Lecture Notes in Earth Sciences 65

Editors:
S. Bhattacharji, Brooklyn
G. M. Friedman, Brooklyn and Troy
H. J. Neugebauer, Bonn
A. Seilacher, Tuebingen and Yale

W0050969

Springer-Verlag Berlin Heidelberg GmbH

Fernando Sansò   Reiner Rummel (Eds.)

Geodetic Boundary Value Problems in View of the One Centimeter Geoid

With 83 Figures and 37 Tables

 Springer

Editors

Prof. Dr. Fernando Sansò
Politecnico di Milano
Dipartimento di Ingegneria Idraulica, Ambientale e del Rilevamento
Piazza Leonardo da Vinci, 32, I-20133 Milano, Italy

Prof. Dr. Reiner Rummel
Technische Universität München
Institut für Astronomie und Physikalische Geodäsie
Arcisstraße 21, D-80333 München, Germany

"For all Lecture Notes in Earth Sciences published till now please see final pages of
the book"

Cataloging-in-Publication data applied for

Die Deutsche Bibliothek - CIP-Einheitsaufnahme

**Geodetic boundary value problems in view of the one centimeter
geoid** / ed.: Fernando Sansò ; Reiner Rummel.

(Lecture notes in earth sciences ; Vol. 65)
ISBN 978-3-540-62636-7   ISBN 978-3-540-68353-7 (eBook)
DOI 10.1007/978-3-540-68353-7

ISSN 0930-0317
ISBN 978-3-540-62636-7

Typesetting: Camera ready by editors
SPIN: 10551354   32/3142-543210 - Printed on acid-free paper

This course of the International Summer School of Theoretical Geodesy, the third of the series, was devoted to a central problem of physical geodesy, namely the determination of the gravity field of the earth from various types of measurements obtained by solving a boundary value problem. Since this problem changes in nature, depending on the resolution and accuracy with which we want to achieve our goal, the environment of the lectures was placed at the 1 cm accuracy level in terms of the geoid.

The course, which has become traditional for the school, consists of three lectures on basic mathematical subjects (potential theory, free boundary value problems, stochastic boundary value problems), three lectures on the classical and advanced theory of boundary value problems in geodesy (on the formulation of geodetic BVPs, on the hierarchy of gravimetric BVPs and on classical numerical methods) and five lectures on relatively new fields devoted to the realization of a geoid at the 1 cm accuracy level. The lectures were complemented by a series of outstanding seminars.

The school has been quite successful, in our opinion, due to the excellent teachers and the vivid interest and participation of the attendees. Also essential for the success of the school has been the support we received from various organizations. This support is gratefully acknowledged: the Politecnico di Milano and the Centro di Cultura Scientifica "A. Volta", on the premises of which the course took place; the European Space Agency (ESA), the Italian National Research Council (CNR), the International Association of Geodesy (IAG), the International Union of Geodesy and Geophysics (IUGG), the Conference of Rectors of Italian Universities (CRUI), under the specific "Programma Vigoni" for the scientific cooperation between Germany and Italy as well as the Municipality of Como which welcomed us warmly in its town. Last but not least, we would like to thank Elena Raguzzoni for the extremely fine organization of all practical aspects of the course including the wonderful social program which was a pleasurable diversion to students and teachers from the heavy program of the school.

Fernando Sansò

TABLE OF CONTENTS

BOUNDARY VALUE PROBLEMS FOR HARMONIC RANDOM FIELDS
Y. Rozanov and F. Sansò

FREE BOUNDARY PROBLEMS
M. Biroli

FORMULATION AND LINEARIZATION OF BOUNDARY VALUE PROBLEMS: FROM OBSERVABLES TO A MATHEMATICAL MODEL
B. Heck

THE HIERARCHY OF GEODETIC BVPs

F. Sansò

GBVP – CLASSICAL SOLUTIONS AND IMPLEMENTATION
H. Sünkel

PART III : FROM THE GBVP TOWARDS
A 1cm GEOID

TOPOGRAPHIC EFFECTS IN GRAVITY FIELD MODELLING FOR BVP
R. Forsberg and C.C. Tscherning

GLOBAL MODELS FOR THE 1cm GEOID – PRESENT STATUS AND NEAR TERM PROSPECTS
R.H. Rapp

AN INTRODUCTION TO AIRBORNE GRAVIMETRY AND ITS BOUNDARY VALUE PROBLEMS
K.-P. Schwarz and Zuofa Li

SPHERICAL SPECTRAL PROPERTIES OF THE EARTH'S GRAVITATIONAL POTENTIAL AND ITS FIRST AND SECOND DERIVATIVES
R. Rummel

SATELLITE ALTIMETRY, OCEAN DYNAMICS AND THE MARINE GEOID
E.J.O. Schrama

STOCHASTIC BOUNDARY VALUE PROBLEM THEORY :
AN ELEMENTARY EXAMPLE

VARIATIONAL METHODS FOR GEODETIC BOUNDARY
VALUE PROBLEMS

TOPICS ON BOUNDARY ELEMENT METHODS
R. Klees 482

SOLVING GEODETIC BOUNDARY VALUE PROBLEMS WITH PARALLEL COMPUTERS
R. Lehmann 532

LIST OF CONTRIBUTORS

Marco Biroli
Dipartimento di Mathematica, Politecnico di Milano
Piazza Leonardo da Vinci. 32, I-20133 Milano
Italy

Martin van Gelderen
Delft University of Technology, Faculty of Geodetic Engineering
Thijsseweg 11, NL-2629 JA Delft
The Netherlands
e-mail: gelderen@geo.tudelft.nl

Bernhard Heck
Geodetic Institute, University of Karlsruhe
Englerstrasse 7, D-76128 Karlsruhe
Germany
e-mail: heck@gik.bau-verm.uni-karlsruhe.de

Petr Holota
Research Institute of Geodesy, Topography and Cartography
250 66 Zdiby 98, Praha-Vychod
Czech Republic
e-mail: vugtk@earn.cvut.cz

Wolfgang Keller
Geodaetisches Institut, Universitaet Stuttgart
Keplerstrasse 11, D-70174 Stuttgart
Germany
e-mail: Wolke@gi5.bauingenieure.uni-stuttgart.de

Roland Klees
Delft University of Technology, Faculty of Geodetic Engineering
Thijsseweg 11, NL-2629 JA Delft
The Netherlands
e-mail: klees@geo.tudelft.nl

Ruediger Lehmann
Geodetic Institute, University of Karlsruhe
Englerstrasse 7, D-76128 Karlsruhe
Germany
e-mail: lehmann@gik.bau-verm.uni-karlsruhe.de

Erich Martensen
Mathematisches Institut II, Universitaet Karlsruhe
D-76128 Karlsruhe
Germany
e-mail: mrt@ma2mar1.mathematik.uni-karlsruhe.de

Stefan Ritter
Mathematisches Institut II, Universitaet Karlsruhe
D-76128 Karlsruhe
Germany
e-mail: Stefan.Ritter@math.uni-karlsruhe.de

Richard H. Rapp
Dept. of Geodetic Science and Surveying, The Ohio State University
1958 Neil Avenue, Columbus, Ohio 43210-1247
USA
e-mail: rhrapp@ohstmvsa.acs.ohio-state.edu

Youri Rozanov
CNR - IAMI
Via Ampere, 56
I-20131 Milano
Italy

Reiner Rummel
Institut fuer Astronomische und Physikalische Geodaesie, TU Muenchen
Arcisstrasse 21, D-80290 Muenchen
Germany
e-mail: rummel@step.iapg.verm.tu-muenchen.de

Fausto Sacerdote
Dipartimento di Ingegneria Idraulica,
Ambientale e del Rilevamento (Sezione Rilevamento), Politecnico di Milano
Piazza Leonardo da Vinci, 32, I-20133 Milano
Italy
e-mail: fausto@ipmtf1.topo.polimi.it

Fernando Sansò
Dipartimento di Ingegneria Idraulica,
Ambientale e del Rilevamento (Sezione Rilevamento), Politecnico di Milano
Piazza Leonardo da Vinci, 32, I-20133 Milano
Italy
e-mail: f.sanso@ipmtf4.topo.polimi.it

Ernst Schrama
Delft University of Technology, Faculty of Geodetic Engineering
Thijsseweg 11, NL-2629 JA Delft
The Netherlands
e-mail: schrama@geo.tudelft.nl

Maurice Schuyer
ESA/ESTEC
Keplerlaan 1, NL - 2201 AZ Noordwijk
The Netherlands
e-mail: mschuyer@estec.esa.nl

Klaus-Peter Schwarz
The University of Calgary
Faculty of Engineering, Department of Geomatics Engineering
2500 University Drive N.W., Calgary, Alberta, T2N 1N4
Canada
e-mail: schwarz@ensu.ucalgary.ca

Nico Sneeuw
Institut fuer Astronomische und Physikalische Geodaesie, TU Muenchen
Arcisstrasse 21, D-80290 Muenchen
Germany
e-mail: sneeuw@step.iapg.verm.tu-muenchen.de

Hans Suenkel
Dept. of Mathematical Geodesy and Geoinformatics
Graz University of Technology
Steyrergasse 30, A-8010 Graz
Austria
e-mail: suenkel@geomatics.tu-graz.ac.at

Carl Christian Tscherning
University of Copenhagen
Geophysical Department
Juliane Maries VEJ 30, DK-2100 Copenhagen OE
Denmark
e-mail: cct@osiris.gfy.ku.dk

Introduction

R. Klees and B. Heck

The lectures given at the International Summer School of Theoretical Geodesy on "Boundary Value Problems and the Modelling of the Earth's Gravity Field in View of the One Centimeter Geoid" can be divided in four groups: mathematical foundations (Martensen & Ritter, Biroli, Rozanov), mathematical geodesy (Heck, Sansò), numerics (Sünkel, Tscherning), and geodetic applications (Rapp, Schwarz, Schrama, Rummel). The lectures have been supplemented by seminars. They may also be assigned to the four categories mentioned before: mathematical foundations (Sacerdote, Martensen), mathematical geodesy (Holota, Grafarend), numerics (Klees, Lehmann), and geodetic applications (Keller, Sneeuw & van Gelderen, Schuyer). In the closing session a list of questions has been compiled in order to point out the most challenging open problems to be solved in future, potentially by the students of the Summer School. The lectures and seminars could only address some of them:

- why do we need boundary value problems at all?
- what is the surface of the Earth?
- is the time of free boundary value problems over?
- how to optimally combine spherical harmonic models with terrestrial data?
- are high degree spherical harmonic expansions needed?
- how to assess the limits of classical geodetic solutions when accuracy and resolution will increase?
- is geophysical information conceptually essential to set up our problems?
- does the classical geodetic approach converge to the boundary element approach and vice versa?
- how to solve the datum problem?
- how to solve mixed boundary value problems efficiently?
- what is an appropriate mathematical representation of sea surface topography and gravity field?
- what are mean anomalies on a rough surface and are there alternatives to working with mean anomalies?
- which theoretical improvements are needed in the estimation of high degree geopotential models and in the determination of geoidal undulations and height anomalies from spherical harmonic models?

1st group: Mathematical Foundations

Lectures:

E. Martensen, W. Ritter: Potential Theory

The series of mathematical lectures is opened by E. Martensen and W.& Ritter with their lecture on "Potential Theory".

The lecture is divided into two parts. The first one, given by E. Martensen, focuses on classical potential theory in the space of continuous and Hölder continuous functions. It consists of three chapters: the first chapter deals with representation formulas and contains the complex and real versions of Cauchy's integral formula, and their generalizations to three dimensions for scalar and vector fields. The second chapter focuses on limit relations and integral equations. Starting with scalar fields, the concept of source distributions on closed surfaces is introduced, the jump relations of the potentials are given and the mapping properties and the spectrum of the corresponding boundary integral operators are discussed. The multiplicity of the eigenvalues ±1 are determined in an elegant way making use of the (second) Betti numbers. The chapter finishes with some extensions to vector fields. The third chapter is devoted to boundary value problems of potential theory and integral equation methods. The (scalar and vector) Dirichlet, Neumann, and Robin boundary value problems for Poisson's and Laplace's equation are introduced and corresponding integral equations are derived making use of the so-called null field method. The second part is given by W. Ritter and aims at providing some mathematical background of boundary element methods, i.e. numerical techniques for solving integral equations making use of trial functions with finite support on the boundary. After discussing the pros and cons of boundary element methods, two discretization methods for boundary integral equations, the (exact) collocation method and the Galerkin method, are explained and some problems related to practical implementations discussed. Most emphasis is put on the definition of Sobolev spaces which is the class of functions normally used in boundary element methods because they provide convergence results and error estimates. Besides, the concept of strong ellipticity of boundary integral operators is introduced, a property that integral operators must have to construct an error and convergence analysis. Finally, the lemma of Céa is discussed which for strongly elliptic integral operators provides the convergence and error estimates of Galerkin boundary element methods.

Y. Rozanov: Random Harmonic Fields

These lectures are motivated by the need of studying various error propagation to the solutions of different B.V.P. (Boundary Value Problems) of geodetic interest, as well as by the new branch of overdetermined B.V.P.'s, because in these cases it is meaningful to ask oneself: what is the part of the solution corresponding to the boundary noise?

In mathematical literature, random fields as solutions of B.V.P.'s are characterized by a suitable, weak formulation, which directly suggests the use of generalized stochastic processes.

This concept however is classically related to linear stochastic functionals defined on spaces of very smooth functions, like the space D of L. Schwarz; the problem is to extend these functionals to much larger spaces in which D is densely embedded and try to characterize the (narrower) space to which the solution has then to belong.

Roughly speaking, for the Dirichlet problem, the reasoning sounds as follows: since a function $\mu \in H^{2,2}(\Omega)$ has a Laplacian in $L^2(\Omega)$ and boundary values in $H^{3/2}(S)$, these can be seen respectively as functionals on $H^{2,2}(\Omega)$ and $H^{-3/2}(S)$; as a matter of fact these functionals can be constructed as limits of sequences of functions $\varphi \in D(\Omega)$; therefore if we provide a source (Δu) and a trace, $u_{|S}$, such that they are linear stochastic functionals continuous on $L^2(S)$ and $H^{-3/2}(S)$ respectively, it is defined a linear stochastic functionals u, continuous $H^{-2,2}(\Omega)$ which is the generalized solution of our problem.

The class of such solution is defined by Rozanov as a vector $\underline{H}^{2,2}(\Omega)$ space, characterized by mapping $H^{-2,2}(\Omega)$ distributions onto random variables with finite variance.

M. Biroli: Free Boundary Problems

The lecture of M. Biroli deals with techniques for solving free boundary value problems that are used in mathematics. Whereas the solution of free geodetic boundary value problems is nowadays based on implicit function theorems, M. Biroli describes frequently used methods based on the calculus of variations which allow to transform the free boundary value problem into a usual minimization problem, thus avoiding the use of an implicit function theorem. Although these methods are not directly applicable to geodetic boundary value problems, they provide more insight into the whole complexity of free boundary value problems. After specifying the conditions for existence of minimum points of a functional making use of the concept of lower semicontinuity, four examples are given to explain the different methods: the equilibrium displacement of a membrane in presence of an obstacle, the so-called dam problem which deals with finding the steady flow of water in a dam, the image segmentation problem, and finally a shape optimization problem for Dirichlet problems. First, each example is described; then a mathematical formulation of the physical problem is given, and finally the problem is transformed into a variational problem (i.e. the problem of minimizing a certain functional on a given set), providing the solution of the free boundary value problem. The question of regularity of the free boundary is left out of consideration.

Seminars:

F. Sacerdote: Stochastic Boundary Value Problem Theory: An Elementary Example

The seminar of F. Sacerdote aims at giving an application of (a part of) the theory developed by Rozanov in his lecture. A Dirichlet problem for a random harmonic field on an open interval is taken as example. First the nonstochastic case is treated and solved in Sobolev spaces. Then, the stochastic case is considered and a rigorous mathematical meaning of the boundary condition and the solution of the problem is given.

E. Martensen: Analogy Between Vectorial and Scalar Formulations of Boundary Value Problems

In the seminar E. Martensen points out the analogies between vectorial and scalar formulations of boundary value problems in potential theory. As a first example the dualism concerning electrostatic and magnetic fields in the context of the Neumann problem is presented, and the properties of the respective eigenvalues are discussed, extending former work of Poincaré and Stakhov. For a further example, where the domain consists of a solid circle torus, the physical properties behind the eigenvalues are identified. Finally the transformation into integral equations is demonstrated by means of the vector Dirichlet problem for a three-axial ellipsoid and a cylinder, magnetized along its axis, which serve as boundary surfaces.

2nd group: Mathematical Geodesy

Lectures:

B. Heck: Formulation and Linearization of Boundary Value Problems: From Observables to a Mathematical Model

The lecture by B. Heck aims at showing the long way from the theory of geodetic boundary value problems to the approximations that underly all practical solutions today, producing a whole hierarchy of approximated problems. Therefore, is may be seen as the core lecture of the Summer School. For each of the three classical formulations of geodetic boundary value problems, namely the scalar and vector Molodensky problem, and the fixed gravimetric boundary value problem, the boundary conditions related to different levels of approximation are derived and thoroughly discussed. First, the boudary conditions in the classical geodetic boundary value problems are derived, and the practical relevance is discussed by comparing them with the real data situation. Then, the boundary conditions are linearized using a reference gravity field and a reference surface, respectively. Thereafter, the linearization errors are evaluated in terms of gravity anomalies and potential quantities, and the differences between the three formulations of the geodetic boundary value problem w.r.t. the impact of nonlinear effects are shown. Finally, starting from the specification of the linearized boundary conditions in several coordinate systems, the corresponding boundary conditions in spherical approximation are given. The fixed gravimetric BVP is taken as example to estimate numerically the influence of the spherical approximation and the constant radius approximation. Finally, the solvability of the first degree harmonic coefficients is discussed, based on the solution of the three basic boundary value problems in nonspherical form in terms of spherical harmonics.

F. Sansò: The Hierarchy of Geodetic Boundary Value Problems

F. Sansò gives in total three lectures which cover various topics of the theory of geodetic boundary value problems. The first lecture on "Wiener measures and the continuous formulation of observation equations" is related to the problem of replacing the original discrete problem of gravity field determination by a continuous problem in a stochastic setting without introducing a nonacceptable bias in the solution of the discrete problem. This leads to the concept of Wiener measures as generalization of the concept of discrete white noise to a continuous white noise. Based on that he introduces the concept of a continuum of noisy measurements as the limit of a large dense data set of measurement points. Then, the question when a switch from continuous to discrete and vice versa is possible, is investigated. A Nyquist type relation is derived stating that the transition may be possible if the discretization error is in some sense "much smaller" than the noise in the data. Only then it is possible to formulate the problem of gravity field determination based on discrete (stochastic) observations in terms of boundary value problems based on a continuum of (stochastic) observations. In a second lecture he discusses the definition of the boundary surface in geodetic boundary value problems and the related problems. He shows that the used boundary is always a (smooth) approximation to the Earth's topography which is located partially inside and partially outside the masses. Therefore, the well known remove restore technique makes it possible to formulate the problem of gravity field determination in terms of boundary value problems. Finally some numerical estimates are presented about the relation between resolution and accuracy of a digital terrain model on the one hand and resolution and accuracy of gravity on the other hand to achieve a 1 cm geoid.

The last lecture is devoted to the analysis of the vector Molodensky problem. The analysis uses a Legendre transform, an approach which is better known in geodesy as the "gravity space approach" worked out by F. Sansò in the seventies. First, the necessary mathematical background such as Legendre transform, Hölder spaces, Schauder estimates for the Dirichlet problem, and local solvability of functional equations are provided. Then, the Legendre transform is applied to the vector Molodensky problem for a non-rotating earth transforming the original BVP in several steps to a fixed boundary value problem for a second order nonlinear partial differential operator. Finally, existence, uniqueness, and regularity of the solution of the transformed BVP are discussed.

Seminars:

P. Holota: Variational Methods for Geodetic Boundary Value Problems

This seminar gives an introduction to variational methods and their application to geodetic boundary value problems. P. Holota starts with a presentation of the functional-analytic foundations of the weak solution concept, concentrating on the construction of a weighted Sobolev space for an unbounded solution domain and estimates for equivalent norms and for traces of functions on a Lipschitz boundary surface. Then he adopts the formulation of the linear Molodensky problem and other geodetic boundary value problems such that the weak solution concept can be applied. With respect to the question of solvability the Lax-Milgram theorem, associated with the ellipticity of bilinear forms, is the basic tool. The ellipticity conditions are interpreted in a geometric way. Finally, numerical solutions in the framework of Galerkin approximations are discussed.

E.W. Grafarend: Three Seminars on Geodetic Boundary Value Problems

The first seminar of E.W. Grafarend is devoted to the determination of the geoid as a (partly) internal level surface. This problem is formulated in the framework of a Cauchy problem for the domain enclosed by the Earth's surface. Special consideration is given to the potential field produced by the topographical and compensation masses, which is calculated by using a slicing technique. A procedure for the determination of the geoidal heights with respect to a given reference surface (sphere, ellipsoid of revolution) is outlined. This procedure is based on (1) Topo-Moho reduction of the potential data given on the Earth's surface, (2) downward continuation by solving an inverse Dirichlet problem making use of Tikhonov regularization, and (3) restoring the effects of topographical-isostatic masses. In the second seminar, this theory is applied to the calculation of the geoid in South-Western Germany from given geopotential numbers. Details about the applied procedure and about the geopotential data and topographic models are presented. The third seminar provides some type of "tool-box" for the representation of scalar, vector, and tensor fields by spherical and ellipsoidal harmonics which are generally used in gravity field analysis. Starting from a description of spherical (polar) and elliptical coordinates in several parameterizations and from the Helmholtz decomposition theorem for differential forms, fields of type scalar, vector and second order (Hesse) tensor are presented. It is shown that vector-valued spheroidal functions are composed of radial, spheroidal and toroidal components, while tensor-valued functions can be decomposed into normal/tangential, antisymmetric/ symmetric and trace/trace-free (deviatoric) parts. The decomposition is illustrated by the aid of the zero, first and second derivatives of the gravitational potential in terms of spherical and ellipsoidal harmonics.

3rd group: Numerics

Lectures:

H. Sünkel: Classical Solutions of GBVP and Their Implementation

This lecture provides a review of the classical analytical approach into the solution of the GBVP, characterized by series expansions with respect to a "small" parameter; as such, it mainly summarizes the work by H. Moritz and the "Graz School". Furthermore, emphasis is given to the numerical implementation of the resulting solution formulae which have the form of spherical integrals. In the first part of the lecture the Stokes problem is treated, aiming at the determination of the geoid as an equipotential surface situated partly inside the earth's surface. After having applied the classical topographic-isostatic and free-air corrections as well as the standard approximations, the geoidal height can be described by the Stokes integral. The Stokes operator is represented in several forms, based on spherical and planar approximation; the effect of the error induced by spherical approximation is counteracted by ellipsoidal corrections. The (spherical/ planar) Stokes integral is evaluated using fast spectral analysis procedures such as one- and two-dimensional Fast Fourier and Fast Hartley Transforms. The second part of the lecture concentrates on the analytical solution of Molodensky's problem in terms of series of spherical integrals. Main emphasis is given to the solution by analytical continuation of the boundary data as well as the evaluation of the integrals by FFT procedures. Finally practical examples and experiences are presented; for the area of Austria the differences between geoidal undulations and height anomalies are reported to amount up to 0.5 m which is about 10% of the geoidal height range in this area.

C.C. Tscherning: Topographic Effects in Gravity Field Modelling for BVP

In methodological respect the lecture by C.C. Tscherning, co-authored by R. Forsberg, is also related to the classical analytical solution approach. Considering geodetic boundary value problems as a subset in the class of gravity modelling approaches, different concepts of terrain reductions are elaborated, including the conventional and the remove-restore technique. The final aim behind terrain reductions in Stokes' as well as in Molodensky's approach is seen in smoothing the residual gravity field, facilitating interpolation, prediction, or the use of collocation. Since the magnitude of the terrain reduction is much larger than the non-level correction (free-air reduction), in general, it is emphasized to use methods which do not require up- or downward movement of the observation points, simultaneously providing the smoothest results. Furthermore he recommends to use density information in terms of density anomalies or isostatic models such as the Airy-Heiskanen compensation model or Helmert's second condensation approach. As an alternative, or in addition, residual terrain modelling (RTM) can be applied, based on a smooth height reference surface. The problem of harmonic continuation from the true topography to the height reference surface is not yet satisfactorily solved in theoretical respect. Another part of the lecture is devoted to the use of terrain reductions in the context of Molodensky's theory. The paper is concluded by a summary of strategies for calculation of the terrain effects, including prism-integration and FFT.

Seminars:

R. Klees: Topics on Boundary Element Methods

The seminar given by R. Klees covers both theoretical and practical aspects of boundary element methods in the context of high resolution gravity field determination. Starting from the mathematical model derived by B. Heck in his lecture, several representation formulas for harmonic functions in an unbounded domain are discussed. Their jump relations provide the boundary integral equations which form the starting point of any boundary element method.Then, the construction of the approximating subspaces where the solution is looked for is discussed and the approximation properties in Sobolev norms of some subspaces relevant for geodetic applications are given. Thereafter, different discretization principles are investigated where most of the time is spent to the Galerkin method, the only discretization method relevant to geodetic problems. The core of the seminar are fast numerical algorithms for Galerkin boundary element methods which are essential for geodetic applications to restrict the numerical costs. Two algorithms are discussed in detail: the panel clustering which exploits the smoothness of the kernel of the integral operator by using certain Taylor series expansions for most of the entries of the equation matrix, and biorthogonal compactly supported wavelets. Both provide sparse linear systems with $O(N \cdot \log^s N)$ nonzero elements where N denotes the number of unknowns and s some positive real number. Thereafter, some of the problems of numerical integration of singular surface integrals are analyzed. They arise when evaluating the presentation formula to get pointwise estimates of the gravity potential and any functionals on the boundary surface. Finally, a detailed state of the art error analysis of Galerkin boundary element methods is given. The three most important types of errors, namely the discretization error, the error of numerical integration, and the error of boundary surface approximation are discussed. Some estimates are given for geodetic boundary value problems making use of the approximation properties of the approximating subspaces.

R. Lehmann: Solving Geodetic Boundary Value Problems with Parallel Computers

This seminar focuses on the use of a special class of parallel computers, the so-called Multiple Instruction Multiple Data Stream (MIMD) computers for gravity field determination. The first part of the seminar deals with a review of modern supercomputer architectures like Single Instruction Multiple Data Stream (SIMD) computers, and MIMD computers with shared, distributed or virtually shared memory. Then some parallel computing concepts for MIMD computers are discussed, for instance interconnection networks, problem decomposition, interprocessor communication, granularity, and load management. Finally, several applications are presented, and analyzed from a parallel computing point of view: the problem of parallel numerical integration which arises for instance when computing terrain corrections, the application of Galerkin boundary element methods to the linearized fixed gravimetric boundary value problem, and some standard problems of numerical linear algebra.

4th group: Geodetic Applications

Lectures:

R. H. Rapp: Global Models for the 1 cm Geoid. Present Status and Near Term Prospects

The lecture given by R. Rapp can be seen as an application of the theory of boundary value problems to the determination of global gravity field models. It mainly focuses on the practical problems related to high resolution gravity field determination in terms of spherical harmonics. First the hierarchy of potential coefficient estimation procedures is outlined: the estimation of a satellite potential coefficient model, a low degree combination with altimeter data and surface gravity anomalies, and two procedures that yield high degree gravity field coefficients, the quadrature approach and the block diagonal procedure. Then, a review of past and current geopotential models is given with special emphasis on the GSFC/DMA joint gravity and geoid improvement project. For the latter, the different types of data are discussed and an evaluation of some preliminary potential coefficient models is given. Finally, some theoretical challenges related to the procdures applied in practice for high resolution global gravity field recovery inview of the 1 cm geoid are discussed.

K.P. Schwarz: An Introduction to Airborne Gravimetry and Its Boundary Value Problems

In the series of lectures presented by K.P. Schwarz (co-author: Zuofa Li) three aspects of airborne gravimetry and gradiometry are elucidated. In the first part the system concepts of airborne gravimetry and gradiometry are explained. The fundamental problems in all approaches to airborne gravimetry are related to the stabilization of a direction in space under motion, and the separation between gravitational and non-gravitational accelerations. After discussing potential solutions to these problems, scalar and vector gravimetry and gravity gradiometry are identified as three basic approaches to the measurement of gravity from a moving platform. The central section of the lecture covers the modelling of the observables and estimation methods.The equations for modelling inertial data are reviewed, and an error model for airborne gravimetry is developed. In particular, the effects of attitude errors and insufficient synchronization between GPS and INS on the derived gravity values are studied, and the dominating role of vibration effects is emphasized. In order to minimize the effects of these and other error sources suitable estimation strategies have been designed. Presently the iterative use of time domain and spectral methods is the most promising way to extract the gravity signal from the noisy data. In the third part several variants of boundary value problems related to airborne gravimetry and gradiometry are formulated. The first group uses airborne data only and includes the Scalar Airborne Gravimetric BVP, the Vector Airborne Gravimetric BVP and the Airborne Gradiometry BVP as special cases, considering continuous or gridded data on a plane at constant flight level. The proposed solution concept makes use of Taylor expansions and Fourier transforms as well as a variational principle. In a second set of BVP airborne data are combined with ground data, stabilizing the downward continuation from flight level to ground level. Finally the solution concept for gridded data is generalized to multiresolution data sets.The lecture is concluded by a presentation of practical results obtained in Greenland and Switzerland and by a discussion of future plans related to airborne gravimetry.

E.J.O. Schrama: Satellite Altimetry, Ocean Dynamics, and the Marine Geoid

This lecture can be seen in some sense as complementary to the lecture given by C.C.Tscherning on the treatment and problems of land topography in the context of the geodetic boundary value problem. However, the time component makes the ocean topography much more challenging although the data coverage is mostly considerably better w.r.t. homogeneity and denseness than on land. The first part of the lecture deals with satellite altimetry as a geodetic measurement technique and is thought for those who are not familiar with satellite altimetry. The principles of altimetric measurements are explained, the corrections that have to be applied to fully exploit the capability of the radar sensor are discussed and the problem of precise orbit determination is briefly outlined. The main part of the lecture focuses on tides and ocean dynamics. The lecture on tides aims at giving an introduction to that subject. It covers such relevant topics like tide generating potential, Earth tides, and Love numbers. Methods for determining ocean tides from altimeter measurements are explained and the problem of load tides is briefly discussed. Finally, a review of recent developments that have dramatically improved the accuracy of deep ocean tide models is given. The last part of the lecture is related to ocean dynamics. It aims at discussing that part of the ocean topography that remains when the tidal effect, the inverse barometer effect, and wind waves have been subtracted: the mean dynamic topography as stationary part and the ocean variability as temporal part. Starting with the equation of motion the concept of a mean dynamic topography is introduced and a physical method for measuring it is explained. Thereafter, comparisons are shown between that method and the results obtained by satellite altimetry. The lecture is finished with some examples showing the temporal variations in the dynamic topography as observed by the French/American Topex/Poseidon altimeter satellite.

R.Rummel: Spaceborne Gradiometry

The last lecture in the group of geodetic applications of boundary value problems is devoted to gravity field determination from spaceborne gradiometry. The course presented by R. Rummel starts with an introduction into the principles of Satellite Gravity Gradiometry. Although the basic concepts of gradiometry are very simple, the technical realization has to struggle with problems like drag effects, calibration and orientation of the devices, and consideration of temporal variations of gravitating masses such as propellant. Due to the construction of gradiometers the error spectrum contains a white noise part as well as a frequency-dependent component. In the second part of the lecture the origin and role of geodetic boundary value problems is explained. Originally the failure of Bruns' polyhedron approach gave rise to the boundary value approach; in the space age the role of the GBVP has been newly defined, e.g. due to the concept of GPS levelling. Satellite gradiometry, providing the second order derivatives of the gravitational potential at satellite height, is embedded into the broader context of gravity sensors and gravity quantities in general and their respective information content. In the third section of the lecture this context is offered by the so-called Meissl scheme, visualized by a "pocket guide" of Physical Geodesy. The structure of the relations based on various self-adjoint (rotational invariant or isotropic) operators is discussed in detail, and extensions to non-self-adjoint operators related to gravity gradiometry observables are pointed out. In the last block of his lectures R. Rummel addresses to spectral analysis of satellite gradiometry which can be considered as an inverse problem of the GBVP, since the disturbing potential on the surface of the earth (or equally well the harmonic coefficients) has to be determined from data given in the exterior. After discussing the signal content of satellite gradiometry and the construction of a noise model (band-limited white noise) several solution concepts are presented, each aiming at a stabilization of the estimation process which proves necessary due to measurement noise propagation. Restricting to a linear functional model the concepts of singular value decomposition, least squares estimation, least squares collocation or regularization, and biased estimation are applied to gravity gradiometry. The lecture is concluded by a synthetic numerical example for spectral estimation based on satellite gradiometry data.

Seminars:

W. Keller: Application of Boundary Value Techniques to Satellite Gradiometry

The seminar given by W. Keller focuses on the gravity field determination using gravity gradients at satellite altitude and terrestrial gravity data. The solution strategy is based on the continuous spacewise approach and the problem is formulated as an overdetermined boundary value problem with noisy data. It is assumed that the data taken along the satellite orbit have been reduced to some boundary surface which in some sense fits the orbit. To set up the functional model, Sobolev spaces on the sphere are used as function spaces and the connection between the error free data and the unknown solution is described by pseudodifferential operators on spheres.

Then, the definition of suitable continuous stochastic models describing the behaviour of the measurement noise, is discussed. He introduces the concept of Hilbert space valued random variables to set up this stochastic model. Thereafter, the solution of the overdetermined boundary value problem is pointed out. First he shows that a simple generalization of the usual least squares principle to infinite dimensional spaces is not possible. As alternative, a generalization of the concept of the best linear unbiased estimation (BLUE) to infinite dimensional spaces is developed and applied to the overdetermined boundary value problem. In addition, a simple a priori error estimate is derived based on the homomorphy of the algebra of pseudodifferential operators with the algebra of their symbols. It can be used in order to investigate the influence of certain mission parameters on the solution. Finally, the theory is applied to the ARISTOTELES and STEP satellite gravity gradiometry missions and some results of a simulation study are presented.

N. Sneeuw & M. van Gelderen: The Polar Gap

The seminar by N. Sneeuw and M. van Gelderen focusses on the effect of nonpolar orbits on the derivation of spherical harmonic coefficients from satellite gradiometry. Based on simulations for a one axis gradiometer (STEP satellite) and a three axis gradiometer (Gravity Explorer Mission), they show what the effect of data (polar) gaps on the estimation of potential coefficients and geoid heights are. Because the problem is ill-posed, regularisation is necessary even though it introduces biases into the potential coefficients. As a result, the polar gap induces a band of low order coefficients which are strongly disturbed and cannot be determined from a combination of gradiometric and GPS data. This could be explained by the support of the Legendre functions in the polar areas and by the condition numbers of the systems of normal equations. If propagated to the space domain (the geoid or the gravity field) the damage caused by the non-global coverage remains limited to an area not much larger than the polar gaps, notwithstanding the global support of the spherical harmonic base functions. Moreover it is demonstrated that the least-squares estimation might contain considerable aliasing errors.

M. Schuyer: European Capabilities and Prospects for a Spaceborne Gravimetry Mission

M. Schuyer presents the goals and the planned mission features of a potential future ESA satellite mission, the Gravity field and Ocean Circulation Earth explorer (GOCE). This mission candidate will have strong benefits to Geodesy, satellite orbit modelling, Solid Earth Geophysics and Geodynamics, and Oceanography (ocean circulation). In technical respect, the GOCE proposal consists of a spacecraft carrying a three-axis gradiometer and a GPS receiver, and flying at 260-270 km altitude on a sun-synchronous dawn-dusk orbit. After a mission length of 8 months the accuracy and resolution requirements (1 mgal in gravity, 2 cm in geoidal height, up to degree and order 180) will be achieved. The gradiometry component is designed to measure the diagonal terms of the tensor of second order gravitational potential derivatives with high accuracy, the non-diagonal terms with less accuracy. The drag-free control of the space vehicle is discussed in detail.

PART I: BOUNDARY VALUE PROBLEMS

Potential theory

E. Martensen and S. Ritter

In the last decades, due to the challenges of applications, problems and methods in potential theory have been extended mainly in two directions. After for a long time potential theory has been a *scalar* theory originated from gravitation and electrostatics, *vector* problems have become more and more of interest. Such problems, for instance, have arisen from magnetostatics, geodesy, and elastomechanics. Secondly, more general *geometries* have occured with problems concerning to disjoint or several-connected domains. The methods developed in this context show a lot of relationship between the scalar and vector problems and provide more understanding of the role of the geometric parameters.

Incited by the rapidly increasing computational power the numerical treatment of boundary value problems in potential theory has become an interesting field of present research. Two powerful numerical methods namely the boundary element method and the finite element method have been developed and successfully applied to various problems in engineering mathematics. The foundations of these techniques such as Sobolev spaces and the concept of strong ellipticity are necessary for proper understanding and working with these methods.

In the first two chapters of this survey the theoretical foundations of scalar and vector problems in potential theory are described. The third chapter is devoted to scalar and vector boundary value problems. In the fourth chapter we present the boundary element method both from the theoretical and the practical point of view. The fifth chapter deals with Sobolev spaces for the sphere which are introduced comprehensively by means of spherical harmonics expansion. Additionally, an idea is given how the basic integral equations in potential theory can be studied in the Sobolev space setting.

1. Representation theorems

Representation theorems are an essential tool for treating boundary value problems in potential theory, but one should be aware that they do not solve such problems by itself.

With some generality, the representation theorems here discussed will be based upon the following geometry. Let $D \subseteq \mathbb{R}^3$ be a finite union of disjoint *compact* domains, say

$$D = D_1 \cup \ldots \cup D_{p'}.$$

For D, the closed complement $D' := \overline{\mathbb{R}^3 \setminus D}$ consists of a finite union of disjoint *closed* domains with exactly one unbounded domain D'_0:

$$D' = D'_0 \cup D'_1 \cup \ldots \cup D'_p.$$

The non-negative integers p, p' occuring here coincide with the so-called *second Betti numbers* for D, D' which specify the number of topologically independent closed surfaces not contractible in D, D', respectively (Fig. 1.1). In the simplest case where D and D' are the domains inside or outside a closed surface, we have $p = 0$ or $p' = 1$. A problem formulated for D or D' is said an *interior* or *exterior* problem, respectively.

The common boundary of D and D' is assumed as a finite number of "sufficiently smooth" closed surfaces S with normal n everywhere directed outward of D (Fig.1). Sufficient smoothness may be understood, roughly spoken, as *piecewise smooth* if the argument points considered are separated from S, and as *continuously curved* in the context of limits where the argument point tends to the boundary. By "$|_+$" or "$|_-$" the limiting values of a function are denoted if the argument point approaches S along the normal from the positive (exterior) or negative (interior) side, respectively.

For the sake of applications which are not concerned with any kind of volume distributions in the exterior, the representation theorems considered will refer to functions satisfying *homogeneous* differential equations in the exterior. Furthermore, all functions are assumed to vanish continuously at infinity.

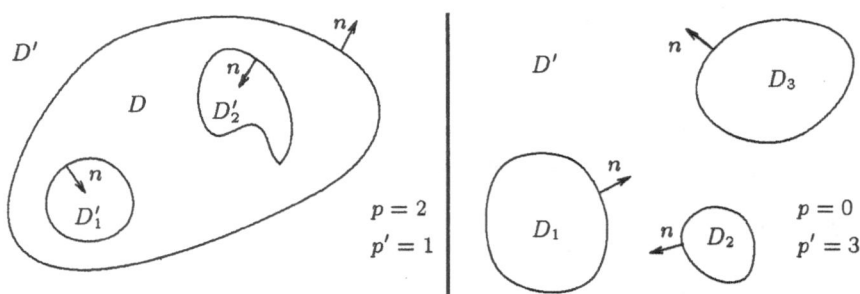

Fig. 1.1. Systems of domains D, D' with second Betti numbers p, p'

1.1 Scalar representation

For a scalar function

$$u \in C^1(D) \cap C^1(D') \cap C^2(\mathbb{R}^3 \setminus S)$$

with properties

$$\Delta u \in C(D), \quad \Delta u = 0 \quad \text{in} \quad \underline{D'}, \quad \lim_{x \to \infty} u(x) = 0,$$

the *scalar representation theorem*

$$u(x) = \frac{1}{4\pi} \left\{ -\int_D \frac{\Delta u(y)}{|x-y|} dV(y) - \int_S \frac{1}{|x-y|} \left(\left.\frac{\partial u}{\partial n}\right|_+ - \left.\frac{\partial u}{\partial n}\right|_- \right) (y) dS(y) \right.$$

$$\left. + \int_S (u|_+ - u|_-)(y) \frac{\partial}{\partial n_y} \frac{1}{|x-y|} dS(y) \right\} \tag{1.1}$$

holds for all argument points $x \in \mathbb{R}^3 \backslash S$. The proof of this classical statement is mainly founded on Green's second integral theorem [22].

By (1.1) the scalar function u is decomposed into a single layer potential of the volume density $-\Delta u$ in D, a single layer potential of the surface density $-\left(\left.\frac{\partial u}{\partial n}\right|_+ - \left.\frac{\partial u}{\partial n}\right|_- \right)$ on S, and a double layer potential of the density $u|_+ - u|_-$ on S. Especially, if u is harmonic in $\mathbb{R}^3 \backslash S$, the volume potential vanishes and (1.1) reduces to what is called *Green's integral formula*.

A representation theorem in potential theory, as a rule, is applied as follows. If a function u is to be represented in D or D', then one is completely free in choosing the values of u in D' or D, respectively. How this should be done, depends on what kind of representation is wanted. Especially, such values can be chosen 0.

Example 1. Let S be a *sphere* with center $x_0 \in \mathbb{R}^3$ and radius $R > 0$, and let inside S a spherical harmonic K_ν of order $\nu \geq 0$ with respect to x_0 be given:

$$u(x) = K_\nu(x), \quad |x - x_0| \leq R. \tag{1.2}$$

In order to represent this function merely by a single layer surface potential, it is seen from (1.1) that we have to choose u outside S as a harmonic function with the same boundary values on S as (1.2) and value 0 at infinity. This is obviously attained by

$$u(x) = R^{2\nu+1} \frac{K_\nu(x)}{|x-x_0|^{2\nu+1}}, \quad |x - x_0| \geq R. \tag{1.3}$$

With regard to S, we calculate from (1.2) and (1.3)

$$\left.\frac{\partial u}{\partial n}\right|_+ (x) - \left.\frac{\partial u}{\partial n}\right|_- (x) = R^{2\nu+1} K_\nu(x) \left.\frac{\partial}{\partial n} \frac{1}{|x-x_0|^{2\nu+1}}\right|_+ = -\frac{2\nu+1}{R} K_\nu(x).$$

Thus Green's integral formula gives us the desired result that

$$K_\nu(x) = \frac{2\nu+1}{4\pi R} \int_S \frac{K_\nu(y)}{|x-y|} dS(y), \quad |x - x_0| < R. \tag{1.4}$$

1.2 First order vector representation

Let now
$$v \in [C(D)]^3 \cap [C(D')]^3 \cap [C^1(\mathbb{R}^3 \backslash S)]^3$$
be a vector function satisfying the conditions
$$\operatorname{div} v \in C(D), \quad \operatorname{rot} v \in [C(D)]^3,$$
$$\operatorname{div} v = 0, \quad \operatorname{rot} v = 0 \quad \text{in} \quad \underline{D'}, \quad \lim_{x \to \infty} v(x) = 0.$$

Then the *vector representation theorem*, otherwise called the *fundamental theorem of vector analysis*,
$$v = -\operatorname{grad} \Phi + \operatorname{rot} A \tag{1.5}$$
with the *scalar* and *vector potentials*

$$\Phi(x) = \frac{1}{4\pi} \left\{ \int_D \frac{\operatorname{div} v(y)}{|x - y|} dV(y) + \int_S \frac{(n, v|_+ - v|_-)(y)}{|x - y|} dS(y) \right\}, \tag{1.6}$$

$$A(x) = \frac{1}{4\pi} \left\{ \int_D \frac{\operatorname{rot} v(y)}{|x - y|} dV(y) + \int_S \frac{[n, v|_+ - v|_-](y)}{|x - y|} dS(y) \right\}, \tag{1.7}$$

is valid for $x \in \mathbb{R}^3 \backslash S$; it refers to v and its first order derivatives. The proof is mainly obtained by the Gauss integral theorem if applied to certain divergence expressions [22].

By (1.5) the vector function v is decomposed into an circulation-free and a flux-free part originated by the Newton-Coulomb law for the source (charge) distributions
$$\varepsilon_0 = \operatorname{div} v \quad \text{in} \quad D, \quad \varepsilon = (u, v|_+ - v|_-) \quad \text{on} \quad S, \tag{1.8}$$
or the Biot-Savart law for the vortex (current) distributions
$$\gamma_0 = \operatorname{rot} v \quad \text{in} \quad D, \quad \gamma = [n, v|_+ - v|_-] \quad \text{on} \quad S, \tag{1.9}$$
respectively. Especially, if v is a harmonic vector field, i.e.,
$$\operatorname{div} v = 0, \quad \operatorname{rot} v = 0 \quad \text{in} \quad \mathbb{R}^3 \backslash S, \tag{1.10}$$
the potentials (1.6) and (1.7) reduce to surface integrals. In this case, the vector representation theorem is called *Cauchy's integral formula*. The background of this notation is as follows.

In the analogous two-dimensional case the common boundary of $D, D' \subseteq \mathbb{R}^2$ consists of a finite number of closed curves C. Then a two-dimensional vector function v for all $x \in \mathbb{R}^2 \backslash C$ is represented by[1]

---

[1] The *$*$-gradient* grad$^* \Psi = (-\Psi_y, \Psi_x)$ arises from grad $\Psi = (\Psi_x, \Psi_y)$ by a rectangular turn in the counter clock sense.

$$v = -\operatorname{grad} \Phi - \operatorname{grad}^* \Psi \tag{1.11}$$

where

$$\Phi(x) = \frac{1}{2\pi} \left\{ \int_D \operatorname{div} v(y) \ln \frac{1}{|x-y|} \, dS(y) \right.$$

$$\left. + \int_C (n, v|_+ - v|_-)(y) \ln \frac{1}{|x-y|} \, dS(y) \right\}, \tag{1.12}$$

$$\Psi(x) = \frac{1}{2\pi} \left\{ \int_D \operatorname{rot} v(y) \ln \frac{1}{|x-y|} \, dS(y) \right.$$

$$\left. + \int_C [n, v|_+ - v|_-](y) \ln \frac{1}{|x-y|} \, dS(y) \right\}, \tag{1.13}$$

are the *logarithmic potential* of the plane source distributions analogous to (1.8) or the *stream function* of the plane vortex distributions analogous to (1.9), respectively; note that in contrast to the three-dimensional case, plane vortex distributions are *scalar* quantities.

Especially, if $v = (v_1, v_2)$ is a harmonic vector field in $\mathbb{R}^2 \backslash C$, the equations analogous to (1.10) have the meaning of the Cauchy-Riemann equations for the complex function

$$f := \bar{v} = v_1 - i v_2 \tag{1.14}$$

thus establishing a holomorphic function in $\mathbb{C} \backslash C$ with value 0 at infinity. Together (1.12) and (1.13) reduce to curve integrals, and it is shown by some computation that (1.11) is identical with the *classical* Cauchy integral formula

$$f(z) = \frac{1}{2\pi i} \int_C \frac{f(\zeta)|_+ - f(\zeta)|_-}{z - \zeta} \, d\zeta \tag{1.15}$$

for the function (1.14), valid for $z \in \mathbb{C} \backslash C$ [26].

1.3 The Lorentz condition

Coming back to three dimensions, we shall now draw attention to the *vortex distributions* (1.9) which arise from the vector representation theorem. Here the volume part γ_0 in D obviously is divergence-free in the weak sense. Furthermore, on S the volume part γ_0 and the surface part γ are linked by

$$\operatorname{Div} \gamma = -\lim_{|\Delta S| \to 0} \frac{1}{|\Delta S|} \int_C (t, v|_+ - v|_-) \, ds$$

$$= -\lim_{|\Delta S| \to 0} \frac{1}{|\Delta S|} \int_{\Delta S} (n, \operatorname{rot} v|_+ - \operatorname{rot} v|_-) \, dS = (n, \gamma_0) \tag{1.16}$$

where the surface divergence Div is calculated in the weak sense by means of some portion $\Delta S \subset S$ with area $|\Delta S|$ and boundary curve $C \subset S$ which shrinks to a point of S; furthermore, the tangent t of C is orientated in the right screw sense with respect to the surface normal n.

After this, from a more general point of view, we consider a volume vortex distribution

$$\gamma_0 \in [C(D)]^3, \quad \operatorname{div} \gamma_0 \in C(D),$$

and a surface vortex distribution

$$\gamma \in [C(S)]^3, \quad (n, \gamma) = 0 \quad \text{on} \quad S, \quad \operatorname{Div} \gamma \in C(S),$$

div and Div to be understood in the weak sense. The vector potential of this distributions

$$A(x) = \frac{1}{4\pi} \left\{ \int_D \frac{\gamma_0(y)}{|x - y|} dV(y) + \int_S \frac{\gamma(y)}{|x - y|} dS(y) \right\} \tag{1.17}$$

is continuous in $\mathbb{R}^3$. A pair of vortex distributions γ_0, γ which has the further properties

$$\operatorname{div} \gamma_0 = 0 \quad \text{in} \quad D, \quad (n, \gamma_0) = \operatorname{Div} \gamma \quad \text{on} \quad S, \tag{1.18}$$

is said to satisfy the *continuity conditions*. Physically spoken, by the second equation (1.18) a volume current γ_0 when entering the boundary is transformed *without lost* into a surface current γ tangential to the boundary.

For the potential (1.17) we calculate by interchanging differentiation with integration and using the Gauss integral theorem for D as well as for S

$$\operatorname{div} A(x) = \frac{1}{4\pi} \left\{ \int_D \left(\operatorname{grad}_x \frac{1}{|x-y|}, \gamma_0(y) \right) dV(y) + \int_S \left(\operatorname{grad}_x \frac{1}{|x-y|}, \gamma(y) \right) dS(y) \right\}$$

$$= \frac{1}{4\pi} \left\{ -\int_D \left(\operatorname{grad}_y \frac{1}{|x-y|}, \gamma_0(y) \right) dV(y) - \int_S \left(\operatorname{Grad}_y \frac{1}{|x-y|}, \gamma(y) \right) dS(y) \right\}$$

$$= \frac{1}{4\pi} \left\{ \int_D \frac{\operatorname{div} \gamma_0(y)}{|x-y|} dV(y) - \int_S \frac{((n, \gamma_0) - \operatorname{Div} \gamma)(y)}{|x-y|} dS(y) \right\}, \tag{1.19}$$

valid for $x \in \mathbb{R}^3 \backslash S$. From this identity we learn that for a pair γ_0, γ the continuity conditions (1.18) are equivalent to the *Lorentz condition*

$$\operatorname{div} A = 0 \quad \text{in} \quad \mathbb{R}^3 \backslash S. \tag{1.20}$$

This condition can therefore serve as a *criterion* whether a vortex pair γ_0, γ is "physical" or not. As can be seen from the special case $\gamma_0 = 0$, the *surface divergence* of a surface vortex distribution γ vanishes if and only if the *divergence* of its vector potential vanishes.

At the beginning, in particular from (1.16), we have seen that the vortex distributions (1.9) satisfy the continuity conditions (1.18). So as an appendix to the vector representation theorem, the vector potential (1.7) always satisfies the Lorentz condition (1.20).

1.4 Second order vector representation

In order to represent a vector function

$$v \in [C^1(D)]^3 \cap [C^1(D')]^3 \cap [C^2(\mathbb{R}^3 \backslash S)]^3$$

which satisfies the conditions

$$\Delta v \in [C(D)]^3, \quad \Delta v = 0 \quad \text{in} \quad \underline{D'}, \quad \lim_{x \to \infty} v(x) = 0,$$

the scalar representation theorem (1.1) can be applied componentwise. But such representation is not yet very relevant to praxis, where vector problems mostly are concerned with other boundary data.

We therefore shall transform (1.1), with u replaced by v, by means of the integral formula[2]

$$\int_S \left\{ \frac{\partial w}{\partial n} + [n, \operatorname{rot} w] - n \operatorname{div} w \right\} dS = 0 \tag{1.21}$$

which holds for any *closed* surface S and a continuously differentiable vector function w defined at least in a one-side vicinity of S. If for fixed $x \in \mathbb{R}^3 \backslash S$, (1.21) is applied to the vector function

$$w(y) = \frac{v(y)}{|x - y|},$$

then some elementary calculation yields

$$\int_S \left\{ \frac{1}{|x - y|} \frac{\partial v(y)}{\partial n} - v(y) \frac{\partial}{\partial n_y} \frac{1}{|x - y|} \right.$$

$$\left. + \frac{[n, \operatorname{rot} v](y)}{|x - y|} + \left(\operatorname{grad}_y \frac{1}{|x - y|} \right)(n, v)(y) \right.$$

---

[2] This is a special case of the Stokes type integral theorem

$$\int_S \left\{ \frac{\partial w}{\partial n} + [n, \operatorname{rot} w] - n \operatorname{div} w \right\} dS = \int_C [t, w] ds,$$

valid for a surface S with boundary curve C where the surface normal n comes out from the curve tangent t in the right screw sense [34].

$$- \frac{(n \operatorname{div} v)(y)}{|x-y|} - \left[\operatorname{grad}_y \frac{1}{|x-y|}, [n,v](y)\right] \Big\} \, dS(y) = 0.$$

By this (1.1) changes to the *second order vector representation theorem* due to Kress [18]

$$v(x) = \frac{1}{4\pi} \Bigg\{ - \int_D \frac{\Delta v(y)}{|x-y|} dV(y)$$

$$+ \int_S \frac{[n, \operatorname{rot} v|_+ - \operatorname{rot} v|_-](y)}{|x-y|} dS(y) - \operatorname{grad} \int_S \frac{(n, v|_+ - v|_-)(y)}{|x-y|} dS(y)$$

$$- \int_S \frac{(n(\operatorname{div} v|_+ - \operatorname{div} v|_-))(y)}{|x-y|} dS(y) + \operatorname{rot} \int_S \frac{[n, v|_+ - v|_-](y)}{|x-y|} dS(y) \Bigg\}$$

where $x \in \mathbb{R}^3 \backslash S$.

2. Limiting relations and integral operators

In the following we consider the geometry as before where the boundary surfaces S now are assumed *continuously curved*.

2.1 Limiting relations for the scalar potential

For a *surface source (charge) distribution* $\varepsilon \in C(S)$ the *scalar potential*

$$\Phi(x) = \frac{1}{4\pi} \int_S \frac{\varepsilon(y)}{|x-y|} dS(y) \tag{2.1}$$

is continuous throughout $\mathbb{R}^3$, harmonic in $\mathbb{R}^3 \backslash S$, but in general not continuously differentiable with respect to each side of S. For the *normal derivatives* of Φ, however, the *limiting relations*

$$\frac{\partial \Phi}{\partial n}\bigg|_\pm = \frac{1}{2}(T\varepsilon \mp \varepsilon) \tag{2.2}$$

with improper scalar parameter integral

$$T\varepsilon(x) := \frac{1}{2\pi} \int_S \varepsilon(y) \frac{\partial}{\partial n_x} \frac{1}{|x-y|} dS(y), \quad x \in S, \tag{2.3}$$

hold in the sense of uniform convergence with regard to S [5, 6, 13]. From (2.2) it follows the *jump relation*

$$\left.\frac{\partial \Phi}{\partial n}\right|_{+} - \left.\frac{\partial \Phi}{\partial n}\right|_{-} = -\varepsilon. \tag{2.4}$$

Example 2. For some illustration of the jump relation (which is also valid for *interior* points of non-closed surfaces) consider the unit disk S in the (x_1, x_2)-plane with normal n in the x_3-direction and a homogeneous single layer distribution $\varepsilon = 1$ on S. The scalar potential (2.1) and its normal derivative on the x_3-axis with the exception of the origin are calculated as

$$\Phi(x) = \frac{1}{4\pi} \int\limits_{S} \frac{dS(y)}{\sqrt{y_1^2 + y_2^2 + x_3^2}} = \frac{1}{4\pi} \int\limits_{0}^{2\pi} \int\limits_{0}^{1} \frac{\varrho \, d\varrho \, d\varphi}{\sqrt{\varrho^2 + x_3^2}} = \frac{1}{2}\left(\sqrt{1 + x_3^2} - |x_3|\right),$$

$$\frac{\partial \Phi(x)}{\partial x_3} = \frac{1}{2}\left(\frac{x_3}{\sqrt{1 + x_3^2}} - \frac{x_3}{|x_3|}\right).$$

By the limits for $x_3 \to \pm 0$ we find that

$$\Phi|_{+}(0) - \Phi|_{-}(0) = 0, \qquad \left.\frac{\partial \Phi}{\partial n}\right|_{+}(0) - \left.\frac{\partial \Phi}{\partial n}\right|_{-}(0) = -1.$$

As an application of the jump relation (2.4), we can conclude from the Lorentz condition (1.20) that the surface source distribution in (1.19) must necessarily vanish. It then follows by standard methods that the remaining volume source distribution in (1.19) must vanish too. Thus the continuity conditions (1.19) follow from (1.20).

2.2 Limiting relations for the dipole potential

For a *dipole (double layer) distribution* $\mu \in C(S)$ the *dipole potential*

$$\Psi(x) = \frac{1}{4\pi} \int\limits_{S} \mu(y) \frac{\partial}{\partial n_y} \frac{1}{|x - y|} dS(y) \tag{2.5}$$

is harmonic in $\mathbb{R}^3 \backslash S$. On S the *limiting relations*

$$\Psi|_{\pm} = \frac{1}{2}(T'\mu \pm \mu) \tag{2.6}$$

with improper scalar parameter integral

$$T'\mu(x) := \frac{1}{2\pi} \int\limits_{S} \mu(y) \frac{\partial}{\partial n_y} \frac{1}{|x - y|} dS(y), \quad x \in S, \tag{2.7}$$

hold in the sense of uniform convergence as to S. This, especially, allows the continuous extension of Ψ to each side of S, so Ψ may be considered as a continuous function in D as well as in D' (but, of course, not in $\mathbb{R}^3$). From (2.6) the *jump relation*

$$\Psi|_+ - \Psi|_- = \mu \tag{2.8}$$

results. Furthermore, for the difference of the normal derivatives of the dipole potential (2.5) it holds

$$\left.\frac{\partial \Psi}{\partial n}\right|_+ - \left.\frac{\partial \Psi}{\partial n}\right|_- = 0 \tag{2.9}$$

uniformly on S with the restriction that the argument points approach to S in the *Cauchy sense*, i.e., with equal distance from both sides [5, 6, 13].

Example 3. Let S be a disk as before bearing now a homogeneous dipole distribution $\mu = 1$. Then for $x = (0, 0, x_3)$, $x_3 \neq 0$, we calculate from (2.5)

$$\Psi(x) = \frac{1}{4\pi} \int_S \left.\frac{\partial}{\partial y_3} \frac{1}{\sqrt{y_1^2 + y_2^2 + (x_3 - y_3)^2}}\right|_{y_3=0} dS(y)$$

$$= \frac{1}{4\pi} \int_0^{2\pi} \int_0^1 \frac{x_3 \varrho d\varrho d\varphi}{\sqrt{\varrho^2 + x_3^2}^3} = \frac{1}{2} \left(\frac{x_3}{|x_3|} - \frac{x_3}{\sqrt{1 + x_3^2}} \right),$$

$$\frac{\partial \Psi(x)}{\partial x_3} = -\frac{1}{2\sqrt{1 + x_3^2}^3};$$

hence

$$\Psi|_+(0) - \Psi|_-(0) = 1, \qquad \left.\frac{\partial \Psi}{\partial n}\right|_+(0) - \left.\frac{\partial \Psi}{\partial n}\right|_-(0) = 0.$$

2.3 The scalar integral operators T and T'

In the context of the limiting relations considered before, the scalar linear integral operators[3]

$$T, T' : C(S) \to C^\alpha(S) \tag{2.10}$$

are established from the improper integrals (2.3) and (2.7) for any $\alpha \in (0, 1)$ [5, 32]. Due to (2.10) there is a "smoothing effect" for the operators T and T' as they map the space $C(S)$ into the *smaller* subspace $C^\alpha(S)$. A property close to this is the *compactness* of the operators T and T'.

The operators T and T' are called the *electrostatic* or *adjoint electrostatic integral operators*, respectively. Here adjointness means

$$\int_S (T\varepsilon)\mu \, dS = \int_S \varepsilon(T'\mu)dS \tag{2.11}$$

which is obtained for any two elements $\varepsilon, \mu \in C(S)$ by interchanging the order of integration. Thus the *Fredholm alternative* can be applied. Next to

---

[3] By $C^\alpha(S)$ the space of **uniformly Hölder continuous** functions on S with Hölder exponent $\alpha \in (0, 1)$ is denoted.

this, possibly with the exception of an eigenvalue 0, the *spectra* $\sigma(T)$ *and* $\sigma(T')$ *coincide* inclusively the *finite multiplicities* of the eigenvalues.

For more understanding of $\sigma(T)$ and by it of $\sigma(T')$, let $\lambda \in \mathbb{C}$ be an eigenvalue of T with corresponding (complex) eigenfunction $\varepsilon \in C(S)\backslash\{0\}$, so

$$T\varepsilon = \lambda\varepsilon. \tag{2.12}$$

For the scalar potential Φ of ε and its conjugate $\overline{\Phi}$ we consider the "energy integrals"

$$I_- := \int_D (\operatorname{grad}\Phi, \operatorname{grad}\overline{\Phi})dV, \quad I_+ := \int_{D'} (\operatorname{grad}\Phi, \operatorname{grad}\overline{\Phi})dV, \tag{2.13}$$

which are obviously *non-negative real* quantities. If we assume $I_- + I_+ = 0$, so it must be $I_- = I_+ = 0$. Then by definiteness, (2.13) yields $\operatorname{grad}\Phi = 0$ in $\mathbb{R}^3\backslash S$ and the jump relation (2.4) leads to the contradiction $\varepsilon = 0$. Hence

$$I_- + I_+ \neq 0. \tag{2.14}$$

For (2.13) it follows by means of Green's first integral theorem, the limiting relations (2.2), and (2.12) that

$$I_{\mp} = \pm \int_S \left.\frac{\partial\Phi}{\partial n}\overline{\Phi}\right|_{\mp} dS = \pm\frac{1}{2}\int_S (T\varepsilon \pm \varepsilon)\overline{\Phi} \, dS = \frac{1\pm\lambda}{2}\int_S \varepsilon\overline{\Phi} \, dS.$$

From this together with (2.14), the eigenvalue λ becomes

$$\lambda = \frac{I_- - I_+}{I_- + I_+}. \tag{2.15}$$

Thus for the spectra of T and T' we obtain

$$\sigma(T), \sigma(T') \subseteq [-1, 1], \tag{2.16}$$

as it has first been shown by Plemelj in 1911 [29]; especially, all eigenvalues of T and T' are *real*.

2.4 The operators T and T' for the sphere

Let now S be a sphere with center $x_0 \in \mathbb{R}^3$ and radius $R > 0$. Using for $x, y \in S$, $x \neq y$ the identities (Fig. 2.1)

$$R^2 = |(x - x_0) - (x - y)|^2 = R^2 - 2R(n(x), x - y) + |x - y|^2,$$

$$R^2 = |(y - x_0) + (x - y)|^2 = R^2 + 2R(n(y), x - y) + |x - y|^2,$$

we calculate

$$\frac{\partial}{\partial n_x}\frac{1}{|x-y|} = -\left(n(x), \frac{x-y}{|x-y|^3}\right)$$

$$\frac{\partial}{\partial n_y}\frac{1}{|x-y|} = \left(n(y), \frac{x-y}{|x-y|^3}\right)$$

$$\left.\vphantom{\frac{\partial}{\partial n_y}}\right\} = -\frac{1}{2R|x-y|}$$

and find from (2.3) and (2.7) that T *and* T' *coincide* as

$$T\varepsilon(x) = T'\varepsilon(x) = -\frac{1}{4\pi R}\int_S \frac{\varepsilon(y)}{|x-y|}dS(y), \quad x \in S. \tag{2.17}$$

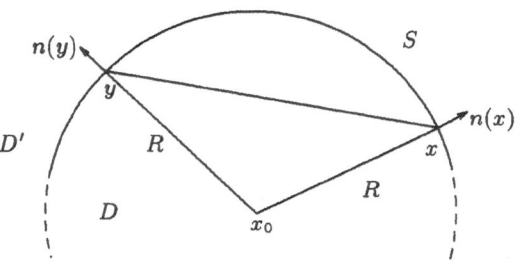

Fig. 2.1. Calculation of T and T' for the sphere

By reasons of continuity, the representation of a spherical harmonic K_ν of order $\nu \ge 0$ by (1.4) remains valid for $|x - x_0| = R$. Thus together with (2.17) we obtain

$$TK_\nu = T'K_\nu = -\frac{1}{2\nu+1}K_\nu, \quad \nu = 0, 1, 2, \ldots, \tag{2.18}$$

and

$$\lambda_\nu = -\frac{1}{2\nu+1}, \quad \nu = 0, 1, 2, \ldots, \tag{2.19}$$

turn out as the eigenvalues of T and T' with corresponding eigenfunctions K_ν. Since there are $2\nu + 1$ linear independent spherical harmonics K_ν of order ν, the eigenvalues (2.19) have *at least* the multiplicities $2\nu + 1$. By completeness arguments it is shown that *all* eigenvalues are given by (2.19) which have *exactly* the multiplicities $2\nu + 1$.

2.5 The eigenvalues $-1, 1$ for T, T'

Again for the general geometry, there are p' linear independent eigenfunctions $\mu_1, \mu_2, \ldots, \mu_{p'}$ to the eigenvalue -1 for T' which are easily verified as (see Fig. 1.1)

$$\begin{aligned}
\mu_1 &= 1 \quad \text{on} \quad \partial D_1, \quad \mu_1 = 0 \quad \text{on} \quad S \backslash \partial D_1, \\
\mu_2 &= 1 \quad \text{on} \quad \partial D_2, \quad \mu_2 = 0 \quad \text{on} \quad S \backslash \partial D_2,
\end{aligned}$$

$$\cdots\cdots\cdots\cdots\cdots\cdots\cdots\cdots\cdots\cdots\cdots\cdots\cdots \tag{2.20}$$

$$\mu_{p'} = 1 \quad \text{on} \quad \partial D_{p'}, \quad \mu_{p'} = 0 \quad \text{on} \quad S \backslash \partial D_{p'}.$$

Analogously, there are p linear independent eigenfunctions $\mu_1, \mu_2, \ldots, \mu_p$ to the eigenvalue 1 for T' with values

$$\begin{aligned}
\mu_1 &= 1 \quad \text{on} \quad \partial D_1', \quad \mu_1 = 0 \quad \text{on} \quad S \backslash \partial D_1', \\
\mu_2 &= 1 \quad \text{on} \quad \partial D_2', \quad \mu_2 = 0 \quad \text{on} \quad S \backslash \partial D_2',
\end{aligned}$$

$$\cdots\cdots\cdots\cdots\cdots\cdots\cdots\cdots\cdots\cdots\cdots\cdots\cdots \tag{2.21}$$

$$\mu_p = 1 \quad \text{on} \quad \partial D_p', \quad \mu_p = 0 \quad \text{on} \quad S \backslash \partial D_p'.$$

Keeping in mind the Fredholm alternative, the eigenvalues $-1, 1$ for the operators T, T' have the multiplicities[4]

$$\dim N(I + T) = \dim N(I + T') = p', \tag{2.22}$$

$$\dim N(I - T) = \dim N(I - T') = p. \tag{2.23}$$

Next we consider an eigenfunction $\varepsilon \in N(I + T)$ to the eigenvalue -1 for T. For the corresponding scalar potential Φ it follows from the limiting relations (2.2) that

$$\left.\frac{\partial \Phi}{\partial n}\right|_{-} = 0 \quad \text{on} \quad S.$$

Hence by Green's first theorem (or, alternatively, by the Hopf maximum principle; see also Fig. 1.1), the potential Φ attains *constant* values in each domain $D_1, \ldots, D_{p'}$ where, however, the constants may vary for different domains. Due to this property, a source (charge) distribution $\varepsilon \in N(I + T)$ is called an *equilibrium distribution (with respect to D)*. It is uniquely determined by given *total sources (total charges)*

$$\int_{\partial D_i} \varepsilon \, dS = E_i, \quad i = 1, \ldots, p', \tag{2.24}$$

on $\partial D_1, \ldots, \partial D_{p'}$, respectively [22, 23]. A physical realization is obtained from electric surface charges distributed on a *system of conductors $D_1, \ldots, D_{p'}$*.

Example 4. Let D be an *elliptical conductor* bounded by a three-axis ellipsoid S with center $x_0 \in \mathbb{R}^3$ and positive half-axis a, b, c. The equilibrium distribution on S uniquely determined by the total charge E (note with regard to (2.24) the Betti number $p' = 1$) can be explicitly given by

$$\varepsilon(x) = \frac{E}{4\pi abc}(n(x), x - x_0), \quad x \in S, \tag{2.25}$$

---

[4] By $N(\cdot)$ the *nullspace* of an operator is denoted.

where $(n(x), x - x_0) > 0$ describes the distance of the tangential plane at x from the center x_0 [13]. It is observed from (2.25) that the charges do not change sign and in absolute value heap up at the points of greatest curvature of S.

Analogously for an eigenfunction $\varepsilon \in N(I - T)$ to the eigenvalue 1 for T, the scalar potential Φ has the property

$$\left.\frac{\partial \Phi}{\partial n}\right|_{+} = 0 \quad \text{on} \quad S$$

and from this *constant* values in each domain $D'_0, D'_1, \ldots, D'_p$; in particular, as Φ tends to 0 at infinity, we have $\Phi = 0$ in D'_0. By this $\varepsilon \in N(I - T)$ is said an *equilibrium distribution (with respect to D')*. For such distribution uniqueness is assured by given *total sources (total charges)*

$$\int_{\partial D'_k} \varepsilon \, dS = E_k, \quad k = 1, \ldots, p, \tag{2.26}$$

on $\partial D'_1, \ldots, \partial D'_p$, respectively, which definitively do not refer to $\partial D'_0$.

Example 5. We consider a *spherical shell* D bounded by two concentric spheres S_1, S_2 with center $x_0 \in \mathbb{R}^3$ and radii $0 < R_1 < R_2$, respectively; for D there is the Betti number $p = 1$. The complement D' splits up into the "cavern" D'_1 inside S_1 and the unbounded domain D'_0 outside S_2. The uniquely determined equilibrium distribution ε on S with total charge E on $S_1 = \partial D'_1$ is easily verified as

$$\varepsilon = \frac{E}{4\pi R_1^2} \quad \text{on} \quad S_1, \quad \varepsilon = -\frac{E}{4\pi R_2^2} \quad \text{on} \quad S_2. \tag{2.27}$$

From (2.27) we calculate the potential in the shell

$$\Phi(x) = \frac{E}{4\pi} \left(\frac{1}{|x - x_0|} - \frac{1}{R_2} \right), \quad R_1 \leq |x - x_0| \leq R_2. \tag{2.28}$$

Because of continuity, (2.28) vanishes on $S_2 = \partial D'_0$. The situation here discussed models a *spherical capacitor*.

2.6 Limiting relations for the vector potential

Let us start from the linear function space

$$F(S) = \left\{ \gamma \in [C(S)]^3 \mid (n, \gamma) = 0 \text{ on } S, \text{ Div } \gamma \in C(S) \right\} \tag{2.29}$$

of continuous *surface vortex (current) distributions* γ directed tangentially to S and provided with a continuous surface divergence in the weak sense. Functions from (2.29) have already played a role in the context of the Lorentz condition. If we take (1.17) and (1.19) for $\gamma_0 = 0$, we see that for every $\gamma \in F(S)$ the *vector potential*

$$A(x) = \frac{1}{4\pi} \int\limits_S \frac{\gamma(y)}{|x - y|} dS(y) \tag{2.30}$$

and its *divergence* are continuous throughout $\mathbb{R}^3$; furthermore, the components of (2.30) are harmonic functions in $\mathbb{R}^3 \backslash S$. For the *tangential components* of rot A the *limiting relations*

$$[n, \mathrm{rot}\, A]|_\pm = -\frac{1}{2}(U\gamma \mp \gamma) \tag{2.31}$$

with improper vector parameter integral

$$U\gamma(x) := -\frac{1}{2\pi} \int\limits_S \left[n(x), \left[\mathrm{grad}_x \frac{1}{|x - y|}, \gamma(y) \right] \right] dS(y), \quad x \in S, \tag{2.32}$$

hold uniformly with respect to S [27]. From (2.31) we have the *jump relation*

$$[n, \mathrm{rot}\, A]|_+ - [n, \mathrm{rot}\, A]|_- = \gamma. \tag{2.33}$$

In addition for reasons which become clear later on, it is appropriate to introduce for $\delta \in F(S)$

$$U'\delta(x) := -U\delta(x), \quad x \in S. \tag{2.34}$$

Because of (2.31) and (2.34) the vector potential A of δ is due to the *limiting relations*

$$[n, \mathrm{rot}\, A]|_\pm = \frac{1}{2}(U'\delta \pm \delta). \tag{2.35}$$

2.7 The vector integral operators U and U'

It is elaborate and lengthy and not the purpose of this survey, to show that from the improper integrals (2.32) and (2.34) the vector linear integral operators

$$U, U' : F(S) \rightarrow F^\alpha(S) \tag{2.36}$$

with arbitrary $\alpha \in (0, 1)$ are established; here $F^\alpha(S)$ denotes the subspace of (2.29) where $\gamma \in [C^\alpha(S)]^3$ and Div $\gamma \in C^\alpha(S)$ [5, 24]. Thus for U and U' we have again a "smoothing effect". Furthermore, U and U' are *compact*.
The operators U and U' are called the *magnetostatic* or *adjoint magnetostatic integral operators*, respectively. Adjointness here means[5]

$$\int\limits_S [n, U\gamma, \delta] dS = \int\limits_S [n, \gamma, U'\delta] dS \tag{2.37}$$

for any two elements $\gamma, \delta \in F(S)$; it is again obtained by interchanging the order of integration. Thus the *Fredholm alternative* is applicable. From this,

---
[5] By $[\cdot, \cdot, \cdot]$ the *spat product* is denoted.

possibly with the exception of an eigenvalue 0, the *spectra $\sigma(U)$ and $\sigma(U')$ coincide* inclusively the *finite multiplicities* of the eigenvalues. Together with (2.34) it follows that if λ is an eigenvalue of U, so $-\lambda$ is also, and λ and $-\lambda$ have the *same multiplicities*. In particular, the spectra of U and U' are *symmetric*.

The electrostatic and magnetostatic integral operators are due to the *interchanging relation*

$$T\operatorname{Div}\gamma = \operatorname{Div}U\gamma, \tag{2.38}$$

valid for $\gamma \in F(S)$ [6]. With the aid of (2.38), for instance, we can conclude that if $\gamma \in F(S)$ satisfies the continuity condition, so $U\gamma \in F(S)$ does also, i.e., from $\operatorname{Div}\gamma = 0$ it follows that $\operatorname{Div}U\gamma = 0$.

In order to study the spectra of U and by it of U', let

$$U\gamma = \lambda\gamma \tag{2.39}$$

hold for an eigenvalue $\lambda \in \mathbb{C}$ with corresponding (complex) eigenfunction $\gamma \in F(S)\backslash\{0\}$. For the first case where the Lorentz condition *holds* for the vector potential (2.30), from $\operatorname{div}A = 0$ and $\Delta A = 0$ it results $\operatorname{rot}\operatorname{rot}A = 0$ in $\mathbb{R}^3\backslash S$. Furthermore, for the *non-negative real* quantities

$$I_- := \int\limits_D (\operatorname{rot}A, \operatorname{rot}\overline{A})dV, \quad I_+ := \int\limits_{D'} (\operatorname{rot}A, \operatorname{rot}\overline{A})dV, \tag{2.40}$$

we have $I_- + I_+ \neq 0$, since otherwise because of $\operatorname{rot}A = 0$ in $\mathbb{R}^3\backslash S$ the jump relation (2.33) would yield the contradiction $\gamma = 0$. Now by the Gauss theorem, the limiting relations (2.31), and (2.39) we can transform (2.40) into

$$I_\mp = \mp \int\limits_S [n, \operatorname{rot}A, \overline{A}]|_\mp \, dS = \pm\frac{1}{2}\int\limits_S (U\gamma \pm \gamma, \overline{A})dS = \frac{1\pm\lambda}{2}\int\limits_S (\gamma, \overline{A})dS,$$

thus arriving with the eigenvalue λ at (2.15).

For the second case where the Lorentz condition *does not hold* for the vector potential (2.30), we first state with regard to (1.19) that

$$\operatorname{div}A(x) = \frac{1}{4\pi}\int\limits_S \frac{\operatorname{Div}\gamma(y)}{|x-y|}dS(y) \tag{2.41}$$

is continuous in $\mathbb{R}^3$, harmonic in $\mathbb{R}^3\backslash S$, and 0 at infinity. Next for the *non-negative real* quantities

$$I_- := \int\limits_D (\operatorname{grad}\operatorname{div}A, \operatorname{grad}\operatorname{div}\overline{A})dV, \quad I_+ := \int\limits_{D'} (\operatorname{grad}\operatorname{div}A, \operatorname{grad}\operatorname{div}\overline{A})dV,$$

$$\tag{2.42}$$

it must be $I_- + I_+ \neq 0$, since otherwise we would get the contradiction $\operatorname{div}A = 0$ in $\mathbb{R}^3\backslash S$. By the Gauss theorem and the limiting relations (2.2) if applied to (2.41), we can transform (2.42) into

$$I_{\mp} = \pm \int\limits_{S} (n, \operatorname{grad} \operatorname{div} A) \operatorname{div} \overline{A}|_{\mp} \, dS = \pm \frac{1}{2} \int\limits_{S} (T \operatorname{Div} \gamma \pm \operatorname{Div} \gamma) \operatorname{div} \overline{A} \, dS.$$

Finally by (2.38) and (2.39) it follows that

$$I_{\mp} = \pm \frac{1}{2} \int\limits_{S} (\operatorname{Div} U \gamma \pm \operatorname{Div} \gamma) \operatorname{div} \overline{A} \, dS = \frac{1 \pm \lambda}{2} \int\limits_{S} \operatorname{Div} \gamma \operatorname{div} \overline{A} \, dS$$

and hence again (2.15). Thus for the spectra of U and U' we have the result first given by Kress in 1968 [16] that

$$\sigma(U), \sigma(U') \subseteq [-1, 1]; \tag{2.43}$$

especially, all eigenvalues of U and U' are *real*. For further results on the spectra of the electrostatic and magnetostatic operators see [20, 25].

Example 6. The eigenvalues and eigenfunctions of the magnetostatic operator U for the *sphere* are completely given by

$$\left. \begin{aligned} \lambda_{-\nu} &= -\frac{1}{2\nu + 1}, & \gamma_{-\nu} &= \operatorname{Grad} K_{\nu} &, \\[2mm] \lambda_{\nu} &= \frac{1}{2\nu + 1}, & \gamma_{\nu} &= [n, \operatorname{Grad} K_{\nu}] &, \end{aligned} \right\} \quad \nu = 1, 2, \ldots, \tag{2.44}$$

K_{ν} denoting a spherical harmonic of order $\nu \geq 1$ [20]. The eigenfunctions with negative indices do not satisfy the continuity condition, those with positive indices, however, do.

2.8 The eigenvalues $-1, 1$ for U, U'

In the following we also need the *first Betti numbers* q, q' for D, D'. These non-negative integers specify the number of topologically independent closed curves which are not contractible in D, D', respectively. Especially, for a *single* domain, bounded or not, the first Betti number is equal to the index of connection minus 1. Thus for the twice-connected interior and exterior of a *torus* we have the first Betti numbers $q = 1$ or $q' = 1$, respectively. As a further example, the *Borromean rings*[6] (Fig. 2.2) show the first Betti numbers $q = 3$ or $q' = 3$, respectively[7].

The multiplicities of the eigenvalues $-1, 1$ for the operators U, U' are given by [23, 36]

$$\dim N(I + U) = \dim N(I + U') = q', \tag{2.45}$$

$$\dim N(I - U) = \dim N(I - U') = q. \tag{2.46}$$

---

[6] The Borromean rings, just so as the Borromean Islands situated in the Lago Maggiore, are named after a Milanese noble family from the 17th and 18th century.

[7] With restriction to the $\mathbb{R}^3$, there is the coincidence $q = q'$.

Fig. 2.2. The Borromean rings

Mentioning that $U' = -U$ and therefore

$$N(I \pm U') = N(I \mp U), \tag{2.47}$$

we may restrict the discussion of the corresponding eigenfunctions to the eigenvalues $-1, 1$ for U.

Let $\gamma \in N(I + U)$ be an eigenfunction to the eigenvalue -1 for U. If the surface divergence Div is applied to $\gamma + U\gamma = 0$, it follows by means of (2.38) that $\text{Div}\,\gamma + T\,\text{Div}\,\gamma = 0$, i.e., $\text{Div}\,\gamma \in N(I + T)$. Since $\text{Div}\,\gamma$ vanishes in the mean over a *closed* surface and therefore is subject to *homogeneous* restraints (2.24), the continuity condition $\text{Div}\,\gamma = 0$ is established. From this the vector potential A of γ satisfies the Lorentz condition $\text{div}\,A = 0$ and so furthermore $\text{rot rot}\,A = 0$ in $\mathbb{R}^3 \backslash S$. Next by the Gauss theorem and the limiting relations (2.31) we calculate

$$\int\limits_D |\text{rot}\,A|^2 dV = -\int\limits_S [n, \text{rot}\,A, A]]_- \, dS = \frac{1}{2}\int\limits_S (U\gamma + \gamma, A) dS = 0$$

and are thus led to $\text{rot}\,A = 0$ in $\underline{D}$. Hence the vector potential A of γ turns out as a *harmonic vector field* in $\underline{D}$ (and thus as the counterpart of a *constant* scalar potential).

In view of the stated harmonicity of the vector potential A in $\underline{D}$, a vortex (current) distribution $\gamma \in N(I + U)$ is called an *equilibrium distribution (with respect to D)*. Uniqueness for such distribution is attained by given *total vortices (total currents)* $\Gamma_1, \ldots, \Gamma_{q'}$ with respect to q' topologically independent "control curves" on S [22, 23]. An equilibrium distribution is physically realized by an electric surface current on a *supra conductor D* where there is no magnetic field within. Due to the continuity of A with regard to S, the magnetic field outside the supra conductor turns out *tangential* to S. The tangential components on S are obtained from the jump relation (2.33).

Example 7. For a *circle torus S* the *toroidal* surface current in equilibrium with respect to the interior is, due to the Betti number $q' = 1$, uniquely determined by the *toroidal* total current. As a numerical result obtained for a torus with radii ratio of 24 per cent, the density at the inner azimuthal circle on S is about *5-times* the density at the outer azimuthal circle [21]. Thus the current is strongly concentrated at the "hole" of the torus.

Analogously, an eigenfunction $\gamma \in N(I - U)$ to the eigenvalue 1 for U is discussed. Since the corresponding vector potential A forms a *harmonic vector field* in $\underline{D'}$, a vortex (current) distribution $\gamma' \in N(I - U)$ is said an *equilibrium distribution (with respect to D')*. Uniqueness is obtained by q analogous restraints.

Example 8. Think of the Borromean rings (Fig. 2.2) as three disjoint *coils*, each of them bearing a *meridional* surface current in equilibrium with respect to the exterior. From these currents, independently from each other, azimuthal magnetic fields are induced within the rings. In the exterior, however, no magnetic field does arise. Due to the Betti number $q = 3$, the equilibrium distribution is uniquely determined by the *meridional* total currents on the three rings.

3. Boundary value problems, the boundary integral method

Representation theorems, limiting relations, and integral operators are the basic elements of the *boundary integral method*. There are mainly two different kinds how to carry out this method for solving boundary value problems in potential theory. For one thing, the solution is extended into the complement (i.e., into D' for an interior or D for an exterior problem, respectively,) in such a way that the representation theorem leads to a "simple" ansatz. The unknown parameters of this ansatz play a *mediate* role for the solution. Secondly, there is the *nullfield method* where the solution is extended into the complement by the values 0. Now the representation theorem yields an ansatz which is possibly "less simple", but refers to unknown parameters which are of *immediate* interest for the problem. In any case, boundary integral equations are obtained from the limiting relations where, in view of the Fredholm theory, one is preferably interested in Fredholm integral equations of the second kind. The essential ideas of the boundary integral method can already be pointed out for a simple geometry. Therefore for the following we shall restrict ourselves to the domains D and D' inside and outside a closed surface S. Here from (2.22) and (2.23) the multiplicities of the eigenvalues $-1, 1$ for the operators T, T' are given by

$$\dim N(I + T) = \dim N(I + T') = 1, \tag{3.1}$$

$$\dim N(I - T) = \dim N(I - T') = 0. \tag{3.2}$$

With regard to (3.1), we remember that an eigenfunction $\varepsilon_0 \in N(I + T)\backslash\{0\}$ has the meaning of an equilibrium distribution (with respect to D). It can be normed by its scalar potential

$$\frac{1}{4\pi}\int\limits_{S}\frac{\varepsilon_0(y)}{|x-y|}dS(y) = 1, \quad x \in D, \tag{3.3}$$

because if ε_0 had the potential 0 in D, then from continuity and the maximum principle the potential must be 0 also in D' and the jump relation (2.4) yields the contradiction $\varepsilon_0 = 0$. Any eigenfunction $\mu_0 \in N(I+T')\backslash\{0\}$ is a constant different from 0 and can trivially be normed to 1.

3.1 The scalar Dirichlet problem

We start with the *interior* scalar Dirichlet problem (first boundary value problem). For this problem a solution $u \in C^1(D) \cap C^2(\underline{D})$ of the Poisson equation

$$\Delta u = -\varrho \quad \text{in} \quad \underline{D} \tag{3.4}$$

satisfying the boundary condition

$$u|_- = -h \quad \text{on} \quad S \tag{3.5}$$

is required where $\varrho \in C(D) \cap C^\alpha(\underline{D})$, $h \in C^{1+\alpha}(S)$ are given for some $\alpha \in (0,1)$. A solution u in D can be harmonically extended into the exterior $\underline{D}'$ with boundary condition $\frac{\partial u}{\partial n}\big|_+ = \frac{\partial u}{\partial n}\big|_-$ on S and limit 0 at infinity (this is done by solving an exterior Neumann problem; see next chapter). Then by (1.1) the solution attains the representation

$$u(x) = \Phi(x) + \frac{1}{4\pi}\int\limits_{S}\mu(y)\frac{\partial}{\partial n_y}\frac{1}{|x-y|}dS(y), \quad x \in \underline{D}, \tag{3.6}$$

as a sum of the volume potential

$$\Phi(x) := \frac{1}{4\pi}\int\limits_{D}\frac{\varrho(y)}{|x-y|}dV(y) \tag{3.7}$$

belonging to $C^{1+\alpha}(\mathbb{R}^3) \cap C^2(\mathbb{R}^3\backslash S)$ [9] and a dipole potential of the distribution $\mu := u|_+ - u|_- \in C(S)$.

In the next step (3.6) and (3.7) are considered as an ansatz with unknown dipole distribution $\mu \in C(S)$ which at once satisfies the Poisson equation (3.4). By the limiting relations (2.6) it follows that the ansatz fulfills the boundary condition (3.5) if and only if $\mu \in C(S)$ satisfies the Fredholm integral equation of the second kind

$$\mu - T'\mu = 2(h + \Phi). \tag{3.8}$$

Due to the Fredholm theory and (3.2), there is exactly one solution $\mu \in C(S)$ of (3.8). Up to now we have merely used that $h \in C(S)$ which for (3.6) guarantees the regularity $u \in C(D) \cap C^2(\underline{D})$. But as we have assumed that $h \in C^{1+\alpha}(S)$, we see from (2.10) and (3.8) that $\mu \in C^\alpha(S)$. From this again together with (3.8), it can be concluded that $\mu \in C^{1+\alpha}(S)$ which finally gives us the regularity $u \in C^1(D)$ of the solution (3.6) [5]. Its uniqueness is assured by the maximum principle.

For the *exterior* scalar Dirichlet problem one looks for a function $u \in C^1(D') \cap C^2(\underline{D}')$ which is *harmonic* in $\underline{D}'$, satisfies boundary condition

$$u|_+ = h \quad \text{on} \quad S \tag{3.9}$$

where $h \in C^{1+\alpha}(S)$, and has the limit 0 at infinity. Here it is generally not possible to extend a solution into the interior as a harmonic function with the same normal derivatives on S, because there is the necessary integrability condition that the normal derivatives must vanish in the mean over S. As it can be seen from the scalar representation theorem (1.1), a solution cannot generally be represented merely by a dipole potential.

We therefore proceed otherwise. Assuming a solution u in D', we introduce the function

$$v := u - C\Phi_0 \in C^1(D') \cap C^2(\underline{D}') \tag{3.10}$$

where

$$\Phi_0(x) = \frac{1}{4\pi|x - x_0|} \tag{3.11}$$

is the potential of an unit source in a fixed chosen point $x_0 \in \underline{D}$ and the constant C is uniquely determined by the condition

$$\int_S \frac{\partial v}{\partial n} dS = \int_S \left(\frac{\partial u}{\partial n} - C \frac{\partial \Phi_0}{\partial n} \right) dS = \int_S \frac{\partial u}{\partial n} dS + C = 0. \tag{3.12}$$

Due to (3.12) the function v now can be harmonically extended into the interior $\underline{D}$ satisfying the boundary condition $\frac{\partial v}{\partial n}\big|_- = \frac{\partial v}{\partial n}\big|_+$ on S (this requires the solution of an interior Neumann problem; see next chapter). Then the scalar representation theorem (1.1) if applied to v, together with (3.10) yields the representation of the original solution

$$u(x) = C\Phi_0(x) + \frac{1}{4\pi} \int_S \mu(y) \frac{\partial}{\partial n_y} \frac{1}{|x - y|} dS(y), \quad x \in \underline{D}', \tag{3.13}$$

where the double layer potential refers to the density $\mu := v|_+ - v|_- \in C(S)$. Next we consider (3.13) together with the potential (3.11) as an ansatz where the constant $C \in \mathbb{R}$ and the distribution $\mu \in C(S)$ are unknown parameters. This ansatz immediately provides a harmonic function with limit 0 at infinity. With regard to the limiting relations (2.6), the boundary condition (3.9) is satisfied if and only if $C \in \mathbb{R}$ and $\mu \in C(S)$ are subject to the Fredholm integral equation of the second kind

$$\mu + T'\mu = 2(h - C\Phi_0). \tag{3.14}$$

By the Fredholm theory, (3.14) is solvable if and only if the right hand side is orthogonal to the general solution of the adjoint homogeneous equation $\varepsilon + T\varepsilon = 0$, or, equivalent to this, orthogonal to the non-trivial equilibrium distribution ε_0 satisfying (3.3). Since from (3.3) and (3.11) it follows that

$$\int_S \Phi_0 \varepsilon_0 \, dS = \frac{1}{4\pi} \int_S \frac{\varepsilon_0(x)}{|x - x_0|} dS(x) = 1, \tag{3.15}$$

the solvability condition just mentioned determines C uniquely by

$$\int_S (h - C\Phi_0)\varepsilon_0 \, dS = \int_S h\varepsilon_0 \, dS - C = 0. \tag{3.16}$$

The general solution $\mu \in C(S)$ of the integral equation (3.14) is now obtained by the sum of a special solution and an arbitrary constant (which is the general solution of the homogeneous equation). But this constant is without influence to the solution (3.13) when μ inserted. The regularity statements $\mu \in C^{1+\alpha}(S)$ and $u \in C^1(D')$ follow like before. Uniqueness again is assured by the maximum principle.

Example 9. We consider the exterior Dirichlet problem for the sphere S with center $x_0 \in \mathbb{R}^3$ and radius $R > 0$. The equilibrium distribution normed by (3.3) is $\varepsilon_0 = \frac{1}{R}$ on S. For the boundary values $u|_+ = 1$ on S the condition (3.16) yields $C = 4\pi R$. From (3.11) we have $\Phi_0 = \frac{1}{4\pi R}$ on S. Thus the right hand side of (3.14) vanishes, and the general solution of (3.14) becomes $\mu = \text{const.}$ By this (3.11) and (3.13) lead to the solution of the boundary value problem

$$u(x) = \frac{R}{|x - x_0|}, \quad |x - x_0| \geq R. \tag{3.17}$$

3.2 The scalar Neumann problem

For the *interior* Neumann problem (second boundary value problem) a solution $u \in C^1(D) \cap C^2(\underline{D})$ of the Poisson equation (3.4) is wanted which satisfies the boundary condition

$$\frac{\partial u}{\partial n}\bigg|_- = k \quad \text{on} \quad S \tag{3.18}$$

for given $k \in C^\alpha(S)$. From the Gauss theorem we get the necessary integrability condition that

$$\int_D \varrho \, dV + \int_S k \, dS = 0. \tag{3.19}$$

The harmonic extension of a solution u in D into the exterior $\underline{D}'$ with boundary values $u|_+ = u|_-$ on S and limit 0 at infinity is done by solving an exterior

Dirichlet problem. Thus the scalar representation theorem (1.1) leads to the ansatz

$$u(x) = \Phi(x) + \frac{1}{4\pi} \int\limits_S \frac{\varepsilon(y)}{|x-y|} dS(y), \quad x \in \underline{D}, \tag{3.20}$$

with the volume potential (3.7) and an unknown single layer distribution $\varepsilon \in C(S)$. By the limiting relations (2.2) together with (3.18), the Fredholm integral equation of the second kind

$$\varepsilon + T\varepsilon = 2\left(k - \frac{\partial\Phi}{\partial n}\right) \tag{3.21}$$

is obtained as a necessary and sufficient condition to $\varepsilon \in C(S)$ that (3.20) solves the boundary value problem. With regard to (2.10) and the right hand side of (3.21), a solution $\varepsilon \in C(S)$ of (3.21) automatically belongs to $C^\alpha(S)$ which for (3.20) guarantees the regularity $u \in C^1(D)$ [9].
In view of (3.1), the Fredholm theory assures that (3.21) is solvable if and only if the right hand side is orthogonal to the general solution of the adjoint homogeneous equation $\mu + T'\mu = 0$, i.e., orthogonal to an arbitrary constant; with other words, the right hand side of (3.21) has to vanish in the mean over S. This just follows from $\Delta\Phi = -\varrho$ in $\underline{D}$ and (3.19). The general solution $\varepsilon \in C(S)$ of (3.21) is given by the sum of a special solution and an arbitrary equilibrium distribution (which is the general solution of the homogeneous equation). From (3.3) and (3.20) it is seen that such equilibrium distribution merely changes (3.20) by an additive constant. The Hopf maximum principle shows that the solution of the interior Neumann problem is unique but a constant.
The *exterior* Neumann problem is simpler, since there is no integrability condition. This problem is formulated for a function $u \in C^1(D') \cap C^2(\underline{D}')$, *harmonic* in $\underline{D}'$, due to the boundary condition

$$\left.\frac{\partial u}{\partial n}\right|_+ = -k \quad \text{on} \quad S \tag{3.22}$$

where $k \in C^\alpha(S)$, and with limit 0 at infinity. By solving an interior Dirichlet problem with boundary condition $u|_- = u|_+$ on S we are led to the ansatz

$$u(x) = \frac{1}{4\pi} \int\limits_S \frac{\varepsilon(y)}{|x-y|} dS(y), \quad x \in \underline{D}', \tag{3.23}$$

with unknown single layer distribution $\varepsilon \in C(S)$.
Observing the limiting relations (2.2), the equivalent Fredholm integral equation of the second kind

$$\varepsilon - T\varepsilon = 2k \tag{3.24}$$

is obtained which with regard to (3.2) has a uniquely determined solution $\varepsilon \in C(S)$. From (2.10) and (3.24) it follows that $\varepsilon \in C^\alpha(S)$, so (3.23) attains the regularity $u \in C^1(D')$. The Hopf maximum principle together with the limit 0 at infinity assures the uniqueness of the solution of the boundary value problem (3.23).

3.3 The Robin problem

We restrict ourselves to the *exterior* Robin problem (third boundary value problem) for a function $u \in C^1(D') \cap C^2\underline{D}')$ which is *harmonic* in $\underline{D}'$, subject to the boundary condition

$$\left.\frac{\partial u}{\partial n}\right|_+ + \lambda u|_+ = -k \quad \text{on} \quad S \tag{3.25}$$

with given $k, \lambda \in C^\alpha(S)$, and has the limit 0 at infinity. By the same argument as for the exterior Neumann problem, the scalar representation theorem (1.1) leads to the ansatz (3.23) for an unknown single layer distribution $\varepsilon \in C(S)$. By means of the limiting relations (2.2) together with (3.25), the equivalent integral equation

$$(\varepsilon - T\varepsilon)(x) - \frac{\lambda(x)}{2\pi} \int_S \frac{\varepsilon(y)}{|x-y|} dS(y) = 2k(x), \quad x \in S, \tag{3.26}$$

is obtained. By the assumptions and (2.10) every solution of (3.26) belongs to $C^\alpha(S)$, so (3.23) turns out with the regularity $C^1(D')$.

Let us take the occasion of the exterior Robin problem also to demonstrate the *nullfield method*. It starts very simply by extending a solution u in D' into the interior $\underline{D}$ by 0. Thus (1.1) and (3.25) lead to the representation

$$u(x) = \Phi(x) + \frac{1}{4\pi} \left\{ \int_S \frac{\lambda(y)\mu(y)}{|x-y|} dS(y) + \int_S \mu(y) \frac{\partial}{\partial n_y} \frac{1}{|x-y|} dS(y) \right\} \tag{3.27}$$

for $x \in \mathbb{R}^3 \backslash S$ with the surface potential

$$\Phi(x) := \frac{1}{4\pi} \int_S \frac{k(y)}{|x-y|} dS(y) \tag{3.28}$$

belonging to $C^{1+\alpha}(D) \cap C^{1+\alpha}(D')$ [9] and $\mu := u|_+ \in C(S)$. Since (3.27) vanishes in $\underline{D}$, the *nullfield condition*[8]

$$u|_- = 0 \quad \text{on} \quad S \tag{3.29}$$

holds which by means of the limiting relations (2.6) is identical with the integral equation

---

[8] This condition has to be chosen in a suitable way dependent on the underlying problem which is treated by the nullfield method.

$$(\mu - T'\mu)(x) - \frac{1}{2\pi} \int_S \frac{\lambda(y)\mu(y)}{|x - y|} dS(y) = 2\Phi(x), \quad x \in S, \tag{3.30}$$

valid for the boundary values μ on S of a solution u in D'.

Vice versa, for a solution $\mu \in C(S)$ of the integral equation (3.30) we can show that (3.27) solves the Robin problem. From (3.30) we successively conclude the regularities $\mu \in C^\alpha(S)$ and $\mu \in C^{1+\alpha}(S)$ thus obtaining for (3.27) the wanted regularity $u \in C^1(D')$. From (3.30) it follows that (3.27) is due to the nullfield condition (3.29), so the maximum principle assures that $u = 0$ in $\underline{D}$. Next by the jump relations (2.4), (2.8), and (2.9), if applied to (3.27) and (3.28), we get

$$u|_+ = \mu, \quad \frac{\partial u}{\partial n}\bigg|_+ = -k - \lambda\mu \quad \text{on} \quad S. \tag{3.31}$$

From this μ gets the meaning of the boundary values of (3.27), and (3.27) satisfies the boundary condition (3.25).

Thus there are two integral equations (3.26) and (3.30) equivalent to the boundary value problem and by this equivalent to each other. This means that in praxis one has the choice between the two methods. The advantages of the nullfield method have become clear from its easy application and the close relation of the solution of the integral equation to the solution of the original boundary value problem. As a disadvantage of the method it may be considered that, contrary to (3.26), the inhomogeneity of (3.30) requires the computation of the potential (3.28) on S.

Both integral equations (3.26) and (3.30) are Fredholm integral equations of the second kind where the corresponding homogeneous integral equations are *adjoint* to each other. From this the Fredholm alternative yields the following. Either both homogeneous equations have *only* the trivial solution or they have the *same finite number* of linear independent solutions. In the first case both inhomogeneous equations and, by equivalence, the Robin problem itself are *uniquely solvable*. Thus the *uniqueness* for the Robin problem already assures the *existence* of a solution. This case, for instance, we have if $\lambda \leq 0$ on S, as it is easily verified from the homogeneous condition (3.25) by Green's first theorem. In the second case the integral equation (3.26) or (3.30) is solvable if and only if the right hand side is orthogonal to the general solution of the corresponding homogeneous equation to (3.30) or (3.26), respectively. The general solutions of (3.26) and (3.30) are given by the sums of a special solution and the general solution of the corresponding homogeneous equation.

Example 10. For the sphere S with radius $R > 0$ we consider the case $\lambda = $ const. From (2.17) the integral equations (3.26) and (3.30) become

$$\varepsilon - (1 - 2R\lambda)T\varepsilon = 2k, \tag{3.32}$$

$$\mu - (1 - 2R\lambda)T'\mu = 2\Phi. \tag{3.33}$$

Mentioning for the operators $T = T'$ the eigenvalues (2.19), there are non-trivial solutions of the homogeneous equations to (3.32) and (3.33) if

$$\lambda = \frac{\nu + 1}{R}, \quad \nu = 0, 1, 2, \ldots . \tag{3.34}$$

Since for fixed $\nu = 0, 1, 2, \ldots$ the corresponding eigenfunctions are given by the spherical harmonics K_ν of order ν, the inhomogeneous equations (3.32) and (3.33) are solvable if and only if

$$\int_S k K_\nu \, dS = 0 \tag{3.35}$$

holds for all spherical harmonics K_ν of order ν. The general solutions of (3.32) and (3.33) are given by a special solution plus an arbitrary spherical harmonic K_ν. Thus uniqueness of a solution of (3.33), for instance, can be attained from the condition

$$\int_S \mu K_\nu \, dS = 0 \tag{3.36}$$

for all spherical harmonics K_ν of order ν. Due to (3.36) and observing that $\mu = u|_+$ on S, all 2^ν-*multipole modes* (1.3) are omitted from the general solution u in D' of the exterior Robin problem.

If, on the other hand, λ differs from (3.34), then both integral equations (3.32) and (3.33) and thus the exterior Robin problem itself are uniquely solvable.

3.4 The vector Dirichlet and Neumann problems

We return to the original geometry with arbitrary Betti numbers. Additionally, we introduce a system of topologically independent closed orientated curves $C_1, \ldots, C_q \subset S$ not contractible in D. For these curves, the tangent t and the normal N lying in S are linked together by $t = [n, N]$. On the other hand we shall restrict ourselves to the interior problems which will be treated only by the nullfield method. Furthermore, no regularity discussion will be done in the following.

The vector Dirichlet or vector Neumann problems refer to solutions $v \in [C(D)]^3 \cap [C^1(\underline{D})]^3$ of the system of partial differential equations

$$\operatorname{div} v = \varepsilon_0, \quad \operatorname{rot} v = \gamma_0 \quad \text{in} \quad \underline{D} \tag{3.37}$$

due to the *Dirichlet boundary condition* and *global Dirichlet restraints*

$$[n, v|_-] = -\gamma \quad \text{on} \quad S, \quad -\int_{\partial D'_k} (n, v|_-) dS = E_k, \quad k = 1, \ldots, p, \tag{3.38}$$

or the *Neumann boundary condition* and *global Neumann restraints*

$$(n, v|_-) = -\varepsilon \quad \text{on} \quad S, \quad \int_{C_k} (t, v|_-)ds = \Gamma_k, \quad k = 1, \ldots, q, \tag{3.39}$$

respectively [17, 23]. For both problems certain integrability conditions have to be fulfilled.

The vector representation theorem (1.5) together with (1.6) and (1.7), the limiting and jump relations concerned, and the nullfield conditions $(n, v|_+) = 0$ or $[n, v|_+] = 0$ on S yield the equivalent Fredholm integral equations of the second kind with corresponding restraints

$$\varepsilon - T\varepsilon = \varphi(\varepsilon_0, \gamma_0, \gamma), \quad \int_{\partial D'_k} \varepsilon \, dS = E_k, \quad k = 1, \ldots, p, \tag{3.40}$$

or

$$\gamma - U\gamma = \psi(\varepsilon_0, \gamma_0, \varepsilon), \quad \int_{C_k} (N, \gamma)ds = \Gamma_k, \quad k = 1, \ldots, q, \tag{3.41}$$

respectively, for the missing boundary values ε or γ on S. The inhomogenities $\varphi \in C(S)$ and $\psi \in F(S)$ occuring in the integral equations are only dependent on the given data. From the Fredholm theory the integral equation problems (3.40) and (3.41) and thus the underlying vector boundary value problems are uniquely solvable.

Example 11. Let $v \in [C(D')]^3 \cap [C^1(\underline{D}')]^3$ be a *flux-free* harmonic vector field in $\underline{D}'$ with limit 0 at infinity. By solving an interior vector Neumann problem it can be extended into the interior $\underline{D}$ as a harmonic vector field with boundary values $(n, v|_-) = (n, v|_+)$ on S. The vector representation theorem (1.5), (1.6), and (1.7) then yields the representation

$$v = \operatorname{rot} A, \quad \operatorname{div} A = 0 \quad \text{in} \quad \underline{D}' \tag{3.42}$$

only by a vector potential A arising from some surface divergence-free vortex distribution on S.

3.5 The vector Poisson equation

From applications, for instance concerning the reflection of electromagnetic waves, one is interested in solutions $v \in [C^1(D)]^3 \cap [C^2(\underline{D})]^3$ of the vector Poisson equation

$$\Delta v = -\omega \quad \text{in} \quad \underline{D} \tag{3.43}$$

which satisfy the *Dirichlet (electric) boundary conditions*

$$\operatorname{div} v|_- = \mu, \quad [n, v|_-] = -\gamma \quad \text{on} \quad S \tag{3.44}$$

or the *Neumann (magnetic) boundary conditions*

$$[n, \operatorname{rot} v|_-] = \delta, \quad (n, v|_-) = -\varepsilon \quad \text{on} \quad S, \tag{3.45}$$

respectively [18, 36].

By applying the second order vector representation theorem (1.22), the null-field method leads to the equivalent systems of integral equations

$$\delta + U'\delta = \varphi_1(\omega, \mu, \gamma), \quad \varepsilon - T\varepsilon = \varphi_2(\omega, \mu, \gamma, \delta), \qquad (3.46)$$

or

$$\mu + T'\mu = \psi_1(\omega, \delta), \quad \gamma - U\gamma = \psi_2(\omega, \delta, \varepsilon, \mu), \qquad (3.47)$$

respectively, for the missing boundary values δ, ε or μ, γ on S. The inhomogeneities for these equations depend on the given data and, additionally with regard to φ_2 or ψ_2, on the solutions δ or μ of the integral equations to be solved in advance, respectively. Furthermore, integrability conditions and restraints have to be taken into consideration.

4. The boundary element method (BEM)

The BEM has been developed originally to treat integral equations that occur in the boundary integral formulation of boundary value problems in mechanics. The historical root goes back to 1963 when Jaswon treated singular integral equations arising in potential theory. BEM is applicable for the classical interior and exterior Dirichlet and Neumann problems for the Laplace equation on a (closed) compact domain $D \subset \mathbb{R}^3$ with smooth boundary $S = \partial D$. By straightforward use of representation theorems and surface potential ansatzes, integral equation formulations have been achieved for a variety of problems in mechanics, electrodynamics, and geodesy. BEM competes with the finite element method (FEM), which seems to be much easier accessible for engineers because of its local differential description rather than the integral approach of BEM. In the last two decades BEM has become more popular in engineering science as the number of publications in this area indicates.

In this chapter we will sketch the principles of BEM. We discuss two methods for solving integral equations for surfaces in $\mathbb{R}^3$, namely the *collocation method* and the *Galerkin method*.

For an exhaustive theoretical treatment of BEM we refer to the monographs of Hackbusch [12] and Kress [19]. Brebbia et al. [3] present a detailed discussion of practical aspects from the engineering point of view. We also want to point out the review articles of Atkinson [2] and Sloan [33].

4.1 General considerations

We discuss Fredholm integral equations of the second kind that arise in the boundary element formulation of the (scalar) interior and exterior Dirichlet and Neumann boundary value problems for the Laplace equation $\Delta u = 0$ for a compact domain $D \subset \mathbb{R}^3$ with smooth boundary $S = \partial D$.

Let X be a Banach space of functions on S equipped with a norm $\| \cdot \|$. By $L(X, X)$ we denote the normed space of bounded linear operators on X with operator-norm

$$\|A\| := \sup_{u \in X \setminus \{0\}} \frac{\|Au\|}{\|u\|}, \quad A \in L(X, X).$$

For instance, consider $X = C(S)$, the linear space of continuous functions on S with the maximum norm.

The second kind integral equations of potential theory have the form

$$(I - K)f = g \tag{4.1}$$

with $g \in X$ given and $f \in X$ unknown.

The operator I denotes the identity operator in X and

$$Kf(x) = \int_S k(x, y)f(y)dS(y), \quad x \in S \tag{4.2}$$

is a compact linear integral operator with weakly singular kernel $k(x, y)$, see [19]. For the classical Dirichlet and Neumann problems for the Laplace equation, K is either $\pm T$ from (2.3) or $\pm T'$ from (2.7). Note that $A := I - K \in L(X, X)$.

It is also possible to derive first kind integral equations

$$Kf = g$$

with K from (4.2) but one encounters the problem of ill-posedness and only the Galerkin method is well understood up to now. We will work out an example for a first kind equation in the next chapter. For a detailed study we refer to [15].

The BEM works with approximations K_n of K, that map X to a finite dimensional subspace X_n, which are considerably simpler than K itself. For the following we assume that $X_n \subset X_{n+1} \subset \ldots \subset X$. The approximated form of (4.1) is

$$(I - K_n)f_n = g_n \tag{4.3}$$

with $g_n \in X_n$ as an approximation to g. The solution $f_n \in X_n$ of (4.3) is taken as an approximation to f.

The questions that arise in this context are:

– Does f_n converge to f ?
– What is the error $\|f - f_n\|$?
– How fast does the error decrease?
– What is the effect of "incorrect" data g_n?

The amount of computation time and storage is rapidly increasing with n. Thus, the above questions are of practical importance. We cannot give satisfying answers to all of these questions — they are far from being cleared in general — but a "qualitative" statement of the behaviour of $\|f - f_n\|$ is possible.

Let $(I - K)^{-1}$, $(I - K_n)^{-1} \in L(X, X)$ for $n \in \mathbb{N}$. Then we have

$$f - f_n = (I - K_n)^{-1}\{(K - K_n)f + g - g_n\} \qquad (4.4)$$

and hence

$$\|f - f_n\| \leq \|(I - K_n)^{-1}\| \{\|K - K_n\|\,\|f\| + \|g - g_n\|\}.$$

The proof is straightforward: from (4.3) we obtain

$$
\begin{aligned}
f - f_n &= (I - K_n)^{-1}\{(I - K_n)f - g_n\} \\
&= (I - K_n)^{-1}\{(K - K_n)f + g - g_n\}
\end{aligned}
$$

where we have used $f = Kf + g$. The triangle inequality and $\|Ax\| \leq \|A\|\,\|x\|$ for $A \in L(X, X)$ yields the error estimate

$$\|f - f_n\| \leq \|(I - K_n)^{-1}\|\{\|K - K_n\|\,\|f\| + \|g - g_n\|\}.$$

Hence the error $\|f - f_n\|$ is decomposed into two components: $\|K - K_n\|$ is the error which is incorporated by approximation to K by K_n, and $\|g - g_n\|$ is the error in the boundary values.

In the following we discuss two methods for BEM namely the *collocation method* and the *Galerkin method*. Before going into details we study what both methods have in common. The collocation and the Galerkin method can be interpreted as *projection methods*.

A *projection* $\Pi_n : X \to X_n$ with $X_n \subset X$ and $\dim X_n < \infty$ is a linear mapping that satisfies $\Pi_n^2 = \Pi_n$.
What we actually want is to find a fairly good approximation $f_n \in X_n$ to the solution $f \in X$ of

$$(I - K)f = g.$$

This produces the defect

$$d_n = (I - K)f_n - g \in X$$

which, in general, does not vanish. The best we can do is to determine f_n such that

$$\Pi_n d_n = 0,$$

from which we obtain

$$(I - K_n)f_n = g_n \tag{4.5}$$

with $K_n f := \Pi_n K f \in X_n$ and $g_n := \Pi_n g \in X_n$. Note that $f_n \in X_n$, which is in general not true for integral equations of the first kind.

For projection methods we have the following result. If $\Pi_n x \to x$ for all $x \in X$ and if K is compact then $\|K - K_n\| \to 0$ for $n \to \infty$. If $(I-K)^{-1}, (I-K_n)^{-1} \in L(X, X)$ for $n \in \mathbb{N}$, then it follows $f_n \to f = (I - K)^{-1}g$. The error $f - f_n$ can be written in terms of the *projection error* of f,

$$f - f_n = (I - K_n)^{-1}(f - \Pi_n f) \tag{4.6}$$

and we obtain

$$\Pi_n f - f_n = (I - K_n)^{-1} K_n (f - \Pi_n f). \tag{4.7}$$

For the first assertion we refer to [12]. To verify (4.6), we consider (4.4),

$$f - f_n = (I - K_n)^{-1}\{(K - K_n)f + g - g_n\}$$

with $K_n = \Pi_n K$, $g_n = \Pi_n g$. From $f = Kf + g$ and $f_n = K_n f_n + g_n$ we deduce

$$
\begin{aligned}
f - f_n &= (I - K_n)^{-1}\{f - K_n f - g_n\} \\
&= (I - K_n)^{-1}\{f - \Pi_n(Kf + g)\} \\
&= (I - K_n)^{-1}(f - \Pi_n f).
\end{aligned}
$$

By $I = (I - K_n) + K_n$ it follows $(I - K_n)^{-1} = I + (I - K_n)^{-1}K_n$ and from the first equation we obtain

$$f - f_n = f - \Pi_n f + (I - K_n)^{-1}K_n(f - \Pi_n f)$$

which implies (4.7).

4.2 The collocation method

Consider, e.g., $X = C(S)$ and a n-dimensional subspace

$$X_n := \operatorname{span}\{\varphi_1, \varphi_2, \ldots, \varphi_n\}, \quad \varphi_1, \varphi_2, \ldots, \varphi_n \in X,$$

and n points $\xi_1, \xi_2, \ldots, \xi_n \in S$.
The *interpolation* $\Pi_n f = \sum_{j=1}^{n} a_j \varphi_j$ of $f \in X$ with respect to X_n and $\xi_1, \xi_2, \ldots, \xi_n \in S$ is defined by

$$(\Pi_n f)(\xi_i) = f(\xi_i), \quad i = 1, \ldots, n. \tag{4.8}$$

Interpolation is a projection from X to X_n. It is well defined if the matrix $(\varphi_i(\xi_k))_{i,k=1,\ldots,n}$ is regular. This is true for the Lagrange-basis with $\varphi_i(\xi_k) = \delta_i^k$.

The integral equation

$$(I - K)f = g \tag{4.9}$$

in the Banach space X is to be solved in X_n in the sense of the above projection. We search for $f_n \in X_n$ that fulfills (4.9) in the *collocation points* $\xi_1, \xi_2, \ldots, \xi_n \in S$, i.e.,

$$(I - K)f_n(\xi_i) = g(\xi_i), \quad i = 1, \ldots, n. \tag{4.10}$$

For $f \in X$ define $K_n f := \Pi_n K f \in X_n$ by $K_n f(\xi_i) = K f(\xi_i)$, $i = 1, \ldots, n$, and for the right-hand side $g_n := \Pi_n g$. Then we have the *collocation equation*

$$(I - K_n)f_n = g_n \tag{4.11}$$

for the unknown function $f_n \in X_n$.

The ansatz

$$f_n = \sum_{k=1}^{n} a_k \varphi_k \tag{4.12}$$

leads to the system of linear equations for the coefficients $a_1, a_2, \ldots, a_n$

$$\sum_{k=1}^{n} a_k \varphi_k(\xi_i) - \sum_{k=1}^{n} a_k K_n \varphi_k(\xi_i) = g(\xi_i), \quad i = 1, \ldots, n$$

or in matrix form

$$(\underline{A} - \underline{B})\underline{a} = \underline{b} \tag{4.13}$$

with

$$
\begin{aligned}
\underline{A} &= ((\varphi_k(\xi_i)))_{i,k=1,\ldots,n}, \\
\underline{B} &= ((K\varphi_k(\xi_i)))_{i,k=1,\ldots,n}, \\
\underline{a} &= (a_i)_{i=1,\ldots,n}, \quad \underline{b} = (g(\xi_i))_{i=1,\ldots,n}.
\end{aligned}
$$

The linear system $(\underline{A} - \underline{B})$ is regular if (4.11) is uniquely solvable. The matrix $\underline{B}$ is, in general, dense. Setting up $\underline{B}$ requires n^2 integrations

$$\underline{B}_{i,j} = K\varphi_j(\xi_i) = \int_S k(\xi_i, y)\varphi_j(y)dS(y), \quad i,j = 1, \ldots, n.$$

Unless known, these integrals have to be evaluated numerically. This is usually the most time consuming part of the procedure and it depends heavily on the choice of the φ_i. The linear system (4.13) can be solved directly, e.g., by a

Gauss-Jordan procedure. For large n iterative linear solvers are required, e.g., GMRES which is a nonsymmetric variant of the CG method, see [31].

Practical implementation. In practice one often works with a triangulation S_N of the surface S rather than with S itself:

$$S \approx S_N = \bigcup_{i=1}^{N} \Delta_i, \quad \Delta_i : \text{ triangle with corners on } S. \tag{4.14}$$

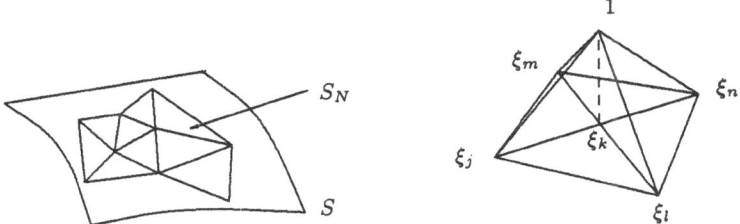

Fig. 4.1. Left: triangulation of S; right: the linear ansatz function φ_k

In Fig. 4.1, S_N is an interpolation of S that consists of straight triangles or *panels*. In general S_N might be a piecewise polynomial surface approximation to S of degree $p \geq 1$.

Take X_N as the set of all piecewise polynomials of degree r over S_N:

Piecewise constant ansatz. For $r = 0$ one can think of piecewise constant interpolation with the *midpoints* of Δ_k as collocation points ξ_k. Here we have

$$\varphi_k = \begin{cases} 1 & \text{on} \quad \Delta_k, \\ 0 & \text{else,} \end{cases} \tag{4.15}$$

and $\dim X_N = N$. Be aware that $\xi_k \notin S$.

Piecewise linear ansatz. Here we have $r = 1$ and the collocation points $\xi_1, \ldots, \xi_n \in S$ are the corners of S_N. For each ξ_k define a *piecewise linear* form function

$$\varphi_k(\xi_j) = \delta_k^j, \quad j = 1, \ldots, n \tag{4.16}$$

with

$$\operatorname{supp} \varphi_k = \{\Delta_j : \xi_k \text{ is corner of } \Delta_j\},$$

see Fig. 4.1. Hence $\dim X_N = n$.

For each Δ_i we introduce *barycentric coordinates* (u, v) by

$$\alpha_1(u, v) = u, \quad \alpha_2(u, v) = v, \quad \alpha_3(u, v) = 1 - u - v.$$

Then we have a parametrization of Δ_i

$$x(u, v) = \sum_{j=1}^{3} \alpha_j(u, v)\xi_{i_j}.$$

For a function $f(u, v)$ in Δ_i we take the linear ansatz

$$f(u, v) = \sum_{j=1}^{3} \alpha_j(u, v)f_{i_j}, \quad f_{i_j} = f(\xi_{i_j}).$$

Another reasonable choice is to take $r = 2$, that are piecewise quadratic polynomials, see [2].

Computation of the matrix. To compute the matrix $\underline{B}$ we have to evaluate

$$K_n\varphi_k(\xi_i) = \int_{S_n} k(\xi_i, y)\varphi_k(y)dS(y) = \int_{\Delta_j \in \text{supp} \{\varphi_k\}} k(\xi_i, y)\varphi_k(y)dS(y). \quad (4.17)$$

Note that for the constant ansatz we have supp $\{\varphi_k\} = \Delta_k$, whereas for the linear ansatz supp $\{\varphi_k\}$ consists of the Δ_j with corner ξ_k.
We sketch the case of the single-layer potential

$$K_n\varphi_k(x) = \int_{\Delta} \frac{\varphi_k(y)}{|x - y|}dS(y)$$

for a triangle Δ with corners $A(a^1, a^2, a^3)$, $B(b^1, b^2, b^3)$, $C(c^1, c^2, c^3)$. The computational problem is to evaluate the above potential for the case that $x = A$ where $k(x, y) = \frac{1}{|x-y|}$ becomes singular.

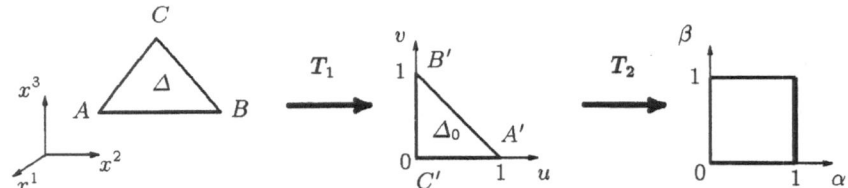

Fig. 4.2. Transformation of Δ to the unit square

First, the cartesian variables (x^1, x^2, x^3) are transformed to a local coordinate system (u, v) by

$$T_1 : \quad \begin{bmatrix} x^1 \\ x^2 \\ x^3 \end{bmatrix} = \begin{bmatrix} \Delta a^1 & \Delta b^1 & c^1 \\ \Delta a^2 & \Delta b^2 & c^2 \\ \Delta a^3 & \Delta b^3 & c^3 \end{bmatrix} \begin{bmatrix} u \\ v \\ 1 \end{bmatrix},$$

with

$$\Delta a^i = a^i - c^i, \quad \Delta b^i = b^i - c^i, \quad i = 1, 2, 3,$$

and Jacobian determinant $|G_1| = 2|\Delta|$, where $|\Delta|$ means the area of Δ, see Fig. 4.2. T_1 maps Δ to the "unit" triangle Δ_0 with corners $(0,0), (1,0), (0,1)$. Hence

$$
\begin{aligned}
\int_\Delta \frac{\varphi_i(y)}{|x-y|} dS(y) &= \int_{\Delta_0} \frac{\varphi_i(u,v)}{|x-y(u,v)|} \cdot 2|\Delta| du\, dv \\
&= 2|\Delta| \int_0^1 \int_0^{1-v} \varphi_i(u,v) \frac{1}{|x-y(u,v)|} du\, dv.
\end{aligned}
$$

A second transformation T_2 from Δ_0 to the unit-square, given by

$$T_2 : \quad u = \alpha, \quad v = \beta(1-\alpha)$$

with Jacobian determinant $|G_2| = (1-\alpha)$ yields

$$\int_\Delta \frac{\varphi_i(y)}{|x-y|} dS(y) = 2|\Delta| \int_0^1 \int_0^1 \varphi_i(\alpha,\beta) \frac{1-\alpha}{|x-y(\alpha,\beta)|} d\alpha\, d\beta.$$

Since

$$|x - y(\alpha, \beta)| = (1-\alpha) \left| \begin{pmatrix} \Delta a^1 - \Delta b^1\, \beta \\ \Delta a^2 - \Delta b^2\, \beta \\ \Delta a^3 - \Delta b^3\, \beta \end{pmatrix} \right|,$$

we have removed the singularity by straightforward transformation and the integral can be evaluated by a standard cubature formula.

For the case of x is another corner of Δ, one has to modify T_1 such that x has local coordinates $u = 1, v = 0$.

Error analysis. For error estimates we refer to [2]. When S is approximated with polynomial degree q, for piecewise polynomial collocation of degree r it has been shown that

$$\|f - f_n\|_\infty = O(h^{(\min\{r+1,q\})})$$

where $h = \max_{k=1,\dots,N} \mathrm{diam}(\Delta_k)$.

An alternative approach is to derive a boundary integral formulation of the boundary value problems for domains and surfaces with corners, e.g., for piecewise smooth surfaces. The integral equations differ from (4.9) in the corner points P by an additive term on the left-hand side which involves the interior solid angle $\Omega(P)$, see [2].

4.3 The Galerkin method

The Galerkin method has been mainly developed for convergence analysis in the variational approach to boundary value problems for partial differential equations. The base is the concept of *weak solutions*.

Let X be a Hilbert space with inner product $\langle \cdot, \cdot \rangle$ and norm $\|u\| = \sqrt{\langle u, u \rangle}$, e.g., $X = L^2(S)$ with $\langle u, v \rangle = \int_S u(y)\overline{v(y)} \, dS(y)$. f is said to be a *weak solution* of $(I - K)f = g$ in X, if

$$\langle (I - K)f, \varphi \rangle = \langle g, \varphi \rangle \quad \text{for all} \quad \varphi \in X. \tag{4.18}$$

Now consider a sequence $X_1 \subset X_2 \subset \ldots \subset X_n \subset X_{n+1} \subset \ldots \subset X$ of finite dimensional subspaces

$$X_n = \text{span}\{\varphi_1, \varphi_2, \ldots, \varphi_n\}$$

of X with $\overline{\lim_{n \to \infty} X_n} = X$. Using a triangulation S_N of S one can choose, for instance, X_n as the linear set of piecewise polynomials of degree $\leq r$ on S_N.

A possible alternative is the so-called *spectral Galerkin method*, where $\varphi_1, \varphi_2, \ldots$ are mutually orthogonal in X.

Let $\Pi_n : X \to X_n$ be the *orthogonal projection* from X to X_n. Then $u \in X$ can be written

$$u = \Pi_n u + (u - \Pi_n u)$$

with $\Pi_n u \in X_n$ and $\langle u - \Pi_n u, u \rangle = 0$.

Note that for $u, v \in X_n$ we have

$$u = v \quad \Longleftrightarrow \quad \langle u, \varphi \rangle = \langle v, \varphi \rangle \quad \text{for all} \quad \varphi \in X_n,$$

$$\Longleftrightarrow \quad \langle u, \varphi_i \rangle = \langle v, \varphi_i \rangle, \quad i = 1, \ldots, n,$$

because from $\varphi = u - v$ we obtain $\langle u - v, u - v \rangle = \|u - v\|^2 = 0$ and hence $u = v$. The second equivalence is because of $\{\varphi_1, \varphi_2, \ldots, \varphi_n\}$ is a basis of X_n.

The Galerkin method is to search an approximation $f_n \in X_n$ to the weak solution $f \in X$ of (4.18), i.e., a solution of

$$\langle (I - K_n)f_n, \varphi \rangle = \langle g_n, \varphi \rangle \quad \text{for all} \quad \varphi \in X_n, \tag{4.19}$$

with $K_n f := \Pi_n K f$ and $g_n := \Pi_n g$.

Because Π_n is self-adjoint with respect to $\langle \cdot, \cdot \rangle$ we have

$$\langle K_n f_n, \varphi_i \rangle = \langle \Pi_n K f_n, \varphi_i \rangle = \langle K f_n, \varphi_i \rangle, \quad i = 1, \ldots, n.$$

Equation (4.19) is equivalent to the *discrete Galerkin equations*

$$\langle f_n, \varphi_i \rangle - \langle K f_n, \varphi_i \rangle = \langle g, \varphi_i \rangle, \quad i = 1, \ldots, n. \tag{4.20}$$

The ansatz

$$f_n = \sum_{j=1}^{n} a_j \varphi_j \tag{4.21}$$

yields the system of linear equations

$$\sum_{j=1}^{n} a_j \langle \varphi_j, \varphi_i \rangle - \sum_{j=1}^{n} a_j \langle K \varphi_j, \varphi_i \rangle = \langle g, \varphi_i \rangle, \quad i = 1, \ldots, n, \tag{4.22}$$

or in matrix form

$$(\underline{A} - \underline{B})\underline{a} = \underline{b} \tag{4.23}$$

with

$$
\begin{aligned}
\underline{A} &= ((\langle \varphi_j, \varphi_i \rangle))_{i,j=1,\ldots,n}, \\
\underline{B} &= ((\langle K \varphi_j, \varphi_i \rangle))_{i,j=1,\ldots,n}, \\
\underline{a} &= (a_i)_{i=1,\ldots,n}, \qquad \underline{b} = (\langle g, \varphi_i \rangle)_{i=1,\ldots,n}.
\end{aligned}
$$

The computation of $\underline{B}$ with components

$$\underline{B}_{ij} = \int_S \int_S k(x, y) \varphi_i(x) \varphi_j(y) dS(y) \, dS(x)$$

is much more expensive than for the collocation method. For each element we have two double integrals versus one evaluation of a double integral in a collocation point. For a symmetric integral kernel $k(x, y) = k(y, x)$ the matrix $\underline{B}$ is obviously symmetric.

Therefore, in most cases, collocation is the method of choice. The advantage of the Galerkin method is that it naturally admits convergence results and error estimates.

Let $X = L^2(S)$, $K \in L(X, X)$ compact, $(I - K)^{-1} \in L(X, X)$ and X_n, $n \in \mathbb{N}$, such that

$$\text{dist}\,(x, X_n) = \inf_{y \in X_n} \|x - y\| \to 0$$

for all $x \in X$. Then the Galerkin method converges, say

$$f_n \to f = (I - K)^{-1} g.$$

We obtain an error estimate of type

$$\|f - f_n\| \leq \|(I - K_n)^{-1}\| \cdot \|f - \Pi_n f\|.$$

It remains to estimate the projection error $\|f - \Pi_n f\|$ for concrete norms. This is one reason for introducing more general function spaces such as Sobolev spaces.

We conclude this chapter with a short list of some facts about BEM and a perspective of present research.

Advantages:

– reduction of dimension, because integral equations are formulated for the boundaries
– exterior problems can be treated easily
– the solution and its derivatives are approximated with high rate of convergence in the interior
– in case of irregular data there is a smoothing effect because the solution is obtained via integration
– numerical computation of matrices (preprocessing) and evaluation of field data (postprocessing) is suited for parallel computers

Drawbacks:

– explicit knowledge of a fundamental solution of the differential equation is required
– for one problem there might exist several boundary integral formulations and to each of them there are different numerical methods
– the matrices are, in general, dense and numerically not well qualified
– the matrix elements are sometimes given as (weakly) singular integrals which causes numerical problems even in the preprocessing step
– non-smooth boundaries still cause problems

The present research of BEM is to overcome this difficulties. *Wavelet methods* are in use to obtain sparse matrix approximations, see, e.g., [8]. Fast summation methods such as the *fast multipole method* [10, 30] and *panel clustering* [11] have been developed to speed up matrix vector multiplication, evaluation of potentials and to avoid the compilation of the matrix.

5. Sobolev spaces – the modern approach to BEM

Sobolev spaces are required for convergence analysis of BEM, especially for the Galerkin method. We provide an elementary approach to Sobolev spaces of scalar functions on the unit sphere. This is less restrictive than it might appear since many integral equations that are formulated for simply connected surfaces can be transformed to the unit sphere.

We discuss mathematical techniques which stem from the theory of partial differential equations. Straightforward use of the concept of strong ellipticity, coercivity, the Lax-Milgram and Cea's lemma lead to both existence results and error estimates for approximate solutions of boundary integral equations. For further study we refer to [4].

5.1 Sobolev spaces on the sphere

Let S_2 be the unit sphere in $\mathbb{R}^3$ and $L^2(S_2)$ denote the Hilbert space of square integrable (complex valued) functions on S_2 with *inner product*

$$\langle u, v \rangle_{L^2} = \int_{S_2} u(y)\overline{v(y)}\, dS(y), \quad u, v \in L^2(S_2) \tag{5.1}$$

and *norm* $\|u\|_{L^2} = \sqrt{\langle u, u \rangle_{L^2}}$.

The *surface spherical harmonics* (SSH)

$$Y_n^m(\theta, \varphi) = \sqrt{\frac{2n+1}{4\pi} \frac{(n-|m|)!}{(n+|m|)!}} P_n^{|m|}(\cos\theta)e^{im\varphi}, \quad \theta \in [0, \pi], \ \varphi \in [0, 2\pi] \tag{5.2}$$

for $m = -n, \ldots, n$, $n \in \mathbb{N}$, form a complete orthonormal system in $L^2(S_2)$. Hence $f \in L^2(S_2)$ can be uniquely represented by a *Fourier series*

$$f = \sum_{n=0}^{\infty} \sum_{m=-n}^{n} f_n^m Y_n^m \quad \text{with} \quad f_n^m = \langle f, Y_n^m \rangle \tag{5.3}$$

where

$$\|f\|_{L^2}^2 = \sum_{n,m} |f_n^m|^2 < \infty.$$

For $s \geq 0$ define the *Sobolev space*

$$H_s(S_2) := \{f \in L^2(S_2) : \sum_{n,m} |f_n^m|^2 (1+n^2)^s < \infty\}, \tag{5.4}$$

or shortly H_s, with *inner product*

$$\langle u, v \rangle_{H_s} := \sum_{n,m} u_n^m \overline{v_n^m} (1 + n^2)^s \qquad (5.5)$$

and *norm*

$$\|u\|_{H_s} = \left(\sum_{n,m} |u_n^m|^2 (1 + n^2)^s \right)^{1/2}. \qquad (5.6)$$

H_s is a Hilbert space and the SSH form a complete orthogonal system in H_s. For $p > s \geq 0$ we have $(1 + n^2)^s \leq (1 + n^2)^p$ for $n \in \mathbb{N}$ and from this the imbedding $H_p \subset H_s$. We obtain the estimate for the norms $\|u\|_{H_p} \geq \|u\|_{H_s}$. In this sense $u \in H_p$ is "*smoother*" than $v \in H_s \setminus H_p$.

Hence a function $f \in H_s$, $s \geq 0$, can be represented by the sequence of its Fourier coefficients $(f_n^m)_{n \in \mathbb{N}, -n \leq m \leq n}$ which satisfies

$$\sum_{n,m} |f_n^m|^2 (1 + n^2)^s < \infty. \qquad (5.7)$$

Vice versa, a sequence $(f_n^m)_{n \in \mathbb{N}, -n \leq m \leq n}$ that satisfies (5.7) represents a function $f \in H_s$, namely the function that is generated by the Fourier series with coefficients (f_n^m). In this sense the function space H_s and the Hilbert space of sequences (f_n^m) satisfying (5.7) with inner product (5.5) and norm (5.6) can be identified.

For $s < 0$ define H_s in like manner as the space of sequences $(f_n^m)_{n \in \mathbb{N}, -n \leq m \leq n}$ by

$$H_s(S_2) = \{ f : \sum_{n,m} |f_n^m|^2 (1 + n^2)^s < \infty \}. \qquad (5.8)$$

This is obviously a larger space than $L^2(S_2)$. The question is whether this makes any sense, i.e., if there are any objects of interest that can be represented by sequences in H_s for $s < 0$. Because of (5.7) this objects cannot be classical functions anymore.

For given $s > 0$, H_{-s} might be interpreted as the space of the *bounded linear forms* from $H_s \to \mathbb{C}$. A linear form F on H_s is given by its values on a basis in H_s, e.g., for the SSH, say

$$F_n^m := F Y_n^m \in \mathbb{C} \quad \text{for all } n, m.$$

Hence for $g \in H_s$ it follows that

$$Fg = F \sum_{n,m} g_n^m Y_n^m = \sum_{n,m} g_n^m F Y_n^m = \sum_{n,m} g_n^m F_n^m.$$

From the triangle and the Cauchy-Schwarz inequality we obtain

$$|Fg|^2 \leq \left(\sum_{n,m} (1+n^2)^{-s/2} |F_n^m| \, (1+n^2)^{s/2} |g_n^m| \right)^2$$

$$\leq \sum_{n,m} (1+n^2)^{-s} |F_n^m|^2 \cdot \sum_{n,m} (1+n^2)^s |g_n^m|^2$$

$$= \|g\|_{H_s}^2 \cdot \sum_{n,m} (1+n^2)^{-s} |F_n^m|^2.$$

Hence for $(F_n^m) \in H_{-s}$ the second factor on the right-hand side of the above term is finite. Therefore $F \in H_{-s}$ is a *bounded linear form* on H_s and for the operator norm we obtain

$$\|F\| = \sup_{g \in H_s \setminus 0} \frac{|Fg|}{\|g\|_{H_s}} < \infty. \tag{5.9}$$

Vice versa, it can be shown that any bounded linear form on H_s can be uniquely represented by an element of H_{-s}, see [28].

Orthogonal Projection Π_N. For practical problems we work with the finite dimensional subspaces

$$X_N := \{ f \in H_s : f_n^m = 0, \quad n \geq N, \quad -n \leq m \leq n \} \tag{5.10}$$

of H_s with $\dim X_N = N^2$. Define the *orthogonal projection* $\Pi_N : H_s \to X_N$ by the cut-off of the Fourier series

$$\Pi_N f := f_N = \sum_{n < N, m} f_n^m Y_n^m. \tag{5.11}$$

The cut-off-error can be estimated in terms of two Sobolev norms: for $f \in H_p$ we obtain

$$\|f - \Pi_N f\|_{H_s} \leq \left(\frac{1}{N} \right)^{p-s} \|f\|_{H_p}, \quad p > s. \tag{5.12}$$

This is to be seen as follows: from $(1+n^2)^{s-p} \leq N^{2(s-p)}$ for $n \geq N$ we get

$$\|f - \Pi_N f\|_{H_s}^2 = \sum_{n \geq N, m} (1+n^2)^s |f_n^m|^2 = \sum_{n \geq N, m} |f_n^m|^2 (1+n^2)^p (1+n^2)^{s-p}$$

$$\leq N^{2(s-p)} \sum_{n \geq N, m} |f_n^m|^2 (1+n^2)^p$$

$$\leq N^{2(s-p)} \|f\|_{H_p}^2.$$

As expected, the cut-off-error decreases with increasing N. The cut-off-error in the H_s-norm decreases more rapidly with N when f is "very smooth", i.e., if $f \in H_p$ with $p \gg s$. The above estimate gives a *quantitative* relation between cut-off errors and smoothness and this will be important for the error analysis of the Galerkin method.

This "friendly" introduction to Sobolev spaces depends heavily on the geometry of S_2 and the properties of the SSH. It has been worked out in detail in [28].

For Sobolev spaces on other domains and surfaces than S_2 the theory of distributions and Fourier integrals are employed, see [4].

We note that for sufficiently smooth S, the piecewise continuous functions on S lie in $H_0(S)$. If $f \in C^{k-1}(S)$ and piecewise $C^k(S)$ then $f \in H_k(S)$ for $k = 1, 2, \ldots$. Using BEM in practice, one often encounters functions of this type. This is one reason because the H_s spaces "fit" to BEM.

5.2 Mapping properties of some integral operators in potential theory

We shall now see that the classical potential integral operators $K = S, T$ or T' are *smoothing* operators, i.e., if $f \in H_s$ then $Kf \in H_{s+1}$. As an example we study the single-layer potential operator

$$Sf(x) = \frac{1}{4\pi} \int_{S_2} \frac{1}{|x-y|} f(y) dS(y), \quad x \in S_2 \tag{5.13}$$

for $f \in H_s$.

From chapter 2 we remind that the Y_n^m, $n \in \mathbb{N}$, $-n \leq m \leq n$, are eigenfunctions of S, say

$$SY_n^m(x) = \frac{1}{2n+1} Y_n^m(x), \quad x \in S_2. \tag{5.14}$$

Let $f \in H_s$ with Fourier expansion (5.3). Then

$$Sf = S\left(\sum_{n,m} f_n^m Y_n^m\right) = \sum_{n,m} f_n^m SY_n^m = \sum_{n,m} \frac{1}{2n+1} f_n^m Y_n^m.$$

For $n \in \mathbb{N}$ we have

$$\frac{1}{16} \leq \frac{1+n^2}{(2n+1)^2} \leq 1,$$

and from this we derive

$$
\begin{aligned}
\|Sf\|_{H_{s+1}}^2 &= \sum_{n,m} \left(\frac{1}{2n+1}\right)^2 |f_n^m|^2 (1+n^2)^{s+1} \\
&\leq \sum_{n,m} |f_n^m|^2 (1+n^2)^s \\
&\leq \|f\|_{H_s}^2.
\end{aligned}
$$

Hence from $f \in H_s$ it follows $Sf \in H_{s+1}$, i.e., S is an *once smoothing* operator, $S : H_s \to H_{s+1}$, $s \in \mathbb{R}$. Speaking in terms of pseudo differential operators, S is of *order* -1.

The classical integral operators of potential theory, namely the double-layer potential operator T' from (2.7) as well as the electrostatic integral operator T from (2.3) map $H_s \to H_{s+1}$ for $s \in \mathbb{R}$.

Note that $S, T, T' : H_s(S) \to H_{s+1}(S)$ for $s \in \mathbb{R}$ when S is a sufficiently smooth surface which is the boundary of a closed bounded simply connected region in $\mathbb{R}^3$. In general this is not true for polyhedral surfaces. A detailed discussion of mapping properties of potential integral operators is to be found in [14] and in [35].

5.3 The concept of strong ellipticity

Many of the boundary integral equations that arise in practice are "strongly elliptic" and therefore "coercive" with respect to some appropriate Hilbert spaces. This provides a simple way of establishing existence and uniqueness of solutions (in weak sense) and yields a framework for analyzing the convergence of the Galerkin method.

To sketch this procedure we restrict ourselves to boundary integral equations of the form

$$Au = f \tag{5.15}$$

with functions u, f on S_2. For instance, consider $A = S$ with S from (5.13), i.e.,

$$\frac{1}{4\pi} \int\limits_{S_2} \frac{1}{|x - y|} u(y) dS(y) = f(x), \quad x \in S_2.$$

(5.15) has the "weak solution" u in the Hilbert space H with norm $||.||_H$, if

$$\langle Au, \varphi \rangle = \langle f, \varphi \rangle \qquad \text{for all} \quad \varphi \in H \tag{5.16}$$

where

$$\langle u, v \rangle := \int\limits_{S_2} u(y) \overline{v(y)} dS(y)$$

denotes the $L^2(S_2)$ inner product.

Lax-Milgram lemma: Existence of a unique weak solution $u \in H$ is guaranteed, if $\langle A\varphi, \psi \rangle$ is *bounded*, say

(i) $$|\langle A\varphi, \psi \rangle| \leq C ||\varphi||_H \cdot ||\psi||_H, \quad \varphi, \psi \in H,$$

and *coercive*, i.e.,

(ii)
$$\operatorname{Re}\langle A\varphi, \varphi\rangle \geq c\,\|\varphi\|_H^2, \quad \varphi \in H$$

with respect to the norm $\|.\|_H$ with constants $C, c > 0$, and, if $\langle f, \cdot\rangle$ is a bounded linear functional on H.

This is a direct consequence of the *Lax-Milgram theorem* which assures existence and uniqueness of u for equations of form

$$a(u, \varphi) = F(\varphi), \quad \varphi \in H \tag{5.17}$$

with a bounded coercive bilinear form $a(\varphi, \psi)$ and a bounded linear functional $F(\varphi)$ on H, see [4] or [9]. The form (5.17) is needed in the study of partial differential equations. In the context of boundary integral equations we have especially $a(\varphi, \psi) = \langle A\varphi, \psi\rangle$.

Now we verify boundedness and coercivity for the operator S from (5.13) in $H_{-1/2}$:

$$
\begin{aligned}
\langle A\varphi, \psi\rangle &= \langle S\varphi, \psi\rangle = \int_{S_2} S\varphi(y)\overline{\psi(y)}\,dS(y) \\[2mm]
&= \int_{S_2} \left\{ \sum_{n,m} \left(\frac{1}{2n+1}\varphi_n^m\, Y_n^m(y) \right) \sum_{k,l} \left(\overline{\psi_k^l}\,\overline{Y_k^l(y)} \right) \right\} dS(y) \\[2mm]
&= \sum_{n,m}\sum_{k,l} \frac{1}{2n+1}\varphi_n^m\, \overline{\psi_k^l}\, \underbrace{\langle Y_n^m, Y_k^l\rangle}_{= \delta_n^m \cdot \delta_k^l} \\[2mm]
&= \sum_{n,m} \frac{1}{\sqrt{2n+1}}\varphi_n^m\, \frac{1}{\sqrt{2n+1}}\overline{\psi_n^m}
\end{aligned}
$$

from which follows

$$
|\langle A\varphi, \psi\rangle| \leq \left(\sum_{n,m} \frac{1}{2n+1}|\varphi_n^m|^2 \right)^{1/2} \left(\sum_{n,m} \frac{1}{2n+1}|\psi_n^m|^2 \right)^{1/2}
$$

by the Cauchy-Schwarz inequality. From

$$\frac{1}{4} \leq \frac{\sqrt{1+n^2}}{2n+1} \leq 1, \quad n \in \mathbb{N}$$

we conclude

$$\frac{1}{4}\sum_{n,m}(1+n^2)^{-1/2}|\varphi_n^m|^2 \le \sum_{n,m}\frac{1}{2n+1}|\varphi_n^m|^2 \le \sum_{n,m}(1+n^2)^{-1/2}|\varphi_n^m|^2$$

from which we obtain the final estimate

$$\langle A\varphi, \psi\rangle \;\le\; \|\varphi\|_{H_{-1/2}}\|\psi\|_{H_{-1/2}} \qquad \text{and} \tag{5.18}$$

$$\langle A\varphi, \varphi\rangle \;\ge\; \frac{1}{4}\|\varphi\|_{H_{-1/2}}^2. \tag{5.19}$$

Hence we have established both boundedness and coercivity of $\langle A\varphi, \psi\rangle$ with $C = 1$, $c = 1/4$ with respect to the Sobolev norm in $H_{-1/2}$. For $f \in H_{1/2}$, the functional $\langle f, \cdot\rangle$ is linear and bounded on $H_{-1/2}$. Therefore we have the following result.

Let $f \in H_{1/2}$ be given. Then $Su = f$ has a weak solution $u \in H_{-1/2}$.

5.4 Galerkin method in Sobolev spaces

Recall that the Galerkin method for $Au = f$ is, to find a weak solution u_n in a finite dimensional subspace X_n of the Hilbert space H with norm $\|.\|_H$:

$$\langle Au_n, \varphi\rangle = \langle f, \varphi\rangle, \quad \varphi \in X_n$$

with

$$\langle u, v\rangle = \int_{S_2} u(y)\overline{v(y)}\,dS(y).$$

Let $\{\varphi_1, \varphi_2, \ldots, \varphi_n\}$ be a basis of X_n. Then the ansatz

$$u_n = \sum_{j=1}^n a_j\varphi_j$$

leads to the system of linear equations for $a_1, \ldots, a_n$

$$\sum_{j=1}^n a_j\langle A\varphi_j, \varphi_i\rangle = \langle f, \varphi_i\rangle, \quad i = 1, \ldots, n \tag{5.20}$$

from which we obtain $u_n \in X_n$ as an approximation to $u \in H$. The questions that naturally arise are:

– What is the error $\|u - u_n\|_H$?
– How is the error to be estimated?

The answer is given by the following lemma.

Céa's lemma: Let $\langle A\varphi, \psi \rangle$ satisfy the boundedness and coercivity properties for H with $C, c > 0$ and let f such that $Au = f$ has a weak solution $u \in H$. Then the Galerkin approximation (5.20) has a unique solution $u_n \in X_n$ which satisfies

$$\|u - u_n\|_H \leq \frac{C}{c} \inf_{v_n \in X_n} \|u - v_n\|_H. \tag{5.21}$$

For the proof we note that existence and uniqueness of u_n follows from the Lax-Milgram lemma, applied to $X_n \subset H$:

$$c\|u_n - u\|_H^2 \leq |\langle A(u_n - u), u_n - u \rangle| = |\langle A(u_n - u), v_n - u \rangle|$$

$$\leq C\|u_n - u\|_H \|v_n - u\|_H.$$

Céa's lemma reduces the *Galerkin error estimation* in the natural norm to a the estimation of the *error of approximation*.

The error of approximation depends on both X_n and the regularity of u. E.g., for X_n being the piecewise linear functions on S and for $u \in H_s$, $s \geq \frac{1}{2}$, it is possible to prove the asymptotic behaviour

$$\inf_{v_n \in X_n} \|u - v_n\|_{H_{1/2}(S)} \leq C\, h^{s-1/2} \|u\|_{H_s(S)}, \quad \frac{1}{2} \leq s \leq 2, \tag{5.22}$$

where C is a positive constant and h denotes the step size of discretization, i.e., $\operatorname{supp} \varphi_i = O(h^2)$, $i = 1, \ldots, n$.

Now at the end of this survey we discuss the spectral Galerkin scheme for our example

$$Su = f \quad \text{with} \quad f \in H_{1/2}, \quad \text{where} \quad u \in H_{-1/2}$$

with S from (5.13) and

$$X_N = \operatorname{span}\{Y_n^m : n < N, \quad -n \leq m \leq n\}. \tag{5.23}$$

From Céa's lemma we obtain the estimate

$$\|u - u_N\|_{H_{-1/2}} \leq 4 \inf_{v_N \in X_N} \|u - v_N\|_{H_{-1/2}}. \tag{5.24}$$

In the Hilbert space setting the approximation error reduces to the projection error. For the spectral Galerkin case of X_N according to (5.23) this is merely the cut-of error of the Fourier sum. Thus for the cases that we know that $u \in H_p$, $p > -\frac{1}{2}$, the right-hand side of (5.24) can be estimated by

$$\inf_{v_N \in X_N} \|u - v_N\|_{H_{-1/2}} = \|u - \Pi_N u\|_{H_{-1/2}} \leq \left(\frac{1}{N}\right)^{p+1/2} \|u\|_{H_p}.$$

Then for the Galerkin error we obtain

$$\|u - u_N\|_{H_{-1/2}} \le 4 \cdot \left(\frac{1}{N}\right)^{p+1/2} \|u\|_{H_p}. \tag{5.25}$$

This result is mainly of theoretical interest.

For the case that $u \in C^{k+\alpha}(S_2)$, i.e., u is k times Hölder-continuously differentiable with Hölder-exponent $\alpha \in (0,1)$, Atkinson [2] has given the following asymptotic formula for the error in the maximum norm

$$\|u - u_N\|_\infty \le O\left(\frac{1}{N^{k+\alpha-0,5}}\right). \tag{5.26}$$

Acknowledgement. One of the authors (S.R.) gives thanks to Professor Andreas Kirsch and to Klaus Giebermann for some good hints and fruitful discussions.

References

1. R. Adams, *Sobolev spaces*, Academic Press, San Diego 1975.
2. K.E. Atkinson, *The numerical solution of boundary integral equations*, The state of the art in numerical Analysis, I. Duff, A. Watson Eds., University Press, Oxford 1996.
3. C.A. Brebbia and J. Dominguez, *Boundary elements an introductory course*, Mc Graw Hill, New York 1989.
4. S.C. Brenner and L.R. Scott, *The mathematical theory of finite element methods*, Springer, New York 1994.
5. D. Colton and R. Kress, *Integral equation methods in scattering theory*, Wiley, New York 1983.
6. D. Colton and R. Kress, *Inverse acoustic and electromagnetic scattering theory*, Springer, Berlin/Heidelberg/New York 1992.
7. M. Costabel, *Principles of boundary element methods*, Computer Physics Reports 6 (1987), 243–274.
8. W. Dahmen, S. Prößdorf, and R. Schneider, *Multiscale methods for pseudo-differential equations on manifolds*, Wavelet Analysis and its Applications 5, C.K. Chui Ed., Academic Press, San Diego 1995.
9. D. Gilbarg and N.S. Trudinger, *Elliptic partial differential equations of second order*, Springer, Berlin/Heidelberg/New York 1977.
10. L. Greengard and V. Rokhlin, *A fast algorithm for particle simulations*, J. Comp. Phys. 73 (1987), 325–348.
11. W. Hackbusch and Z. Nowak, *On the fast matrix multiplication in the boundary element method by panel clustering*, Numer. Math. 54 (1989), 436–491.
12. W. Hackbusch, *Integral equations: theory and numerical treatment*, Birkhäuser, Basel 1994.
13. O.D. Kellogg, *Foundations of potential theory*, Springer, Berlin/Heidelberg/New York 1967.
14. A. Kirsch, *Surface gradients and continuity properties for some integral operators in classical scattering theory*, Math. Meth. Appl. Sci. 11 (1989), 789–804.

15. A. Kirsch, *An introduction to the mathematical theory of inverse problems*, Springer, Berlin/Heidelberg 1996.
16. R. Kress, *Über die Integralgleichung des Pragerschen Problems*, Arch. Rational. Mech. Anal. **30** (1968), 381–400.
17. R. Kress, *Grundzüge einer Theorie der verallgemeinerten harmonischen Vektorfelder*, Meth. Verf. Math. Phys. **2** (1969), 49–83.
18. R. Kress, *Die Behandlung zweier Randwertprobleme für die vektorielle Poissongleichung nach einer Integralgleichungsmethode*, Arch. Rational Mech. Anal. **39** (1970), 202–226.
19. R. Kress, *Linear integral equations*, Springer, Berlin/Heidelberg 1989.
20. R. Kress, *On the spectrum of the magnetostatic integral operator*, Inverse scattering and potential problems in mathematical physics, R.E. Kleinman, R. Kress, and E. Martensen Eds., Peter Lang, Frankfurt a.M./Bern 1995, 95–105.
21. E. Martensen, *Numerische Auflösung der Integralgleichung des Robinschen Problems für eine torusartige Berandung*, Symposium on the numerical treatment of ordinary differential equations, integral and integro-differential equations, Birkhäuser, Basel/Stuttgart 1960, 129–150.
22. E. Martensen, *Potentialtheorie*, Teubner, Stuttgart 1968.
23. E. Martensen, *Die Dualität des Robinschen und Pragerschen Problems in drei Dimensionen*, Arch. Rational Mech. Anal. **30** (1968), 360–380; **34** (1969), 405.
24. E. Martensen, *On geometric inequalities in potential theory*, Complex analysis and its applications, Nauka, Moscow 1978, 367–374.
25. E. Martensen, *A relationship between the electrostatic and magnetostatic integral operators*, Inverse scattering and potential problems in mathematical physics, R.E. Kleinman, R. Kress, and E. Martensen Eds., Peter Lang, Frankfurt a.M./Bern 1995, 119–127.
26. E. Martensen, *Analysis IV — Funktionentheorie mit Differentialgleichungen im Komplexen —*, Spektrum Akademischer Verlag, Heidelberg/Berlin/Oxford 1995.
27. C. Müller, *Foundations of the mathematical theory of electromagnetic waves*, Springer, Berlin/Heidelberg/New York 1969.
28. F. Nestel, *Zur numerischen Integration einiger Integralgleichungen auf der Sphäre*, Dissertation, Erlangen 1996.
29. J. Plemelj, *Potentialtheoretische Untersuchungen*, Preisschriften der Fürstlich Jablonowskischen Gesellschaft zu Leipzig, Teubner, Leipzig 1911.
30. V. Rokhlin, *Rapid solution of integral equations of classic potential theory*, J. Comp. Phys. **60** (1983), 187–207.
31. Y. Saad and M.H. Schulz, *GMRES: A generalized minimal residual algorithm for solving nonsymmetric linear systems*, SIAM J. Sci. Comp. **7** (1986), 856–869.
32. J. Schauder, *Potentialtheoretische Untersuchungen I*, Math. Z. **33** (1931), 602–640; **35** (1932), 536–538.
33. I. Sloan, *Error analysis for boundary integral methods*, Acta Numerica **1** (1992), 287–339.
34. M.R. Spiegel, *Vector analysis*, Mc Graw Hill, New York 1959.
35. W. Wendland, *Die Behandlung von Randwertaufgaben im $\mathbb{R}^3$ mit Hilfe von Einfach- und Doppelschichtpotentialen*, Numer. Math. **11** (1968), 380–404.
36. P. Werner, *On an integral equation in electromagnetic diffraction theory*, J. Math. Anal. Appl. **14** (1966), 445–462.

Boundary Value Problems for Harmonic Random Fields

Y. Rozanov and F. Sansò

1 Introduction

We consider Laplace equation

$$\Delta\xi = 0 \qquad (1.1)$$

and boundary value problems for harmonic functions $\xi \in \underline{W}(G)$ of a certain class $\underline{W}(G)$ in a bounded region $G \subseteq R^d$, whose behaviour near the boundary $\Gamma = \partial G$ is as much chaotic as we see in a case of generalized random fields ξ characterized by the stochastic equation

$$\Delta\xi = \eta \qquad (1.2)$$

with a stochastic source η of white noise type in G. The choice of $\underline{W}(G)$ is motivated by a few important reasons as follows.

We would like to have the corresponding sample $\xi \in \underline{W}(G)$ of chaotic boundary behaviour in the case of (1.2) with any random Schwartz distribution

$$\eta = (\varphi, \eta), \varphi \in C_0^\infty(G),$$

which is meansquare continuous over Schwartz test functions φ with respect to L_2-norm:

$$E|(\varphi, \eta)|^2 \le C\|\varphi\|_2^2 \ , \quad \varphi \in C_0^\infty(G) \ . \qquad (1.3)$$

We would like also to get a variety of stochastic boundary conditions such that the corresponding boundary data, given arbitrarily, determine a unique solution $\xi \in \underline{W}(G)$ of equation (1.1)/(1.2). And apart from of that, we would like not to fail in case, that the region G considered has inside a lot of singularities (totally of zero Lebesgue measure in R^d). These motivations lead us to *stochastic Sobolev spaces*

$$\underline{W}(G) = \underline{W}_2^p(G) \ , \quad p = 2 \ ,$$

of random Schwartz distributions

$$\xi = (\varphi, \xi) \ , \quad \varphi \in C_0^\infty(G) \ ,$$

which are meansquare continuous over Schwartz test functions φ with respect to the corresponding norm

$$\|\varphi\|_{-p} = \left(\int |\tilde{\varphi}(\lambda)|^2 (1 + |\lambda|^2)^{-p} d\lambda \right)^{1/2} \tag{1.4}$$

given by means of the corresponding Fourier transform [1].

Our approach to stochastic boundary problems for harmonic functions $\xi \in \underline{W}_2^p(G)$, $p = 2$, is based on the application of an appropriate *test function space*

$$X(G) = [C_0^\infty(G)]$$

which appears as a closure of all Schwartz test functions $\varphi \in C_0^\infty(G)$ with respect to norm (1.4).

---

[1] A material on classical (deterministic) Sobolev spaces can be found in nearly any book on Functional Analise and PDE-see, for example, 1 e 2.

It occurs that in this way we get

$$X(G) = \underline{W}_2^{-p}(G) \ , \quad p = 2 \ ,$$

as a space of all Schwartz distributions

$$x = (x, u) \ , u \in C_0^\infty(R^d) \ ,$$

with supports $\mathrm{supp}\,x \subseteq [G]$ in the closure $[G]$ of G, having a finite norm

$$\|x\|_X \asymp \|x\|_{-p} = \left(\int |\tilde{x}(\lambda)|^2 \left(1 + |\lambda|^2\right)^{-p} d\lambda \right)^{1/2} \ , \quad p = 2 \ . \quad (1.5)$$

Dealing with any $\xi \in \underline{W}(G)$, by meansquare continuity we can determine

$$(x, \xi) = \lim_{\varphi \to x} (\varphi, \xi)$$

for all $x \in X(G)$ and treat $x \in X(G)$ as test functions for

$$\xi = (x, \xi) \ , \quad x \in X(G) \ , \quad (1.6)$$

with the corresponding values (x, ξ) of ξ at x.

Speaking of boundary conditions for the random Schwartz distributions $\xi = (\varphi, \xi)$, $\varphi \in C_0^\infty(G)$, we have in mind to condition appropriate limit values $\lim(\varphi, \xi)$ which appear by means of appropriate test functions $\varphi \in C_0^\infty(G)$ with support in an infinitely small neighbourhoods Γ^ϵ of the boundary Γ, ($\Gamma^\epsilon \searcharrow \Gamma$ of course). Having in mind boundary conditions for possibly any $\xi \in \underline{W}(G)$, we have to verify the existence of these limit values $\lim(\varphi, \xi)$ for all ξ, what implies the existence of the limit $\lim \varphi = x$ in $X(G)$ for the corresponding φ. Thus the limit values we are speaking about are just certain values $(x, \xi) = \lim_{\varphi \to x}(\varphi, \xi)$ where the limits $x = \lim \varphi$ can be obtained by means of $\varphi \in C_0^\infty(G)$ with supports in $\Gamma^\epsilon \downarrow \Gamma$, and therefore $\mathrm{supp}\, x \subseteq \Gamma$.

We introduce a *boundary test functions* subspace

$$X(\Gamma) = \{x \in X(G) : \mathrm{supp}\, x \subseteq \Gamma\}$$

which consists of all Schwartz distributions $x \in X(G)$ with support on the boundary Γ.

As the multiplication by $w \in C_0^\infty(R^d)$ is well defined in our space $X(G)$, we see that for any $x = \lim \varphi$ in $X(G)$, with $\mathrm{supp}\, x \subseteq \Gamma$ we also have $x = \lim w\varphi$ for appropriate $w \in C_0^\infty(\Gamma^\epsilon)$, $wx = x$, and this shows that limit values $\lim(w\varphi, \xi)$ which can appear in boundary conditions are just *boundary values* (x, ξ), $x \in X(\Gamma)$ of the generalized function (1.6). Then we can consider any set of limit boundary values as

$$(x, \xi) = (x, \xi_+) \ , \quad x \in X^+(\Gamma) \ , \tag{1.7}$$

by choosing the corresponding set $X^+(\Gamma) \subseteq X(\Gamma)$ of the boundary test functions and appropriate values (x, ξ_+), $x \in X^+(\Gamma)$, which can be obtained by testing an appropriate stochastic sample $\xi_+ \in \underline{W}(G)$, (in this way we follow a certain general approach to stochastic boundary problems, see for example, [3]-[5].)

To conclude our introduction, we have to emphasize that $\xi \in \underline{W}(G)$ has to be considered as a generalized vector function having values $(\varphi, \xi), \varphi \in C_0^\infty(G)$, in the well know Hilbert space of random variables with the meansquare norm $(E|(\varphi, \xi)|^2)$. Any such function has an equivalent modification with all its realizations ξ_ω, $\omega \in \Omega$ (on a probability space Ω) being Schwartz distributions

$$\xi_\omega = (\varphi, \xi_\omega) \ , \quad \varphi \in C_0^\infty(G) \ .$$

Equation (1.1) can be treated as for the generalized vector functions ξ and, on the same time, as for all realizations ξ_ω of a proper modification of ξ. In the latter case, dealing with harmonic Schwartz distributions ξ_ω, we find them as classical (analytical) harmonic functions

$$\xi_\omega = \xi_\omega(t) \ , \quad t \in G \ ,$$

in the region G. However, our approach to stochastic boundary problems led us to determine just random variables

$$(\varphi, \xi_\omega) = \int \varphi(t)\xi_\omega(t)dt \ , \quad \omega \in \Omega \ ,$$

with probability 1 for every $\varphi \in C_0^\infty(G)$; then we can employ a delta-sequence $\varphi \to \delta_s$ to determine $\xi_\omega(s) = \lim \int \varphi(t)\xi_\omega(t)dt$ with probability 1 for every $s \in G$.

2 The stochastic Laplace equation

Let $\underline{L}_2(G)$ be a variety of random Schwartz distributions $\eta = (\varphi, \eta)$, $\varphi \in C_0^\infty(G)$, of type (1.3). A typical sample from this *stochastic L_2 space* can be represented by the so-called *white noise*, satisfying

$$E|(\varphi, \eta)|^2 = \|\varphi\|_{L_2}^2 \ , \quad \varphi \in C_0^\infty(G) \ ,$$

i.e., roughly speaking, having $|(\varphi, \eta)|^2$ in the average like $\|\varphi\|_{L_2}^2$ over $\varphi \in C_0^\infty(G)$ (this obviously shows a drastic difference from the deterministic case when $\eta \in L_2(G)$). Note, that with probability 1 realizations of general representives $\eta \in \underline{L}_2(G)$ are not better than Schwartz distributions of the Sobolev space $W_2^{-q}(G)$, $q > d/2$, depending on the dimension of $R^d \supseteq G$.

Let us show that with the Laplace operator Δ we have

$$\Delta \underline{W}_2^2(G) = \underline{L}_2(G) \ . \tag{2.8}$$

Of course, this statement concerns the Laplace (Poisson) equation (1.2) which means that

$$(\Delta \varphi, \xi) = (\varphi, \eta) \ , \quad \varphi \in C_0^\infty(G) \ . \tag{2.9}$$

Dealing with $\xi \in \underline{W}_2^2(G)$ and $\eta \in \underline{L}_2(G)$, we have the norm equivalence

$$\|\Delta \varphi\|_X^2 \asymp \int |\tilde{\varphi}(\lambda)|^2 \frac{|\lambda|^4}{(1+|\lambda|^2)^2} d\lambda \asymp \int |\tilde{\varphi}(\lambda)|^2 d\lambda \asymp \|\varphi\|_{L_2}^2 \ ;$$

this holds true thanks to the general fact that, for bounded and non-negative F_1, F_2,

$$\int |\tilde{\varphi}(\lambda)|^2 F_1(\lambda) \, d\lambda \asymp \int |\tilde{\varphi}(\lambda)|^2 F_2(\lambda) d\lambda \ , \quad \varphi \in C_0^\infty(G) \ ,$$

in any bounded region $G \subseteq R^d$ when

$$0 < \underline{\lim}_{\lambda \to \infty} \frac{F_1(\lambda)}{F_2(\lambda)} \leq \overline{\lim}_{\lambda \to \infty} \frac{F_1(\lambda)}{F_2(\lambda)} < \infty$$

see, for example [6]. Hence, for any $\xi \in \underline{W}_2^2(G)$ we have $\Delta \xi \in \underline{L}_2(G)$. On the other hand, for any $\eta \in \underline{L}_2(G)$ equation (1.2)/(2.9) determines ξ on a closure of $\Delta C_0^\infty(G)$ in our test function space $X(G) = W_2^{-2}(G)$,

$$[\Delta C_0^\infty(G)] = \Delta L_2(G) \subseteq X(G) \ ;$$

namely, we have

$$(x,\xi) = (f,\eta) \ , \quad x = \Delta f \ , \quad f \in L_2(G) \ . \tag{2.10}$$

And this ξ can be extended on $X(G)$ with

$$(x,\xi) = (x,\xi_+) \ , \quad x \in X^+ \tag{2.11}$$

by means of an appropriate sample $\xi_+ \in \underline{W}_2^2(G)$ on some direct complement X^+ to the subspace $\Delta L_2(G)$ in $X(G)$, as

$$(\varphi,\xi) = (\Delta f,\xi) + (x,\xi) = (f,\eta) + (x,\xi_+) \ , \quad \varphi \in X(G) \ , \tag{2.12}$$

according to the direct sum decomposition

$$\varphi = \Delta f + x \ , \quad f \in L_2(G) \ , \quad x \in X^+$$

see Fig. 1

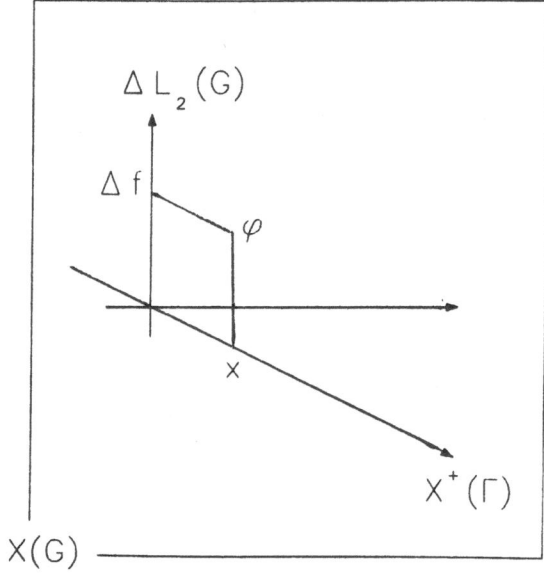

Fig. 1

We are to see that equation $(1.2)/(1.7)$ gives us nothing but ξ on the subspace $\Delta L_2(G) \subseteq X(G)$, see (2.10), so we need an additional information to determine $\xi \in \underline{W}_2^2(G)$; that information equivalently gives ξ on the corresponding direct complement X^+ to $\Delta L_2(G)$ in $X(G)$ by appropriate conditions (2.11) as *boundary conditions* by choosing some *boundary subspace*

$$X^+ = X^+(\Gamma) \subseteq X(\Gamma) \tag{2.13}$$

as in (1.7).

3   Direct complements to Laplace equation

To describe direct complements (2.13) to the subspace $\Delta L_2(G)$ in our test functions space $X(G) = [C_0^\infty(G)]$ we apply a certain Riesz representation of the dual space $W(G) = X(G)^*$- cfr. [5].

Let

$$X(R^d) = \left[C_0^\infty(R^d)\right] = W_2^{-p}(R^d) \ , \quad p = 2 \ ,$$

be a Hilbert space of all Schwartz distributions of type (1.4). Its dual space

$$W(R^d) = \left[C_0^\infty(R^d)\right] = W_2^p(R^d) \ , \quad p = 2 \ ,$$

is the closure of $C_0^\infty(R^d)$ with respect to the corresponding norm

$$\|u\|_W^2 \asymp \|u\|_p^2 = \int |\tilde{u}(\lambda)|^2 (1 + |\lambda|^2)^p d\lambda \ . \tag{3.14}$$

Our test functions space

$$X(G) = [C_0^\infty(G)] \subseteq X(R^d)$$

consists of Schwartz distributions with supports in the closure $[G]$, and we can consider them just in some (bounded) neighbourhood $\mathcal{O} \supseteq [G]$ as

$$x = (x, u) \ , \quad u \in C_0^\infty(\mathcal{O}) \ .$$

Let us turn to the closure

$$\overset{\circ}{W}(\mathcal{O}) = [C_0^\infty(\mathcal{O})] = \overset{\circ}{W}{}_2^p(\mathcal{O}) \ , \quad p = 2 \ ,$$

of $C_0^\infty(\mathcal{O})$ with respect to norm (3.14).

Considering $x \in C_0^\infty(\mathcal{O})$ as an element of the dual space

$$\overset{\circ}{X}(\mathcal{O}) = [C_0^\infty(\mathcal{O})] = \overset{\circ}{W}_2^{-p}(\mathcal{O}) \ , \quad p = 2 \ ,$$

by the application of an appropriate multiplier $w \in C_0^\infty(\mathcal{O})$, $wx = x$, we have

$$
\begin{aligned}
\|x\|_{\overset{\circ}{X}} \ &\asymp \ \sup_{u \in C_0^\infty(\mathcal{O}) : \|u\|_{\overset{\circ}{W}} \leq 1} |(x, u)| \leq && (3.15) \\
&\leq \ \sup_{u \in C_0^\infty(R^d) : \|u\|_W \leq 1} |(x, u)| \asymp \|x\|_X \asymp \\
&\asymp \ \sup_{u \in C_0^\infty(R^d) : \|u\|_W \leq 1} |(x, wu)| \leq \\
&\leq \ C \sup_{u \in C_0^\infty(\mathcal{O}) : \|u\|_{\overset{\circ}{W}} \leq 1} |(x, u)| \leq C \|x\|_{\overset{\circ}{X}} \ .
\end{aligned}
$$

Thus, we have the norm equivalence

$$\|x\|_X \asymp \|x\|_{\overset{\circ}{X}} \ , \quad x \in C_0^\infty(\mathcal{O}) \ . \tag{3.16}$$

Let us employ for $\overset{\circ}{W}(G)$ the following equivalent norm:

$$\|u\|_{\overset{\circ}{W}}^2 = \|\Delta u\|_{L_2}^2 \asymp \int |\tilde{u}(\lambda)|^2 |\lambda|^4 d\lambda \asymp \|u\|_p^2 \ , \quad p = 2 \ .$$

Then, with $u \in \overset{\circ}{W}(\mathcal{O})$ and $\mathcal{P} = \Delta^2$, the corresponding scalar product

$$\langle \varphi, u \rangle_{\overset{\circ}{W}} = (\mathcal{P}u, \varphi) \ , \varphi \in C_0^\infty(\mathcal{O}) \ ,$$

actually gives the *Riesz representation*

$$\mathcal{P} \overset{\circ}{W}(\mathcal{O}) = \overset{\circ}{X}(\mathcal{O}) \tag{3.17}$$

of the dual space $\overset{\circ}{X}(\mathcal{O}) = \overset{\circ}{W}(\mathcal{O})^*$ with its scalar product

$$\langle x, \mathcal{P}u \rangle_{\overset{\circ}{X}} = (x, u) , \quad x \in \overset{\circ}{X}(\mathcal{O}) , \quad u \in C_0^\infty(\mathcal{O}) .$$

In particular, this shows that an orthogonal complement to $\mathcal{P}C_0^\infty(G)$ in $\overset{\circ}{X}(\mathcal{O})$ consists of all Schwartz distributions with supports supp $x \subseteq G^c$, in the complement of G

$$\langle x, \mathcal{P}u \rangle_{\overset{\circ}{X}} = (x, u) = 0 , \quad u \in C_0^\infty(G) , \text{supp } x \subset \overline{\mathcal{O}} - \overline{G} .$$

Hence, the corresponding orthogonal complement to

$$\mathcal{P}C_0^\infty(G) \subseteq X(G) = [C_0^\infty(G)] \subseteq \overset{\circ}{X}(\mathcal{O})$$

in our test function space $X(G)$ consists of all boundary test functions x, supp $x \subseteq \Gamma$, since supp $x \subseteq [G]$ for all $x \in X(G)$, so we have

$$X(G) = [\mathcal{P}C_0^\infty(G)] + X(\Gamma) , \quad \mathcal{P} = \Delta^2 .$$

Taking in account that $\Delta^2 C_0^\infty(G) \subseteq \Delta L_2(G)$, we obtain a representation

$$X(G) = \Delta L_2(G) + X_0^+(\Gamma)$$

with the corresponding orthogonal complement $X_0^+(\Gamma) \subseteq X(\Gamma)$ to the subspace $\Delta L_2(G)$ in $X(G)$.

Hence, we see that there is a variety of direct complements $X^+(\Gamma) \subseteq X(\Gamma)$ to $\Delta L_2(G)$:

$$X(G) = \Delta L_2(G) + X^+(\Gamma) \qquad (3.18)$$

see fig. 2, which we can employ in boundary conditions (1.7)/(2.11).

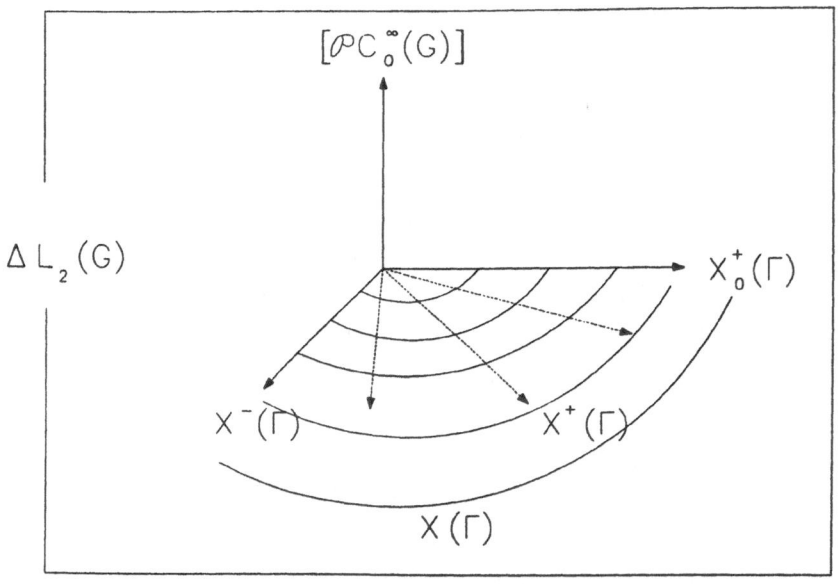

Fig. 2

Let us return to $\overset{\circ}{W}(\mathcal{O}) = \overset{\circ}{W}_2^p(\mathcal{O})$, $p = 2$, considering it as the dual space $\overset{\circ}{W}(\mathcal{O}) = \overset{\circ}{X}(\mathcal{O})^*$. A restriction of $u \in \overset{\circ}{W}(\mathcal{O})$ to the region G, that is on the subspace $[C_0^\infty(G)] = X(G)$ in $\overset{\circ}{X}(\mathcal{O})$, is a function of the Sobolev space $W(G) = W_2^p(G)$, $p = 2$, with the known norm

$$\|u\|_W^2 \asymp \sum_{|k| \le p} \int_G |\partial^k u|^2 dt \ .$$

Moreover, in the case of a regular boundary Γ any function from this space $W_2^p(G)$ can be extended up to some element $u \in \overset{\circ}{W}(\mathcal{O})$, so this space can be treated as dual to our test functions space $X(G) = W_2^{-p}(G)$, $p = 2$.

Hence, the following criterion for $X^+(\Gamma)$ to be a direct complement in representation (3.18) can be applied:

$X^+(\Gamma)$ is a direct complement of $\Delta L^2(G)$, if and only if the *deterministic Laplace equation*

$$\Delta u = f \ , \tag{3.19}$$

for any $f \in L_2(G)$ and zero boundary conditions

$$(x, u) = 0 \ , \quad x \in X^+(\Gamma) \ , \tag{3.20}$$

has a unique solution $u \in W_2^2(G)$.

Note that we can represent as in (3.20), with the corresponding boundary subspace $X^+(\Gamma) \subseteq X(\Gamma)$, any boundary condition

$$u|_\Gamma = 0 \tag{3.21}$$

given by means of a linear operators on $W_2^2(G)$, with a closed subspace $[u \in W_2^2(G) : u|_\Gamma = 0] \supseteq C_0^\infty(G)$, see [4].

4 General solution of the Laplace equation

Considering the homogeneous equation (1.1) with arbitrarily given stochastic boundary conditions (1.7)/(2.11) on $X^+ = X^+(\Gamma)$ we get a unique solution $\xi \in \underline{W}_2^2(G)$ if and only if $X^+(\Gamma)$ is a direct complement to the subspace $\Delta L_2(G)$ in our test function space $X(G) = W_2^{-2}(G)$.

The solution can be described as

$$(\varphi, \xi) = (x, \xi_+) \;, \quad \varphi \in X(G) = [C_0^\infty(G)] \;, \qquad (4.22)$$

where $x \in X^+(\Gamma)$ is a proper projection of $\varphi \in X(G)$ onto the boundary subspace $X^+(\Gamma)$ - see fig. 1.

Of course, for general boundary conditions $(1.7)/(2.11)$ with any boundary sample $\xi_+ \in \underline{W}_2^2(G)$, we can employ just a complete system of elements $x \in X^+(\Gamma)$. Let us take a basis x_n in the subspace $X^+(\Gamma)$, so that any $x \in X^+(\Gamma)$ can be represented as a strong limit

$$x = \lim_{n \to \infty} \sum_{k \leq n} c_k x_k = \sum_k c_k x_k \;. \qquad (4.23)$$

Then we can set our boundary conditions as

$$(x_k, \xi) = \xi_k \;, \quad k = 1, 2, \ldots \qquad (4.24)$$

by means of appropriate random variables ξ_k, $k = 1, 2, \ldots$, such that

$$E\left| \sum_k c_k \xi_k \right|^2 \leq C \sum_k |c_k|^2 \qquad (4.25)$$

for all linear combinations. To explain in more detail, we have

$$\sum_k c_k \xi_k = (x, \xi_+) \;, \quad x = \sum_k c_k x_k \;,$$

with some representative $\xi_+ \in \underline{W}_2^2(G)$ for which

$$E|(x, \xi_+)|^2 \leq C \|x\|_X^2 \;, \quad \|x\|_X^2 \asymp \sum_k |c_k|^2 \;.$$

Let us employ a *dual system* u_j of deterministic harmonic functions $u_j \in W_2^2(G)$ with

$$(x_k, u_j) = \begin{cases} 1, & j = k \\ \\ 0, & j \neq k \end{cases} \quad ; \qquad (4.26)$$

for each u_j, $\Delta u_j = 0$, this is a boundary condition of type (4.24). Then the coefficients c_k in representation (4.23) of any $x \in X^+(\Gamma)$ can be determined as

$$c_k = (x, u_k) \quad , \quad k = 1, 2, ...,$$

and, according to (4.22)-(4.25), we have

$$(\varphi, \xi) = (x, \xi_+) = \sum_k c_k \xi_k = \lim_{n \to \infty} \sum_{k \leq n} c_k \xi_k = \qquad (4.27)$$

$$= \lim_{n \to \infty} \sum_{k \leq n} (x, u_k) \xi_k = \lim_{n \to \infty} \left(x, \sum_{k \leq n} \xi_k u_k \right) =$$

$$= \lim_{n \to \infty} \left(\varphi, \sum_{k \leq n} \xi_k u_k \right) \quad , \quad \varphi \in X(G) \quad .$$

Thus, the following result holds true

Theorem: A general solution $\xi \in \underline{W_2^2}(G)$ of Laplace equation (1.1) can be obtained by boundary conditions of type (4.24) according to a limit formula

$$\xi = \sum_k \xi_k u_k = \lim_{n \to \infty} \sum_{k \leq n} \xi_k u_k \qquad (4.28)$$

as a series of the corresponding harmonic functions $u_k \in W_2^2(G)$ - see (4.26), with the (random) coefficients ξ_k, $k = 1, 2, ...$, defined in the boundary conditions (4.24).

We conclude here with the following remark. Let U be a subspace of all (deterministic) harmonic functions $u \in W_2^2(G)$. Considering $W_2^2(G)$ as the dual space to $X(G)$, we have U as the annihilator of the subspace $\Delta L_2(G)$ in $X(G)$. On the other hand, any direct complement X^+ to $\Delta L_2(G)$ in $X(G)$ can serve as a dual space to U, if we identify elements $x \in X^+$ and the corresponding function $x = (x, u)$, $u \in U$. Recall, that for the dual space $X(G) = W_2^{-2}(G)$ to $W_2^2(G)$ we have the direct sum representation (3.18) and $x = 0$ if $(x, u) = 0$, $u \in U$, since for any $x \in X^+$, $x \neq 0$, there is a harmonic function $u \in W_2^2(G)$ with $(x, u) \neq 0$. Hence *dealing with an appropriate basis u_j of harmonic functions in $U \leq W_2^2(G)$ we can identify it by boundary conditions of type (4.26) with the corresponding dual basis x_k in the dual space $X^+ = X^+(\Gamma)$.*

5 Boundary values characterization

We refer to [5] where it is shown that in a case of regular boundary Γ the test function subspace $X(\Gamma)$ is a direct sum

$$X(\Gamma) = \left[W_2^{-3/2}(\Gamma) \times \delta^{(0)} \right] + \left[W_2^{-1/2}(\Gamma) \times \delta^{(1)} \right] \ . \tag{5.29}$$

Here the subspaces

$$\left[W_2^{-(p-k-1/2)}(\Gamma) \times \delta^{(k)} \right] \quad ; \quad k = 0, 1; \quad p = 2$$

are correspondingly generated by the Schwartz distributions

$$x \times \delta^{(k)} = \int x(s) \delta_s^{(k)} ds \ :$$

$$(x \times \delta^{(k)}, u) = \int_\Gamma x(s) u^{(k)}(s) ds \ , \quad u \in C_0^\infty(R^d) \ . \tag{5.30}$$

which represent distributed delta-function $\delta_s^{(0)}$ and its derivative $\delta_s^{(1)}$ along a normal $\ell = \ell(s)$, $s \in \Gamma$, with weight-functions $x \in C(\Gamma)$, say. Moreover, we have the norm equivalence

$$\|x \times \delta^{(k)}\|_X^2 \;\asymp\; \|x \times \delta^{(k)}\|_{-p}^2 = \tag{5.31}$$

$$= \int_{R^d} \Big| \int_\Gamma x(s) e^{i\lambda s} (\lambda, \ell)^k ds \Big|^2 \left(1 + |\lambda|^2\right)^{-p} d\lambda \;\asymp\;$$

$$\asymp \|x\|_{-(p-k-1/2)}^2 \;, \quad x \in C(\Gamma), \quad p = 2 \;\; .$$

By means of boundary test functions (5.29) we can define a *boundary trace*

$$\xi^{(k)} = (x, \xi^{(k)}) \stackrel{def}{=} (x \times \delta^{(k)}, \xi), x \in C(\Gamma) \tag{5.32}$$

of the function $\xi = \xi^{(0)}$ ($k = 0$) and its generalized normal derivative $\partial \xi = \xi^{(1)}$ ($k = 1$) which are characterized by meansquare continuity over $x \in C(\Gamma)$ with respect to the corresponding norm $\|x\|_{-(p-k-1/2)}$ in Sobolev spaces $W_2^{-(p-k-1/2)}(\Gamma)$, $p = 2$, on the boundary Γ. We shall indicate this characterization as

$$\xi^{(k)} \in \underline{W}_2^{p-k-1/2}(\Gamma) \;, \quad p = 2 \;\; k = 0, 1 \;\; . \tag{5.33}$$

Thus, *all boundary values (x, ξ), $x \in X(\Gamma)$, are linear combinations of the corresponding values of $\xi^{(k)}$, $k = 0, 1$.*

6   The stochastic Dirichlet problem

In our general scheme (1.1), (2.11) we can set a *stochastic Dirichlet problem* with boundary conditions (2.11) on the boundary subspace

$$X^+(\Gamma) = W_2^{-3/2}(\Gamma) \times \delta^{(0)} \tag{6.34}$$

which can be given by means of an arbitrary stochastic sample $\xi_+^{(0)} \in \underline{W}_2^{3/2}(\Gamma)$ as

$$\xi^{(0)} = \xi_+^{(0)} \quad :$$

$$(x, \xi^{(0)}) = (x, \xi_+^{(0)}), \quad x \in C(\Gamma) \ . \tag{6.35}$$

In form (2.11) this means that $(x \times \delta^{(0)}, \xi) = (x \times \delta^{(0)}, \xi_+)$, $x \times \delta^{(0)} \in X^+(\Gamma)$, - see (5.29)-(5.33). To explain it further, we note that we can state the direct sum representation (3.18), according to our general criterion (3.19)-(3.20), since the classical Dirichlet problem:

$$\Delta u = f \ , \quad u^{(0)} = 0 \ ,$$

for any (deterministic) $f \in L_2(G)$ has a unique solution $u \in W_2^2(G)$, see for example, [2].

By the general formula (4.22) *we can describe a unique solution of our stochastic Dirichlet problem as*

$$(\varphi, \xi) = (x, \xi^{(0)}) \ , \quad \varphi \in X(G) \tag{6.36}$$

with corresponding $x \in W_2^{-3/2}(\Gamma)$ from the direct sum decomposition

$$\varphi = \Delta f + (x \times \delta^{(0)}) \ , \quad f \in L_2(G) \ , \quad x \times \delta^{(0)} \in X^+(G)$$

Note that in case of small dimensions $d \leq 3$ (recall, that $G \subseteq R^d$) all the stochastic Schwartz distributions $\xi \in \underline{W}_2^2(G)$ are regular in a sense that we can treat as point-wise functions

$$\xi = \xi(t) \ , \quad t \in [G]$$

meansquare continuous up to the boundary Γ. Indeed, in this case we have at our disposal delta-functions $\delta_t \in X(G)$ at all points $t \in [G]$, since $\tilde{\delta}_t(\lambda) = e^{i(\lambda, t)}$ and

$$\int_{R^d} (1 + |\lambda|^2)^{-2} d\lambda < \infty , \quad d \leq 3 ,$$

so that, according to (1.4), $\delta_t \in W_2^{-2}(G)$. Therefore we can set

$$\xi(t) = (\delta_t, \xi) , \quad t \in [G] , \tag{6.37}$$

getting

$$E|\xi(t) - \xi(s)|^2 \leq C \|\delta_t - \delta_s\|_X^2$$

with

$$\|\delta_t - \delta_s\|_X^2 \asymp \int_{R^d} |e^{i\lambda(t-s)} - 1|^2 (1 + |\lambda|^2)^{-2} d\lambda \to 0 , \quad t - s \to 0 .$$

After this specification, we find as we might expect,

$$\int \varphi(t)\xi(t)dt = (\varphi, \xi) , \quad \varphi \in C_0^\infty(G) .$$

For any harmonic $\xi \in \underline{W}_2^2(G)$, by the general formula (6.36), we have

$$\xi(t) = (x_t, \xi^{(0)}) = \int_\Gamma x_t(s)\xi(s)ds , \quad t \in [G] , \tag{6.38}$$

where $x_t \in W_2^{-3/2}(\Gamma)$ is derived from the direct sum decomposition

$$\delta_t = \Delta f + (x_t \times \delta^{(0)}) , \quad f \in L_2(G) , \quad x \times \delta^{(0)} \in X^+(G) ,$$

which is indeed applicable as $\delta_t \in X(G)$ as we have already explained. Along this line one can rederive some classical formulas for harmonic functions - as Poisson integral

$$\xi(r, \alpha) = \frac{1}{2\pi} \int_0^{2\pi} \frac{1 - r^2}{1 + r^2 - 2r\cos(\alpha - \beta)} \xi(\beta)d\beta$$

with polar coordinates (r, α) in the unit circle, and so on.

7 Negligible singularities

Suppose that, dealing with the function $\xi \in \underline{W}_2^2(G)$, we know that it is harmonic everywhere in the region G but for some singular (closed) set $\gamma \subseteq G$. Then, *if this set γ is of Lebesgue measure zero, the function ξ is harmonic in the whole region G.*

Indeed, the equation

$$(\Delta\varphi, \xi) = 0 \quad, \quad \varphi \in C_0^\infty(G - \gamma) \quad,$$

implies that $\xi = 0$ on the whole subspace $\Delta L_2(G)$, since $(\Delta\varphi, \xi)$ is meansquare continuous with respect to $\|\varphi\|_{L_2}$ and $C_0^\infty(G - \gamma)$ is dense in

$$L_2(G - \gamma) \equiv L_2(G) \quad.$$

As an example let us take G as an interval (a, b) with a singular point γ. In general a harmonic function ξ on $G - \gamma$ is represented by a broken line - see Fig. 3. However if $\xi \in \underline{W}_2^2(G)$, then it must be endowed with a continuous derivative, so ξ must be represented just by a straight-line.

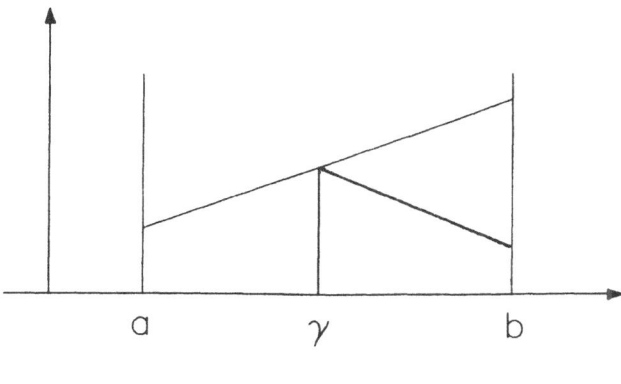

Fig. 3

8 Expectation and correlation

Suppose, now we are dealing with a harmonic random function $\xi \in \underline{W}_2^2(G)$, having a Gaussian (Normal) probability distribution. As it is well known, ξ is completely characterized by its *expectation function*

$$\mathcal{A} = (\varphi, \mathcal{A}) = E(\varphi, \xi) \ , \quad \varphi \in C_0^\infty(G) \ ,$$

and its *correlation function*

$$B = (\varphi, B\psi) = E\left[(\varphi, \xi) - (\varphi, \mathcal{A})\right]\left[(\psi, \xi) - (\psi, \mathcal{A})\right] \ , \quad \varphi, \psi \in C_0^\infty(G) \ ;$$

as it is apparent, $\mathcal{A}$ is a linear form on $C_0^\infty(G)$ while $\mathcal{B}$ is a quadratic form on $C_0^\infty(G) \times C_0^\infty(G)$. Dealing with any harmonic realization ξ_ω of the harmonic random function ξ, for any $t \in G$ we can define a delta-sequence $\varphi \to \delta_t$ and use it to get

$$\xi_\omega(t) = \lim_{\varphi \to \delta_t} \int \varphi(s)\xi_\omega(s)ds \ .$$

with probability 1. For Gaussian random variables

$$(\varphi, \xi) = \int \varphi(s)\xi_\omega(s)ds \quad , \quad \omega \in \Omega \quad ,$$

this limit holds also in meansquare sense and, therefore, we actually have ξ as a Gaussian harmonic function

$$\xi = \xi(t) \quad , \quad t \in G \quad ,$$

with its *expectation*

$$\mathcal{A}(t) = E\xi(t) \quad , \quad t \in G \quad ,$$

and *correlation*

$$B(s,t) = E\left[\xi(s) - \mathcal{A}(s)\right]\left[\xi(t) - \mathcal{A}(t)\right] \quad , \quad s,t \in G \quad .$$

defined pointwise in G. Naturally, for ξ as a generalized random function $\xi \in \underline{W_2^2}(G)$ we have its expectation in terms of $\mathcal{A}(t)$ as

$$\mathcal{A} = (\varphi, \mathcal{A}) = \int \varphi(t)\mathcal{A}(t)dt \quad , \quad \varphi \in C_0^\infty(G) \quad ,$$

and its correlation, in terms of $B(t,s)$, given by

$$B = (\varphi, B\psi) = \int \varphi(s)\psi(t)B(s,t)dsdt \quad , \quad \varphi, \psi \in C_0^\infty(G) \quad .$$

Suppose now we have to do with a $\xi \in \underline{W_2^2}(G)$, as a solution of Laplace equation determined by given boundary data

$$(x, \xi) = (x, \xi_+) \ , \quad x \in X^+(\Gamma) \ ,$$

of type (1.7)/(2.11). These now have to be treated as Gaussian random variables with expectation

$$\mathcal{A}_+ = (x, \mathcal{A}_+) = E(x, \xi_+) \ , \quad x \in X^+(\Gamma) \ ,$$

and correlation

$$B_+ = (x, B_+ y) = E\left[(x, \xi_+) - (x, \mathcal{A}_+)\right]\left[(y, \xi_+) - (y, \mathcal{A}_+)\right] \ , \quad x, y \in X^+(\Gamma)$$

Then we have both $\mathcal{A}, B$ as solutions $u \in W_2^2(G)$ of the Laplace equation

$$\Delta u = 0$$

with the respective boundary conditions

$$(x, u) = \left\{ \begin{array}{c} (x, \mathcal{A}_+) \\ \\ (x, B_+ y) \end{array} \right\} \ , \quad x \in X^+(\Gamma) \ ; \tag{8.39}$$

to explain, we have

$$(\varphi, \mathcal{A}) = (x, \mathcal{A}_+) \ , \quad \varphi \in X(G) \ ,$$

and

$$(\varphi, B\psi) = (x, B_+ y) \quad, \varphi, \psi \in X(G) \quad,$$

over our test function space $X(G) = [C_0^\infty(G)]$ with the direct sum decomposition

$$\varphi = \Delta f + x \ , \quad \psi = \Delta g + y \ , \quad f, g \in L_2(G) \ , \quad x, y \in X^+(\Gamma) \ ,$$

as explained in (3.18).

9 A filtering problem

Suppose, we have to determine a certain solution $\theta \in W_2^2(G)$ of the Laplace equation

$$\Delta \theta = 0$$

by means of appropriate boundary values

$$(x, \theta) \ , \quad x \in X^+(\Gamma)$$

of type (1.7)/(2.11). However instead of these values we have actually some other data which are represented by Gaussian random variables

$$(x, \xi) \ , \quad x \in X^+(\Gamma) \ ,$$

so that instead of the function we are looking for, we can just determine a $\xi \in \underline{W}_2^2(G)$ as a solution of the Laplace equation (1.1) with the corresponding boundary conditions (1.7)/(2.11). If we have some knowledge of some relation of $\xi|_\Gamma$ with θ, can we do something better?

Let us formulate this question more precisely. Suppose, we expect that $\theta \in \Theta$ i.e.that θ belongs to a certain set $\Theta \subseteq W_2^2(G)$ of harmonic functions.

Suppose further that the boundary data

$$(x, \xi) = (x, \theta) + (x, \xi_0) \tag{9.40}$$

differ from the values (x, θ) by Gaussian random variables (x, ξ_0), $x \in X^+(\Gamma)$, having zero expectation. Then we can treat (x, ξ), $x \in X^+(\Gamma)$, as random variables with a Gaussian distribution P_θ, having an unknown expectation

$$(x, \theta) = E_\theta(x, \xi) \ , \quad x \in X^+(\Gamma) \ ,$$

and a given correlation

$$E_0(x, \xi_0)(y, \xi_0) = E_\theta \left[(x, \xi) - (x, \theta) \right] \left[(y, \xi) - (y, \theta) \right] \ , \quad x, y \in X^+(\Gamma) \ .$$

The question we want to address, is to determine a harmonic function $\hat{\theta} \in \underline{W}_2^2(G)$ such that

$$E_\theta \hat{\theta}(t) = \theta(t) \ , \quad E_\theta |\hat{\theta}(t) - \theta(t)|^2 = \min \tag{9.41}$$

at every point $t \in G$, say, or equivalently,

$$E_\theta(\varphi, \hat{\theta}) = (\varphi, \theta) \ , \quad E_\theta |(\varphi, \hat{\theta}) - (\varphi, \theta)|^2 = \min \tag{9.42}$$

for every test function $\varphi \in X(G)$.

Here it is natural to assume that the Gaussian distributions P_θ are not singular with respect to each other for different $\theta \in \Theta$, including $\theta \equiv 0$, and this implies that

$$E_0\left[(x,\xi)\eta_\theta\right] = (x,\theta) \ , \quad x \in X^+(\Gamma) \ , \tag{9.43}$$

for some element η_θ, $(\theta \in \Theta)$, belonging to the linear meansquare closure H of all the random variables $(x,\xi), x \in X^+(\Gamma)$, with respect to the probability distribution P_0, see for example [6]. Formula (9.43) with $(x,\theta) = E_\theta(x,\xi)$ can be extended over all $\eta \in H$ as

$$E_0\left[\eta \cdot \eta_\theta\right] = E_\theta\eta \ , \quad \theta \in \Theta \ . \tag{9.44}$$

Let $\hat{H}$ be the linear closure of all the elements η_θ, $(\theta \in \Theta)$, in H.

Then its orthogonal complement

$$H_0 = H \ominus \hat{H} \tag{9.45}$$

represents the variety of all random variables $\eta \in H$ with expectation

$$E_\theta\eta \equiv 0 \ , \quad \theta \in \Theta \ ,$$

as it follows from formula (9.44).

Hence, *the best unbiased estimate of the unknown value $\theta(t)$ at $t \in G$, can be obtained as the orthogonal projection $\hat{\theta}(t)$ of the corresponding random variable $\xi(t) \in H$ on the subspace $\hat{H}$ in H*; in fact thanks to the orthogonal decomposition $\xi(t) = \hat{\theta}(t) + \eta$ with $\eta \in H_0$, this estimate $\hat{\theta}(t)$ satisfies conditions (9.41). In the same way we can get the best unbiased estimate $(\varphi, \hat{\theta})$ of the unknown value (φ, θ). In this case we have to use the projection of $(\varphi, \xi) \in H$ on $\hat{H}$, which satisfies conditions (9.41). As an illustration, let us assume for example that the variance

$$E_0|(x,\xi)|^2 \asymp \|x\|_X^2 \ , \quad x \in X^+(\Gamma) \tag{9.46}$$

is non-degenerate over the boundary test functions subspace $X^+(\Gamma)$ and use the left hand side in (9.46) to determine a new scalar product on $X^+(\Gamma)$ as

$$\langle x, y \rangle = E_0(x,\xi)(y,\xi) \ , \quad x, y \in X^+(\Gamma) \tag{9.47}$$

so as to claim that ξ is a "white noise" over $X^+(\Gamma)$.

We shall now consider the space U of deterministic harmonic functions in $W_2^2(G)$ and its dual $X^+(\Gamma)$.

We choose in U an orthonormal basis $\{u_i, \theta_j\}$, done in such a way that $\{\theta_j\}$ is also a basis of Θ.

Let x_i, y_j be the corresponding dual basis in $X^+(\Gamma)$, so as to have

$$(x, \theta_j) = \langle x, y_j \rangle = E_0(x,\xi)(y_j,\xi) \ , \quad x \in X^+(\Gamma) \ ,$$

We see that the elements $\{\eta_{\theta_j}\}$, defined through the equation (9.43), are just $\eta_{\theta_j}(= \eta_j) = (y_j, \xi)$ and all these $\{\eta_j\}$ form an orthonormal basis in the subspace $\hat{H}$; correspondingly the elements

$$\xi_i = (x_i, \xi)$$

form an orthonormal basis in H_0 - see (9.45). Let us set

$$\hat{\theta} = \sum_j \eta_j \theta_j \ , \tag{9.48}$$

representing one of the two parts in which the representation (4.28), which here reads

$$\xi = \sum_i \xi_i u_i + \sum_j \eta_j \theta_j \ , \tag{9.49}$$

is naturally split.

Then the harmonic (random) function $\hat{\theta} \in \underline{W}_2^2(G)$ represents *the best unbiased estimate of the unknown* θ, or, to be more precise, its point-wise values $\hat{\theta}(t)$, $t \in G$, or

$$(\varphi, \hat{\theta}) = \sum_j \eta_j(\varphi, \theta_j) \ , \quad \varphi \in X(G) = [C_0^\infty(G)] \ ;$$

represent the best unbiased estimates of the corresponding values of the harmonic function $\theta(t)$ or (φ, θ), we are looking for.

10　Generalized least square method

Let us suppose that, considering a boundary problem of type $(1.7)/(2.11)$, we prefer to deal with some appropriate scalar product on the corresponding boundary test function subspace $X^+(\Gamma)$; this can be any scalar product such that

$$\langle x, x \rangle \asymp \|x\|_X^2 \ , \quad x \in X^+(\Gamma) \ .$$

Let's assume that, according to (9.40), we can obtain as a solution just

$$\xi = \theta + \xi_0 \in \underline{W}_2^2(G) \ , \tag{10.50}$$

though we are actually looking for its deterministic component $\theta \in \Theta \subseteq U$, from a linear subspace Θ of a certain space of harmonic (deterministic) functions.

Considering the subspace $U \subseteq W_2^2(G)$ of all harmonic (deterministic) functions as dual to $X^+(\Gamma)$ with our scalar product, we can take the orthonormal basis $\{u_i, \theta_j\}$ in U and the dual basis $\{x_i, y_j\}$ in $X^+(\Gamma)$ as it was suggested in the previous paragraph; namely $\{\theta_j\}$ is a basis in Θ and $\{u_i\}$ is a basis in its orthogonal complement $U \ominus \Theta$. Accordingly we can use formula (9.48) giving us the best estimate $\hat{\theta}$ of θ, $\hat{\theta} = \sum_j (y_j, \theta) \theta_j$.

According to formula (9.49) for the solution ξ, its part $\hat{\theta} = \sum_j \eta_j \theta_j$ represents some kind of "generalized projection" of ξ on Θ.

Indeed, taking in the functional series with random coefficients (9.49) just finite sums, for any realization we get $\sum_j \eta_j(\omega) \theta_j$ as the orthogonal projection on Θ, since $u_i \perp \Theta$ and $\theta_j \in \Theta$. Recall, that the functional series (9.49) describes ξ as the harmonic random function with values on $X(G)$,

$$(\varphi, \xi) = \sum_i \xi_i(\varphi, u_i) + \sum_j \eta_j(\varphi, \theta_j) , \quad \varphi \in X(G) = [C_0^\infty(G)] \quad ;$$

the series is converging in mean square sense (cfr. [6]).

Suppose, we apply formula (9.48) and get the estimate $\hat{\theta}$ related to some scalar product of type (9.46), (9.47) with respect to a certain probability distribution P_0. Then, as we know, it represents the best unbiased estimate (related to P_0). One can imagine, that this P_0 differs from the true probability distribution - let it be P_0^*, say.

Thus, we actually apply some kind of the *pseudobest unbiased estimate* $\hat{\theta}$ of θ, still unbiased with respect to the true probability distribution P_0^* giving the corresponding expectation

$$E_\theta^*(\varphi, \hat{\theta}) \equiv (\varphi, \theta)$$

Suppose for instance, we can obtain the probability distributioin P_0 only from some sampled data, and in this way we overestimate the true variance, i.e. we obtain

$$E_0|(x,\xi)|^2 \geq E_0^*|(x,\xi)|^2 , \quad x \in X^+(\Gamma) .$$

In this case *the pseudobest unbiased estimate $\hat{\theta}$ is asymptotically efficient* when

$$E_0|(x,\xi)|^2 \to E_0^*|(x,\xi)|^2 , \quad x \in X^+(\Gamma) ; \qquad (10.51)$$

namely, taking into account the true best unbiased estimate $\hat{\theta}^*$, we have

$$E_{\hat{\theta}}^*|(\varphi,\hat{\theta}) - (\varphi,\theta)|^2 \to E_{\hat{\theta}}^*|(\varphi,\hat{\theta}^*) - (\varphi,\theta)|^2 , \quad \varphi \in X(G) = [C_0^\infty(G)] .$$
$$(10.52)$$

Indeed, we have $(\varphi,\hat{\theta})$ and $(\varphi,\hat{\theta}^*)$ in the space H of all random variables (x,ξ), $x \in X^+(\Gamma)$, and therefore

$$E_0^*|(\varphi,\hat{\theta}^*)|^2 \leq E_0^*|(\varphi,\hat{\theta})|^2 \leq E_0|(\varphi,\hat{\theta})|^2 \leq E_0|(\varphi,\hat{\theta}^*)|^2$$

where, according to (10.51),

$$E_0|(\varphi,\hat{\theta}^*)|^2 \to E_0^*|(\varphi,\hat{\theta}^*)|^2$$

and

$$\begin{aligned}
E_{\hat{\theta}}^*|(\varphi,\hat{\theta}) - (\varphi,\theta)|^2 &= E_0^*|(\varphi,\hat{\theta})|^2 \to \qquad (10.53)\\
&\to E_0^*|(\varphi,\hat{\theta}^*)|^2 = E_{\hat{\theta}}^*|(\varphi,\hat{\theta}^*) - (\varphi,\theta)|^2
\end{aligned}$$

References

[1] I.R. Aubin, *Applied Functional Analysis*, John Wiley e Sons., 1979.

[2] J.L. Lions, E. Magenes, *Problemes aux limites non homogenes*, Dunod, 1968.

[3] Y.A. Rozanov, *On generalized Dirichlet problem*, Dokl. Acad. Nauk of USSR, 1982.

[4] Y.A. Rozanov. *General boundary problems for stochastic partial differential equations*, Proceedings of Steklov Math. Inst., 200, 1991.

[5] Y.A. Rozanov *Stochastic Sobolev spaces*, Theory Prob. and Appl. n. 1, 1995.

[6] Y.A. Rozanov, *Gaussian infinite dimensional distributions*, Proceedings of Steklov Math. Inst., 108, 1968 (AMS English transl., 1971).

Free Boundary Problems

M. Biroli

1. Introduction

In this paper we are concerned with free boundary problems and in paricular with free boundary problems arriving from Calculus of Variations or at least can be solved by methods from Calculus of Variations.

Roughly speaking a free boundary problem is a problem with not only a function but also a set as unknown. Our effort in such a case will be transform our problem in a problem where the free boundary disappears and the only unknown is a function or, eventually, a measure. The free boundary is contained implicitly in the problem and we can construct it when we know the solution of the transformed problem.

We are interested here in method from Calculus of variations which allows us to transform a free boundary problem in an usual minimization problem; then we do not deal with methods using the abstract implicit function theorem (see for that method the lectures of F. Sansó).

We present here four examples that describe the methods used most frequently to transform the free boundary problem in an usual minimization problem.

The first example is the equilibrium displacement of membrane in presence of an obstacle. If we tray to formulate the problem as a free problem we have the reaction (and the set where the reaction acts) as unknown; so we have a free boundary type problem. Using the method of the minimum of the energy we can transform our problem into a usual minimization problem. The price to pay is that we have to minimize the energy only on the displacements consistent with the obstacle; then the set of functions on which we minimize the energy is no more linear (as it is in the free case) but (as we shall see) convex.

The second example we deal with is the dam problem: we have a dam between to water basins; we have to find the (steady) flow of the water in the dam. Here the unknown set is the wet part of the dam. We formulate at first the problem as a free boundary one and then we will see that a suitable change of the unknown function transform our problem into a problem mathematically equivalent to the one arising in the first example.

In the third example we deal with the image segmentation problem; roughly speaking we have to minimize a functional that depends also on a set. Intuitively the set is the set of jumping discontinuities of the unknow function; so the method we choose to solve our problem, the use in the minimization of a suitable function space, whose function can have, roughly speaking, only jumping discontinuities, giving suitable compactness and semicontinuity results.

In the fourth and last example we deal with a shape optimization problem, that in general has no solutions. We use a relaxation method: the method consists in transforming a minimization problem, that we are not able to solve, in a second problem with solutions such that:

(a) the infimum of the two functionals are the same; then a minimum point for the first problem is also minimum point for the second

(b) a minimizing sequence of the first problem converges at least after extraction of subsequence to a minimum point of the second problem.

So we plunge our initial problem in a more general class of problem that have solutions.

Finally we remark that we do not touch a field in which the mathematical research is still intensely active, that is the regularity problem for the free boundary.

For sake of brevity we do not recall here some notions that are generally known by a beginner in the research in Partial Differential Equations, but we refer to [8] for the few notion on distribution theory we will use, to [6] for Sobolev spaces and partial differential equations of elliptic type, to [5] for measures and finally to [7] for the functional space $BV(\Omega)$.

2. The Weierstrass Theorem revisited

First we are concerned with the classic Weierstrass Theorem; we observe that, if we have to prove only the existence of minimum points, the assumption on the continuity of the function is too strong and we can substitute the previous assumption with the one of lower semicontinuity.

Definition 1 *A real function f(x) defined on a set $E \subseteq R^N$ is lower semicontinuous at a point $x_0 \in E$ if x_0 is an isolated point of E or if*

$$liminf_{x \to x_0} f(x) \geq f(x_0).$$

The function f is lower semicontinuous on E if it is lower semicontinuous at every point in E.

Theorem 1. (Refinement of the Wierstrass Theorem) Let f(x) be a real function defined and lower semicontinuous on a bounded closed set $E \in R^n$; then f has at least a minimum point in E.

Proof. From Heine Borel's Theorem thhe set E is a compact set in R^N. We prove first that $l = inf_E f$ is finite.

If $inf_E f = -\infty$, there exists a sequence x_n in E such that $lim_{n \to +\infty} f(x_n) = -\infty$. By the compactness of E we can assume, without loss of generality, that $lim_{n \to +\infty} x_n = x_0 \in E$. From the lower semicontinuity of f on E we obtain $-\infty = lim_{n \to +\infty} f(x_n) \geq f(x_0)$. So we have a contradiction and our result is proved.

Now we prove the existence of a minimum point of f in E. Let $l = inf_E f$; there exists a sequence x_n in E such that $lim_{n \to +\infty} f(x_n) = l$. By the compactness of E we can assume, without loss of generality, that $lim_{n \to +\infty} x_n = x_0 \in E$. From the lower semicontinuity of f on E we obtain $l = lim_{n \to +\infty} f(x_n) \geq f(x_0)$. Then $f(x_0) = l$, so x_0 is a minimum point for f in E.

Our aim is to extend the Theorem 1 in an abstract framework, that will allow us to deal with many different problems in the Calculus of Variations.

Let X be a set with a notion of convergence (an Hausdorff topology); a map $J : X \to R \cup +\infty$ (not identically $+\infty$) is called a *functional* (in the following we assume the usual conventions on infinity).

Definition 2. *The functional $J : X \to R$ is lower semicontinuous at a point $x_0 \in E$ if x_0 is an isolated point or if*

$$liminf f(x) \geq f(x_0)$$

whenever $x \to x_0$. We say that J is lower semicontinuous on X if it is lower semicontinuous at every $x \in X$.

We remark that the lower semicontinuity for a weaker convergence implies the lower semicontinuity for the initial one. Then is usually easier in the applications prove the lower semicontinuity for the stronger convergence of J than the lower semicontinuity for the weaker one

Theorem 2. *If $J : X \to R$ is lower semicontinuos and there exists a non empty level set of J*

$$E_\alpha = \{x \in X; J(x) \leq \alpha\}$$

that is relatively (sequentially) compact, then there exists at least a minimum point of J in E.

The proof of the result is the same as the one of Theorem 1 observing that is equivalent minimize J on X or on E_α and that, due to the lower semicontinuity of J, E_α is closed.

We assume now that X is a separable Banach space. We observe that in the following will be usefull for us consider functionals defined on a subset E of X; in such a case we have an equivalent problem by extending J to X by $+\infty$.

We can consider on X two relevant notion of convergence: the strong convergence (or convergence in norm) and the weak convergence. The difficulty to apply directly the result of Theorem 2 in the case of the strong topology is the proof of the relative compactness of E_α; on the contrary, in the case of weak topology, the difficulty to apply directly the result of Theorem 2 is the proof of the lower semicontinuity (for the weak topology). Concerning the second case we recall that, if X is reflexive, every bounded set is relatively compact for the weak topology; so it seems that the best is use the weak topology having some assumptions (easy to verify in the applications) assuring the lower semicontinuity of J for the weak topology.

To solve the last problem we have to study a little of the connections between lower semicontinuity and geometrical type properties.

Consider the functional $J : E \subseteq X \to R$ the epigraph of J, denoted $epi(J)$, is the set

$$\{(x, \alpha); x \in E \quad J(x) \leq \alpha\}.$$

Theorem 3. *Let E be a closed (for the strong, weak topology) set in X and $J : K \to R$ a functional on K; J is lower semicontinuous (for the strong, weak topology) iff the set $epi(J)$ is closed (for the strong, weak topology).*

bf Proof. Let J be lower semicontinuous (for the strong, weak topology) and $(x_n, \alpha_n) \in epi(J)$ a sequence converging (for the strong, weak topology) on X and for the usual one on R) to x_0, α_0. We have

$$liminf_{n \to +\infty} J(x_n) \geq J(x_0)$$

then $\alpha_0 \geq J(x_0)$; this implies $(x_0, \alpha_0) \in epi(J)$. So $epi(J)$ is closed (for the strong, weak topology).

Let now $epi(J)$ be closed (for the strong, weak topology), $x_n \to x_0$, $x_n, x_0 \in K$ (for the strong, weak topology). We have $(x_n, J(x_n)) \in epi(J)$. By the closedness of $epi(J)$ we obtain $(x_0, \liminf_{n \to +\infty} J(x_n)) \in epi(J)$; so J is lower semicontinuous (for the strong, weak topology).

Let $K \subseteq X$ a convex set and $J : K \to R$ be a functional, J is said to be convex iff
$$J(\lambda x_1 + (1 - \lambda)x_2) \leq \lambda J(x_1) + (1 - \lambda)J(x_2)$$

for every $x_1, x_2 \in K$ and $0 \leq \lambda \leq 1$. It is easy to see that the functional J is convex iff $epi(J)$ is a convex set.

We now recall a fundamental (and difficult) result on convex sets in Banach spaces, that we don't prove:

Theorem 4. *Every convex set closed in the strong topology is also closed for the weak topology.*

An easy consequence of Theorem 4 is:

Theorem 5. *Let X be a reflexive Banach space, $K \subseteq X$ be a closed convex set and $J : K \to R$ be a convex functional lower semicontinuous for the strong topology and that there exists a non empty bounded level set of J*

$$K_\alpha = \{x \in K; J(x) \leq \alpha\};$$

then there exists at least a minimum point for J in K.

The assumption in Theorem 5 is easily verfied under a condition of *coercivity*, i.e.

$$\lim_{\|u\|_X \to +\infty} J(u) = +\infty$$

Theorem 6. *Let X be a reflexive Banach space, $K \subseteq X$ be a closed convex set and $J : K \to R$ be a convex coercive functional lower semicontinuous for the strong topology; then there exists at least a minimum point for J in K.*

To deal with the uniqueness of the minimum points is usefull to introduce the notion of strict convexity. Let $K \subseteq X$ a convex set and $J : K \to R$ be a functional, J is said to be strictly convex iff

$$J(\lambda x_1 + (1 - \lambda)x_2) < \lambda J(x_1) + (1 - \lambda)J(x_2)$$

for every $x_1, x_2 \in K$ and $0 < \lambda < 1$.

Theorem 7. *Let X be a Banach space, $K \subseteq X$ be a convex set and $J : K \to R$ be a strictly convex functional; then there exists at most one minimum point for J in K.*
Proof. Let $x_1 \neq x_2$ be minimum points of J in K and denote $J(x_1) = J(x_2) = m$. By the strict convexity we have

$$\frac{x_1 + x_2}{2} \in K; \quad J(\frac{x_1 + x_2}{2}) < \frac{J(x_1) + J(x_2)}{2} = m.$$

So we have a contradiction and the result is proved.

We end the section giving a differential conditions for minima.

Definition 3. *Let X be a Banach space, $J : X \to R$ be functional; assume that for a fixed $u \in X$ the limit*

$$lim_{t \to 0} \frac{J(u + tv) - J(u)}{t} = < J'(u), v >$$

exists, for every $v \in X$, where $< ., . >$ denotes the duality between X and its dual X^*. The vector $J'(u) \in X^*$ is the derivative at u of J. We say that J is derivable on X, if it is derivable at every point of X.

We have easily:

Theorem 8. *Let X be a Banach space, $J : X \to R$ be a functional derivable on X. Let u be a (local) minimum point for J on X, then $J'(u) = 0$.*

Now we consider the particular case of the minima of convex functionals on a convex.

Theorem 9. *Let X be a Banach space, $J : X \to R$ be a convex functional derivable on X. Let K be a convex set in X; then u is a minimum point for J in X iff*

$$< J'(u), v - u > \geq 0, \quad \forall v \in K,$$

where $< ., . >$ denotes the duality between X and its dual X^*.
Proof. Let u be a minimum point, $v \in K$ and $w = v - u$; then $u + tw = (1 - t)u + tv, 0 < t < 1$, is in K.

We have $J(u + tw) - J(u) \geq 0$ so

$$< J'(u), w > = < J'(u), v - u > \geq 0.$$

Let now $u \in K$ be such that

$$< J'(u), v - u > \geq 0, \quad \forall v \in K.$$

Since J is convex, choosen $u, v \in K$ and denoted $w = v - u$ the function $f(t) = J(u + tw)$ is convex; so

$$f(1) - f(0) \geq f'(0)$$

that is

$$J(v) - J(u) \geq < J'(u), v - u > \geq 0$$

for every $v \in K$. The above relation say that u is a miimum point for J in K.

3. The obstacle problem

We will now give a first example of free boundary problem, that can be solved by means of the Calculus of Variations.

We consider an elastic homogeneous isotropic membrane clamped at the boundary; the membrane, in absence of exterior forces, has an equilibrium configuration given by an open set $\Omega \subseteq R^2$ with a smooth boundary Γ. We will find the equilibrium configuration of the membrane under the action of forces directed ortogonally to Ω and with density f and with an obstacle on the displacement given on Ω. If we describe the obstacle by the condition $u(x) \leq \psi(x)$, $\psi \in H^1(\Omega)$ with $\psi|_\Gamma \geq 0$.

The above problem can be described as a free boundary problem. We have to consider as unknown also the set where we have a reaction μ of the obstacle.

Our problem, if we arrive to know μ, can be described as a problem without obstacle (free):

$$\Delta u = f + \mu \ \ in \ \Omega; \quad u = 0 \ \ on \ \Gamma$$

$$supp[\mu] \subseteq \{x \in \Omega; \ \ u(x) = \psi(x)\}.$$

If we use the principle of minimization of the energy between all possible configurations of the membrane we arrive to a different formulation of our problem: minimize the integral

$$\int_\Omega |\nabla v|^2 dx - 2 \int_\Omega fv dx,$$

where we can assume $f \in L^2(\Omega)$, on the convex

$$K = \{v \in H_0^1(\Omega), \ \ v(x) \leq \psi(x) \ a.e. \ in \ \Omega\} \subseteq H_0^1(\Omega).$$

We now precise the abstract framework; we choose $X = H_0^1(\Omega)$ and K as above. Then our problem is minimize $J(v) = ||v||^2$, where $||.||$ denotes the norm in $H_0^1(\Omega)$.

Our functional is strictly convex coercive and continuous (for the strong topology) on K: then

Theorem 1. *There exisists a unique minimum point u for $J(v)$ in K.*

The functional $J(v)$ is differentiable then an easy computation gives that u satisfies the following Euler's condition

$$((u, v - u)) \geq (f, v - u)$$

$$\forall v \in K; \quad u \in K$$

where $((.,.))$ denotes the scalar product in $H_0^1(\Omega)$.

We recall that the first line of the Euler's condition can be written as

(1) $< \Delta u - f, v - u >_{H^{-1}, H_0^1} \geq 0;$

so we obtain

$$< \Delta u - f, v >_{H^{-1}, H_0^1} \geq 0$$

$\forall v \leq 0$ in $H_0^1(\Omega)$. Then we have

$$\Delta u = f + \mu, \quad u \in H_0^1(\Omega)$$

where μ is a negative Radon measure with $supp[\mu] \subseteq \{x \in \Omega; \; u(x) = \psi(x)\}$, that gives the reaction of the obstacle ψ. We oberve that so we have, at the end, the reaction μ of the obstacle and consequently the set $supp[\mu]$ where the reaction acts.

Finally we will prove that

Theorem 2. *Let $\Delta \psi, f \in L^2(\Omega)$; we have $\Delta u \in L^2(\Omega)$ (then u in$H^2(\Omega)$) where u is the solution of (1).*

We give only the sketch of the proof.

We consider at first the "penalized" problem

(2_n) $\Delta u_n + n(u_n - \psi)^+ = f \;\; in \; \Omega, \quad u \in H_0^1(\Omega).$

The problem (2_n) has a unique solution u_n.

Using $n(u_n - \psi)^+$ as test function in (2_n) we obtain easily

(3) $n^2 \int_\Omega [(u_n - \psi)^+]^2 dx \leq C$

where C denotes possibly different constants independent of n.

The estimate (3) implies

$$\int_\Omega |\Delta u_n|^2 dx \leq C$$

so u_n is a sequence bounded in $H^2(\Omega)$: then, at least after extraction of a subsequence, we obtain that u_n weakly converges to u in $H^2(\Omega)$ and that u_n strongly converges to u in $H_0^1(\Omega)$ and in $L^2(\Omega)$. Moreover Δu_n weakly converges to Δu in $L^2(\Omega)$.

Consider now the problem (2_n) and use $(v - u_n)$ with $v \in K$ as test function; we obtain

(4_n) $\int_\Omega (\Delta u_n - f)(v - u_n) dx \geq 0$

Passing to the limit in (4_n) as $n \to +\infty$ we obtain that u is the solution of the problem (1) and the result is proved.

We observe that the above result give a regularity result for the reaction, that, under the assumptions of Theorem 2, is not only a measure, but a function in $L^2(\Omega)$. Moreover by the result of Theorem 2 we realize that, roughly speaking, the regularity of u can increase with the regularity of ψ. We conclude this section remarking that one can prove that there is an upper limit on the regularity of u; also if the obstacle is $C^\infty(\overline{\Omega})$ we can only assure the boundness of the second derivatives of u.

4. The rectangular dam problem, [2]

We consider a dam with rectangular section, that in the plane can be described as th rectangle $D = (0, c) \times (0, d)$.

On the two side of the dam we have two water basins, where the water attains different heights y_1 in the left hand side and y_2 in the right hand side and we assume $d > y_1 > y_2 > 0$. The water seeps trough the dam from the left basin to the right one. We assume that the dam is homogeneous incompressible and isotropic and that the basement of the dam $\Gamma_0 = (0, c) \times \{0\}$ is water-proof; the flow of the water is steady, incompressible, irrotational and bidimensional; we suppose also that capillarity and evaporation phenomena are negligible. Moreover we assume that the basis $(0, c) \times \{0\}$ of the dam is water-proof.

We have in the dam two parts: the wet part and the dry one.

The two parts are divided by a line, the *free* line denoted by $\phi(x)$, and the pressure is 0 in the dry part and positive in the wet one. We assume, as experimentally reasonable, that $\phi(c) > y_2$ and we denote by Γ_σ the line $\{c\} \times (y_2, \phi(c))$, called in the following the seepage line. Moreover we assume that the water does not come in or out trough the seepage line.

Denote by Ω the wet region ($\Omega = \{(x, y) : x \in (0, c), \ 0 < y < \phi(x)\}$.

The problem of characterization of the flow of of the water in the dam is a free boundary problem why we don't know the free line separating the wet and the dry part of the dam.

To have a mathematical formulation of the problem we assume, as verified by experiments, that the flux of the water follows the Darcy's law:

$$\vec{V} = -k \ grad(y + \frac{p}{\gamma}) \quad in \ \Omega$$

where $\vec{V}$ is the velocity of the water in the wet region, $p = p(x, y)$ is the pressure, γ is the specific weight (for the water) and k is the permeability coefficient of the material, that makes up the dam, with respect to the water. In the following we choose, without loss of generality $k, \gamma = 1$ and we denote

$$u = y + p.$$

Then the Darcy's law can be rewritten as

$$\vec{V} = grad\ u$$

so u is the potential of the velocity.

From the continuity equation, taking into account that the mass density is constant due to incompressibility of the water, we obtain

$$div\ \vec{V} = 0$$

and that implies

$$\Delta u = 0\ in\ \Omega$$

The above relation means that u is harmonic in Ω. Then there exists an harmonic function v (called stream function) such that $\phi = u + iv$ is an function (the complex potential). The level lines of u (equipotential lines) and the level lines of v (streamlines) are mutually orthogonal. If we consider a region bounded by two streamlines (a stream tube) the water does not come in or out from the region through the streamlines, then the free line and the basement of the dam have to be a streamline for the flow. Moreover due to the incompressibility condition we have that if L is the section of a stream tube then

$$q_L = \int_L \vec{V}.\vec{n}_L dl$$

depends only on the stream tube but not on the section.

Considering now the tube bounded by the free line and by the basement of the dam we can define the costant quantity:

$$q = q(x) = -\int_0^{\phi(x)} u_x dy.$$

Moreover recalling that $u = y + p$, we can also write

$$q = -\int_0^{\phi(x)} p_x dy = -\int_0^d p_x dy$$

We recall now that u and so also p are harmonic, i.e.

(1) $$\Delta p = 0\ \ in\ \Omega$$

Now we will deal with the boundary conditions.

1) *The water – proof basement denoted by* Γ_0.

Taking into account that Γ_0 is water-proof we have

$$0 = \vec{V}.\vec{n} = u_n = u_y\ \ on\ \Gamma_0$$

then

(2)
$$p_y = -1.$$

2) *The boundary wet by the basins* $\Gamma_1 = \{0\} \times (0, y_1)$, $\Gamma_2 = \{c\} \times (c, y_2)$.
If we assume that the velocity of the water in the two basins is negligible in comparison with the one in the dam then $u = cst$ in the two basins. In the two points $(0, y_1)$, (c, y_2) we have $p = 0$ (where we take $p = 0$ at the contact with the atmosphere) so we have

$$u = y_1 \text{ on } \Gamma_1, \quad u = y_2 \text{ on } \Gamma_2.$$

Then

(3)
$$p = (y_1 - y) \text{ on } \Gamma_1, \quad p = (y_2 - y) \text{ on } \Gamma_2.$$

3) *The seepage line denoted by* Γ_σ.
On the line Γ_σ there is contact with the atmosphere then

(4)
$$p = 0 \text{ on } \Gamma_\sigma.$$

The condition that the water can does not come in or out through the seepage line, can be analytically written as $u_n = 0$ on Γ_σ or equivalently

(5)
$$(p + y)_n = 0, \text{ on } \Gamma_\sigma$$

4) *The free line denoted by* Γ_λ.
The free line is a stream line in contact with the atmosphere, then

(6)
$$v = cst, \ p = 0$$

on Γ_λ. The first relation can be written as

(7).
$$u_n = (p + y)_n = 0, \text{ on } \Gamma_\lambda$$

We summarize the formulation of the free boundary problem in the following way (F1):
We seek for a function $y = \phi(x)$, $x \in [0, c]$, *with ϕ nondecreasing and a function p defined in* $\Omega = \{(x, y) : 0 < x < c, 0 < y < c\}$ *such that:*
$\Delta p = 0$ in Ω
$p(0, y) = y_1 - y$ for $0 \leq y \leq y_1$
$p(c, y) = y_2 - y$ for $0 \leq y \leq y_2$
$p(c, y) = 0$ for $y_2 \leq y \leq \phi(c)$
$p(x, \phi(x)) = 0$ for $0 \leq x \leq c$

$(p+y)_n(x, \phi(x)) = 0$ for $0 \leq x \leq c$, where n is the normal unit vector to Γ_λ
$(p+y)_y(x, 0) = 0$ for $0 \leq x \leq c$

We observe that from minimal assumptions on the regularity of p and ϕ the maximum principle gives that $p > 0$ in Ω.

Now we will transform the above formulation in the following one (F2):
We seek for a function p defined in D such that:
$p \geq 0$ in D
$p(0, y) = [y_1 - y]^+$ for $0 \leq y \leq d$
$p(c, y) = [y_2 - y]^+$ for $0 \leq y \leq d$
Let Ω be the set $\{(x, y) \in D : p(x, y) > 0\}$; Ω is an open relatively compact open set such that if $(x_0, y_0) \in \Omega$, then $(x_0, y) \in \Omega$ for $0 < y < y_0$ and

$$\int_\Omega grad(p+y) \, grad\psi \, dxdy = 0$$

$\forall \psi \in C_0^1((0, c) \times R)$.

The last relation can be written also as

(8) $$\int_D gradp \, grad\psi \, dxdy = \int_D \chi_\Omega \psi_y dxdy$$

$\forall \psi \in C_0^1((0, c) \times R)$, where χ_Ω denotes the characteristic function of Ω.

The above relation implies

(9) $$\Delta p = (\chi_\Omega)_y$$

in distribution sense. We remark that it is difficult to deal with (9), why the function χ_Ω is discontinuous and then $(\chi_\Omega)_y$ is a distribution but isn't a function.

We will prove also that (8) implies

(10) $$\int_0^d p_x(x, y)dy \text{ is linear on } [0, c].$$

Choose $\psi(x, y) = \psi_1(x)\psi_2(y)$ where $\psi_1 \in C_0^1((0, c))$ and $\psi_2 \in C_0^1((0, d))$ with $\psi_2 = 1$ on $[0, d]$; we obtain

$$\int_0^c [\int_0^d p_x(x, y)dy]\psi_1' \, dx = 0$$

Then

$$(\int_0^d p_x(x, y)dy)_x = 0$$

in distribution sense on (c,d) and (10) is proved.

To avoid the above remarked difficulty in (9) we change the unknow function choosing

$$w(x,y) = \int_y^d p(x,t)dt$$

The formulation of the problem becomes (F3):

$w \le 0$

$\Delta w = \chi_\Omega$ in D where $\Omega = \{(x,y) \in D : w(x,y) > 0\}$

$w(c,y) = \int_y^d (y_2 - t)^+ dt$

$w(0,y) = \int_y^d (y_1 - t)^+ dt$

$w(x,0) = \frac{y_1^2 - y_2^2}{2c}(c - x) + \frac{y_2^2}{2}$

where the last relation derives easily from (10).

We observe that again in this last formulation of our problem the unknown set Ω appears explicitly.

Now we arrive to the final formulation, that will allow us to hide Ω.

We observe that from the previous formulation we obtain

(11) $(\Delta w - 1)w = 0, \ \Delta w - 1 \ge 0, \ w \ge 0 \ in \ D$

and

(12) $w = g \ on \ \partial D$

where g is defined by

$$g(x,d) = 0, \quad g(x,0) = \frac{y_1^2 - y_2^2}{2c}(c - x) + \frac{y_2^2}{2} \ for \ x \in [0,c]$$

$$g(0,y) = \int_y^d (y_1 - t)^+ dt, \ g(c,y) = \int_y^d (y_2 - t)^+ dt \ for \ y \in [0,d].$$

Let K be the convex in $H^1(D)$ defined by

$$K = \{v \in H^1(D) : \ v \ge 0 \ in \ D, \ v = g \ on \ \partial D.$$

From (11) we have (F4):

(13) $\int_D grad w \ grad(z - w) \ dxdy \ge - \int_D (z - w)dxdy$

$\forall z \in K$, moreover $w \in K$.

Then w is the solution of the variational inequality (13), that is equivalent to the problem ((F5)) of the minimization on K of the functional

$$(14) \qquad \int_D (\frac{1}{2}|gradw|^2 - w)dxdy.$$

From the abstract results we obtain:

Theorem 1. *The functional (14) has a unique minimization point w.*

From a more subtle analysis of the minimization (or variational inequality) inequality) problem, [2], we also obtain:

Theorem 2. *Let w as in Theorem 1; we have w_{xx}, w_{xy}, $w_{yy} \in L^r(D)$ for every $r \in [1, +\infty)$, then w, w_x, w_y are continuous in D. Moreover we have w_x, $w_y \leq 0$ in D.*

We observe that the inequalities on w_x, w_y allow use to represent the set $\Omega = \{(x, y) \in D : w(x, y) > 0\}$ as $\Omega = \{(x, y) \in D : y < \phi(x)\}$ where ϕ is a bounded decreasing function.

Theorem 3. *The function ϕ is continuous in [0,c] and analytic in (0,c). Moreover ϕ is concave and $\phi'(0) = 0, \phi'(c) = +\infty$.*

Finally we remark that the results in Theorem 2, 3 allow us to affirm that w or $p = -w_y$ are also solution of every of the previous formulations of our problem.

5. The image segmentation problem, [1]

Let Ω be an open bounded set in R^2 or R^3; we assume that a distribution of intensity of gray is given in Ω by a function g. We will seek for an image such that the intensity of gray in the different objects in the image has a "minimum" gradient. We also assume that the perimeter of the objects is also of "minimal" lenght.

If we try to have a mathematical formulation of the problem we naturally arise to a minimization problem:

minimize the following functional

$$(1) \qquad F(v, K) = \int_{\Omega-K} |gradv|^2 dx + \alpha H^{n-1}(K) + \beta \int_\Omega |v - g|^2 dx$$

with $K \in \Omega$ closed set, n=2,3, $v \in H^1(\Omega - K)$ and $g \in L^\infty(\Omega)$.
The set K is the set where v is possibly discontinuous and from the meaning of the problem we can admit only *jumping discontinuities*. In the following we will deal with the problem in R^n, why there is no supplementary difficulties

We observe that the difficulty to deal with the functional $F(u, K)$ is the dependence of the functional from the set K and the assumption that on K we can have only jumping discontinuities. This last assumption prevent us to try to use the space $BV(\Omega)$; in fact a function in $BV(\Omega)$ may have any type of discontinuities (not only the jumping ones).

Moreover we can define for function in $BV(\Omega)$ the point of jumping discontinuity:

Definition 1. *Let u be a function defined in the open bounded set $\Omega \subseteq R^n$; we say that $u^+(x)$ is the approximate superior limit of u at x iff*

$$u^+(x) = inf\{t \in (-\infty, +\infty);\ lim_{\rho \to 0} \frac{m(\{x \in \Omega;\ u(x) < t\} \cup B(x, \rho))}{\rho^n} = 0\}$$

We say that $u^-(x)$ is the approximate inferior limit of u at x iff

$$u^-(x) = inf\{t \in (-\infty, +\infty);\ lim_{\rho \to 0} \frac{m(\{x \in \Omega;\ u(x) < t\} \cup B(x, \rho))}{\rho^n} = 0\}$$

If $u^+(x) = u^-(x)$ we say that x is a point of approximate continuity.

In the following we denote by S_u the set $\{x \in \Omega;\ u^+(x) > u^-(x)\}$; then u is approximately continuos in $\Omega - S_u$. We observe that if u is a characteristic function χ_E then $S_u = \partial^* E$.

Definition 2. *Let $x \in \Omega - S_u$, p is the approximate differential of u at x iff:*

$$V(y) = \frac{|u(y) - u^+(y) - <p, y - x>|}{|y - x|}$$

has an approximate limit equal 0 in x.

If we consider a function u, which is in $BV(\Omega)$ we obtain, [1]:

Theorem 1. *Let $u \in BV(\Omega)$; then u is approximately differentiable almost everywhere and the approximate gradient ∇u is in $L^1(\Omega)$ and $\nabla u . L^n|_\Omega$ ($L^n|_\Omega$ is the Lebesgue measure restricted to Ω) is the absolutely continuous part of Du (Du denotes the distributional gradient of u) with respect to $L^n|_\Omega$ and the following inequality holds:*

$$\int_\Omega |\nabla u| dx + \int_{S_u} |u^+ - u^-| dH^{n-1} \le |Du|(\Omega)$$

Moreover S_u is the union of a sequence of C^1 ipersurfaces with the execption of of the part of the boundaries, that belongs to two different ipersurfaces, denoted by N and $H^{n-1}(N) = 0$.

If we consider now a function $u \in BV(\Omega)$ the disributional gradient can be divided into three parts:

$$Du = \nabla u . L^n|_\Omega + Ju + Cu$$

where J is the term corresponding to jumping discontinuities and is defined as

$$Ju(B) = \int_{S_u \cap B} |u^+ - u^-| dH^{n-1}$$

for every Borel set B in Ω.

The third part Cu is called the Cantor part is singular with respect to $L^n|_\Omega$ and such that $|Cu(B)| = 0$ for every Borel set B in Ω such that $H^{n-1}(B) = +\infty$.

Definition 3. *A function $u \in BV(\Omega)$ is in $SBV(\Omega)$ iff $Cu = 0$.*

Now we try to transform our initial free boundary problem into a variational problem in $SBV(\Omega)$.

The first result interesting for us is the following, [1] :

Theorem 2. *Let K be a closed set with $H^{n-1}(K) < +\infty$ and $u \in H^{1,1}(\Omega - K) \cap L^\infty(\Omega)$. Then $u \in SBV(\Omega)$ and $S_u \subseteq K \cup N$, where $H^{n-1}(N) = 0$.*

The above result allow us to affirm the equivalence of the minimization problem (1) with the following different minimization problem

minimize the following functional

$$(2) \qquad G(v) = \int_\Omega |\nabla v|^2 dx + \alpha H^{n-1}(S_v) + \beta \int_\Omega |v - g|^2 dx$$

in $SBV(\Omega)$

We observe that in the new formulation (2) of our problem the functional depends only on the function v and the dependence on a set (present in (1)) has disappeared. This last fact allow us to use for the minimization of J the abstract results that we have seen before.

Theorem 3. *Let $\Phi(t)$ be an increasing function with $\lim_{t \to +\infty} \frac{\Phi(t)}{t} = +\infty$ and assume that u_h is a sequence in $SBV(\Omega) \cap L^\infty(\Omega)$ such that*

$$||u_h||_{L^\infty(\Omega)} + \int_\Omega \Phi(|\nabla u_h|) dx + H^{n-1}(S_{u_h}) \leq C$$

where C is a constant independent on h. Then there exists a subsequence u_{h_k} converging strongly in $L^1(\Omega)$ to $u \in SBV(\Omega)$. Moreover

$$\nabla u_{h_k} \to \nabla u \text{ weakly in } L^1(\Omega; R^n)$$

$$Ju_{h_k} \to Ju \text{ weakly}^* \text{ as Radon measures}$$

The proof of Theorem 3 uses a slicing method and is given in [1].

We observe now that the functional $G(v)$ is decrasing by truncature on v then:

Theorem 4. Let M = $||g||_{L^\infty(\Omega)}$; then

$$inf_{u \in SBV(\Omega)} G(v) = inf_{\{u \in SBV(\Omega); \, ||u||_{L^\infty(\Omega)} \leq M\}} G(v)$$

Denote by $W \subseteq L^1(\Omega)$ the set $\{u \in L^1(\Omega); \, ||u||_{L^\infty(\Omega)} \leq M\}$. Using the results and the methods of Theorem 3 we can prove:

Theorem 5. *The set W is closed for the strong topology of $L^1(\Omega)$; the level sets of G in W are compact for the strong topology of $L^1(\Omega)$.*

Theorem 6. *The functional G is lower semicontinuous on W for the strong topology of $L^1(\Omega)$.*

The abstract result and Theorems 5,6 prove that G has a minimum point in W, that due to Theorem 4 is a minimum for G in $SBV(\Omega)$. Then:

Theorem 7. *The minimization problem (2), and then also the minimization problem (1), has at least a solution.*

We have so solved our original free boundary problem transforming it in a usual variational problem by a good choise of the functional space related to the problem.

We end the section by giving a regularity result, which is a consequence of well known regularity results on Laplace equation and of Theorem 2:

Theorem 8. *Let u be a solution of the minimization problem (2) (or (1)) then:*

$$u \in H^{2,p}(\Omega - \overline{S}_u) \cap C^1(\Omega - \overline{S}_u)$$
$$H^{n-1}(\Omega \cap (\overline{S}_u - S_u)) = 0.$$

6. The shape optimization for Dirichlet problems, [4]

In this section we give an example of a method, the relaxation method, that allow us to deal with problems without solution.

Consider a functional J and the minimization problem for J on a set $Y \subseteq Z$; suppose that we are not able to solve our problem; we can consider a set X, such that Y is dense in X for a suitable convergence in Z. Define an extension J^* of J to X The minimization problem for J^* on X is called a relaxation of the original problem.

We observe that, in general, we can relax a problem in different ways; the ground of the methods is choose the relaxation in such a way as to have that from every minimizing sequence in X we can extract a sequence converging to a minimum point. Under the previous assumption we have

(1) $$inf_Y J = min_X J^*$$

and that from every minimizing sequence in Y we can extract a subsequence converging to a minimum of J^* in X. Moreover from (1) we have that if the problem of minimization on Y has a solution, that solution is also a solution of the minimization problem on X.

So the method consists in trnsforming a minimmization problem, that we are not able to solve, in a problem with solutions such that:

(a) (1) holds; then a minimum point for the first problem is also minimum point for the second

(b) a minimizing sequence of the first problem converges at least after extraction of subsequence to a minimum point of the second problem.

We have now to introduce some preliminary results and notations. The capacity of a subset of R^N is defined by

$$cap(E) = inf_{U_E} ||u||_{H^1(R^N)}$$

where $U_E = \{u \in H^1(\Omega); \ u \geq 1 \ a.e. \ on \ a \ neighbourhood \ of \ E\}$.

We say that a property $P(x)$ holds *quasi everywhere* (q.e.) in an open set A if $P(x)$ holds except for a set of zero capacity.

A set A is *quasi open* if for every $\epsilon > 0$ there exists an open set A_ϵ containing A and such that $cap(A_\epsilon - A) < \epsilon$.

We say that a function $u : A \to R$, where $A \subseteq R^N$ ia an open set, is *quasi continuous* in A, if for every $\epsilon > 0$ there exists a subset E of A with $cap(A - E) \leq \epsilon$ such that the restriction $u|_E$ of u to E is continuous on E.

We recall that if $u \in H^1(A)$ then the limit

$$lim_{r \to 0+} \frac{1}{m(B(x,r))} \int_{B(x,r)} u(y)dy = \tilde{u}(x)$$

exists and is finite q.e. in A. Moreover $\tilde{u}(x)$ is a quasi continuous representative of u.

Another property we will recall is the following: if u_n is a sequence in $H^1(A)$ converging to u, then u_n converges to u q.e. in A.

Let Ω be a bounded open set in R^N.

By $\mathcal{B}(\Omega)$ we denote the σ-field of all Borel subsets of Ω and by $\mathcal{B}_c(\Omega)$ the δ-ring of all Borel subsets such that $B \subseteq\subseteq \Omega$. By a *Borel measure* on Ω we mean a countably additive set function $\mu : \mathcal{B}(\Omega) \to] -\infty, +\infty]$, not necessarily finite or σ-finite. By a *Radon measure* we mean a countably additive set function $\mu : \mathcal{B}_c(\Omega) \to R$.

For a Borel (Radon) measure we can consider a positive and a negative part denoted by μ^+ and μ^-, where μ^+ and μ^- are (orthogonal and uniquely defined) Borel (Radon) positive measures and $\mu = \mu^+ - \mu^-$ (Dunford - Schwartz, Vol I pg. 130); we denote $|\mu| = \mu^+ + \mu^-$ and the Borel (Radon) nonegative measure $|\mu|$ is called the *total variation* of μ.

By $\mathcal{M}_0(\Omega)$ we denote the set of all nonegative Borel measures on Ω such that $\mu(B) = 0$ for every $B \in \mathcal{B}(\Omega)$ with $cap(B) = 0$. By $\mathcal{M}_0^+(\Omega)$ we denote the set of nonegative Borel measures on Ω in $\mathcal{M}_0(\Omega)$.

An important example of measures in $\mathcal{M}_0^+(\Omega)$, which will play an important role in the following, is, for every set $S \subseteq \Omega$, the Borel measure ∞_S defined as

$$\infty_S(B) = 0 \ \ if \ cap(B \cap S) = 0$$

$$\infty_S(B) = +\infty \ \ if \ cap(B \cap S) > 0.$$

We observe that for every μ in $\mathcal{M}_0^+(\Omega)$ there exists a quasi open subset $A(\mu)$ Borel set in Ω such that $\mu(A)$ is finite for every Borel set A with closure contained in $A(\mu)$ and $\mu(A) = +\infty$ if A is a quasi-open set with $cap(A \cap S(\mu) > 0$, where $S(\mu) = \Omega - A(\mu)$. We say that $A(\mu)$ is the *regular set* of μ and that $S(\mu)$ is the *singular set* of μ. So the measure μ can be decompsed in a mesure of the type $\infty_{S(\mu)}$ and in a mesure with support in $\overline{A}(\mu)$ which is finite on the compacts contained in $A(\mu)$.

We will now consider problems of the type

(2) $$-\Delta u + \mu u = f, \ u = 0 \ on \ \partial\Omega.$$

where $\mu \in \mathcal{M}_0^+(\Omega)$ and $f \in L^2(\Omega)$. First we have to give the exact formulation for problem (2).

We say that u is a weak solution of (2) iff

(3) $$\int_\Omega gradu \ gradvdx + \int_\Omega uv \ d\mu = \int_\Omega fv \ dx$$

for every $v \in H_0^1(\Omega) \cap L^2(\Omega; \mu)$, where $u \in H_0^1(\Omega) \cap L^2(\Omega; \mu)$.

By standard methods of Functional Analysis we can prove :

Theorem 1. *The problem (3) has a unique solution.*

We denote by $R_\mu(f)$ the solution u of (3).

The problem (3) has an important particular case. Let A be an open set in Ω and $S = \Omega - A$; if we have $\mu = \infty_S$ the problem (3) is equivalent to the problem

(4) $$-\Delta u = f|_A. \ u \in H_0^1(A).$$

The result in Theorem 1 allow us to define a notion of convergence on $\mathcal{M}_0(\Omega)$.

Definition 1. *Let μ_n be a sequence in $\mathcal{M}_0^+(\Omega)$; the sequence μ_n γ-converges to $\mu \in \mathcal{M}_0^+(\Omega)$ iff*

(5) $$R_{\mu_n}(f) \to R_\mu(f)$$

in $L^2(\Omega)$. Let now μ_n be a sequence in $\mathcal{M}_0(\Omega)$; the sequence μ_n γ-converges to $\mu \in \mathcal{M}_0(\Omega)$ iff μ_n^+ and μ_n^- γ-converge to μ^+ and μ^-.

We have the following compactness and density results that we do not prove:

Theorem 2. For every sequence μ_n in $\mathcal{M}_0^+(\Omega)$ there exists a subsequence which γ converges to a measure μ in $\mathcal{M}_0^+(\Omega)$.

Theorem 3. For every μ in $\mathcal{M}_0^+(\Omega)$ there exists a sequence S_n of compacts subsets of Ω such that the sequence ∞_{S_n} γ-converges to μ.

We now give the optimal shape probleme we are concerned with.

Let u_A be the solution of the problem

$$(6) \qquad -\Delta u_A = f \ in \ A; \ u_A \in H_0^1(A)$$

where A is an open subset of Ω; we have to minimize the functional

$$(7) \qquad J(A) = \int_\Omega |u_A - g|^2 dx$$

with respect to the open subsets $A \subseteq \Omega$, where $g \in L^2(\Omega)$ and the extension of u_A to Ω by 0 is again denoted by u_A.

We observe that the minimization problem (7) does not have a solution for any $g \in L^2(\Omega)$. Moreover we can give an equivalent formulation for (6)(7). Let u_A be the weak solution of the problem

$$(6') \qquad -\Delta u_A + \infty_S u = f \ in \ \Omega, \ u_A \in H_0^1(\Omega)$$

where S is an arbitrary compact subset of Ω and $A = \Omega - S$ is an arbitrary open subset of Ω; we have to minimize the functional

$$(7') \qquad J(\infty_S) = \int_\Omega |u_A - g|^2 dx$$

with respect to the measures ∞_S.

We have now to relax the problem (6')(7') taking into account the results of Theorems 2,3. Let u_μ be the weak solution of the problem

$$(6'') \qquad -\Delta u_\mu + \mu u = f \ in \ \Omega, \ u_\mu \in H_0^1(\Omega)$$

where μ is a Borel measure in $\mathcal{M}_0^+(\Omega)$; we have to minimize the functional

$$(7'') \qquad J(\mu) = \int_\Omega |u_\mu - g|^2 dx$$

with respect to the measures in $\mathcal{M}_0^+(\Omega)$. We observe That $J(\mu)$ is continuous on $\mathcal{M}_0^+(\Omega)$ for the γ-convergence. From Theorem 2 we have that $\mathcal{M}_0^+(\Omega)$

is (sequentially) compact for the γ-convergence; then every level set of $J(\mu) : \mathcal{M}_0^+(\Omega) \to R$ is (sequentially) compact. We can conclude that the problem (6")(7") has a solution.

The continuity of $J(\mu)$ on $\mathcal{M}_0^+(\Omega)$ for the γ-convergence and the result in Theorem 3 give that the infimum of the functional in (7') is equal to the minimum of the functional in (7"); then if there exists a minimum point for (7') the sama measure is also a minimum point for (7"). Moreover from the continuity of $J(\mu)$ on $\mathcal{M}_0^+(\Omega)$ for the γ-convergence we have that from any minimizing sequence for (7') it is possible to extract a subsequence γ-convergent to a minimum point for (7").

We can so conclude that (6")(7") is the relaxed problem of (6')(7').

REFERENCES

1. L. Ambrosio, Variational methods in *SBV*, *Actae Applicandae Math.*, **17** (1989), pp. 1-40.

2. C. Baiocchi, *Disequazioni variazionali e quasivariazionali: applicazioni a problemi di frontiera libera.* Vol.I,II, Pitagora, Bologna, 1985.

3. M. Brelot, *On topologies and boundaries in Potential Theory*, Springer V., Lectures Notes in Math. 175, Berlin - Heidelberg - New York, 1971.

4. G. Buttazzo, G. Dal Maso, Shape optimization for Dirichlet problems: relaxed formulation and optimality conditions, *ppl. Math. Opt.*, **23** (1991), 17–49.

5. N. Dunford, J.T. Schwartz, *Linear operators*, Wiley, New York, 1957.

6. Gilbarg T., Trudinger N.S., *Elliptic partial differential equations of second order*, Springer V., Grundlehren der Math. Wiss. 224, Berlin - Heidelberg - New York, 1982.

7. E. Giusti, *Minimal surfaces and functions of bounded variation*, Birkhäuser, Boston - Basel - Stuttgart, 1984.

8. L. Schwartz, *Théorie des distributions*, Hermann, Paris, 1966.

PART II: GEODETIC BOUNDARY VALUE PROBLEM (GBVP)

Formulation and Linearization of Boundary Value Problems : From Observables to a Mathematical Model

B. Heck

1   Introduction

Since the beginning of Geodesy in a modern sense, gravity field determination has been a basic task. The determination of the external gravity field of the earth from terrestrial observational data is related to the formulation of boundary value problems with respect to the Laplace-Poisson differential equation. Depending on the type of boundary data and on the type and number of unknown functions to be solved for from geodetic observational data, several versions of the geodetic boundary value problem (GBVP) can be formulated. Besides the classical observational data, such as potential (differences) , gravity and astronomical latitudes and longitudes, new types of boundary data have become available from satellite techniques in recent years, involving new versions of the GBVP, like mixed and overdetermined ones. Any of these formulations deserves an analysis of its own, since the properties are strongly different.

This series of lectures is restricted to three classical formulations of the GBVP, related to terrestrial geodetic observables acting as boundary data. Only external problems are treated, the solution domain corresponding to the space outside the earth's surface which is considered as a closed, sufficiently smooth surface. The topic of these lecture notes is the derivation and discussion of the boundary conditions in several levels, arising from several steps of approximation, the most important ones being formed by linearization, spherical and constant radius approximation.

The presented material is organized as follows: In *section 2* we will derive the boundary conditions of the so-called fixed, vectorial free and scalar free versions of the classical GBVP, starting from a common basic physical-mathematical model; furthermore we will discuss the "degree of practicability" by comparing the assumptions with the availability of the observational data. In *section 3* the originally non-linear relationships between observables and unknown functions are linearized. The resulting linearization errors are estimated and compared with the practical requirements of (quasi-)geoid determination. In *section 4* the linearized boundary conditions are specified with respect to several coordinate systems; the coefficients of the disturbing potential and its functionals are simplified using the so-called spherical approximation. The last step in the hierarchy of approximations is the constant radius approximation which formally transfers the boundary data to a sphere. The respective approximation errors for the fixed GBVP will be presented in detail. Finally *section 5* provides some considerations about the solution of the three formulations of the linearized GBVP using a Hilbert space approach in terms of spherical harmonics; a comparison of the resulting algebraical systems of equations gives valuable insight into the uniqueness properties of the solution.

2 The boundary conditions in the classical GBVPs

2.1 Basic model and assumptions

It is a common feature of the classical GBVPs that they are based upon Newtonian mechanics. The respective physical-mathematical model can be formulated as follows:

(1) The earth is assumed to behave like a rigid, non-deformable body B, uniformly rotating about a space- and body-fixed axis in Newtonian absolute space which is by definition three-dimensional Euclidean. All attracting mass elements are located in the interior of the closed boundary surface $S = \partial B$ which represents the earth's surface. The integrated attraction of mass elements on a test particle generates the Newtonian *gravitational potential* V which is regular at infinity and fulfills Poisson's partial differential equation (Lap: Laplace operator)

$$\text{Lap}(V) = -4\pi G\rho \qquad (2\text{-}1)$$

in the whole Euclidean space E^3 except at the points and surfaces of discontinuity of the volume density $\rho^{(1)}$; the gravitational constant G ($G = 6.6726 \cdot 10^{-11}$ m^3 s^{-2} kg^{-1}) is a universal, time-independent constant. In the mass-free exterior domain $\Omega_e = E^3 - \bar{B}$ (B closure of B) the Poisson equation is identical with Laplace's differential equation

$$\text{Lap}(V) = 0. \qquad (2\text{-}2)$$

---

[1]

More precisely, the existence of second derivatives of V implies that the density function ρ is piecewise continuously differentiable or at least Hölder continuous.

The rotational motion produces the *centrifugal potential Z*,

$$Z = \tfrac{1}{2}\,\omega^2 p^2 \,, \tag{2-3}$$

depending on the (constant) angular velocity ω and the orthogonal distance p of the test particle from the space- and body-fixed rotation axis. The centrifugal potential Z fulfills the Poisson equation

$$\text{Lap } Z = 2\omega^2 \tag{2-4}$$

in any point fixed with respect to the rotating massive body.
The sum $W := V + Z$ is the *gravity potential* which fulfills the Poisson equation

$$\text{Lap } (W) = -\,4\pi G\rho + 2\omega^2 \tag{2-5}$$

at any point in E^3 except the surfaces of discontinuity of ρ. In the mass-free external space (2-5) transforms into the extended Laplace equation

$$\text{Lap } W = 2\omega^2. \tag{2-6}$$

The gradient of the gravity potential

$$\Gamma = \text{grad } W \tag{2-7}$$

is the (vectorial) gravity field intensity, which is called the *vector of gravity* in Geodesy and Geophysics. Equations (2-5) and (2-7) can be replaced by the vectorial field equations for Γ

$$\text{rot } \Gamma = \mathbf{0}$$
$$\text{div } \Gamma = -\,4\pi G\rho + 2\omega^2. \tag{2-8}$$

The gravity vector Γ at some point $P \in E^3$ is always perpendicular to the *equipotential surface* (or *level surface*) $W = \text{const.}$ running through the same point in space (Heiskanen and Moritz, 1967, p. 49). The orthogonal trajectories of the set of level surfaces are the *plumb lines* which are complicated curves in space, possessing non-vanishing curvature and torsion in general. The gravity vector Γ at any point is tangent to the plumb line at the same point. One specific level surface is formed by the *geoid* which approximates the mean sea surface in some specific sense not discussed here in more detail.

(2) In order to describe positions in space and field quantities like the gravity vector Γ we introduce an earth-fixed, orthonormal *coordinate frame* $\{0, \mathbf{f}_i\}$, $i = 1, 2, 3$. It is postulated that the origin 0 is situated at or near the earth's centre of mass, the $\mathbf{f}_3$ basis vector is directed along the rotational axis, $\mathbf{f}_1$ is parallel to the Greenwich meridian plane, and $\mathbf{f}_2$ completes the orthonormal global geocentric triad, oriented in a mathematically positive sense.[2]

The position of an arbitrary point $P \in E^3$, e.g. a boundary point $P \in S$, is described by the Cartesian coordinates X_i, which are related to the *position vector* $\mathbf{X}$

$$\mathbf{X} = X_i\,\mathbf{f}_i\,. \tag{2-9}$$

In (2-9) we have applied Einstein's summation convention, i.e. summation over pairs of indices occurring in a product.

Correspondingly, the gravity vector Γ can be represented by its Cartesian coordinates Γ_i

$$\Gamma = \Gamma_i\,\mathbf{f}_i\,. \tag{2-10}$$

2

Realizations of conventional terrestrial reference frames affixed to the (deformable) earth are provided by the International Earth Rotation Service (IERS), e.g. the International Terrestrial Reference Frame 1994 (ITRF 94).

Instead of the rectilinear coordinates X_i, Γ_i curvilinear coordinates are often preferable. Concerning the position coordinates X_i, we will consider three alternatives, namely spherical, geographical (geodetic) and elliptical coordinates. The functional relationship between Cartesian and *spherical (polar) coordinates* is provided by the equations

$$X_1 = r \cdot \cos \varphi \cdot \cos \lambda$$
$$X_2 = r \cdot \cos \varphi \cdot \sin \lambda \qquad (2\text{-}11)$$
$$X_3 = r \cdot \sin \varphi \,,$$

where (r, φ, λ) are (geo)centric radius, (geo)centric latitude and longitude.

Furthermore, we introduce an ellipsoid of revolution with semi-major axis a and semi-minor axis $\overline{b}$ by fixing the centre of the ellipsoid at 0 and aligning the semi-minor (symmetry) axis along $\mathbf{f}_3$. Then the Cartesian coordinates are related to the *ellipsoidal (geodetic) coordinates* (H, b, l) in the following way:

$$X_1 = (N + H) \cdot \cos b \cdot \cos l$$
$$X_2 = (N + H) \cdot \cos b \cdot \sin l \qquad (2\text{-}12)$$
$$X_3 = (N (1\text{-}e^2) + H) \cdot \sin b;$$

(H, b, l) denote the ellipsoidal height, the geodetic (geographical) latitude and longitude, respectively, e the first numerical eccentricity and N the prime vertical radius of curvature of the ellipsoid

$$e^2 = \frac{a^2 - \overline{b}^2}{a^2} \quad , \quad N = \frac{a}{\sqrt{1 - e^2 \sin^2 b}} \qquad (2\text{-}13)$$

Without restricting the general validity we can identify the geodetic longitude l with the geocentric longitude λ. For a visualization of the coordinates (r, φ, λ) and (H, b, l) see Fig. 2.1.

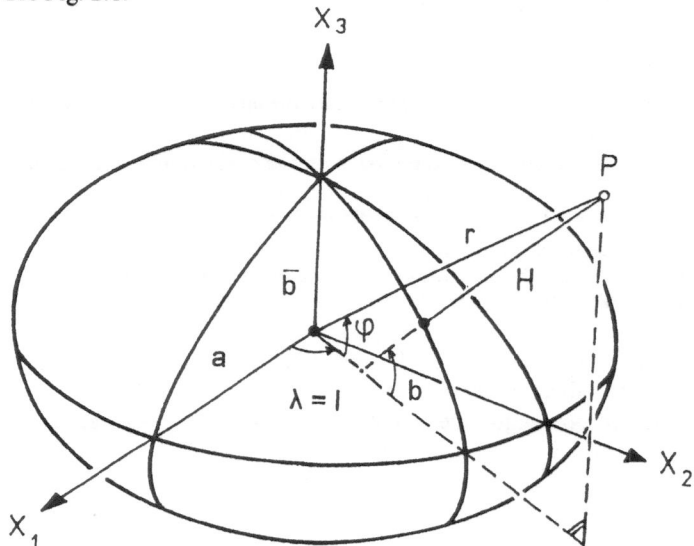

Figure 2.1:   Spherical (polar) coordinates (r, φ, λ) and ellipsoidal (geodetic) coordinates (H, b, l)

Finally *elliptical coordinates*, also related to the ellipsoid of revolution defined above, are introduced, which are connected to the Cartesian coordinates by the formulae (see Heiskanen and Moritz, 1967, p. 40; Thong and Grafarend, 1989)

$$X_1 = \sqrt{u^2 + E^2} \cdot \cos\beta \cdot \cos l$$
$$X_2 = \sqrt{u^2 + E^2} \cdot \cos\beta \cdot \sin l \qquad (2\text{-}14)$$
$$X_3 = u \cdot \sin\beta \ ;$$

$E = \sqrt{a^2 - \bar{b}^2}$ = const. is the linear eccentricity of the set of confocal ellipsoidal coordinate surfaces u = const., (u, β, l) denote the semi-minor axis of the coordinate surface, the reduced latitude and longitude, respectively, where again $l \equiv \lambda^{(3)}$.

Equivalently, the Cartesian coordinates Γ_i of the gravity vector can be expressed in terms of polar coordinates

$$\Gamma_1 = -\Gamma \cdot \cos\Phi \cdot \cos\Lambda$$
$$\Gamma_2 = -\Gamma \cdot \cos\Phi \cdot \sin\Lambda \qquad (2\text{-}15)$$
$$\Gamma_3 = -\Gamma \cdot \sin\Phi,$$

where $\Gamma = \|\Gamma\|^{(4)}$, Φ, Λ denote the *modulus of gravity* and the *astronomical latitude* and *longitude*, respectively. The astronomical coordinates (Φ, Λ) describe the direction of the gravity vector with respect to the earth-fixed reference frame $\mathbf{f}_i$. In general, the astronomical coordinates Φ, Λ have to be distinguished from the "geometrical" coordinates (φ, λ), (b, l) and (β, l) defined above.

(3) The boundary surface S = ∂B is a closed, star-shaped (with respect to the origin 0) and sufficiently smooth regular surface. Mathematically, the degree of smoothness is often expressed in terms of Hölder spaces. On this boundary surface the *boundary data* are given in the form of continuous functions. The boundary data are presumed as observable quantities which can be determined from geodetic observations on the earth's surface and which are related to some functionals of the gravity potential W. Examples are potential differences between surface points in the form of geopotential numbers C which can be determined by combining geodetic levelling and gravity observations along the levelling lines, or the modulus of gravity Γ which is derived from absolute and relative gravity observations (ballistic instruments, spring gravity meters), or the astronomical latitudes and longitudes which can be determined from astronomical observations of stars; in addition, some components of the gravity gradient tensor, i.e. second order derivatives of the gravity potential, can be obtained from terrestrial gradiometry or satellite gravity gradiometry. It is evident that for a detailed mathematical analysis the properties of the boundary data have to be defined in terms of suitable function spaces, e.g. Sobolev spaces.

---

[3]

If the flattening of the ellipsoid decreases, the eccentricities e and E approach zero, and the three definitions of curvilinear coordinates will coincide, i.e. $\varphi = b = \beta$ and r = R+h = u, where R is the radius of a mean earth sphere.

[4]

$\|\Gamma\|$ denotes the Euclidean norm of the vector Γ

Here it will be assumed that the boundary data are "sufficiently" smooth, whatever this notion may mean in a special case. The relationship between any boundary data $\tilde{L}$ and the gravity potential W is formally expressed as

$$BW = \tilde{L} \qquad (2\text{-}16)$$

where B denotes the *boundary operator* acting on W. This operator is composed of an operator D related to a spatial differential operator of order 0 or 1 (or 2 for the gradiometric observables) and an evaluation operator E restricting the spatial function to surface points $P \in S$. In general form, the combined operation is expressed as

$$E \circ DW = DW \big|_s = \tilde{L} \qquad (2\text{-}17)$$

Generally, D is a *non-linear* operator, related to a non-linear mapping between two function spaces. For external GBVPs, which are exclusively treated here, the spatial function DW is formed in the spatial domain on the external side of S.

(4) The primary *unknown* to be solved for in the framework of the GBVP is the gravity potential W on and outside the boundary surface S, i.e. in the external, mass-free space Ω_e. As discussed above, the behaviour of W in the external space Ω_e is ruled by the equations

$$W(\mathbf{X}) = V(\mathbf{X}) + Z(\mathbf{X})$$
$$\text{Lap } V(\mathbf{X}) = 0 \text{ for } \mathbf{X} \in \Omega_e \qquad (2\text{-}18)$$
$$V(\mathbf{X}) \to 0 \text{ for } r \to \infty, r = |\mathbf{X}|$$
$$Z(\mathbf{X}) = \tfrac{1}{2}\,\omega^2\,(X_1^2 + X_2^2).$$

If the geometry of the boundary surface S is completely known, then W($\mathbf{X}$) for $\mathbf{X} \in \Omega_e$ is the only unknown to be solved for. The respective BVPs based on arbitrary types of boundary data $\tilde{L}$ are denoted as *fixed GBVPs*. On the other hand, the geometric information about the boundary may be incomplete, such that the determination of the position vector of the boundary surface becomes part of the problem. These problems, dealing with an unknown geometry of the boundary are called *free GBVPs*. One might be forced to assume that no information at all about the positions of the surface points $\mathbf{X} \in S$ are given; then the number of unknowns in the respective free GBVP is four, namely the potential function W and the three coordinate functions of $\mathbf{X}$. This type of GBVP is called the *vectorial free GBVP*, or the *astronomical variant of Molodensky's problem*. As an alternative one can imagine that the "horizontal" coordinates of $\mathbf{X} \in S$ are given, e.g. the geodetic coordinates b and l. Since the vertical or height coordinate of $\mathbf{X}$ remains the only geometrical unknown, the total number of unknowns is two. This setup of the GBVP is called the *scalar free GBVP*, or the *geodetic variant of Molodensky's problem*. As a matter of fact, the larger number of unknowns in the free GBVP has to be counterbalanced by additional boundary functions.

Obviously, the type of non-linearity in the boundary conditions differs strongly, considering free and fixed GBVPs: For the fixed GBVP the non-linearity of the operator D is decisive, while for the free GBVP there exists an additional non-linear dependence on the (unknown) boundary surface, since the relationship between W($\mathbf{X}$) and $\Gamma(\mathbf{X})$ and $\mathbf{X}$ is a non-linear one.

Besides the unknown functions W and $\mathbf{X}$, other (secondary) unknowns may be existent in the GBVP, too. Since it is impossible to observe absolute potentials or to determine potential differences with respect to an infinitely distant point, the absolute potential value W_0 at some vertical datum reference point P_0 (e.g.

coinciding with a tide gaugue) may be considered as an additional discrete unknown. Obviously, the occurrence of such an unknown has to be counteracted by at least one additional observation, e.g. by an observed distance between two specified boundary points (see e.g. Witsch, 1987). This case can be generalized into a multi-datum GBVP, assuming that on the boundary surface S a total number of k>1 datum zones exist, each of them being related to a different vertical datum reference point. This problem can be handled assuming that the "absolute" positions of the k reference points are known from e.g. satellite positioning (Rummel and Teunissen, 1988; Sansò und Usai, 1995).

The assumptions (1) - (4) describe an idealized situation which differs in many respects from the real world. Since no external masses (earth's atmosphere, moon, sun, planets) have been admitted, their effects must have been reduced in the boundary data. Moreover the GBVP is considered as a stationary problem, neglecting (or reducing for) geodynamic phenomena; for the treatment of the space-time GBVP see e.g. Heck (1982, 1984). Another artificial feature is the assumption of continuously distributed boundary data, given over the whole boundary surface, while in geodetic practice the observations are provided in discrete form, containing gaps in some regions and a surplus of data in other regions. Considering mixed data on different subdomains of the boundary surface or additional data of different types one is lead to mixed and overdetermined GBVPs. For the latter type of BVPs the stochastic behaviour of the boundary data is another crucial item, while the randomness of observational errors etc. plays no role in the formulation and solution of the classical (uniquely determined) GBVP. In the following, the boundary conditions related to the classical GBVPs of purely fixed and free vectorial/scalar types will be discussed in detail; mixed, overdetermined and time-dependent GBVPs will be left out of consideration.

2.2 The fixed gravimetric GBVP

The practical background of the fixed gravimetric GBVP is provided by recent advances in satellite positioning techniques using SLR (Satellite Laser Ranging) and VLBI (Very Long Baseline Interferometry) for fixing a highly precise global reference frame which is densified and interpolated by GPS (Global Positioning System) carrier phase observations. In this way it is possible to determine the position of any point on the earth's surface with respect to a uniform global reference frame with an accuracy of a few centimetres or better. This precision justifies the assumption that the (geocentric) position vector $\mathbf{X}$ of any surface point $P \in S$ is completely known.

In order to determine the only unknown function, namely the gravity potential W in the external space outside S, it is sufficient to postulate that one type of continuous boundary data is given on S. If it was possible to observe the values of the gravity potential W on S, a *Dirichlet problem* in the sense of classical potential theory would result. Unfortunately the observability of (relative) potential values by combining levelling and gravimetry is restricted to the solid parts of the earth's surface. On the other hand, the ocean surface - after reduction for tidal effects and wave heights - differs by no more than about 2 metres from an equipotential surface; thus potential differences in oceanic areas can be determined by oceanographic methods, see e.g.

Rummel and Ilk (1995). Considering the drawback that geodetic as well as oceanographic levelling is extremely labourious this solution is impracticable.

On the other hand the *modulus of gravity*, i.e. the length of the gravity vector Γ =grad W,

$$\tilde{\Gamma} = \| \Gamma \|_S = \| \operatorname{grad} W \|_S$$

(2-19)

is an observable both on the solid and fluid parts of the earth's surface using land-based and shipborne gravimetry. Assuming that a complete, sufficiently dense and precise survey of the modulus of gravity $\tilde{\Gamma}$ has been performed on the whole (fixed) surface, and that the discrete distribution has been interpolated and transformed into a continuous function on the boundary S, the *fixed gravimetric GBVP* arises. Since only the absolute value of Γ is given, while the direction of Γ remains unknown, the boundary condition (2-19) gives a non-linear relationship between the observable $\tilde{L} = \tilde{\Gamma}$ and the unknown function W, the boundary condition referring to the Laplace-Poisson equation (2-5) or (2-6). The reason for the non-linearity property of the boundary equation is the *non-linearity* of the (Euclidean) norm operator.

The non-linear fixed gravimetric GBVP first has been discussed by Backus (1968). Local existence, uniqueness and stability results, referring to some standard, approximate solution, have been derived by Bjerhammar and Svensson (1983); in contrast to the boundary condition (2-19) these authors have used the squared norm $\tilde{\Gamma}^2$ as boundary data. For a survey of recent work done in the field of the fixed gravimetric GBVP see Klees (1992).

2.3 The vectorial free GBVP

Before the advent of space techniques it was impossible to determine the geometry of the earth's surface as a whole with high enough precision. The difficulties with vertical refraction prevented Geodesy from establishing three-dimensional networks of global dimensions by pure terrestrial measurements of distances, horizontal angles, azimuths and zenith distances. This unpleasant situation gave rise to the formulation of *free* GBVPs, requiring that the position vector **X** of the boundary surface S has to be determined from a set of suitable boundary data as a part of the solution.

If no information about the position vector is known, then the number of inherent unknown functions increases to four, namely the three coordinate functions in **X** and the potential function W. It is evident that (at least) four functionals of W have to be given on the boundary in order to solve the respective GBVP related to the Laplace-Poisson equation (2-5) or (2-6). In the classical formulation of the vectorial free GBVP the boundary data consist of the potential function and the three components of the gravity vector Γ. This problem had first been defined by M.S. Molodenskii (Molodenskii et al., 1962) and is called *Molodensky's problem*; relevant steps in its analysis have been made by the work of Krarup (1973), Hörmander (1976), Sansò (1978), Moritz (1980) and others. Inspecting (2-15) it becomes clear that Γ can be expressed by the polar coordinates $\Gamma = \| \Gamma \|$, Φ, Λ. Since this GBVP is based on the astronomical coordinates Φ, Λ acting as boundary data, the problem is sometimes denoted as the *astronomical variant* of Molodensky's problem.

The relationship between the observable boundary data and the unknown functions is provided by the formulae

$$\tilde{W} - W|_S = W(\mathbf{X})$$

$$\tilde{\Gamma} - \text{grad } W|_S = \text{grad } W(\mathbf{X})$$

(2-20a)

(2-20b)

or equivalently

$$\tilde{W} - W|_S = W(\mathbf{X}) \qquad (2\text{-}21\text{a})$$

$$\tilde{\Gamma} - \sqrt{\Gamma_1^2 + \Gamma_2^2 + \Gamma_3^2}\Big|_S = \sqrt{(\Gamma_1(\mathbf{X}))^2 + (\Gamma_2(\mathbf{X}))^2 + (\Gamma_3(\mathbf{X}))^2} \qquad (2\text{-}21\text{b})$$

$$\tilde{\Phi} - \arctan \frac{-\Gamma_3}{\sqrt{\Gamma_1^2 + \Gamma_2^2}}\Big|_S = \arctan \frac{-\Gamma_3(\mathbf{X})}{\sqrt{(\Gamma_1(\mathbf{X}))^2 + (\Gamma_2(\mathbf{X}))^2}} \qquad (2\text{-}21\text{c})$$

$$\tilde{\Lambda} - \arctan \frac{-\Gamma_2}{-\Gamma_1}\Big|_S = \arctan \frac{-\Gamma_2(\mathbf{X})}{-\Gamma_1(\mathbf{X})} \qquad (2\text{-}21\text{d})$$

where $\Gamma_i = \partial W/\partial X_i$. Obviously the relationships (2-20) and (2-21) are non-linear, the *non-linearity* being mainly induced by the free boundary. The postulate of prescribing absolute potential values W can be weakened by assuming potential differences $W - W_0$ as boundary data; the additional degree of freedom due to W_0 can be counterbalanced e.g. by one observed distance.

It should be noted that the astronomical observables $\tilde{\Phi}, \tilde{\Lambda}$ mainly influence the "horizontal" components of the position vector $\mathbf{X}$ (Rummel and Teunissen, 1982). A precision of say 0.3 for the astronomical observations corresponds to an accuracy of about 10 m in the horizontal position. As a matter of fact, this data has neither the required density of distribution nor a sufficient precision to justify the boundary value approach for the determination of the horizontal components of the position vector $\mathbf{X}$. This property reduces the practical significance of the vectorial free GBVP. Nevertheless its analysis gives a deep insight into the properties of the whole class of GBVPs.

2.4 The scalar free GBVP

The practical difficulties with the determination of the horizontal components of the position vector have motivated the *scalar free* formulation of the GBVP. In fact, the information about horizontal positions of surface points have never been extracted from the solution of a GBVP in classical Geodesy. On the other hand, the vertical component of the position vector can be rather accurately derived from the solution of the GBVP, essentially depending on the quality of the gravity data.

In geodetic practice the horizontal coordinates of survey points have traditionally been the result of terrestrial trigonometric measurements arranged in (horizontal) geodetic networks. Generally the regional and continental networks have been referred to an ellipsoid of revolution with given geometrical parameters, position and orientation. The reference ellipsoid is affixed to the geodetic network by astronomical observations at some fundamental point. Since the geodetic control points are projected onto the chosen reference ellipsoid along the ellipsoidal normal, the horizontal coordinates can be identified with the geographical (geodetic) longitude l and latitude b. As a matter of fact, the ellipsoidal height H (distance measured along the ellipsoidal normal) of the horizontal control points is only approximately determined. It is evident that a change in the ellipsoidal parameters a and e^2 (or the flattening f) or in the orientation and position of the reference ellipsoid affects the geodetic coordinates H, b, l of an arbitrary point in space.

In the following we will presuppose that the *geographical coordinates* b, l of the points on the boundary are absolutely known. Although this assumption is made plausible by the above considerations, some difficulties have to be mentioned. Since in the past the geodetic datum of any national geodetic network has been independently fixed, the horizontal coordinates of survey points do not refer to a unique reference frame, producing a bias in the national geographical coordinates up to about 10 arcseconds (corresponding to a horizontal displacement of about 300 metres), see e.g. Heck (1990). This systematic effect propagates into gravity related products like gravity anomalies, and into the numerical solution of the GBVP. Transformations of independent national horizontal networks into a global reference frame have become feasible only recently by satellite positioning procedures, e.g. by the products of the International Earth Rotation Service (IERS). Thus, in principle it is possible nowadays to transform all local datum-geographical coordinates into a global, "absolute" reference frame. Unfortunately the precision of the classical terrestrial geodetic networks is disparaged by the propagation of random, systematic and gross errors in the observational data producing distortions in the network's geometry. Nevertheless the horizontal coordinates of the points of the earth's surface are known with much better precision than the vertical coordinate, i.e. the ellipsoidal height H. This property holds for satellite positioning, e.g. by GPS observations, too, justifying the assumptions related to the scalar free GBVP.

Fixing the horizontal coordinates b, l of the boundary surface, the pair of unknowns W, H remains which have to be determined from suitable boundary data. As in section 2.3 we will assume that the modulus of gravity and potential differences with respect to a fundamental levelling point are known. The relationship between this boundary data and the unknown functions is provided by the formulae

$$\tilde{W} - \tilde{W}_0 - (W - W_0)|_S - W(\mathbf{X}) - W(\mathbf{X}_0) \qquad (2\text{-}22a)$$

$$\tilde{\Gamma} - \| \operatorname{grad} \ W \||_S - \sqrt{(\Gamma_1(\mathbf{X}))^2 + (\Gamma_2(\mathbf{X}))^2 + (\Gamma_3(\mathbf{X}))^2} \qquad (2\text{-}22b)$$

The dependence on $\mathbf{X}$ of the right hand side has to be understood in the sense that $\mathbf{X}$ (b, l ; H) depends on the given parameters b, l and the unknown parameter H which itself is a function of b, l.

W_0 denotes the unknown gravity potential at a fundamental datum point P_0, the position of which is described by the position vector $\mathbf{X}_0$ (b$_0$, l$_0$; H$_0$), the ellipsoidal height H_0 being unknown. The relationships between $\tilde{W}$, $\tilde{\Gamma}$ and W, H are non-linear, the *non-linearity* being induced by the non-linear (Euclidean) norm operator in (2-22b) and by the "vertically free" boundary surface.

Due to the strong relationship to the geodetic coordinates b, l the GBVP based on the boundary conditions (2-22) is sometimes denoted as the *geodetic variant* of

Molodensky's problem. The first formulation of this GBVP can be found in Molodenskii et al. (1962); relevant steps in the analysis of this problem have recently been made by the work of Sacerdote and Sansò (1986) and Otero (1987).

Obviously the scalar formulation based upon boundary data of type potential (differences) and modulus of gravity is much closer to geodetic practice than the vectorial version: It is general custom to describe the position of some point where gravity has been measured by its geographical coordinates b, l, which either are taken from a topographic map or from GPS positioning. On the other hand the "height" of the gravity observation point is derived from levelling, implicitly applying some type of telluroid mapping.

3. Linearization of the boundary conditions

3.1 Approximate and difference quantities

In section 2 it has been demonstrated that the boundary conditions for various formulations of the GBVP are generally non-linear, i.e. the (continuous) boundary data $\tilde{L}$ given on the earth's surface S depend in a non-linear way on the respective unknown functions. Since the gravity potential W in the external space Ω_e acts as the primary unknown, the observables $\tilde{L}$ can be understood as (non-linear) functionals of W, restricted to the boundary surface S, of the general form (2-17). It is evident from the discussion in section 2 that - in the case of the free GBVP - the boundary data depend on additional unknown functions like the Cartesian coordinates of the boundary surface or the ellipsoidal height.

Until now there exist no mathematical tools for solving the non-linear GBVP directly. All analytical and numerical methods developed in the past are related to linear BVPs, the solution of the respective non-linear problems being based on a linearization of the original problem around an approximate solution and on the construction of a convergent iteration process. In this way the solution of the non-linear GBVP is achieved by solving a family of related linear problems (see e.g. Klees, 1995). As a matter of fact, the speed of convergence depends on the degree of "closeness" of the approximation with respect to the true solution. On the other hand, the functions used as reference in the solution of the GBVP should be as simple as possible from the mathematical-algebraic point of view for numerical reasons.

As a compromise between these two requirements the gravity potential W is often approximated by a Somigliana-Pizzetti field (Heiskanen and Moritz, 1967, p. 64f.) acting as a normal or *reference field*. As an alternative one may choose some (finite) expansion of the gravity potential in spherical or ellipsoidal harmonics as reference field. In any case, the reference potential w is formed by an analytical expression which allows the evaluation of the normal potential and its derivatives at arbitrary points in the solution domain. The properties of the normal potential w are required to be similar to those of W, including regularity at infinity of the gravitational part and validity of the Laplace-Poisson equation in Ω_e

$$\text{Lap } w = 2 \, \omega^2. \tag{3-1}$$

The angular velocities in (2-4) and (3-1) are assumed to coincide.

The gradient of the reference potential w is the *normal gravity* vector

$$\gamma = \text{grad } w \tag{3-2}$$

which can be represented in the earth-fixed coordinate frame $\{0, f_i\}$ by its Cartesian coordinates γ_i. Equivalently to (2-15) the Cartesian coordinates can be expressed in terms of polar coordinates

$$\gamma_1 = -\gamma \cdot \cos \varphi_\gamma \cdot \cos \lambda_\gamma$$
$$\gamma_2 = -\gamma \cdot \cos \varphi_\gamma \cdot \sin \lambda_\gamma \tag{3-3}$$
$$\gamma_3 = -\gamma \cdot \sin \varphi_\gamma,$$

where $\gamma = \|\gamma\|$, φ_γ, λ_γ denote *normal gravity* and the *normal latitude and longitude*, respectively. The parameters (φ_γ, λ_γ) describe the direction of the normal gravity vector (with respect to the earth-fixed reference frame f_i), which is tangential to the normal plumb line running through the respective point in space. In general, (φ_γ, λ_γ) do not coincide with the "geometrical" coordinates (φ, λ), (b, l) or (β, l); only in the case of a rotationally symmetric field, e.g. a Somigliana-Pizzetti normal field, the longitudes are equal, i.e. $\quad \lambda_\gamma = \lambda = 1$.

In addition to the reference potential, a *reference surface* has to be adopted in the case of the free GBVP. In principle, any surface s approximating the earth's surface S in some sense can be used, as long as the surface points p ∈ s and P ∈ S are in a one-to-one correspondence. Geometrically simple surfaces like spheres and ellipsoids of revolution will not fulfill the requirements of closeness, since the distances between corresponding points p ∈ s and P ∈ S may be as large as 10-20 km. A surface in close vicinity of the earth's surface is obtained via a *telluroid mapping* using a suitable subset of the boundary data. For the vectorial free GBVP a natural choice is an isozenithal telluroid mapping of type (Grafarend, 1978)

$$\frac{\Gamma(P)}{\Gamma(P)} = \frac{\gamma(p)}{\gamma(p)} \tag{3-4}$$

which is completed by one of the following conditions

$$W(P) = w(p) \qquad \text{(Marussi telluroid)} \tag{3-5a}$$
$$\Gamma(P) = \gamma(p). \qquad \text{(Gravimetric telluroid)} \tag{3-5b}$$

For a discussion of the numerical properties of telluroid mappings of different kinds see e.g. Heck (1986). An appropriate definition of the telluroid in the case of the scalar free GBVP is provided by the mapping equations (Molodensky's telluroid)

$$b(P) = b(p)$$
$$l(P) = l(p) \tag{3-6}$$
$$W(P) - W_0 = w(p) - w_0$$

where w_0 may be identified with the (constant) potential on the surface of the Somigliana- Pizzetti reference ellipsoid; by the first and second condition in (3-6) the associated points p and P are restricted to the same ellipsoidal normal. It is known from practical results that the radial (or vertical) distance of the telluroid from the earth's surface is of the order 100 metres only, in the case of Marussi's and Molodensky's telluroid mappings based on a Somigliana-Pizzetti normal field. It should be noted that choosing the simple isotropic potential as normal potential, i.e. $w = \mu/r$, will produce maximum distances between s and S on the order of about 10 km. As a consequence, it is not advisable to apply the isotropic normal potential for telluroid definition[5].

Denoting the position vector of an arbitrary telluroid point $p \in s$ by $\mathbf{x}$ we can identify the position correction vector or *position anomaly* by the difference

$$\Delta \mathbf{x} := \mathbf{X}(P) - \mathbf{x}(p) = \Delta x_i \cdot \mathbf{f}_i. \tag{3-7}$$

The position anomaly $\Delta \mathbf{x}$ may be expressed in terms of either Cartesian coordinates Δx_i (referring to the global reference frame $\{0, \mathbf{f}_i\}$ or curvilinear, e.g. geodetic coordinates

$$\Delta h := H(P) - h(p)$$
$$\Delta b := b(P) - b(p) \tag{3-8}$$
$$\Delta l := l(P) - l(p);$$

for obvious reasons the difference Δh is often called the *height anomaly*. If Molodensky's telluroid mapping is applied to the scalar free GBVP then $\Delta b = \Delta l = 0$ holds due to (3-6) and (3-8). In addition, Δh vanishes for the fixed GBVP.

In a similar way the *disturbing potential* δw at an arbitrary point in space $\mathbf{y}$ is defined by the difference

$$\delta w(\mathbf{y}) = W(\mathbf{y}) - w(\mathbf{y}). \tag{3-9}$$

If the centrifugal parts in W and w are identical, then the disturbing potential is harmonic outside the boundary surface S, i.e.

$$\text{Lap } \delta w(\mathbf{y}) = 0 \qquad \text{for } \mathbf{y} \in \Omega_e. \tag{3-10}$$

The domain of harmonicity of δw is often extended down to the surface of the telluroid[6]; this presumption will not be necessary in the case of the fixed GBVP where $\Delta \mathbf{x} \equiv \mathbf{0}$. A second property of the disturbing potential is regularity at infinity, i.e.

$$\delta w = \frac{c}{r} + 0\,(r^{-k}). \tag{3-11}$$

If the origin of the reference system coincides with the earth's centre of mass, then $k = 3$, otherwise $k = 2$.

The reference potential w can also be used to define model quantities $\tilde{\mathbf{l}}$ as a counterpart of the boundary data $\tilde{L}$. Applying the differential operator D in the boundary conditions (2-17) to the reference potential w, as well as the evaluation operator e wich restricts the resulting functional to the model surface s, we obtain the model quantity $\tilde{\mathbf{l}}$

$$\tilde{l} = e \circ Dw = Dw|_s \tag{3-12}$$

which can be exactly calculated. In the case of the fixed GBVP the surfaces s and S, and thus the operators e and E, are identical.

The difference between (2-17) and (3-12)

$$\tilde{L} - \tilde{l} = E \circ DW - e \circ Dw = DW|_S - Dw|_s \tag{3-13}$$

is a two-point function in the context of the free GBVP and is denoted by the term

---

We will not further discuss the problem of a stochastic reference surface considering the geodetic observations on the boundary surface as stochastic quantities. The reference surface s is here constructed on the basis of non-stochastic prior information e.g. by simply assigning numerical values as fixed quantities.

This extension of the domain of harmonicity still forms an open problem since neither the theorem of Keldych-Lavrentieff (1937, see Moritz, 1980, p. 69) nor the theorem of Yamabe (1950) give a definite answer.

anomaly

$$\Delta l := \tilde{L}(P) - \tilde{l}(p);$$ (3-14)

in contrast, the one-point function

$$\delta l := \tilde{L}(P) - \tilde{l}(P)$$ (3-15)

occurring in the case of the fixed GBVP is called a *disturbance*. Throughout the paper the symbols δ and Δ will be used in the sense of disturbances and anomalies, respectively. In particular, we can formulate the difference equations for

the *potential anomaly*

$$\Delta w := W(P) - w(p) = \delta w(P) + [w(P) - w(p)],$$ (3-16)

the *vectorial gravity anomaly*

$$\Delta \gamma := \Gamma(P) - \gamma(p) = \delta\gamma(P) + [\gamma(P) - \gamma(p)],$$ (3-17)

and the *vectorial gravity disturbance*

$$\delta\gamma := \Gamma(P) - \gamma(P) = \text{grad } \delta w(P).$$ (3-18)

In terms of curvilinear gravity space coordinates (3-17) is equivalent to the scalar equations

$$\Delta\gamma := \Gamma(P) - \gamma(p)$$ (3-19a)

$$\Delta\varphi_\gamma := \Phi(P) - \varphi_\gamma(p)$$ (3-19b)

$$\Delta\lambda_\gamma := \Lambda(P) - \lambda_\gamma(p).$$ (3-19c)

$\Delta\gamma$ is the *scalar gravity anomaly*. In a similar way the vectorial gravity disturbance (3-18) can be expressed by differences in curvilinear gravity space coordinates:

$$\delta\varphi_\gamma := \Phi(P) - \varphi_\gamma(P) = \xi(P)$$ (3-20a)

$$\delta\lambda_\gamma := \Lambda(P) - \lambda_\gamma(P) = \eta(P)/\cos\varphi_\gamma(P)$$ (3-20b)

$$\delta\gamma := \Gamma(P) - \gamma(P);$$ (3-20c)

$\delta\gamma$ is the *scalar gravity disturbance*, while the directional components are related to the *deflections of the vertical ξ, η*.

Referring to the third equation of (3-6) the potential anomaly Δw can be transformed into the anomaly of the geopotential number

$$\Delta c := [W_0 - W(P)] - [w_0 - w(p)] = \Delta w_0 - \Delta w$$ (3-21)

where $\Delta w_0 = [W_0 - w_0]$ is a constant potential unknown related to the fundamental vertical datum point P_0, and $C(P) := W_0 - W(P)$ is the observable geopotential number, corresponding to the model quantity $c(p) = w_0 - w(p)$ which can be calculated from the known position of the point p. Obviously Δc vanishes for Molodensky's telluroid mapping (3-6).

3.2 Linearization of the fixed GBVP

Without restricting the general validity it can be assumed that the reference surface s coincides with the actual boundary S in the context of the fixed GBVP. As a consequence, the boundary condition (3-13), specified for the observable Γ (2-19), takes the difference form

$$(\|\text{grad }(w + \delta w)\| - \|\text{grad } w\|)|_s = \tilde{\Gamma}(P) - \tilde{\gamma}(P) = \delta\gamma(P)$$ (3-22)

making use of (3-9) and (3-20). Since the norm operator is non-linear, the left hand side of   (3-22) is not simply $B \delta w = E \circ D\delta w$, D linear, but contains additional terms non-linear in the unknown function δw. Neglecting these non-linear terms corresponds to the operation of linearization, and the linearized boundary condition takes the form

$$B^{(1)} \delta w^{(1)} = E \circ D^{(1)} \delta w^{(1)} = \delta l$$ (3-23)

where $D^{(1)}$ is a linear operator. Higher order approximations can be constructed by series expansions, as will be shown below.

By the aid of (3-2) and $\|\gamma\| = \gamma$ the differential operator in (3-22) can be expressed in the following form($<\cdot, \cdot>$ denotes the scalar product of vectors):

$$\| \text{grad } (w + \delta w)\| - \| \text{grad } w\| =$$

$$= (\|\gamma\|^2 + 2 < \gamma, \text{grad } \delta w > + \| \text{grad } \delta w\|^2)^{\frac{1}{2}} - \|\gamma\| = \quad (3\text{-}24)$$

$$= \gamma \cdot (1 + \frac{2}{\gamma^2} \cdot < \gamma, \text{grad } \delta w > + \frac{1}{\gamma^2} \| \text{grad } \delta w\|^2)^{\frac{1}{2}} - \gamma.$$

Since it has been assumed that the normal potential w approximates the true potential W "sufficiently well", the term under the square root $(1 + x)^{\frac{1}{2}}$ will contain the "small" quantity x, $|x| \ll 1$, so that the binomial series can be applied. Neglecting the terms non-linear in δw results in the linearized boundary condition

$$< \frac{\gamma}{\gamma}, \text{grad } \delta w^{(1)} > |_s = \delta \gamma. \quad (3\text{-}25)$$

Obviously the boundary operator of the linearized gravimetric GBVP is identical with the derivative of $\delta w^{(1)}$ in the direction of the normal gravity vector, i.e.

$$< \frac{\gamma}{\gamma}, \text{grad } \delta w^{(1)} > = \frac{\partial \delta w^{(1)}}{\partial \tau}$$

$$\tau = \gamma / \gamma. \quad (3\text{-}26)$$

(The direction of $-\gamma$ for a Somigliana-Pizzetti normal field differs only slightly from the geocentric radial direction, 11 arcmin at most. For this reason it is common practice to introduce further approximations in the linearized problem, namely the so-called spherical approximation replacing $\partial/\partial\tau$ by $-\partial/\partial r$. These approximations will be discussed in section 4).

The linearized boundary condition (3-25) has to be combined with the field equation for $\delta w^{(1)}$

$$\text{Lap } \delta w^{(1)} (y) = 0 \quad \text{for } y \in \Omega_e \quad (3\text{-}27)$$

which is linear from the beginning. The respective solution $\delta w^{(1)}$ can be taken to construct an improved reference potential

$$w \leftarrow w + \delta w^{(1)}.$$

Repetition of the above procedure yields higher order solutions of the gravimetric GBVP.

Using the binomial series expansion of (3-24) an iterative approach, derived in detail by Heck (1989a), can easily be constructed:

$$\delta w^{(0)} = 0$$

$$< \frac{\gamma}{\gamma}, \text{grad } \delta w^{(n)} > = \delta \gamma - \sum_{i=2}^{n} \delta \gamma_i (\delta w^{(n-1)}), n \geq 1 \quad (3\text{-}28)$$

$$\text{Lap } \delta w^{(n)} (y) = 0 \quad \text{for } y \in \Omega_e,$$

where the correction terms on the right hand side of the boundary condition depend on the solutions in former steps. The first terms are

$$\delta \gamma_2 = \frac{1}{2\gamma} [\| \text{grad } \delta w^{(1)}\|^2 - (< \frac{\gamma}{\gamma}, \text{grad } \delta w^{(1)} >)^2]$$

$$\delta \gamma_3 = -\frac{1}{\gamma} \cdot < \frac{\gamma}{\gamma}, \text{grad } \delta w^{(2)} > \delta \gamma_2, \quad (3\text{-}29)$$

$\delta\gamma_2$ being strongly related with the deflections of the vertical.

It should be noted that the boundary operator of the linearized problem is identical in every iteration step, a property which is well-known from the simple Newton method. The convergence of this procedure in suitable function spaces can be proved using a "soft" implicit function theorem (Zeidler 1986). A similar procedure, based on the simple Newton iteration has been proposed by Bjerhammar and Svensson (1983). The functional analytic background of these procedures is provided by the concept of the Fréchet derivative, see also Meissl (1971). A different approach, based on power series expansion with respect to a "small" parameter ϵ, has been investigated by Klees (1995). In Klees (1992) a synopsis of several iteration procedures for the solution of the non-linear fixed gravimetric GBVP is given.

The degree of non-linearity can easily be changed for the fixed gravimetric GBVP. Squaring the boundary condition (13) and considering Γ^2 instead of Γ as the fundamental boundary data, a strictly quadratic boundary condition in W is achieved. This procedure, proposed by Bjerhammar and Svensson (1983), gives an essential simplification for the analysis and the numerical treatment of the non-linear fixed GBVP, see also Sacerdote and Sansò (1990).

3.3   Linearization of the vectorial free GBVP

As noted above, the free formulations of the GBVP are related to another type of non-linearity which is induced by the fact that the position vector of the boundary surface is unknown. In this case, a *continuation* of the unknown potential function and its derivatives has to be performed implicitly which involves the problem of extending the domain of harmonicity from the exterior of S to the exterior of the telluroid s. Starting from the original boundary conditions (2-20 a,b) we obtain the difference equations (3-16), (3-17) for the potential anomaly Δw and the vectorial gravity anomaly $\Delta\gamma$ which are related to the unknown disturbing potential δw and the position anomaly Δx:

$$\Delta w = \delta w\,(P) + [w\,(P) - w\,(p)] \tag{3-30a}$$
$$\Delta\gamma = \text{grad}\,\delta w\,(P) + [\gamma\,(P) - \gamma(p)]. \tag{3-30b}$$

Symbolically, the boundary condition takes the form

$$D\delta w|_s + (Dw|_s - Dw|_s) = \Delta l \tag{3-31}$$

where Δl is composed of the potential anomaly Δw and the vectorial gravity anomaly $\Delta\gamma$, while the differential operator D consists of the identity operator I and the gradient operator ∇

$$\Delta l = \begin{pmatrix} \Delta w \\ \Delta \gamma \end{pmatrix} \quad, \quad D = \begin{pmatrix} I \\ \nabla \end{pmatrix}; \tag{3-32}$$

obviously the differential operator D is *linear* in the case of the vectorial free GBVP. In the boundary condition (3-30) or (3-31) the first term is induced by the disturbing potential , while the second one is related to the "displacement" between the boundary surface S and the telluroid surface s, the position anomaly Δx.

In the process of linearization the continuation problem can be formally tackled by Taylor expansion on the known approximate surface s, the telluroid. Assuming that δw and Δx are first-order quantities, the respective expansions up to second order terms are

$$w(P) = w(p)+\nabla w(p)\cdot\Delta x + \tfrac{1}{2}\nabla\nabla w(p)\cdot\Delta x\cdot\Delta x +...$$
$$\delta w(P) = \delta w(p)+\nabla\delta w(p)\cdot\Delta x + ... \qquad (3\text{-}33)$$
$$\nabla w(P) = \nabla w(p)+\nabla\nabla w(p)\cdot\Delta x+\tfrac{1}{2}\nabla\nabla\nabla w(p)\cdot\Delta x\cdot\Delta x+...$$
$$\nabla\delta w(P) = \nabla\delta w(p)+\nabla\nabla\delta w(p)\cdot\Delta x+.....$$

These expansions can be employed in order to derive second order terms in the boundary conditions (3-30). It should be noticed that the (first, second and third order) covariant derivatives have to be applied in the calculation of the gradient operators if curvi-linear coordinates q^i, $i\in\{1,2,3\}$ are used for the description of the telluroid surface $s\ni p$ (Smeets 1991). Introducing the covariant base vectors g_j, $j\in\{1,2,3\}$

$$g_j := \frac{\partial x}{\partial q^j} \qquad (3\text{-}34)$$

the position correction vector Δx can be expressed by the corresponding contravariant coordinates $\Delta\xi^j$

$$\Delta x := \Delta\xi^j\cdot g_j \qquad (3\text{-}35)$$

where we have applied Einstein's summation rule. The contravariant coordinates $\Delta\xi^j$ are related to the differences of curvilinear coordinates

$$\Delta q^i := q^i(P) - q^i(p) \qquad (3\text{-}36)$$

by the Taylor expansion

$$\Delta\xi^j = \Delta q^j + \tfrac{1}{2}\Gamma^j_{kl}\Delta q^k\Delta q^l +....., \qquad (3\text{-}37)$$

involving the Christoffel symbols of second kind.

Now the representation (3-35) as well as the respective expressions for the gradients can be inserted in (3-30), (3-33), resulting in expansions of the boundary conditions up to second order terms in δw and $\Delta\xi^i$ (Heck 1988; Smeets 1991). Neglecting second and higher order terms the linearized (vectorial free) GBVP is based on the boundary conditions

$$\delta w_{(1)} + w_{|i}\cdot\Delta\xi^i_{(1)} = \Delta w \qquad (3\text{-}38a)$$

$$\frac{\partial\delta w_{(1)}}{\partial q^i} + w_{|i|j}\cdot\Delta\xi^j_{(1)} = \Delta\gamma_i , \; i\in\{1,2,3\} \qquad (3\text{-}38b)$$

where $w_{|i}$ and $w_{|i|j}$ denote the first and second order covariant derivatives of the reference potential with respect to the curvilinear coordinates at the telluroid point $p\in s$. The terms $\Delta\gamma_i$ denote the covariant coordinates of the vectorial gravity anomaly $\Delta\gamma$ with respect to the contravariant basis g^i, i.e.

$$\Delta\gamma = \Delta\gamma_i\cdot g^i = \Delta\gamma_i\cdot g^{ij}\cdot g_j, \qquad (3\text{-}39)$$

where g^{ij} denotes the inverse of the metric tensor related to the coordinates q^i.

In general, the contravariant coordinates $\Delta\xi^i$ are not identical with the coordinate differences Δq^i due to non-linear terms in (3-37). For this reason, a *second linearization* has to be performed in (3-38a,b), neglecting the non-linear terms in Δq^i in (3-37). As a final result we end up with the linearized boundary conditions

$$\delta w_{(1)} + w_{|i}\cdot\Delta q^i_{(1)} = \Delta w \qquad (3\text{-}40a)$$

$$\frac{\partial\delta w_{(1)}}{\partial q^i} + w_{|i|j}\cdot\Delta q^j_{(1)} = \Delta\gamma_i. \qquad (3\text{-}40b)$$

It should be noted that the second linearization is a peculiarity of the description in curvilinear coordinates, see Heck (1995); no second step of linearization will be necessary if rectilinear (e.g. Cartesian) coordinates are employed.

The boundary conditions (3-40) can be combined in such a way that only the unknown function δw is left in the reduced boundary condition. We first solve (3-

40b) for $\Delta q^j_{(1)}$ by multiplying this equation with the inverse of the matrix

$$m_{ij}: = w_{|i|j} \; ; \tag{3-41}$$

this inverse will be denoted by m^{ri} and fulfills the relationship

$$m^{ri} m_{ij} = \delta^r_j \tag{3-42}$$

with δ^r_j the Kronecker tensor. Hence we have

$$\Delta q^r_{(1)} = m^{ri} \{\Delta \gamma_i - \frac{\partial \delta w_{(1)}}{\partial q^i}\}, \tag{3-43}$$

and inserting (3-43) into (3-40a) we find the reduced linearized boundary condition

$$\delta w_{(1)} - \frac{\partial w}{\partial q^r} \cdot m^{ri} \frac{\partial \delta w_{(1)}}{\partial q^i} = \Delta w - \frac{\partial w}{\partial q^r} \cdot m^{ri} \cdot \Delta \gamma_i. \tag{3-44}$$

For an isozenithal telluroid mapping of type (3-4) the vectorial gravity anomaly can be expressed by the aid of the scalar gravity anomaly $\Delta \gamma$

$$\Delta \boldsymbol{\gamma} = \frac{\boldsymbol{\gamma}}{\gamma} \cdot \Delta \gamma \quad ; \tag{3-45}$$

the boundary condition (3-44) based on (3-45) can rigorously be transformed into the form (Moritz, 1980, p. 348)

$$- \frac{\partial \delta w_{(1)}}{\partial \tau} + \frac{1}{\gamma} \cdot \frac{\partial \gamma}{\partial \tau} \cdot \delta w_{(1)} = \Delta \gamma^* + \frac{1}{\gamma} \cdot \frac{\partial \gamma}{\partial \tau} \cdot \Delta w \quad . \tag{3-46}$$

Here $\partial/\partial \tau$ denotes the derivative along the isozenithal line.

The explicit form of the reduced boundary condition depends on the choice of the reference potential as well as on the type of coordinates. For the simple *isotropic reference potential* $w = \mu/r$ we obtain

$$\delta w_{(1)} + \frac{1}{2} < \mathbf{x} , \text{ grad } \delta w_{(1)} > = \Delta w + \frac{1}{2} < \mathbf{x} , \Delta \boldsymbol{\gamma} > , \tag{3-47}$$

which in spherical coordinates equals

$$- \frac{\partial \delta w_{(1)}}{\partial r} - \frac{2}{r} \delta w_{(1)} = \Delta \gamma - \frac{2}{r} \Delta w \tag{3-48}$$

for an isozenithal telluroid mapping. As mentioned above, it is not advisable to choose the isotropic potential as reference due to the large distances between the surfaces s and S. Respective boundary conditions for the more complicated Somigliana-Pizzetti normal field and several parameterizations in curvilinear coordinates have been derived by Dermanis (1984, 1987), Heck (1991) and Holota (1988), see also section 4.1.

Again the linearized boundary conditions can be used as starting point for an iterative solution of the non-linear problem. In contrast to the fixed GBVP a "hard" implicit function theorem of Moser-Nash type has to be applied for analyzing the convergence of the (Nash-Hörmander) iteration process, see Moritz (1980, p. 434 ff) and Sansò (1981).

Concluding this section we will derive the non-linear, second-order terms neglected in the linearized boundary conditions. For the sake of simplicity the derivation will be based on Cartesian coordinates x_i. Inserting the second-order expansions (3-33) into the boundary conditions (3-30) yields

$$\Delta w = \delta w(p) + <\gamma, \Delta \mathbf{x}> + c_w \qquad (3\text{-}49a)$$
$$\Delta \gamma = \text{grad } \delta w \,(p) + M \cdot \Delta \mathbf{x} + \mathbf{c}_\gamma, \qquad (3\text{-}49b)$$

$$M = \text{grad } \gamma = \left(\frac{\partial^2 w}{\partial x^i \, \partial x^j} \right) \cdot \mathbf{f}_i \otimes \mathbf{f}_j$$

where the second order terms are abbreviated by

$$c_w := \partial_i \delta w \cdot \Delta x_i + \frac{1}{2} \partial_j w \cdot \Delta x_i \, \Delta x_j$$

$$\mathbf{c}_\gamma := \partial_i \delta \gamma \cdot \Delta x_i + \frac{1}{2} \partial_j \gamma \cdot \Delta x_i \, \Delta x_j \quad . \qquad (3\text{-}50)$$

All quantities in (3-49) and (3-50) refer to the known telluroid point p.

Neglecting the non-linear terms c_w, $\mathbf{c}_\gamma$ in (3-49) results in the linearized problem
$$\Delta w = \delta w^0 + <\gamma, \Delta \mathbf{x}^0>$$
$$\Delta \gamma = \text{grad } \delta w^0 + M \cdot \Delta \mathbf{x}^0 . \qquad (3\text{-}51)$$

A second-order solution can be obtained in the following way: Insert the solution δw^0 and $\Delta \mathbf{x}^0$ of the linear problem in c_w, $\mathbf{c}_\gamma$ (3-50) and decompose the solution into the linear term and a second-order incremental part, respectively, i.e.

$$\delta w = \delta w^0 + \delta u \qquad (3\text{-}52a)$$
$$\Delta \mathbf{x} = \Delta \mathbf{x}^0 + \mathbf{d}. \qquad (3\text{-}52b)$$

Since δw and δw^0 are regular harmonic functions, δu is harmonic, too, and regular at infinity.

Solving (3-49b) for $\Delta \mathbf{x}$ and inserting into (3-49a) yields the reduced boundary condition
$$\delta w - <\gamma, M^{-1} \text{ grad } \delta w> = \qquad (3\text{-}53)$$
$$= \Delta w - <\gamma, M^{-1} \Delta \gamma> - \kappa,$$
$$\kappa = c_w - <\gamma, M^{-1} \mathbf{c}_\gamma>, \qquad (3\text{-}54)$$

which by (3-52a) and (3-51) gives the boundary condition for the second-order incremental potential term δu
$$\delta u - <\gamma, M^{-1} \text{ grad } \delta u> = -\kappa. \qquad (3\text{-}55)$$

Accordingly, the second-order incremental part $\mathbf{d}$ of the position correction vector is expressed by the formula
$$\mathbf{d} = -M^{-1} (\text{grad } \delta u + \mathbf{c}_\gamma). \qquad (3\text{-}56)$$

Obviously the non-linear effects in the disturbing potential and in the position correction vector are governed by the second-order correction terms c_w, $\mathbf{c}_\gamma$ and its combination term κ. In consistence with the second-order approximation, c_w and $\mathbf{c}_\gamma$ can be evaluated by replacing $\Delta \mathbf{x}$, δw in (3-50) by the linear solution $\Delta \mathbf{x}^0$, δw^0. In index notation we obtain the expressions

$$d_i = -m_{ij} (\partial_{jk} \delta w^0 \cdot \Delta x^0_k + \tfrac{1}{2} \partial_{jkl} w \cdot \Delta x^0_k \cdot \Delta x^0_l + \partial_j \delta u) \qquad (3\text{-}57)$$
$$\kappa = -m_{ij} \, \gamma_i (\partial_{jk} \delta w^0 \cdot \Delta x^0_k + \tfrac{1}{2} \partial_{jkl} w \cdot \Delta x^0_k \cdot \Delta x^0_l) +$$
$$+ \partial_i \delta w^0 \cdot \Delta x^0_i + \tfrac{1}{2} \partial_{ij} w \cdot \Delta x^0_i \cdot \Delta x^0_j . \qquad (3\text{-}58)$$

Using spherical approximation, i.e. replacing the normal potential by the isotropic field in evaluating the coefficients in (3-57), (3-58), see section 4.2, finally results in lengthy expressions compiled in Heck (1988) and Smeets (1991), which can be used for estimating second-order effects in the boundary condition, the position anomaly and the potential function.

3.4 Linearization of the scalar free GBVP

The linearization of the boundary equations (2-22a,b) can be performed along the same lines of reasoning. Difference equations for the anomaly of the geopotential number Δc (equivalent to the potential anomaly) and the scalar gravity anomaly $\Delta \gamma$ can be constructed on the basis of (3-21), (3-16) and (3-19a)

$$\Delta c = - \delta w (P) - [w (P) - w (p)] + \Delta w_0 \qquad (3-59a)$$

$$\Delta \gamma = \| \text{ grad } w (P) + \text{grad } \delta w (P) \| - \| \text{ grad } w (p) \|. \qquad (3-59b)$$

In consistency with practical procedures we will assume in the following that the relationship between corresponding points $p \in s$ and $P \in S$ is provided by Molodensky's telluroid mapping. Due to the fact that any related pair of points p and P is situated on the same ellipsoidal normal the "horizontal" components of the position anomaly vanish, i.e. $\Delta b = \Delta l = 0$ due to (3-6) and (3-8), such that $\Delta \mathbf{x} = \Delta h \cdot \mathbf{n}$ leaving the "vertical displacement" Δh between P and p as an unknown. This unknown is hidden in the right hand sides of (3-59a, b), therefore the boundary data Δc, $\Delta \gamma$ depend in a non-linear way on Δh. In addition, the norm operator in (3-59b) is responsible for another type of non-linearity.

Assuming that w and δw are four times continuously differentiable functions of position we can formally expand the boundary conditions (3-59a, b) into Taylor series. The resulting expressions including second-order terms have been derived in Heck (1989b):

$$\Delta c = - \delta w (p) + \Delta w_0 - <\gamma, \mathbf{n}> \cdot \Delta h + 0_c \qquad (3-60a)$$

$$\Delta \gamma = \frac{1}{\gamma} <\gamma , \text{ grad } \delta w (p) > + \frac{1}{\gamma} <\gamma, M \cdot \mathbf{n}> \cdot \Delta h + 0_\gamma \qquad (3-60b)$$

where the second-order expressions are denoted by

$$0_c = -\tfrac{1}{2} (\partial_{ij} w \cdot n_i \cdot n_j) \cdot (\Delta h)^2 - (\partial_i \delta w \cdot n_i) \cdot \Delta h \qquad (3-61a)$$

$$0_\gamma = \frac{1}{2\gamma} [\partial_i \delta w \cdot \partial_i \delta w - (\frac{1}{\gamma} \partial_i w \cdot \partial_i \delta w)^2] +$$

$$+ \frac{1}{2\gamma} [(\partial_{ij} w \cdot \partial_{ik} w + \partial_i w \cdot \partial_{ijk} w) \cdot n_j \cdot n_k -$$

$$- (\frac{1}{\gamma} \partial_i w \cdot \partial_{ij} w \cdot n_j)^2] \cdot (\Delta h)^2 + \qquad (3-61b)$$

$$+ \frac{1}{\gamma} [\partial_{ij} w \cdot \partial_i \delta w + \partial_i w \cdot \partial_{ij} \delta w - \frac{1}{\gamma^2} \partial_i w \cdot \partial_{ij} w \cdot \partial_k w \cdot \partial_k \delta w] \cdot$$

$$\cdot n_j \cdot \Delta h ;$$

n_i are the coordinates of the ellipsoidal unit normal vector $\mathbf{n}$ with respect to the earth-fixed reference frame

$$\mathbf{n} = n_i \cdot \mathbf{f}_i = \cos b \cdot \cos l \cdot \mathbf{f}_1 + \cos b \cdot \sin l \cdot \mathbf{f}_2 + \sin b \cdot \mathbf{f}_3 . \qquad (3-62)$$

Neglecting the non-linear terms in (3-60a, b) we get the boundary conditions for the linearized problem, corresponding to the solution functions δw^0, Δh^0

$$\Delta c = - \delta w^0 (p) + \Delta w_0 - <\gamma, \mathbf{n}> \cdot \Delta h^0 \qquad (3-63a)$$

$$\Delta \gamma = \frac{1}{\gamma} <\gamma , \text{ grad } \delta w^0 (p) > + \frac{1}{\gamma} <\gamma, M \cdot \mathbf{n}> \cdot \Delta h^0 . \qquad (3-63b)$$

As shown above, a second-order solution of the non-linear problem can be achieved by partitioning the solution functions into the linear components δw^0, Δh^0 and second-order increments δu, d

$$\delta w = \delta w^0 + \delta u \tag{3-64}$$

$$\Delta h = \Delta h^0 + d. \tag{3-65}$$

Differencing (3-60) and (3-63) the increments δu, d can be shown to fulfill the boundary conditions

$$\delta u + <\gamma, \mathbf{n}> \cdot d = 0_c \tag{3-66a}$$

$$\frac{1}{\gamma} <\gamma, \text{grad } \delta u> + \frac{1}{\gamma} <\gamma, M \cdot \mathbf{n}> \cdot d = - 0_\gamma \tag{3-66b}$$

where 0_c and 0_γ can be evaluated using the respective linear solutions δw^0, Δh^0 instead of δw, Δh. From the harmonicity of δw and δw^0 follows the harmonicity of δu.

The incremental terms 0_c, 0_γ as well as δu, d can be assumed to be small in comparison with the respective first order terms, at least for a suitably chosen reference potential. Thus it will be justified to apply spherical approximation of the coefficients in (3-66a, b), admitting a relative error of the order of the earth's flattening. Omitting some lengthy but elementary derivations results in (see Heck, 1989b)

$$\delta u - \bar{\gamma} \cdot d = 0_c \tag{3-67a}$$

$$- \frac{\partial \delta u}{\partial r} - \frac{2\bar{\gamma}}{r} \cdot d = - 0_\gamma \tag{3-67b}$$

$$0_c = - \frac{\bar{\gamma}}{r} \cdot (\Delta h^0)^2 - \frac{\partial \delta w^0}{\partial r} \cdot \Delta h^0 \tag{3-68a}$$

$$0_\gamma = \frac{3\bar{\gamma}}{r^2} (\Delta h^0)^2 - \frac{\partial^2 \delta w^0}{\partial r^2} \cdot \Delta h^0 + \tag{3-68b}$$

$$+ \frac{1}{2\bar{\gamma}} [(\frac{\partial \delta w^0}{r \partial \varphi})^2 + (\frac{\partial \delta w^0}{r \cos \varphi \partial \lambda})^2] \quad ;$$

$\bar{\gamma}$ is a global mean value of gravity, and r can be approximated by the radius of a mean earth sphere.

Solving (3-67a) for d

$$d = \frac{\delta u - 0_c}{\bar{\gamma}} \tag{3-69a}$$

and inserting in (3-67b) yields the reduced boundary condition

$$- \frac{\partial \delta u}{\partial r} - \frac{2}{r} \delta u = v \tag{3-69b}$$

containing the incremental boundary data

$$v := - \frac{2}{r} 0_c - 0_\gamma$$

$$= - \frac{\bar{\gamma}}{r^2} \cdot (\Delta h^0)^2 + [\frac{\partial^2 \delta w^0}{\partial r^2} + \frac{2}{r} \frac{\partial \delta w^0}{\partial r}] \cdot \Delta h^0 - \tag{3-70}$$

$$- \frac{1}{2\bar{\gamma}} [(\frac{\partial \delta w^0}{r \partial \varphi})^2 + (\frac{\partial \delta w^0}{r \cos \varphi \partial \lambda})^2]$$

Obviously the reduced boundary condition (3-69) has formally the same structure as the respective boundary condition of the linear GBVP in spherical and constant radius approximation (Rummel, 1984); the solution is the well-known Stokes' formula, leaving the first degree terms in the spherical harmonic expansion of δu indeterminate. It is common practice in Physical Geodesy to enforce the first degree terms to vanish; as noted by Rummel (1995), this procedure does not correspond to placing the origin of the coordinate system into the earth's centre of mass.

3.5 Evaluation of linearization errors

The relationships for the second-order incremental terms in the boundary conditions can be employed to give some impression on the order of magnitude of the terms neglected in the linearized GBVPs. Since the true disturbing potential is unknown it has to be approximated in the incremental boundary conditions in a suitable form. In Heck and Seitz (1993) a truncated spherical harmonic series of degree N has been used. The harmonic coefficients of δw^0 are based on the geopotential model OSU91A; a Somigliana-Pizzetti model field has been chosen for the reference potential w.

An analysis of the incremental terms in the boundary condition reveals strong differences between the three formulations of the GBVP with respect to the impact of non-linear effects. The smallest order of magnitude is obtained for the *fixed GBVP*, the incremental term $\delta\gamma_2$ (3-29) amounting up to $20 \cdot 10^{-8}$ ms^{-2}. In contrast, in the *vectorial free GBVP* the respective term κ (3-58), transformed into an impact on gravity anomalies, takes maximum values up to $3 \cdot 10^{-5}$ ms^{-2} which is two orders of magnitude larger than the observational precision. A view to (3-58) shows that κ is dominated by the term $\partial_{jk} \, \delta w^0 \cdot \Delta x^0_k$ involving the products between the horizontal components of Δx and the mixed radial-horizontal derivatives $\partial^2 \, \delta w^0/\partial r \, \partial\varphi$, $\partial^2 \delta w^0/\partial r \, \partial\lambda$. Since these second-order derivatives of the disturbing potential show strong variations in mountainous regions, and the horizontal components of Δx (corresponding to the deflections of the vertical) are large, too, in these regions, the absolute values of κ may take even larger values in these extreme situations. The *scalar free GBVP* takes a mediate position between the vectorial free and the fixed GBVP. The non-linear terms ν in the reduced boundary condition (3-69) amount up to $3 \cdot 10^{-6}$ ms^{-2} in some regions which is larger than the precision of gravity observations; the main impact in (3-70) is provided by the term $\partial^2 \, \delta w^0/\partial r^2 \cdot \Delta h^0$, involving the second vertical derivative of the disturbing potential, which is a strongly oscillating quantity on the earth's surface. In contrast, $\delta\gamma_2$ contains only first order derivatives of the disturbing potential; thus $\delta\gamma_2$ is a smoother function on the earth's surface. The behaviour of the non-linear terms $\delta\gamma_2$ and ν in the reduced boundary conditions of the fixed and scalar free GBVP are illustrated in figs. 3.1 and 3.2.

If the Somigliana-Pizzetti field is replaced by a higher degree reference field, e.g. a low-degree geopotential model containing the harmonic coefficients up to degree N_0, say $N_0 \le 20$, then the respective incremental terms in the boundary conditions will decrease. For the scalar free GBVP it has been shown (Seitz et al., 1994) that increasing the degree of the reference potential from $N_0 = 2$ to 20 results in a decrease of the amplitude of ν from $0.3 \cdot 10^{-5}$ ms^{-2} to about $0.06 \cdot 10^{-5}$ ms^{-2}.

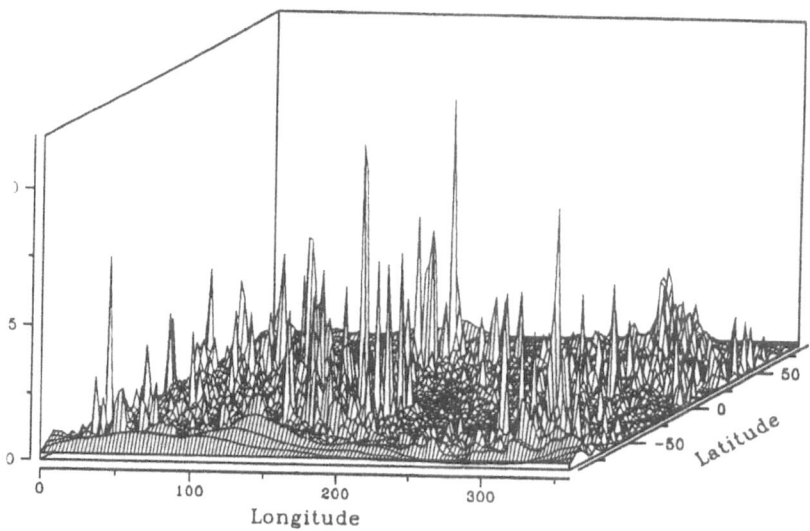

Fig. 3.1:   Non-linear term $\delta\gamma_2$ [10^{-6} ms^{-2}] in the boundary condition of the fixed
              GBVP (δw^0 based on GEM 10C and GRS 80)

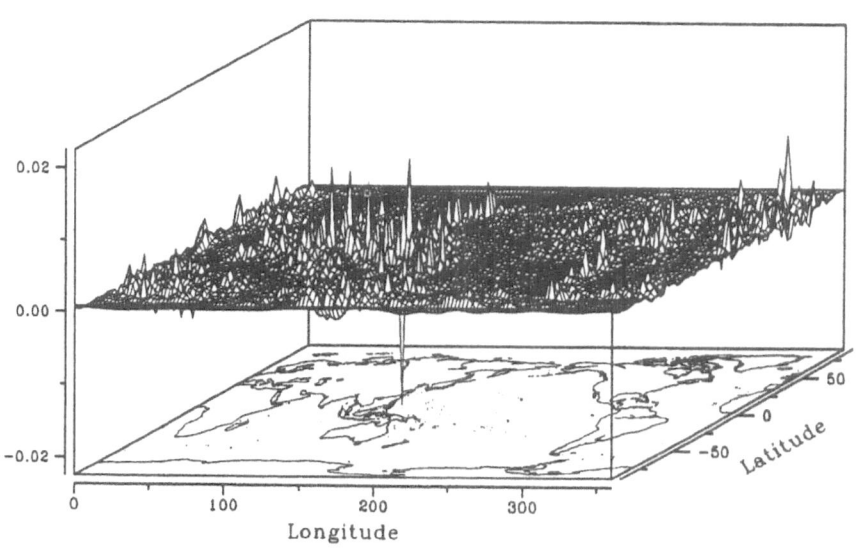

Fig. 3.2:   Non-linear term ν [10^{-4} ms^{-2}] in the reduced boundary condition of the
              scalar free GBVP (δw^0 based on GEM 10C and GRS 80)

In Heck and Seitz (1993) and Seitz et al. (1994) the effects of non-linear terms on the unknown functions to be solved for in the GBVPs have been estimated. In the *fixed* GBVP an incremental effect in the gravity potential of about 0.04 m^2s^{-2} is produced, corresponding to a maximum vertical "shift" of equipotential surfaces of about 4 mm which is negligible in all practical cases. In contrast, δu in the *vectorial free* GBVP amounts up to 2 m^2s^{-2}, corresponding to an effect of 0.2 m in the geoidal height. A similar order of magnitude is reached for the impact on the position correction vector Δx: the effects on the radial component amount up to 0.4 m, and up to 3 metres in the tangential components. These results prove that linear solutions of the vectorial free GBVP are useless from a practical point of view. For the *scalar free* GBVP the effect on the potential δu amounts up to 0.1 $m^2 s^{-2}$ corresponding to a vertical "shift" of up to 1 cm in the geoidal heights. Since the very long wavelength constituents in δu are prevailing, this effect is negligible in nearly any practical application. Similarly, the effect on the vertical position correction is rather small, showing amplitudes up to about 3 mm only.

4. Further approximations of the boundary conditions

4.1 Specification of the boundary conditions in several coordinate systems

Considering the (reduced) linearized boundary conditions (3-25), (3-44), (3-63) of the three standard forms of the GBVP it can easily be recognized that the direction of the derivative of δw generally is not normal to the boundary surface; the boundary surface in the free GBVP is now fixed, namely the telluroid s. Thus, the linearized GBVPs are formally equivalent to the classical *oblique-derivative BVP* of potential theory which first has been analyzed by Giraud in the thirties. Symbolically, the (reduced) boundary conditions for the disturbing potential δw read

$$e \circ D_L \, \delta w \equiv D_L \, \delta w \mid_s = \Delta \, l, \qquad (4\text{-}1)$$

where D_L is a linear differential operator of order one, and e is the evaluation operator restricting the spatial function $D_L \, \delta w$ to the telluroid surface s. In the case of the fixed GBVP s coincides with S (e = E), and Δl is equal to $\delta l = \delta \gamma$.

The coefficients of the disturbing potential δw and its gradient in the boundary conditions depend on the first and second order gradients of the reference potential w. The reference potential as well as the reference surface (telluroid) can be arbitrarily chosen apart from requiring that they are "close" to the actual gravity potential and boundary surface, such that the linearization errors are small enough. In view of high-precision practical applications at least the central field component, the quadrupole momentum of the earth's gravitational field, the centrifugal potential and the ellipticity of the boundary surface have to be taken into account for the explicit evaluation of the differential operator D_L in (4-1).

The most general choice of a reference potential is a (infinite or finite degree) spherical harmonic series extended by the centrifugal term. For the sake of simplicity, we will choose a simple, but very efficient approximation for the evaluation and representation of D_L

$$w = \frac{\mu}{r} \cdot \left\{ 1 - J_2 \cdot \left(\frac{a}{r} \right)^2 \cdot P_2(\sin \varphi) + \dots \right\} + \frac{1}{2} \omega^2 r^2 \cos^2 \varphi$$

$$= \frac{\mu}{r} \left\{ 1 - J_2 \cdot \left(\frac{a}{r} \right)^2 \cdot P_2(\sin \varphi) + \frac{1}{2} \overline{m} \left(\frac{r}{a} \right)^3 \cos^2 \varphi + \dots \right\} \tag{4-2}$$

where we have put

$$\overline{m} = \frac{\omega^2 a^3}{\mu} \tag{4-3}$$

For the expressions of the first and second order gradients of w see e.g. Heck (1991), Appendix A.1. Since for the earth any other spherical harmonic coefficient is at least two orders of magnitude smaller than J_2, the expression (4-2) already provides a fairly good approximation for any choice of reference potential.

The *linearized fixed GBVP* is characterized by the linear differential operator

$$D_L \, \delta w := < \frac{\gamma}{\gamma} , \text{ grad } \delta w >. \tag{4-4}$$

Inserting the expression for γ / γ in spherical coordinates into (4-4), neglecting non-linear terms in J_2 and $\overline{m}$, yields the explicit form

$$D_L \, \delta w = - \frac{\partial \delta w}{\partial r} - \sin\varphi \, \cos \varphi \, [3 \, J_2 \left(\frac{a}{r} \right)^2 + \overline{m} \left(\frac{r}{a} \right)^3] \cdot \frac{\partial \delta w}{r \partial \varphi}. \tag{4-5}$$

The second term depends on the latitudinal derivative of the disturbing potential and possesses the order of magnitude $0(J_2) \sim 0(\overline{m}) \sim 0(e^2)$. Since for any point on the earth's surface the ratio (a/r) differs from 1 by no more than 0.003, this difference can be neglected in consistency with the above requirements. As a result, we obtain the boundary condition for the linearized fixed GBVP in terms of *spherical coordinates*

$$\left(- \frac{\partial \delta w}{\partial r} - [3 \, J_2 + \overline{m}] \cdot \sin \varphi \cdot \cos \varphi \cdot \frac{\partial \delta w}{a \partial \varphi} \right) \Big|_s = \delta \gamma \quad . \tag{4-6a}$$

Other forms of this boundary condition can be constructed by transformations from spherical to *ellipsoidal* or *elliptical coordinates*

$$\left(- \frac{\partial \delta w}{\partial H} - [3 \, J_2 + \overline{m} - e^2] \sin b \cdot \cos b \cdot \frac{\partial \delta w}{a \partial b} \right) \Big|_s = \delta \gamma \tag{4-6b}$$

$$\left(- [1 + \frac{e^2}{2} \cos^2 \beta] \cdot \frac{\partial \delta w}{\partial u} - [3J_2 + \overline{m} - e^2] \sin \beta \cdot \cos \beta \cdot \frac{\partial \delta w}{u \partial \beta} \right) \Big|_s = \delta \gamma \quad , \tag{4-6c}$$

using the relationships derived in Heck (1991), Appendix A.2.

If a *Somigliana-Pizzetti* normal field is applied, then the parameters J_2, $\overline{m}$, e^2 are no more independent. In consistence with the first-order theory the relationship

$$3 \, J_2 + \overline{m} - e^2 + \dots = 0 \tag{4-7}$$

holds (Heiskanen and Moritz, 1967, p. 78), and the boundary conditions (4-6a, b, c) take the simple form (Heck, 1991)

$$\left(- \frac{\partial \delta w}{\partial r} - e^2 \sin\varphi \cdot \cos \varphi \cdot \frac{\partial \delta w}{a \partial \varphi} \right) \Big|_s = \delta \gamma \tag{4-8a}$$

$$\left(- \frac{\partial \delta w}{\partial H} \right) \Big|_s = \delta \gamma \tag{4-8b}$$

$$\left(- [1 + \frac{e^2}{2} \cos^2 \beta] \cdot \frac{\partial \delta w}{\partial u} \right) \Big|_s = \delta \gamma. \tag{4-8c}$$

Analogously the boundary conditions for the *vectorial free GBVP* can be derived, starting from equation (3-44) and using the first and second order derivatives of (4-2). For a *Somigliana-Pizzetti* normal field the explicit form of the boundary conditions in the various coordinate systems reads

$$\left(- \frac{2}{a} [1 + \overline{m} + \frac{1}{2} e^2 \cos 2\varphi - \frac{h}{a}] \delta w^0 - \frac{\partial \delta w^0}{\partial r} \right.$$

$$\left. + [5\overline{m} - 2 e^2] \cdot \sin \varphi \cdot \cos \varphi \cdot \frac{\partial \delta w^0}{a \partial \varphi} \right) \Big|_s = \Delta \gamma \tag{4-9a}$$

$$\left(- \frac{2}{a} [1 + \overline{m} + \frac{1}{2} e^2 \cos 2b - \frac{h}{a}] \delta w^0 - \frac{\partial \delta w^0}{\partial h} \right.$$

$$\left. + [5\overline{m} - e^2] \cdot \sin b \cdot \cos b \cdot \frac{\partial \delta w^0}{a \partial b} \right) \Big|_s = \Delta \gamma \tag{4-9b}$$

$$\left(- \frac{2}{u} [1 + \overline{m} - e^2 \sin^2 \beta] \cdot \delta w^0 - [1 + \frac{e^2}{2} \cos^2 \beta] \cdot \frac{\partial \delta w^0}{\partial u} \right.$$

$$\left. + [5\overline{m} - e^2] \sin \beta \cdot \cos \beta \cdot \frac{\partial \delta w^0}{u \partial \beta} \right) \Big|_s = \Delta \gamma \tag{4-9c}$$

On the right hand side of the boundary conditions (4-9) we have assumed that the potential anomaly Δw vanishes, which holds for Marussi's telluroid mapping. For the derivation of the respective relationships for the components of the position anomaly Δx in the framework of the linear approximation see Heck (1991).

Finally the reduced boundary condition for the *scalar free GBVP* results from eliminating Δh^0 in (3-63a, b). For a Somigliana-Pizzetti normal field the boundary condition in the considered coordinate systems takes one of the following forms (Heck 1991):

$$\left(- \frac{2}{a} [1 + \overline{m} + \frac{e^2}{2} \cos 2\varphi - \frac{h}{a}] \cdot \delta w^0 - \frac{\partial \delta w^0}{\partial r} \right.$$

$$\left. - e^2 \sin \varphi \cdot \cos \varphi \cdot \frac{\partial \delta w^0}{a \partial \varphi} \right) \Big|_s = \Delta \gamma - \frac{2}{a} \Delta w_0 \tag{4-10a}$$

$$\left(- \frac{2}{a} [1 + \overline{m} + \frac{e^2}{2} \cos 2b - \frac{h}{a}] \delta w^0 - \frac{\partial \delta w^0}{\partial h} \right) \Big|_s = \Delta \gamma - \frac{2}{a} \Delta w_0 \tag{4-10b}$$

$$\left(- \frac{2}{u} [1 + \bar{m} - e^2 \sin^2 \beta] \cdot \delta w^0 - [1 + \frac{e^2}{2} \cos^2 \beta] \cdot \frac{\partial \delta w^0}{\partial u} \right) \Big|_s = \Delta \gamma - \frac{2}{a} \Delta w_0$$

$$(4\text{-}10c)$$

In (4-10a, b, c) it has been assumed that $\Delta c \equiv 0$ (i.e. Molodensky's telluroid mapping holds) and Δw_0 is a small quantity reflecting the vertical datum unknown. The respective relationships for the height anomaly in linear approximation (Δh^0) have been derived in Heck (1991).

Considering the numerical values of e^2 and $\bar{m}$ for a Somigliana-Pizzetti field of the earth, e.g. the Geodetic Reference System 1980,

$$e^2 = 0.006\ 694\ 380, \quad \bar{m} = 0.003\ 438\ 220,$$

the relationship

$$\bar{m} \approx e^2/2 \qquad\qquad (4\text{-}11)$$

is fulfilled with a high degree of precision. Utilizing (4-11), the formulae (4-9) and (4-10) can be given an even simpler form which is often refered to in geodetic literature, see e.g. Holota (1988), Moritz (1980, p. 317). But it should be kept in mind that the order of error is now $O(e^3)$ instead of $O(e^4)$.

It should be noticed that the boundary conditions in the different forms compiled above still refer to the boundary surface, i.e. to the true boundary S in the case of the fixed GBVP and to the telluroid s for the free GBVPs. In the following these boundary conditions will be simplified furthermore, introducing two other approximations, namely the so-called "spherical approximation" and the "constant radius approximation". The way of operation of these simplifications will be demonstrated for the boundary condition of the fixed GBVP, formulated in spherical coordinates.

4.2   Spherical approximation

From equation (4-6a) it becomes clear that the boundary condition of the linearized fixed GBVP has the general form

$$E \circ D_L \delta w \equiv E \circ \left\{ - \frac{\partial \delta w}{\partial r} + \delta a_\varphi \cdot \frac{\partial \delta w}{\partial \varphi} \right\} = \delta \gamma \qquad\qquad (4\text{-}12)$$

The procedure of "spherical approximation" consists in simply neglecting the terms of order $O(e^2)$, i.e. the terms proportional to J_2 and $\bar{m}$ in the representation of the differential operator D_L. Formally, this procedure is equivalent to calculating the coefficients of the operator D_L by the aid of the isotropic potential approximation w $\approx \mu/r$. It must be emphasized that the isotropic potential approximation is only used in the calculation of the coefficients in the differential operator D_L, not in the computation of the boundary data $\delta \gamma$; here of course the exact form of the reference potential has to be applied for evaluating γ (P) in $\delta \gamma$.

After spherical approximation the boundary condition of the linearized fixed GBVP takes the simplified form

$$E \circ \left\{ - \frac{\partial \delta w}{\partial r} \right\} \equiv - \frac{\partial \delta w}{\partial r} \Big|_s = \delta \gamma \qquad\qquad (4\text{-}13)$$

The terms of order J_2 and $\bar{m}$ (and e^2) in the boundary condition are often called *ellipsoidal effects*. A rough estimate, assuming $|\partial \delta w/(\gamma \cdot a \cdot \partial r)| \leq 30$ arcsec.

(latitude component of the deflection of the vertical), gives maximum effects on the order of $0.5 \cdot 10^{-5}$ ms^{-2}. This order of magnitude is corroborated by a detailed analysis (Eichhorn, 1996), based on a representation of the global gravitational field by the spherical harmonic model OSU91A (max. degree 360), see fig. 4.1. Due to the dependence of δa_φ on $\sin \varphi \cdot \cos \varphi$ the maximum ellipsoidal effects can be found in mid-latitudes. Obviously these ellipsoidal effects cannot be neglected in highly precise modelling of the disturbing potential. The effect on the solution of the GBVP is further discussed in Heck (1991), see also Rapp (1991) and Cruz (1986).

Fig. 4.2 illustrates the effects of the higher order terms (J_4, J_2^2, $\bar{m} J_2$ etc.) in the boundary condition, which have been neglected in (4-6a). Considering the order of magnitude of these effects (less than $5 \cdot 10^{-9}$ ms^{-2}) it is obvious that they are negligible in practice.

It should be stressed that the GBVP in spherical approximation still refers to the surface of the earth S (or the telluroid s in the case of the free GBVP, respectively) which is by no means spherical. In this respect a lot of confusion can be found in literature, since some authors apply the expression "spherical approximation" to a formal transition from S (or s) to a sphere Σ. For this reason the notion "isotropic potential approximation", acting on the coefficients of the linearized boundary operator, would be better suited. The GBVP at this level of approximation is often called the "simple" GBVP, e.g. the "simple Molodensky problem" (equivalent to the free GBVP in linear and spherical approximation). The application of the isotropic potential approximation is justified by the fact that - due to the smallness of the disturbing potential δw - it will not be necessary, from a numerical point of view, to calculate the coefficients of the boundary operator with highest precision. By the isotropic potential approximation a relative error of the order $3 \cdot 10^{-3}$ is introduced, leaving at least two leading digits of the coefficients correct.

It is remarkable that the (reduced) boundary conditions of the vectorial and scalar free GBVP coincide at the level of isotropic potential approximation; a minor distinction remains due to the slightly different telluroid definitions (Heck, 1991). On the other hand, a view to (4-9) and (4-10) shows that both problems will differ at the level of linear approximation, resulting in different ellipsoidal corrections. The formal equality of the boundary conditions of the vectorial and scalar free GBVP in spherical approximation has become another source of confusion in literature, where sometimes both formulations are mixed in an inadmissible way.

In standard analytical procedures for the solution of the GBVP another approximation, the so-called *planar approximation*, is introduced (Moritz, 1980, p.358), neglecting terms of order $|h/a| \leq 1.5 \cdot 10^{-3}$ in the coefficients of the boundary operator. After linear, spherical and planar approximation the resulting form of the boundary condition is simple enough to admit analytical solutions via perturbation approaches, e.g. Molodensky's series solution or corresponding alternatives, see Moritz (1980, pp. 360-401).

Once again, a solution for the general linearized boundary condition can be constructed by an iterative approach. For the linearized fixed GBVP the solution in spherical approximation (isotropic potential approximation), based on the boundary condition (4-13), can be considered as starting approximation δw^0, which can be introduced into the ellipsoidal term in (4-6a). Solving the boundary condition for the "reduced" boundary data

$$\delta \gamma^{(1)} := \delta \gamma + [3J_2 + \bar{m}] \cdot \sin \varphi \cdot \cos \varphi \cdot \frac{\partial \, \delta w^0}{a \, \partial \varphi} \Big|_s$$

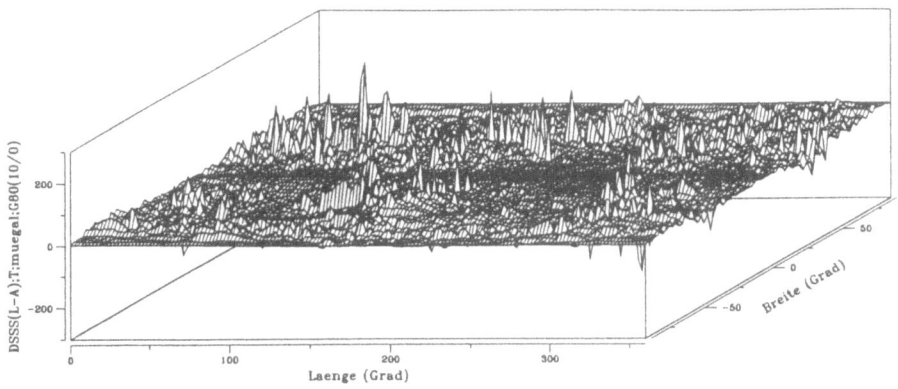

Fig. 4.1:   Ellipsoidal effects in the boundary condition of the linearized fixed GBVP ("True" field: OSU91A, Reference field: GRS 80). Units: 10^{-8} ms^{-2}

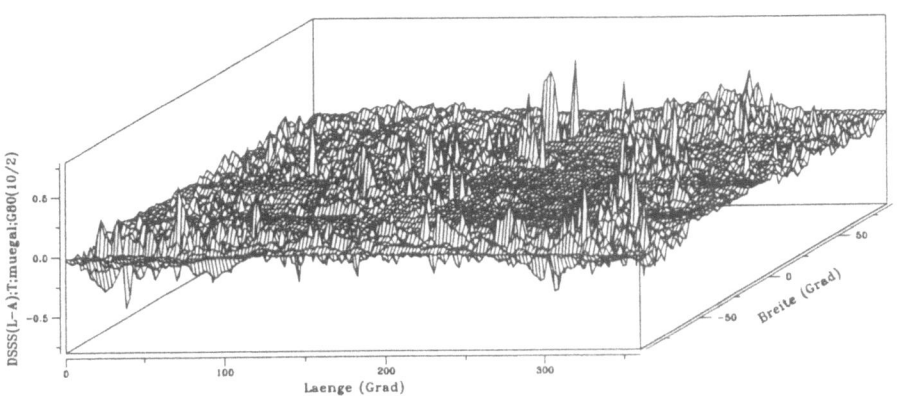

Fig. 4.2:   Higher order effects $0(J_4)$, $0(J_2^2)$, $0(\bar{m}J_2)$ etc neglected in the boundary condition (4-6a). ("True" field: OSU91A, Reference field: GRS 80). Units: 10^{-8} ms^{-2}

provides an improved solution δw^1. This iteration process can be continued until the differences between the solutions in successive iteration steps are "small enough".

4.3 Constant radius approximation

The linearized fixed GBVP in spherical approximation, based on the boundary condition (4-13), permits no rigorous analytical solution in closed form as long as the boundary is not spherically shaped. In the past, perturbative solutions have been proposed and constructed in various ways which basically can be interpreted in terms of a decomposition of the evaluation operator E. Assuming that the spatial function $g = D_L \, \delta w$ (or its spherical approximation $g = - \, \partial \delta w / \partial r$) is sufficiently smooth, a Taylor series can be set up, the expansion taking place at the surface of the sphere Σ with radius a centered at the coordinate origin

$$Eg = E_\Sigma g + (E_\Sigma \, \frac{\partial g}{\partial r}) \cdot (r - a) + \ldots\ldots \qquad (4\text{-}14)$$

E_Σ denotes evaluation on the surface of the sphere, the Taylor point having the same horizontal coordinates φ, λ as the boundary point $(r = r\,(\varphi, \lambda))$ under consideration. It should be kept in mind that $(r - a)$ is composed of terms of order $0(e^2)$ and $0(h/a)$ due to the ellipticity of the height reference surface and the topography of the boundary surface. Neglecting the non-spherical parts in (4-14) is equivalent to the procedure of *constant radius approximation* which is often interpreted as a mapping of the boundary data on the sphere of radius a (Moritz, 1980, p. 351). The solution of the linearized fixed GBVP in spherical and constant radius approximation, based on the simple spherical boundary condition[7]

$$- \frac{\partial \delta w}{\partial r} \, |_\Sigma - \delta \gamma \qquad (4\text{-}15)$$

is given by Hotine's integral formula

$$\delta w - \frac{a}{4\pi} \iint_\sigma \delta \gamma \cdot H(r, \psi) \cdot d\sigma \qquad (4\text{-}16)$$

$$H(r, \psi) - \sum_{n=0}^{\infty} \left(\frac{a}{r} \right)^{n+1} \cdot \frac{2n+1}{n+1} \cdot P_n(\cos \psi)$$

$$- \frac{2a}{1} - \ln \frac{1 + a - r \cdot \cos \psi}{r\,(1 - \cos \psi)} \qquad (4\text{-}17)$$

$$1 = (r^2 + a^2 - 2\,r \cdot a \cdot \cos \psi)^{\frac{1}{2}},$$
$$\cos \psi = \sin \varphi \cdot \sin \varphi' + \cos \varphi \, \cdot \cos \varphi' \cdot \cos (\lambda' - \lambda).$$

The spherical coordinates (r, φ, λ) and (a, φ', λ') refer to the evaluation point and the variable integration point, respectively; $\delta \gamma$ and $d\sigma$ in (4-16) depend on (φ', λ'). The space domain representation (4-16) of the disturbing potential is fully equivalent with the spectral domain form

---

[7]

It is remarkable that this BVP is formally equivalent to a Neumann problem for a spherical boundary.

$$\delta w_{lm} = \frac{a^2}{\mu (l+1)} \cdot \delta \gamma_{lm}$$

(4-18)

where δw_{lm} are the spherical harmonic coefficients in the expansion of the disturbing potential

$$\delta w \, (r, \varphi, \lambda) = \frac{\mu}{a} \cdot \sum_{l=0}^{\infty} \left(\frac{a}{r}\right)^{l+1} \cdot \sum_{m=-l}^{+l} \delta w_{lm} \cdot Y_{lm}(\varphi, \lambda),$$

(4-19)

and Y_{lm} (φ, λ) are the (non-normalized) surface spherical harmonics of degree l and order $|m|$

$$Y_{lm} (\varphi, \lambda) = \begin{cases} P_{lm}(\sin \varphi) \cdot \cos m \lambda & m \geq 0 \\ P_{l|m|} (\sin \varphi) \cdot \sin |m| \lambda & m < 0 \end{cases} .$$

(4-20)

Correspondingly $\delta \gamma_{lm}$ are the coefficients in the surface spherical harmonic expansion of the gravity disturbance $\delta \gamma$

$$\delta \gamma \, (\varphi, \lambda) = \sum_{l=0}^{\infty} \sum_{m=-l}^{+l} \delta \gamma_{lm} \cdot Y_{lm} (\varphi, \lambda) .$$

(4-21)

It should be kept in mind that the simple proportionality relationship (4-18) between the spectra of the disturbing potential and the boundary data is valid only for a spherical boundary surface.

In general, the higher order terms in (4-14) are not negligible. This is evident from fig. 4.3 where the influence of the non-sphericity of the boundary surface is illustrated: the error induced by constant radius approximation amounts up to $5 \cdot 10^{-4}$ ms^{-2}, which is more than three orders of magnitude larger than the observational errors in gravimetry. Maximum effects are visible for the polar regions, since the largest distances between S and Σ occur there; in the equatorial region the effects are minimal.

The series expansion of the evaluation operator can also be interpreted in terms of analytical continuation: the data on the boundary S are obtained by analytical continuation from the spherical surface Σ. Due to the choice of the spherical radius, the continuation is directed downward, in general, except some regions in the vicinity of the equator. It is well-known that downward continuation is strongly related to a Cauchy initial value problem the solution of which is unstable for harmonic functions. This instability manifests itself e.g. in the high-frequency behaviour of the error induced by constant radius approximation (see fig. 4.3) which can also be interpreted as the effect of downward continuation from Σ to S.

The main part of the error induced by constant radius approximation can be corrected by a reduction term Δ

$$\Delta = E_{\Sigma} \left\{ \delta a_{\varphi} \cdot \frac{\partial \delta w}{\partial \varphi} \right\} - \sum_{k=1}^{K} E_{\Sigma} \left\{ \frac{\partial^{k+1} \delta w}{\partial r^{k+1}} \, (r-a)^k \right\}$$

(4-22)

taking the analytical continuation of the function into account. Expanding the second term in (4-22) up to $K = 8$ the residual error in the boundary condition decreases to less than $5 \cdot 10^{-9}$ ms^{-2} for the synthetic example ("True" field: OSU91A, reference field GRS 80) mentioned above (Eichhorn, 1996), see fig. 4.4.

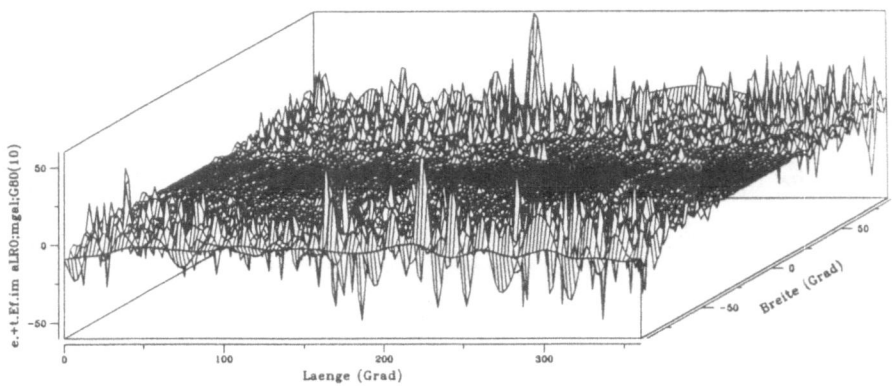

Fig. 4.3: Error in the boundary condition of the linearized fixed GBVP due to constant radius approximation ("True" field: OSU91A, reference field: GRS 80). Units: 10^{-5} ms^{-2}

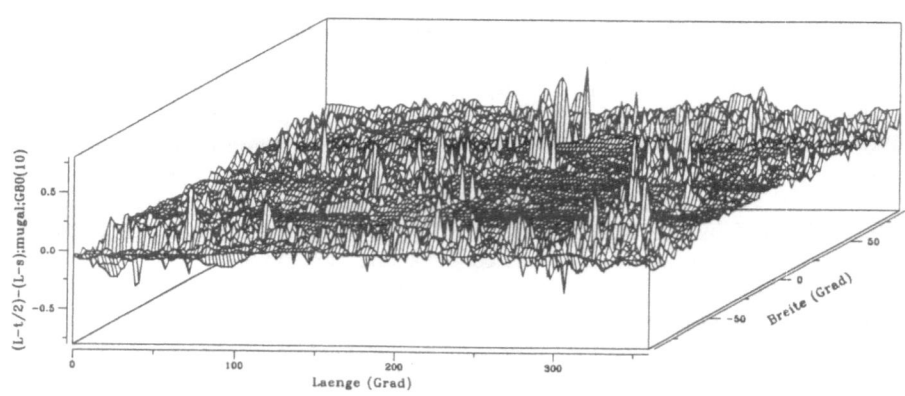

Fig. 4.4: Residual error in the boundary condition of the linearized fixed GBVP after taking the Taylor terms (4-18) into account. Units: 10^{-8}ms^{-2}

Due to the fact that the dominating constituent in (4-14) is provided by the spherical evaluation operator E_Σ, it suggests itself to solve the non-spherical problem by iteration. For this purpose the boundary condition (4-12) is transformed, using (4-14) and (4-22):

$$E_\Sigma \left\{ -\frac{\partial\, \delta w}{\partial\, r} \right\} \equiv -\frac{\partial\, \delta w}{\partial\, r} \bigg|_\Sigma = \delta\gamma - \Delta \tag{4-23}$$

Again the solution of this BVP based on the "reduced" boundary data $\delta\gamma - \Delta$ is given by Hotine's integral formula, calculating Δ by the aid of the simple solution (4-16). Retaining the first order ($k = 1$) component in the correction term Δ corresponds to the so-called "gradient solution" (Moritz, 1980); similarly as in Molodensky's problem, the convergence of the series (4-22) cannot be guaranteed for the real topographic surface of the earth.

Finally the iteration idea can be used to build up a complete hierarchy of iteration solutions (Rummel, 1988). At the lowest level - solution of the linearized fixed GBVP in spherical and constant radius approximation - the solution is analytically given by the simple integral formula (4-16). This is the starting point for the iterative solution of the linearized GBVP in spherical approximation. This more rigorous second level result provides the starting point for the iterative process aiming at the general solution of the linearized GBVP. At the highest level, the non-linear problem is solved along the same lines of reasoning. The same concept can be transferred to the solution of the free GBVP, accordingly. Until now there exist no practical experiences with this "complete" solution of the GBVP, due to the large amount of computational work.

Another concept for the solution of the general linearized problem in spherical (i.e. isotropic potential) approximation has been proposed by Holota (1991), based on a"deformation" of the boundary surface into a sphere and a transformation of the boundary condition. In this way the classical Green's function method can be applied; as a matter of fact, the field equation (Laplace's equation) has to be transformed, too. The general problem can be solved by means of the method of successive approximations. Considering the first step in the successive scheme, the resulting correction terms resemble strongly the well-known topographic reduction (Holota, 1993). As a conclusion, one might conjecture that the iterative solution described above and Holota's approach - although more appealing from a theoretical point of view - give the same solution, at least numerically.

5 Solution of the non-spherical GBVP in terms of spherical harmonics

In this paragraph a concept for the solution of the linearized GBVP in spherical (isotropic potential) approximation for non-spherical boundary surfaces is presented, based on series expansions in spherical harmonics. As a preparation, we express the geocentric radius of a surface point in the form (Heck, 1991, p.57)

$$r = a \left(1 - \frac{1}{2}e^2 \sin^2 \varphi + \frac{h}{a} + \ldots \right) ,$$

(5-1)

neglecting higher order terms $O(e^4)$, $O(e^2 h/a)$, $O((h/a)^2)$. Since $a/r \ll 1$ the $(1 + 1)^{th}$ power of this expression can be approximated by

$$\left(\frac{a}{r}\right)^{1+1} = 1 + \frac{1+1}{2} e^2 \sin^2 \varphi - (1+1) \cdot \frac{h}{a} + \ldots .$$

(5-2)

which is sufficient at least for "small" degrees l; for increasing l the neglected higher order terms of order $O(e^4)$ may become more and more important. Inserting (5-2) in (4-19) is equivalent to evaluating δw on the boundary surface, i.e. to applying the evaluation operator E:

$$E \circ \delta w = \frac{\mu}{a} \sum_{l=0}^{\infty} \left(1 + \frac{1+1}{2}e^2 \sin^2 \varphi - (1+1) \cdot \frac{h}{a}\right) \cdot \sum_{m=-1}^{+1} \delta w_{lm} \cdot Y_{lm}(\varphi, \lambda) .$$

(5-3)

In a similar way the derivatives of δw are evaluated on the boundary surface (Heck, 1991, p. 27), e.g.:

$$E \circ \left(\frac{\partial \delta w}{\partial r}\right) = -\frac{\mu}{a^2} \cdot \sum_{l=0}^{\infty} \left(1 + \frac{1+2}{2}e^2 \sin^2 \varphi - (1+2) \cdot \frac{h}{a}\right) \cdot$$

$$\cdot (1+1) \cdot \sum_{m=-1}^{+1} \delta w_{lm} \cdot Y_{lm}(\varphi, \lambda) .$$

(5-4)

Due to the property of surface spherical harmonics, that products of harmonic series can be "written back" into a spherical harmonic series, we can express the products $\sin^2 \varphi \cdot Y_{lm}(\varphi, \lambda)$ in the following way (Moritz, 1980)

$$\sin^2 \varphi \cdot Y_{lm}(\varphi, \lambda) = \alpha_{lm} \cdot Y_{l+2,m}(\varphi, \lambda) + \beta_{lm} \cdot Y_{lm}(\varphi, \lambda) +$$
$$+ \gamma_{lm} \cdot Y_{l-2,m}(\varphi, \lambda)$$

(5-5)

$$\alpha_{lm} = \frac{(1-k+1)(1-k+2)}{(21+1)(21+3)}$$

$$\beta_{lm} = \frac{21^2 - 2k^2 + 21 - 1}{(21-1)(21+3)}$$

$$\gamma_{lm} = \frac{(1+k)(1+k-1)}{(21-1)(21+1)} , \quad k = |m| .$$

Similarly we can write (Moritz, 1980, p. 321)

$$\sin \varphi \cdot \cos \varphi \cdot \frac{\partial Y_{lm}(\varphi, \lambda)}{\partial \varphi} = -1 \cdot \alpha_{lm} \cdot Y_{l+2,m}(\varphi, \lambda) +$$

$$+ \frac{1}{2}(3\beta_{lm} - 1) \cdot Y_{lm}(\varphi, \lambda) + (1+1) \cdot \gamma_{lm} \cdot Y_{l-2,m}(\varphi, \lambda) .$$

(5-6)

These expressions can be introduced in the boundary conditions for non-spherical boundaries. On the other hand, the boundary data f (gravity anomalies $\Delta \gamma$ or gravity disturbances $\delta\gamma$) can be expanded in surface spherical harmonics since they depend on the horizontal coordinates (φ, λ):

$$f(\varphi, \lambda) = \sum_{l=0}^{\infty} \sum_{m=-l}^{+l} f_{lm} \cdot Y_{lm}(\varphi, \lambda) \tag{5-7a}$$

$$f_{lm} = \frac{2l + 1}{2\pi (1 + \delta_{lo})} \cdot \frac{(l - m)!}{(l + m)!} \iint_{\sigma} f(\varphi', \lambda') \cdot Y_{lm}(\varphi', \lambda') \cdot d\sigma(\varphi', \lambda') \; . \tag{5-7b}$$

Inserting in the boundary conditions provides a system of linear algebraic equations for the determination of the unknown harmonic coefficients δw_{lm} from the known coefficients f_{lm} ($\Delta\gamma_{lm}$ or $\delta\gamma_{lm}$ respectively). We will demonstrate this approach by the example of the *fixed GBVP*; for the sake of simplicity we will neglect the terms of order $0(h/a)$, formally solving the linearized fixed GBVP approximately for an ellipsoidal boundary surface.

Starting from the boundary condition (4-8a) and inserting the harmonic series expansions for $\partial\delta w/\partial r$, $\sin\varphi \cdot \cos\varphi \cdot \partial\delta w/\partial\varphi$ and $f = \delta\gamma$ we obtain the system of algebraical equations

$$\frac{\mu}{a^2} \cdot \left\{ (l+1) \cdot \delta w_{lm} + \frac{e^2}{2} \left[(l^2 + l - 4) \cdot \alpha_{l-2,m} \cdot \delta w_{l-2,m} + \right. \right.$$

$$\left. + \left[\beta_{lm} (l^2 + 3l - 1) + 1 \right] \cdot \delta w_{lm} + (l+2) \, (l+3) \, \gamma_{l+2,m} \cdot \delta w_{l+2,m} \right] \Big\}$$

$$= \delta\gamma_{lm} \; . \tag{5-8}$$

Considering (5-8) it becomes obvious that by constant radius approximation, i.e. by neglecting terms of order $0(e^2)$, the system of equations reduces to the simple spectral form (4-18). Due to the dominance of this term - at least for low degrees l - the equation system (5-8) can be solved iteratively, starting with the zero order solution

$$\delta w_{lm}^{(o)} = \frac{a^2}{\mu} \cdot \frac{\delta\gamma_{lm}}{l + 1} \; . \tag{5-9}$$

Inserting (5-9) in the terms of order $0(e^2)$ in (5-8) and again solving for δw_{lm} yields

$$\delta w_{lm}^{(1)} = \frac{a^2}{\mu} \left\{ \frac{\delta\gamma_{lm}}{l + 1} - \frac{e^2}{2} \cdot \left[\frac{l^2 + l - 4}{(l-1) \, (l+1)} \cdot \alpha_{l-2,m} \cdot \delta\gamma_{l-2,m} + \right. \right.$$

$$\left. + \left[\beta_{lm} \cdot \frac{l^2 + 3l - 1}{(l + 1)^2} + \frac{1}{(l + 1)^2} \right] \cdot \delta\gamma_{lm} + \frac{l+2}{l+1} \cdot \gamma_{l+2,m} \cdot \delta\gamma_{l+2,m} \right] \Big\} \tag{5-10}$$

where terms of order $0(e^4)$ have been omitted. No problem arises with the factor $(l-1)^{-1}$ since the corresponding term is dropping out for $l=1$. It should be mentioned that in (5-8) the terms depending on e^2 are proportional to l^2, while the first term is proportional to l. As a matter of fact, the omitted terms of order $0(e^4)$ are proportional to l^3 for large l, so the convergence will not be guaranteed. This property is a consequence of the (unstable) process of continuation from the sphere of radius a to the boundary surface. Equivalently the neglected terms in (5-10) are of order $0(e^2 \cdot l)$, which grow up for increasing l. For further details, including the formulae describing the disturbing potential in space and on the boundary surface, see Heck (1991, section 5.3).

It is a well-known property of the fixed GBVP that the first-degree harmonics of the disturbing potential can be fully determined in this concept. This uniqueness manifests itself in the equation system (5-8) which is solvable for arbitrary degrees, including l=1. This property is no more valid in the case of the free GBVPs which are characterized by a zero order solution (equivalent to the solution of the free GBVP in spherical and constant radius approximation)

$$\delta w_{lm}^{(o)} = \frac{a^2}{\mu} \cdot \frac{\Delta \gamma_{lm}}{1 - 1} \quad ,$$
(5-11)

leaving the first degree terms indeterminate. In Heck (1991) the procedure described above is applied to the scalar and vectorial variants of the free GBVP. For the *vectorial free GBVP* the following equation system for l=1 results:

$$\frac{3}{2}(5 \beta_{1m} - 3) \cdot \overline{m} \cdot \delta w_{1m} + 5(-3J_2 + 3\overline{m} + e^2) \cdot \gamma_{3m} \cdot \delta w_{3m} =$$

$$= -\frac{a^2}{\mu} \cdot \Delta \gamma_{1m} \quad .$$
(5-12)

For m= 0 we obtain

$$5\beta_{10} - 3 = 0$$
(5-13)

which leaves the coefficient δw_{10} indeterminate and gives rise to a consistency condition which the boundary data must fulfill

$$\Delta \gamma_{10} = \frac{3}{7}(3\overline{m} - 3J_2 + e^2) \cdot \Delta \gamma_{30} \quad .$$
(5-14)

On the other hand, the coefficients of degree l=1 and orders |m|=1 can be determined at the level of the general linearized vectorial free problem:

$$\delta w_{1, \pm 1} = -\frac{a^2}{\mu} \cdot \frac{7\Delta \gamma_{1, \pm 1} - 6(-3J_2 + 3\overline{m} + e^2) \cdot \Delta \gamma_{3, \pm 1}}{21 \overline{m}} \quad .$$
(5-15)

It becomes clear from (5-15) that these coefficients are determinable due to the fact that $\overline{m} \neq 0$, i.e. due to the centrifugal term in the gravity potential. For a non-rotating earth the terms of order m = 1 remain indeterminate.

For the *scalar free GBVP* we obtain the following equation system for l = 1 (Heck, 1991)

$$\left[\frac{3}{2}(3 J_2 + \overline{m}) \beta_{lm} - \frac{1}{2}(3 J_2 + 5\overline{m})\right] \delta w_{1m} +$$

$$+ \left[3 J_2 - \overline{m} + 5 e^2\right] \gamma_{3m} \cdot \delta w_{3m} = \frac{a^2}{\mu} \cdot \Delta \gamma_{1m} \quad .$$
(5-16)

It can easily be shown that here all three first-degree coefficients can be determined, i.e. the translational invariance existing in (5-11) has disappeared. In detail, we obtain from (5-16)

$$\delta w_{10} = -\frac{a^2}{\mu} \cdot \frac{35\Delta\gamma_{10} - 3(5\,e^2 - 3\,J_2 - \overline{m}) \cdot \Delta\gamma_{30}}{7(-6\,J_2 + 8\overline{m})} \tag{5-17a}$$

$$\delta w_{1,\pm 1} = -\frac{a^2}{\mu} \cdot \frac{35\,\Delta\gamma_{1,\pm 1} - 6(5\,e^2 - 3\,J_2 - \overline{m}) \cdot \Delta\gamma_{3,\pm 1}}{7(3\,J_2 + 11\overline{m})} \tag{5-17b}$$

Another difference with respect to the vectorial free GBVP consists in the denominator which now depends on J_2, too; due to this property the first degree terms are determinable even for a non-rotating earth model. It should be evident, however, that the small denominators in the first-degree terms will produce instabilities and large numerical errors in the calculated values of δw_{1m} for the free GBVP. On the other hand it should be kept in mind that forcing $\delta w_{1m} = 0$ will result in a bias in the calculated coefficients $\delta w_{3,\,m}$, $\delta w_{5,\,m}$, etc for $|m| \le 1$ due to the coupling of these coefficients in the system of algebraical equations.

6   Concluding remarks

The considerations presented in this paper have proved that the modelling aspect in the GBVP is not a trivial one. The transition from one level of approximation (linearization, spherical approximation, constant radius approximation) to another one may even change the fundamental properties of the solution. In any case, the effort put into modelling should correspond to the quality of the boundary data in a reasonable proportion; this procedure has been called the "engineering principle".

There are many other facets of gravity field modelling, not covered in these lecture notes, e.g. the combination of terrestrial gravity data with potential coefficients derived from satellite orbit analysis, residual terrain modelling or the use of fast numerical techniques for evaluating the spherical surface integrals. These aspects have been considered in other lectures. Since emphasis has been put on the modelling of the boundary conditions, solution concepts could only be touched. Besides the classical analytical solutions in terms of spherical integral formulae, modern numerical solutions based on the Boundary Element Method are more and more promising; these approaches make no use of the classical corrections or reductions. Nevertheless, modelling the boundary conditions in a reasonable, cost-effective way will remain a specific task of the geodesist also in future.

References

Backus GE (1968) Application of a non-linear boundary value problem for Laplace's equation to gravity and geomagnetic intensity surveys. Quart. Journ. Mech. and Applied Math., Vol. XXI, pt. 2, 195-221

Bjerhammar A, Svensson L (1983) On the geodetic boundary value problem for a fixed boundary surface - a satellite approach. Bull. Géod., 57, 382-393

Cruz J Y (1986) Ellipsoidal corrections to potential coefficients obtained from gravity anomaly data on the ellipsoid. Reports of the Ohio State University, Dept. of Geodetic Science and Surveying, Report No. 371

Dermanis A (1984) The geodetic boundary value problem linearized with respect to the Somigliana-Pizzetti normal field. Manuscr. geodaetica, 9, 77-92

Dermanis A (1987) The Bruns formula in three dimensions. Bull.Géod. , 61, 297-309

Eichhorn A (1996) Ellipsoidische Effekte beim fixen Geodätischen Randwertproblem. Diploma thesis, University of Karlsruhe (unpublished)

Grafarend E (1978) The definition of the telluroid. Bull.Géod., 52, 25-37

Heck B (1982) Combination of leveling and gravity data for detecting real crustal movements. Proc. Int. Symp. Geod. Networks and Computations, Munich 1981. Deutsche Geodätische Kommission, B 258/VII, 20-30

Heck B (1984) Zur Bestimmung vertikaler rezenter Erdkrustenbewegungen und zeitlicher Änderungen des Schwerefeldes aus wiederholten Schweremessungen und Nivellements. Deutsche Geodätische Kommission, C 302, Munich

Heck B (1986) A numerical comparison of some telluroid mappings. In: F. Sansò (ed.), Proc. I. Hotine-Marussi Symposium on Mathematical Geodesy, Rome 1985. Milano 1986, 19-38

Heck B (1988) The non-linear geodetic boundary value problem in quadratic approximation. Manuscr. geodaetica, 13, 337-348

Heck B (1989a) On the non-linear geodetic boundary value problem for a fixed boundary surface. Bull. Géod., 63, 57-67

Heck B (1989b) A contribution to the scalar free boundary value problem of Physical Geodesy. Manuscr. geodaetica, 14, 87-99

Heck B (1990) An evaluation of some systematic error sources affecting terrestrial gravity anomalies. Bull.Géod., 64, 88-108

Heck B (1991) On the linearized boundary value problems of Physical Geodesy. Reports of the Ohio State University, Department of Geodetic Science and Surveying, Report No. 407

Heck B (1995) Linearization procedures. Manuscr. geodaetica, 20, 386-392

Heck B, Seitz K (1993) Effects of non-linearity in the geodetic boundary value problems. Deutsche Geodätische Kommission, A 109, Munich

Heiskanen W, Moritz H (1967) Physical Geodesy. W.H.Freeman, San Francisco, Calif.

Holota P (1988) Isozenithals in the neighbourhood of an earth's model and the boundary condition for the disturbing potential. Manuscr. geodaetica, 12 , 257-266

Holota P (1991) On the iteration solution of the geodetic boundary-value problem and some model refinements. Travaux de l'Association Internationale de Géodésie, Tome 29, Paris 1992, 260-289

Holota P (1993) The mathematics of gravity field and geoid modelling: Green's function method. Paper, General Meeting of the IAG, Beijing, China

Hörmander L (1976) The boundary problems of physical geodesy. Arch. Rat. Mech. Anal., 62, 1-52

Klees R (1992) Lösung des fixen geodätischen Randwertproblems mit Hilfe der Randelementmethode. Deutsche Geodätische Kommission, C 382, Munich

Klees R (1995). Perturbation expansion for solving the fixed gravimetric boundary value problem. In: Sanso F (ed.), Geodetic theory today. IAG Symposium No.

114, L'Aquila, Italy 1994. Springer-Verlag, 340-349

Krarup T (1973) Letters on Molodensky's problem. Communication to the members of the IAG SSG 4.31 (unpublished)

Meissl P (1971) On the linearization of the geodetic boundary value problem. Reports of the Ohio State University, Dept. of Geodetic Science and Surveying, Report No. 152

Molodenskii M S, Eremeev V F, Yurkina M I (1962) Methods for study of the external gravitational field and figure of the earth. Israel Program for Scientific Translations, Jerusalem

Moritz H (1980) Advanced Physical Geodesy. H.Wichmann Verlag Karlsruhe, Abacus Press, Tunbridge Wells Kent

Otero J (1987) An approach to the scalar boundary value problem of Physical Geodesy by means of Nash-Hörmander theorem. Manuscr. geodaetica, 12, 245-252

Rapp R H (1981) Ellipsoidal corrections for geoid undulation computations. Reports of the Ohio State University, Dept. of Geodetic Science and Surveying. Report No. 308

Rummel R (1984) From the observation model to gravity parameter estimation. In: Schwarz K.P. (ed.), Proc. Beijing Int. Summer School on Local Gravity Field Approximation, Div. of Surveying Engineering, The Univ. of Calgary, 67-106

Rummel R (1988) Zur iterativen Lösung der geodätischen Randwertaufgabe. Deutsche Geodätische Kommission, B 287, Munich, 175-181

Rummel R (1995) The first degree harmonics of the Stokes problem - what are the practical implications? In: Festschrift E. Groten on the occasion of his 60th anniversary, Munich, 98-106

Rummel R , Ilk K H (1995) Height datum connection - the ocean part. Allgemeine Vermessungs-Nachrichten, 102, 321-330

Rummel R, Teunissen P (1982) A connection between geometric and gravimetric geodesy - Some remarks on the role of the gravity field. In: Feestbundel ter gelegenheid van de 65ste verjaardag van Prof. Baarda. Deel II, 603-623. Dept. of Geodetic Science, Delft University of Technology

Rummel R, Teunissen P (1988) Height datum definition, height datum connection and the role of the geodetic boundary value problem. Bull. Géod., 62, 477-498

Sacerdote F, Sansò F (1986) The scalar boundary value problem of Physical Geodesy. Manuscr. geodaetica, 11, 15-28

Sansò F (1978) Molodensky's problem in gravity space: a review of the first results. Bull.Géod., 52, 59-70

Sansò F (1981) Recent advances in the theory of the geodetic boundary value problem. Rev. Geophys. and Space Physics, 19, 437-449

Sansò F, Usai S (1995) Height datum and local geodetic datums in the theory of geodetic boundary value problems. Allgemeine Vermessungs-Nachrichten, 102, 343-355

Seitz K, Schramm B, Heck B (1994) Non-linear effects in the scalar free GBVP based on reference fields of various degrees. Manuscr. geodaetica, 19, 327-338

Smeets I (1991) Note on the second-order theory of the geodetic boundary value problem. Manuscr. geodaetica, 16, 173-176

Thong N C , Grafarend E W (1989) A spheroidal harmonic model of the terrestrial gravitational field. Manuscr. geodaetica 14, 285-304

Witsch K J (1987) Über eine modifizierte Version der freien geodätischen Randwertaufgabe. Abschlußberichtsband des SFB 72, Universität Bonn

Yamabe H (1950) On an extension of the Helly's theorem. Osaka Mathem. Journal, 2, 15-17

Zeidler E (1986): Non-linear functional analysis and its application. Vol.I: Fixed-point theorems. Springer, New York

Exercises

E1: Show that for an isozenithal telluroid mapping of type (3-4) the vectorial gravity anomaly $\Delta\gamma$ can be expressed by the aid of the scalar gravity anomaly $\Delta\gamma$, formula (3-45)

E2: Derive the linearized boundary condition (3-47) of the vectorial free GBVP for an isotropic reference potential $w = \mu/r$. Hint: Start from (3-38 a, b) and use Cartesian coordinates

E3: Find the linearized boundary condition for the vectorial GBVP in geographical coordinates (H, b, l) in first-order approximation for a Somigliana-Pizzetti reference potential

The Hierarchy of Geodetic BVPs

F. Sansò

On Wiener measures and the continuous formulation of observation equations

1   Wiener measures and Wiener integral

The description of a discrete finite white noise is elementary and imme-. diately perceived by scientists interested in measurements, like geodesists; it is just an n-dimensional variate $\underline{\nu} = [\nu_1, ..., \nu_n]^+$ with independent components, identically distributed, with zero mean. This is indeed a natural model for the errors of n measurements, performed under independent but statistically similar conditions; since we are thinking of measurements, let us restrict ourselves to normal variates, because this makes the whole theory easier.

The generalization of this concept to sequences of white noise $\underline{\nu} = [\nu_i, \ i = 1, 2, ...]^+$ in R^∞ is less elementary but still intuitive; its mathematical background lies in measure theory and more precisely in the theorems of Caratheodory and Kolmogorov, which allow us to extend a sequence of finite dimensional measures to a measure in R^∞.

This measure is normal and hence characterized by its mean and covariance; this is easy to express because it implies considering the components of $\underline{\nu}$ one or two at the time, so

$$E\{\underline{\nu}_i\} = 0 \quad , \quad C_{ik} = E\{\nu_i \nu_k\} = \sigma_\nu^2 \delta_{ik} \quad . \tag{1.1}$$

As usual we can construct a Hilbert space, $\mathcal{L}_\nu^2$, of random variables on the above probability space, namely those functions $X(\underline{\nu})$ such that

$$\|X\|_2^2 = E\{X^2(\underline{\nu})\} < +\infty \quad ; \tag{1.2}$$

this is a Hilbert space with scalar product

$$(X, Y)_2 = E\{X(\underline{\nu})Y(\underline{\nu})\} \quad , \tag{1.3}$$

inducing the norm (1.2).

Dealing with linear problems, we can limit ourselves to considering a closed subspace of $\mathcal{L}^2_\nu$, namely that of homogeneous linear functions of $\underline{\nu}$

$$X(\underline{\nu}) = \sum_{i=1}^{+\infty} \lambda_i \nu_i = \underline{\lambda}^+ \underline{\nu} \ .$$

(1.4)

(λ_i constant)

It is immediate to verify that

$$X = \underline{\lambda}^+ \underline{\nu} \in \mathcal{L}^2 \nu \iff \|\underline{\lambda}\|_2^2 = \sum_{i=1}^{+\infty} \lambda_i^2 < +\infty \ ,$$

(1.5)

in fact in this case

$$\|X\|_2^2 = E\{X^2\} = \underline{\lambda}^+ C_{\nu\nu} \underline{\lambda} = \sigma_\nu^2 \|\underline{\lambda}\|_2^2 \ .$$

(1.6)

As a matter of fact it is obvious that under condition (1.5) the series (1.4) is convergent in $\mathcal{L}^2_\nu$ norm (i.e. in quadratic mean) and it is even possible to see that it converges with probability $P = 1$ (i.e. for almost all realizations of ν).

The space of such homogeneous linear functions will be called L^2_ν and it is obvious through (1.6) that this is not only in biunivocal relation with ℓ^2, but even isometric to this space, the correspondence between $\underline{\lambda} \in \ell^2$ and $X \in L^2_\nu$ being defined through (1.4). We immediately recognize that all r.v. in L^2_ν have zero average.

It is in L^2_ν that we can solve the problem of building a stochastic model for continuous white noise.

Remark 1.1: It might seem that a natural generalization of the concept of discrete white noise might be a process $\nu(t)$ where each $\nu(t)$, $t \in \Omega$, is an independent, finite variance, variable.

However it is easy to see that this model leads nowhere and in particular we could consistently define a basic linear operation like

$$\int f(t)\nu(t)dt$$

only for functions $f(t)$ which are identically zero almost everywhere. So let us first define a Wiener measure and then justify why this is a good model to represent a continuous white noise.

Definition 1.1: Let Ω be a domain in R^n, which we consider as bounded for the moment; let us consider $L^2(\Omega)$ and a complete O.N.

(Ortho-Normal) sequence in it, $\{\varphi_n\}$; let us consider any measurable set $A \subseteq \Omega$ and let us define its indicator function

$$\lambda(A,t) = \begin{cases} 1 & t \in A \\ 0 & t \in \Omega\backslash A \end{cases} ; \tag{1.7}$$

moreover let us take any discrete normal white noise $\underline{\nu} = [\nu_1, \nu_2, ...]^+$; we define the r.v. (random variable)

$$\mu_A = \sum_{n=1}^{+\infty} \chi_n(A)\nu_n ,$$
$$\tag{1.8}$$
$$\chi_n(A) = (\chi(A,t), \varphi_n(t))_2 = \int_A \varphi_n(t)dt ;$$

Theorem 1.1: The following properties hold for μ_A

a) $\mu_A \in L^2_\nu, \forall A \subseteq \Omega$ measurable;

b) μ_A is a normal family, with index A;

c) μ_A, μ_B are independent r.v. when $A \cap B = 0$; this is also phrased by saying that μ_A has independent increments;

d) μ_A is an additive function of A, namely

$$\forall A, B , \quad A \cap B = 0 ; \quad \mu_{A \cup B} = \mu_A + \mu_B ; \tag{1.9}$$

e) μ_A is a σ-additive function of A, i.e.

$$\forall \{A_n\} , \quad A_n \cap A_m = 0 , \quad n \neq m ; \quad \mu_{\cup A_n} = \sum \mu_{A_n} ; \tag{1.10}$$

therefore μ_A is what is called a normal stochastic measure with independent increments or a Wiener measure on Ω , with constant intensity (or variance density).
We start the proof;

a) $\chi(A,t) \in L^2(\Omega) , \quad \|\chi(A,t)\|_2^2 = \int_\Omega \chi^2(A,t)dt = \int_A dt = m(A)$
    (Lebesgue measure);

$$\Rightarrow \sum_{n=1}^{+\infty} \chi_n(A)^2 = m(A) < +\infty ; \tag{1.11}$$

$$\Rightarrow \mu_A \in L^2_\nu , \quad \|\mu_A\|_2^2 = \sigma_\nu^2 m(A) , \tag{1.12}$$

by (1.6), (1.11);

b) we only note that $\sum_{n=1}^{N} \chi_n(A)\nu_n$ is normal for $\forall N$ and converges in quadratic mean to μ_A, what implies that it converges in law to μ_A, so μ_A is normal. The same reasoning can be repeated for any vector $[\mu_{A_1}, \mu_{A_2}, ..., \mu_{A_n}]$, what is enough for us;

c) let us put $\underline{\chi}_A = [\chi_1(A), \chi_2(A)...]^+$, and compute $\forall A, B$, measurable,

$$E\{\mu_A \mu_B\} = E\{\underline{\chi}_A^+ \underline{\nu} \underline{\nu}^+ \underline{\chi}_B\} = \sigma_\nu^2 \underline{\chi}_A^+ \underline{\chi}_B =$$

$$= \sigma_\nu^2 \sum_{n=1}^{+\infty} \chi_n(A)\chi_n(B) = \sigma_\nu^2 (\chi(A,t), \chi(B.t)) =$$

$$= \sigma_\nu^2 \int_\Omega \chi(A,t)\chi(B,t)dt = \sigma_\nu^2 \int_{A \cap B} dt = \sigma_\nu^2 m(A \cap B) \tag{1.13}$$

Since $E\{\mu_A\} = E\{\mu_B\} = 0$, (1.13) is the covariance of μ_A, μ_B and it is zero when $A \cap B = 0$; since μ_A, μ_B are jointly normal, this implies independence;

d) compute $\forall A, B$, $A \cap B = 0$, by using (1.13),

$$\|\mu_{A \cup B} - \mu_A - \mu_B\|_2^2 = E\{\mu_{A \cup B}^2\} + E\{\mu_A^2\} + E\{\mu_B^2\} +$$

$$-2E\{\mu_{A \cup B}\mu_A\} - 2E\{\mu_{A \cup B}\mu_B\} - 2E\{\mu_A \mu_B\} =$$

$$= \sigma_\nu^2[m(A \cup B) + m(A) + m(B) - 2m(A) - 2m(B)] = 0 \;\;;$$

(1.19) follows;

e) first observe that $m(A) = \sum_{n=1}^{+\infty} m(A_n)$ so that $\sum_{n=N+1}^{+\infty} m(A_n) \to 0$ when $N \to \infty$; but then, using d),

$$\|\mu_A - \sum_{n=1}^{N} \mu_{A_n}\|_2^2 = \|\mu_A - \mu\left(\bigcup^N \mu_{A_n}\right)\|_2^2 = \sigma_\nu^2 m(\bigcup_{n=N+1}^{\infty} \mu_{A_n})$$

$$= \sigma_\nu^2 \sum_{n=N+1}^{+\infty} m(A_n) \to 0 \;\;,$$

what implies (1.1 - 1.10).

Now we want to show that when a stochastic measure μ_A is given satisfying

$$E\{\mu(A)\mu(B)\} = \sigma_0^2 m(A \cap B) \;\;, \tag{1.14}$$

then one can define an isometric operator J mapping (one to one) $L^2(\Omega)$ into a closed subspace of $\mathcal{L}^2_\mu$, which we call L^2_μ; this isometric operator is called the *Wiener integral*.

In fact let us put by definition

$$J[\chi(A,t)] = \mu(A) \qquad (1.15)$$

and let us extend J by linearity, in the sense that for any sequence $\{A_i, i = 1, 2..N\}$, of disjoint sets, and $\underline{\lambda} = \{\lambda_1, \lambda_2, ...\lambda_N\}$ we put

$$J[\sum \lambda_i \chi(A_i, t)] = \sum \lambda_i J[\chi(A_i, t)] = \sum \lambda_i \mu(A_i) \qquad (1.16)$$

Now the piecewise constant functions

$$f(t) = \sum_{i=1}^N \lambda_i \chi(A_i, t) \qquad (1.17)$$

are first of all contained in $L^2(\Omega)$, as

$$\|f\|_2^2 = \sum_{i=1}^N \lambda_i^2 m(A_i) \ , \qquad (1.18)$$

since the sets A_i are by hypothesis disjoint and bounded; moreover the set of $f(t)$ of the form (1.17), $\forall N$, $\forall \underline{\lambda}$, $\forall \{A_i\}$ is a set of functions dense in $L^2(\Omega)$. Furthermore the r.v. $\sum_i \lambda_i \mu(A_i)$ has norm in $\mathcal{L}^2_\mu$

$$\|Jf\|_2^2 = \|\sum_i \lambda_i \mu(A_i)\|_2^2 = \sum_i \lambda_i^2 \sigma_0^2 m(A_i) = \sigma_o^2 \|f\|_2^2 \ , \qquad (1.19)$$

because by virtue of (1.14)

$$E\{\mu(A_i)\mu(A_k)\} = \sigma_0^2 m(A_i)\delta_{ik} \ . \qquad (1.20)$$

As we see, apart from the inessential constant σ_0^2, (1.19) proves that J is an isometry between the set $\{f$ piecewise constant$\}$ and some subspace of $\mathcal{L}^2_\mu$.

Let us observe here that an isometry preserves the norm and hence the scalar product too.

This isometry can therefore be extended by continuity since

$$\sum_{i=1}^N \lambda_i^N \chi(A_i^N, t) \to g(t) \text{ in } L^2(\Omega)$$

implies by (1.19) that also

$$\sum_{i=1}^{N} \lambda_i^N \mu(A_i^N)$$

is a sequence convergent in mean square sense so that there must be an element $X \in \mathcal{L}_\mu^2$ such that

$$\sum_{i=1}^{N} \lambda_i^N \mu(A_i^N) \to X \quad , \quad \text{in } \mathcal{L}_\mu^2 \;\; ;$$

but then we can set up a suitable
Definition 1.2: $\forall g \in L^2$, take f_N (piecewise constant) $f_N \to g$ in L^2 and put

$$Jg = \lim_{N \to \infty} Jf_N \tag{1.21}$$

the limit being in $\mathcal{L}_\nu^2$.
Remark 1.2: The symbol Jg is usually written

$$Jg = \int_\Omega g(t)d\nu \tag{1.22}$$

Remark 1.3: The image of $L^2(\Omega)$ under J is necessarily a closed subspace of $\mathcal{L}_\nu^2$ and we will call it L_ν^2.
Now we are in a position to prove a theorem which is in a sense the converse of Theorem 1.1.
Theorem 1.2: Assume $\mu(A)$ to be a normal stochastic measure with independent increments and uniform variance density, i.e.

$$E\{\mu(A)\} = 0 \quad E\{\mu(A)^2\} = \sigma_0^2 m(A) \tag{1.23}$$

or, equivalently,

$$E\{\mu(A)\mu(B)\} = \sigma_0^2 m(A \cap B) \tag{1.24}$$

then $\mu(A)$ is a Wiener measure in the sense of Definition 1.1, namely given any complete O.N. sequence $\{\varphi_n\}$ in $L^2(\Omega)$, there is a discrete white noise $\underline{\nu} = [\nu_1, \nu_2, ...]^+$, with $\sigma_{\nu_i}^2 = \sigma_0^2$, such that

$$\mu(A) = \sum_{n=1}^{+\infty} \chi_n(A)\nu_n \;\; . \tag{1.25}$$

In fact let us put

$$\nu_n = J\varphi_n = \int_\Omega \bigcup \varphi_n(t) d\mu \ ; \tag{1.26}$$

then $E\{\nu_n\} = 0$ and

$$E\{\nu_n \nu_m\} = (\nu_n, \nu_m)_{\mathcal{L}^2} = (J\varphi_n, J\varphi_m)_{\mathcal{L}^2} = \sigma_0^2(\varphi_n, \varphi_m)_{L^2} = \sigma_0^2 \delta_{nm} \tag{1.27}$$

because J is an isometry and therefore, apart from multiplication by σ_0^2, it preserves the scalar product too. But then $\{\nu_n\}$ is a normal sequence of independent variables with the same variance. Moreover from the identity

$$\chi(A,t) = \sum_n (\chi(A,t), \varphi_n(t))\varphi_n(t) = \sum_n \chi_n(A)\varphi_n(t) \ ,$$

by applying the isometry J one obtains

$$\mu(A) = \sum_{n=1}^{+\infty} \chi_n(A)\nu_n$$

as we wanted to prove.

Corollary 1.1: The Wiener integral J satisfies the identity, $\forall f, g \in L^2(\Omega)$

$$E\left\{ \int_\Omega f(t)d\mu \int_\Omega g(t)d\mu \right\} = (Jf, Jg)_{\mathcal{L}\epsilon} = \sigma_0^2(f, g)_{L^2} = \sigma_0^2 \int_\Omega f(t)g(t)dt \ . \tag{1.28}$$

Remark 1.4: Let us take (1.28) with $g = f$ and we get the most characteristic relation of Wiener integral, namely

$$E\left\{ \left(\int_\Omega f(t)d\mu \right)^2 \right\} = \sigma_0^2 \int_\Omega f^2(t)dt \tag{1.29}$$

$$(dt = dm \text{Lebesgue measure element});$$

now let us choose $f(t) = \sqrt{\gamma(t)}\chi(A,t)$, with $\gamma \geq 0$, so that (1.29) yields

$$E\left\{ \left(\int_A \sqrt{\gamma(t)}d\mu \right)^2 \right\} = \sigma_0^2 \int_A \gamma(t)dm \tag{1.30}$$

this proves that

$$\bar{\nu}(A) = \int_A \sqrt{\gamma(t)}d\mu \ , \tag{1.31}$$

which is indeed a zero mean variable, can be taken to model a non uniform stochastic measure with variance

$$M(A) = E\{\bar{\mu}(A)^2\} = \sigma_0^2 \int_A \gamma(t)dm$$

and than with variance density, with respect to Lebesgue measure,

$$\frac{d\mu}{dm} = \sigma_0^2 \gamma(t)$$

This new measure indeed models an area spread noise, which, like the Wiener measure, is independent on non overlapping areas, with the difference however that the variance $\bar{\mu}$ is now varying from area to area and it is not any more constant.

Of course we can define a "Wiener integral" on the stochastic measure $\bar{\mu}$ too, obtaining in this way a r.v. satisfying the characteristic relation

$$E\left\{\int_\Omega f(t)d\bar{\mu} \int_\Omega g(t)d\bar{\mu}\right\} = \sigma_0^2 \int_\Omega f(t)g(t)\gamma(t)dm \ ,$$

$$\forall f,g \ , \ \int_\Omega f^2(t)\gamma(t)dm \ , \ \int_\Omega g^2(t)\gamma(t)dm < +\infty \ . \tag{1.32}$$

2 Continuous observation equations

We are now in a position to start our discussion on the representation of a "continuum" of measurements of some function $u(t)$ with noise. Since our idea is to describe a situation in which we have a very large and dense set of measurement points, it seems natural to think of a limit process such that each point t is surrounded by an infinitesimal set dt and to write

$$\mu_0(dt) = u(t)dt + \mu_\nu(dt)$$

$$(t \in \Omega \ ; \ \int dt = \Omega) \tag{2.1}$$

where $u(t)$ is the function we want to measure while $\mu_\nu(dt)$ is the continuous noise represented by a Wiener measure such that

$$E\{\mu_\nu(dt)\} = 0 \quad E\{\mu_\nu(dt)^2\} = \sigma_0^2 \gamma(t) m(dt); \quad (2.2)$$

the sum of the two actually represents the measurement which is apparently attached to the infinitesimal volume dt.

Remark 2.5: In this way the representation of the measurement is rather extensive than intensive, i.e. pointwise; this is however the only way to represent reasonably the white noise part. On the other hand (2.1) is an equality between measures and as such its "signal" part $u(t)dt$ is also a measure; more precisely a signed measure with density $u(t)$ with respect to the Lebesgue element $dt = m(dt)$. As such $u(t)$ needs not to be a continuous function but it can be for instance an $L^2(\Omega)$ function, which is often the case. For those functions on the other hand a pointwise measurement would not be meaningful, while (2.1) still has a clear meaning.

Willing to go to a finite formulation we could multiply (2.1) by a smooth function $\varphi(t)$ (test function) and integrate over the set Ω where the measurements are available.

So we find

$$\int_\Omega \varphi(t)\mu_0(dt) = \int_\Omega \varphi(t)u(t)dt \, t \int_\Omega \varphi(t)\mu_\nu(t) \quad , \quad (2.3)$$

where the last term is indeed a Wiener integral.

At this point it is also natural to ask ourselves how large could be the space of test functions $\{\varphi\}$; in our case this is dictated by the requirement that the disturbance part of (2.3), namely

$$\eta = \int_\Omega \varphi(t)\mu_\nu(dt)$$

be finite in variance, i.e.

$$E\{\mu^2\} = \sigma_0^2 \int_\Omega \varphi^2(t)\gamma(t)dt < +\infty \quad . \quad (2.4)$$

hence we must have $\varphi \in L_\gamma^2(\Omega)$, i.e. the space of functions square summable over Ω with weight $\gamma(t)$; only in case that

$$0 < \alpha \leq \gamma(t) \leq \beta \quad (2.5)$$

this space coincides with $L^2(\Omega)$, what we shall assume from now on.

It is interesting to note that if (2.3) has to be meaningful $\forall \varphi \in L^2(\Omega)$, then we must also have $u \in L^2(\Omega)$, i.e. we must put on $u(t)$ the requirement to belong to a space contained in (or coinciding with) $L^2(\Omega)$.

Now let us try to create a connection between (2.3) and the usual discrete observational model

$$v_{0i} = u(t_i) + \nu_i \; ; \tag{2.6}$$

an idea is to make a partition of Ω into sets Δ_i,

$$\Omega = \cup_i \Delta_i \; ; \quad \Delta_i \cap \Delta_j = 0 \; ; \quad t_i \in \Delta_i$$

and take $\varphi = \frac{1}{\Delta_i} \chi(\Delta_i, t)^1$ in (2.3), thus getting

$$\frac{1}{\Delta_i} \int_{\Delta_i} \mu_0(dt) = U_{0i} = \frac{1}{\Delta_i} \int_{\Delta_i} u(t) dt t \frac{1}{\Delta_i} \mu_\nu(\Delta_i) = \overline{u}_i + \nu_i^0 \; , \quad i = 1, 2, ..., N \; . \tag{2.7}$$

We shall use (2.7) as a bridge between (2.3) and (2.6), in the sense that first of all we shall prove that from (2.7), by taking a limit with Max Diam $(\Delta_i) \to 0$ we can go back to (2.3), than by studying under what conditions (2.6) and (2.7) can be consider as "equivalent".

Just for the sake of brevity let us assume that $\gamma \equiv 1$; the generalization to an arbitrary $\gamma(t)$, satisfying (2.5), is straightforward.

Let us now start from any set $\Delta_i = \Delta_i^0$ and divide it in two

$$\Delta_i^0 = \Delta_{i0}^1 \cup \Delta_{i1}^1 \; , \quad m(\Delta_{ik}^1) = \frac{1}{2} m(\Delta_i^0) \; ;$$

accordingly define

$$\nu_{ik}^1 = \frac{\mu_\nu(\Delta_{ik}^1)}{\Delta_{ik}^1} \quad (k = 0, 1)$$

and note that, since $\gamma \equiv 1$ by hypothesis,

---

[1] For the sake of simplicity we shall use one and the same symbol Δ_i to indicate both the set Δ_i and its measure, $m(\Delta_i) = \Delta_i$.

$$\sigma^2(\nu_i^0) \quad = \frac{\sigma_0^2}{\Delta_i}$$

$$\sigma^2(\nu_{ik}^1) \quad = \frac{\sigma_0^2}{\Delta_{ik}^1} = 2\frac{\sigma_0^2}{\Delta_i} = 2\sigma^2(\nu_i^0) \ .$$

It is clear that we could repeat the same process N times, obtaining 2^N subsets of Δ_i and corresponding noises such that

$$\begin{cases} \Delta_i = \bigcup_{k=0}^{2^N-1} \Delta_{ik}^N \ , \quad m(\Delta_{ik}^N) = \frac{1}{2^N} m(\Delta_i^0) \\[2mm] \sigma^2(\nu_{ik}^N) = 2^N \sigma^2(\nu_i^0) \ . \end{cases} \tag{2.8}$$

If we choose a point t_{ik}^N for each Δ_{ik}^N we can write

$$U_{ik}^N = \frac{1}{\Delta_{ik}^N} \int_{\Delta_{ik}^N} u(t)dt + \nu_{ik}^N$$

which multiplied by Δ_{ik}^N and by $\varphi(t_{ik}^N)$, with φ any smooth function, and added over all k and over all i gives

$$\sum_{ik} \varphi(t_{ik}^N) U_{ik}^N \Delta_{ik}^N = \sum_{ik} \varphi(t_{ik}^N) \int_{\Delta_{ik}^N} u(t)dt + \sum_{ik} \varphi(t_{ik}^N) \mu_\nu(\Delta_{ik}^N) \tag{2.9}$$

In (2.8) it is clear that for $N \to \infty$ (Max Diam $\Delta_{ik}^N \to 0$)

$$\sum_{ik} \varphi(t_{ik}^N) \int_{\Delta_{ik}^N} u(t)dt \to \int_\Omega \varphi(t)u(t)dt$$

and that

$$\sum_{ik} \varphi(t_{ik}^N) \mu_\nu(\Delta_{ik}^N) \to \int_\Omega \varphi(t) \mu_\nu dt$$

by definition of Wiener integral.
So from (2.9) we find that the second member is the same as (2.3) in the limit for $N \to \infty$, so the same must be true for the first members, at least for smooth functions, but this is enough to claim that (2.9) converges to (2.3) for any $\varphi \in L^2(\Omega)$.

This shows that we can consider (2.3) as the limit of discrete equations of type (2.9) obtained by increasing the resolution of observations, i.e. lething Max Diam $\Delta_{ik}^N \to 0$, and on the same time increasing the variance of the noise on each domain, according to the rule (2.8).

Yet the crucial question here is when can we consider (2.7) close enough to (2.3)?

To answer to this question we rewrite (2.7) as

$$\sum U_{0i}\varphi_i\Delta_i = \int_\Omega \left[\sum \varphi_i\chi(\Delta_i,t)\right] u(t)dt + \sum \varphi_i\mu_\nu(\Delta_i) \qquad (2.10)$$

and we note that this corresponds basically to using the test function

$$\tilde{\varphi}(t) = \sum_i \varphi_i\chi(\Delta_i,t) \qquad (2.11)$$

in (2.3). Now (2.3) says that we can estimate through $\int_\Omega \varphi(t)\mu_0(dt)$ the projection of u in any direction φ, committing the error η, while (2.10) tells us that if we can use only the observables U_{0i} instead of μ_0, we have information about the projection of u on directions $\tilde{\varphi}$ of the type (2.11). This restriction of information will not be particularly significant only if the component of u orthogonal to $\tilde{\varphi}$ is small compared to its total norm. More precisely if we call $P = \sum \frac{1}{\Delta_i}\chi(\Delta_i,t)(\chi(\Delta_i,\cdot),\cdot)_2$ the projector on the manifold of the $\{\tilde{\varphi}\}$, we must have

$$\|(J - P)u\| \ll \|u\| \quad . \qquad (2.12)$$

Just to make (2.12) more handable let us observe that

$$\|(J - P)u\|^2 = \sum_i \int_{\Delta_i} [u(t) - \overline{u}_i]^2 \, dt \quad ; \qquad (2.13)$$

then if we assume $u(t)$ to be regular and we call

$$\sigma_{\Delta_i}^2[u] = \tfrac{1}{\Delta_i} \int [u(t) - \overline{u}_i]^2 \, dt \simeq \nabla u(\theta_i)^+ C_i' \nabla u(\theta_i)$$

$$C_i = \tfrac{1}{\Delta_i} \int_{\Delta_i} (t - \theta_i)(t - \theta_i)^+ dt \qquad (2.14)$$

$$\theta_i = \tfrac{1}{\Delta_i} \int_{\Delta_i} t dt \quad ,$$

namely the variances of $u(t)$ over a uniform distribution of t on Δ_i, we can rewrite (2.13) as

$$\|(J - P)u\|^2 \simeq \sum_i \sigma_{\Delta_i}^2[u]\Delta_i \simeq \int \sigma_{\Delta_t}^2[u]dt \qquad (2.15)$$

Now if Δ_i are constant in shape, e.g. all cubes then

$$C_i = C = H \cdot I$$

with H a suitable shape factor, which is

$$H = \frac{\Delta^2}{12} \quad , \quad H = \frac{\Delta}{12} \qquad (2.16)$$

respectively for intervals in R^1 and for squares in R^2.
Then using (2.12) through (2.16) one gets

$$H \cdot \int_\Omega |\nabla u|^2 dt \ll \int_\Omega u^2 dt \quad , \qquad (2.17)$$

which is a relation easy enough to verify.
Example 2.1: With $t \in R^1$, $\Omega = (2\pi)$, $\Delta = \frac{2\pi}{N}$, $u = A \sin nt$ (2.17) becomes just

$$\frac{n}{N} \ll \frac{\sqrt{3}}{\pi} \sim \frac{1}{2} \quad ,$$

while with $t = [t_1 t_2]^+ \in R^2$, $\Omega = (2\pi) \times (2\pi)$, $\Delta = \frac{4\pi^2}{N^2}$, $u = A \sin nt_1 \sin mt_2$, $k^2 = n^2 + m^2$, we get again

$$\frac{k}{N} \ll \frac{\sqrt{3}}{\pi} \sim \frac{1}{2}$$

thus showing that (2.17) is basically the usual deterministic Nysquit relation.
So up to now we can maintain that if a condition like (2.17) holds, than the use of a fully continuous equation like (2.3) $\forall \varphi$ or the use of reduced equation like (2.10) $\forall \tilde{\varphi}$, gives about the same information on u.
Now we have to compare (2.6) with (2.7); since the noise part of the two equations can be made identical, the question is whether we can substitute $u_i = u(t_i)$ with $\bar{u}_i = \frac{1}{\Delta_i} \int_{\Delta_i} u(t)dt$ without changing too

much the observational scheme. A traditional criterion for that to happen is just that

$$|u_i - \overline{u}_i| \ll \sigma_{\nu_i} \tag{2.18}$$

because in this case the bias introduced in each equation is much smaller than the noise.

Let us make the restrictive hypothesis that t_i is just the barycentre of Δ_i; this is of course acceptable when the Δ_i are all similar, but this is not the only case.

Then we have

$$\int_{\Delta_i} (t - t_i)dt \equiv 0$$

so that, calling $U(t)$ the matrix of second derivatives of u,

$$|\overline{u}_i - u_i| = \tfrac{1}{\Delta_i}|\int [u(t) - u(t_i)]\,dt| \simeq$$

$$\simeq \tfrac{1}{2}\tfrac{1}{\Delta_i}|\int TrU(t_i)(t - t_i)(t - t_i)^+\,dt| =$$

$$= \tfrac{1}{2}|TrU(t_i)C_i| \;;$$

if Δ_i are all of the same type so that

$$C_i = H \cdot I$$

then we get, recalling (2.18),

$$|\overline{u}_i - u_i| \simeq \frac{1}{2}H|\Delta u(t_i)| \ll \sigma_{\nu_i} \;. \tag{2.19}$$

Multiplying (2.19) by $|\Delta|$ and adding over i one gets

$$\frac{H}{2}\int_\Omega |\Delta u(t)|dt \ll \sigma_\nu \cdot m(\Omega) \tag{2.20}$$

Example 2.1 (continued): With $t \in R^1$, $\Omega = 2\pi$, $\Delta = 2\pi$, $\Delta = \frac{2\pi}{N}$, $u(t) = A \sin nt$, $H = \frac{\Delta^2}{12} = \frac{4\pi^2}{12N^2}$ we find as a consequence of (2.20),

$$\frac{n}{N} \ll \sqrt{\frac{\sigma_\nu}{A}}\sqrt{\frac{6}{2\pi}} \;; \tag{2.21}$$

with $t \in R^2$, $\Omega = (2\pi) \times (2\pi)$, $\Delta = \frac{4\pi^2}{N^2}$, $u = A \sin nt_1 \sin mt_2$, $k^2 = n^2 + m^2$, $H = \frac{\Delta}{12} = \frac{4\pi^2}{12N^2}$ we get

$$\frac{k}{N} \ll \sqrt{\frac{\sigma_\nu}{A}} \sqrt{\frac{3}{2}} . \tag{2.22}$$

Remark 2.6: The interest of this new Nyquist type relations, like (2.20), (2.21), (2.22) is that they look at the resolution parameter $\frac{n}{N}$, including the stochastic behaviour of the model, summarized by the signal to noise ratio $\frac{A}{\sigma_\nu}$. It is clear that when $\frac{A}{\sigma_\nu} \sim 1$, (2.21), (2.22) are back again to the usual Nyquist relation, while when $\frac{A}{\sigma_\nu} \ll 1$, (2.21), (2.22) become more stringent than Nyquist relations because when the signal is much higher than the noise we have to use smaller intervals between measurement points so that the variation of u from point to point is again small, compared to noise, and we can still describe the situation as a continuum of measurements.

We conclude this chapter trying to apply formula (2.22) to a case relevant to geodesy and to the formulation of geodetic problems in terms of Boundary Value Problems.

Example 2.2: Let us represent the earth surface as a rectangle with sides $(2\pi) \times (2\pi)$, covered by a regular grid of measured gravity anomalies.

Repeating the same reasoning as in Example 2.1 one gets the relation

$$\frac{k}{N} \ll \sqrt{\frac{\sigma_\nu}{A} 3} ;$$

however considering that $k = \sqrt{n^2 + m^2}$ and assuming that both n and m have the same maximum value n_{max}, we derive

$$n_{max} \ll N \sqrt{\frac{\sigma_\nu \cdot 3}{A \cdot 2}} . \tag{2.23}$$

If we apply this relation to the case that

$$A = 30\text{mGal} , \quad \sigma_\nu = 0,3\text{mGal} , \quad N = 10800 ,$$

corresponding to 1 measurement per each $(1') \times (1')$ square, we get $n_{max} \sim 720$. This seems to indicate that with such a data set we could think of reaching the determination of a signal with maximum frequency (degree) up to 722. Of course this is not the case today. Now if we imagine to average our signal over squares $(0, 5°) \times (0, 5°)$, so that the power is decreased and on the same time the noise increases because we just estimate the mean value from, may be few, point values, we could guess

$$A = 15\text{mGal} \ , \quad \sigma_\nu = 5\text{mGal} \ , \quad N = 360 \ ,$$

so getting $n_{max} \ll 254$, for instance $n_{max} \sim 180$, which seems a reasonable figure in the described situation.

What is a geodetic B.V.P.?

3 Purely gravitational B.V.P's of geodesy

We shall consider the earth as a rigid body in uniform rotation around a direction fixed in space; moreover we shall assume that all time dependent gravity effects, like the influence of the moon and sun, tidal effects, the effect of the atmosphere, have been computed and subtracted from data.

We shall work with an ordinary cartesian triad with the origin placed at the barycentre of the masses, the z axis coinciding with the rotation axis and the X, Y axes fixed with respect to the earth, i.e. rotating with the same angular velocity with respect to an inertial frame.

Under these conditions the gravity potential $W(P)$, is given by

$$W(P) = u(P) + \frac{1}{2}\omega^2(x_P^2 + y_P^2) \tag{3.1}$$

with $u(P)$ the pure Newtonian potential, ω the angular velocity of the earth; $u(P)$ is a harmonic function outside the masses and the idea is to try to determine this function by using suitable measurements on the boundary, i.e. by solving the corresponding boundary value problem (B.V.P.)

$$\begin{cases} \Delta u = 0 & \text{in } \Omega \\ \Gamma(w) = f & \text{on } S \\ u(P) = 0(r_P^{-1}) & r_P \to \infty \end{cases} \tag{3.2}$$

with B the body of the earth, $\Omega = B^C$, $S = \partial B$.

Remark 3.1: Sometimes (3.2) is substituted with the equations

$$\begin{cases} \Delta w = 2\omega^2 & \text{in } \Omega \\ \Gamma(w) = f & \text{on } S \end{cases},$$

which do not define uniquely a B.V.P., unless it is specified that w has the form (3.1) and that $u(P) = 0(r_P^{-1})$ for $r_P \to \infty$.

Remark 3.2: In order to make the treatment of geodetic B.V.P.'s neater and more appealing it is useful to consider the elimination of centrifugal terms as one of the many corrections of the data performed before coming to the core of the subject.

We try to justify that on the basis of today's achievable information. Namely let us assume that the position of points where we perform measurements is known a priori with an error of the order of $\rho \sim 10^2 m$ and let us see how this affects the two basic quantities that are related to the potential W, namely the gravity modulus g as well as the potential W itself.

As for g we observe that the centrifugal potential contributes to it through

$$g = |\nabla u + \underline{c}|$$

$$\underline{c} = \omega^2(x_P \underline{e}_x + y_P \underline{e}_y) \ ,$$

(3.3)

with $\underline{e}_x$, $\underline{e}_y$ unit coordinate vectors; then

$$|\delta \underline{c}| \leq \omega^2 \rho \simeq 5 \cdot 10^{-2} \text{mGal}$$

(3.4)

which favourably compares with noise level, that we fix here as $\sigma_\nu \sim 0.1$mGal.

As for the potential itself one can use (3.1); yet to make the result more readable it is convenient to divide the error δw by γ ($\gamma \sim 10 \ m \ sec^{-2}$), in order to get a (linear) geoid error through $\delta N = \frac{\delta W}{\gamma}$,

$$|\delta N| = |\frac{\delta W}{\gamma}| \leq \frac{\omega^2 \sqrt{x_P^2 + y_P^2}\rho}{\gamma} \lesssim 30 \text{ cm} \ .$$

(3.5)

This seems to be a troublesome figure, probably not acceptable as such. Yet one has to consider that we never get hold of $w(P)$ by a direct measurement; rather we compute potential variations between points[2]

$$\Delta w = w(P) - w(Q)$$

from leveling measurements and some (interpolated) gravity information.

But then (3.1) should be more properly substituted by

$$\Delta w \simeq \Delta u + \omega^2(\overline{x}\Delta x + \overline{y}\Delta y) \ ,$$

(3.6)

---

[2]It is clear that here Δ means variation between points P and Q and not Laplacian.

with $\bar{x}, \bar{y}$ cordinates of the middle point between P and Q, so that the error in $\Delta\omega$, $\delta\Delta w$, becomes

$$|\delta\Delta w| \simeq \omega^2 |\delta x \Delta x + \delta y \Delta y|$$

or

$$|\delta\Delta N| = |\frac{\delta\Delta w}{\gamma}| \lesssim \frac{\omega^2}{\gamma}\rho\sqrt{\Delta x^2 + \Delta y^2} \ . \tag{3.7}$$

For two points P, Q, 1 Km far one from the other, we get from (3.7)

$$|\delta\Delta N| \lessgtr 5 \cdot 10^{-5} \ m$$

which is fairly below the measurement noise, along a 1 Km leveling line.

Based on this consideration we shall, from now on, consider only B.V.P.'s for the purely gravitational potential u; these will be typically of the form

$$\begin{cases} \Delta u = 0 & \text{in } \Omega \\ \Gamma(u) = f & \text{on } S \\ u = 0(r_P^{-1}) & r_P \to \infty \ . \end{cases} \tag{3.8}$$

Example 3.1: This is the so called fixed boundary gravimetric[3] B.V.P. and it consists in assuming as known the geometry of the boundary and the modulus of gravity on it

$$\begin{cases} \Delta u = 0 \\ |\nabla u||_S = g(\sigma) \\ u = 0(r_P^{-1}) & (r_P \to \infty) \end{cases} \tag{3.9}$$

with

_____

[3] Here we use the customary word gravity which however, in this context, should be substituted by gravitation.

$\sigma \quad = (\varphi, \lambda),$ geodetic latitude and longitude

$S \quad = \{\underline{r}(\sigma) = \underline{R}(\sigma) + h\underline{\nu}(\sigma)\}$

$$\underline{r}(\sigma) = \frac{a}{\sqrt{1 - e^2 \sin^2 \varphi}} \begin{bmatrix} \cos \varphi \cos \lambda \\ \cos \varphi \sin \lambda \\ (1 - e^2) \sin \varphi \end{bmatrix} \text{(ellipsoidal point)}$$

$$\underline{\nu}(\sigma) = \begin{bmatrix} \cos \varphi \cos \lambda \\ \cos \varphi \sin \lambda \\ \sin \varphi \end{bmatrix} \text{(ellipsoidal normal)}$$

$h \quad =$ ellipsoidal height

This problem, though already difficult in terms of mathematical analysis[4], yet is considerably simpler than the other two gravimetric B.V.P. where the geometry of the boundary, namely S, is in itself partly or completely unknown.

Example 3.2: The scalar gravimetric B.V.P. or scalar Molodensky's problem; in this case we assume for each point P on the surface S (cf. Fig. 1), that it is known the projection of P on the ellipsoid E, i.e. the point P_0; moreover we assume to know $g = |\nabla u|$ in P. On the other hand the ellipsoidal height h is assumed to be unknown, so that some other information on the "height" of P has to be supplied; in this case this is the value of the potential itself in P, which can be obtained, apart from an additive unknown constant, from leveling and gravimetry.

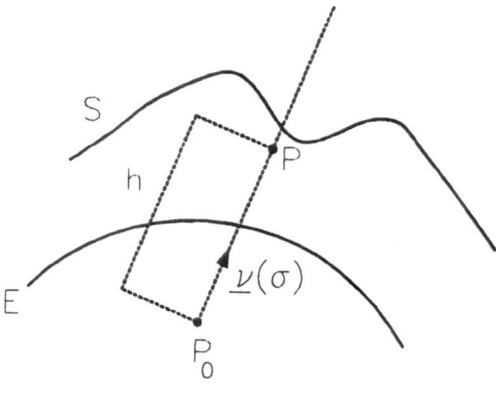

Fig. 1

[4] Some analyses on this problem can be found in [Backus 1968], [Sacerdote-Sansó 1989].

Formally the problem becomes

$$
\begin{cases}
\Delta u = 0 & \text{in } \Omega \\
u[h(\sigma), \sigma] = v(\sigma) & \text{on } S \\
|\nabla u[h(\sigma), \sigma]| = g(\sigma) & \text{on } S \\
u(r) = 0(r^{-1}) & r \to \infty
\end{cases}
\tag{3.10}
$$

with $u(h, \sigma)$ unknown $(h \geq h(\sigma))$ as well as $h(\sigma)$, i.e. the function that defines the boundary S.

It has to be stressed that implicit in the formulation (3.10) is that the ellipsoid E is given and fixed a priori.

Remark 3.3: The scalar Molodensky's problem is considered nowadays as the most important B.V.P. in geodesy; yet its mathematical analysis is still far from being complete.

One of the most striking features of this problem is that, if we linearize it starting from the purely central field $\tilde{u} = \frac{\mu}{r}$, we get a so-called simple Molodensky's problem which has a null space at least of dimension 3, while the original non linear problem has a null space at most of dimension 1 (cf. ?).

In any way we shall come back to this problem in a quasi-linear version, namely

$$
\begin{cases}
\Delta u = 0 \\
u[h(\sigma), \sigma] = v(\sigma) \\
-\frac{\partial u}{\partial \nu}[h(\sigma), \sigma] = g(\sigma) \\
u = 0(r^{-1}) \ ,
\end{cases}
\tag{3.11}
$$

to illustrate a particular technique called partial Legendre transform

Remark 3.4: The vector gravimetric B.V.P. or vector Molodensky's problem. In this case no information at all is given on the coordinates of boundary points, so that beyond the "altimetric type" information on the potential, more data have to be added to compensate for the lack of horizontal coordinates. For this purpose the astronomical coordinates $\alpha = (\Phi, \Lambda)$ of each point P on S are assumed to be known. These are related to the gravity field through the equation

$$
-\frac{\nabla u}{g} = \begin{bmatrix} \cos \Phi \cos \Lambda \\ \cos \Phi \sin \Lambda \\ \sin \Phi \end{bmatrix}
\tag{3.12}
$$

Since g is known at P, we see that this provides us with the knowledge of the full vector $\underline{g} = \nabla u$ at any point P of S.

So what is known now on S is u and $\underline{g} = \nabla u$; yet what does it mean exactly to know the 4 quantities (u, g_x, γ_y, g_z) (or alternatively (u, g, Φ, Λ)) on the unknown S?

It means precisely that we know that the 4 given numbers refer to the same (unknown) point P; another way of expressing the same thing would be to use $\alpha = (\Phi, \Lambda)$ as a 2-D parameter, maintaining that $v(\alpha), \underline{g}(\alpha)$ are known for every α.

Formally the vector Molodensky's problem writes

$$\begin{cases} \Delta u = 0 & \text{in } \Omega \\ u[\underline{r}(\alpha)] = v(\alpha) & \text{on } S \\ \nabla u[\underline{r}(\alpha)] = \underline{g}(\alpha) & \text{on } S \\ u(r) = 0(r^{-1}) & r \to \infty \end{cases}$$

with $\underline{r} = \underline{r}(\alpha)$ the parametric representation of the unknown boundary S.

This problem is nowadays well analyzed from the mathematical point of view and we shall report some results in the last lecture.

4   The physical surface of the earth

Before starting the study of B.V.P.'s it seems reasonable to ask oneself, what is the boundary in these problems. A first answer might seem reasonable; what we see around as the "ground". This answer however is absolutely wrong; what we can call the visible topographic surface, T, first of all includes a number of objects like trees, buildings etc which have such a low density that they are irrelevant from the gravimetric point of view, or better they appear as a highly oscillating, low intensity noise in the gravimetric signal.

Moreover, even going down to the floor so to say, if we try to get a realistic digital model of the terrain we have at least to reach the 10 m horizontal resolution level.

How can we hope to get this resolution level with gravity (and potential) data that in the best areas can reach a density of one measurement per kilometer? The answer is that we simply cannot.

So we must imagine that the boundary is "some relatively smooth surface" for instance going through the measurement points or wandering close to them, in such a way that we can reduce the measurements to S without introducing model errors, large with respect to noise.

However if we move the boundary S away from T we loose the harmonicity of the potential u, because S will be partly buried in the masses.

Furthermore, and maybe even more important, the gravity signal S due to exceeding or lacking masses will be strongly oscillating, though S is smoother than T (cfr. Fig. 2).

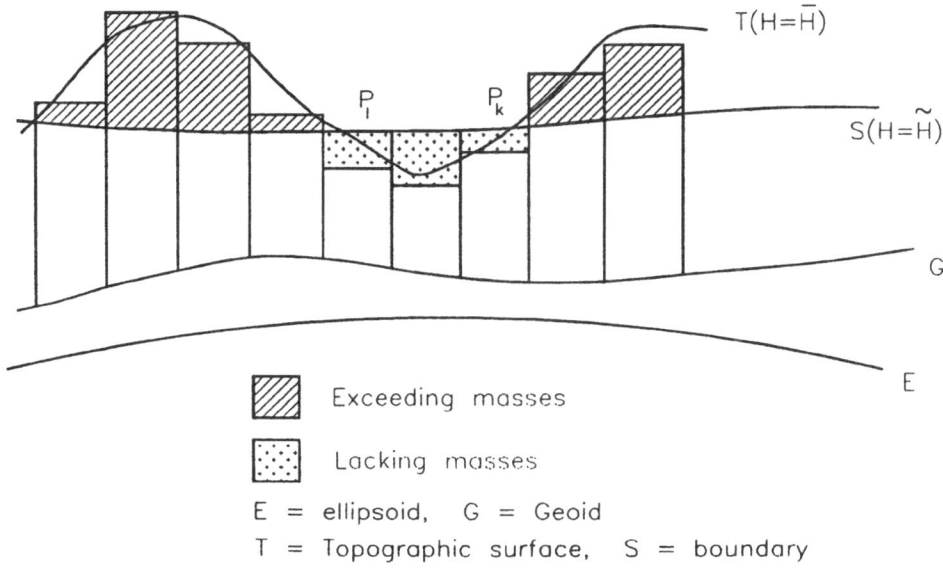

Fig. 2

The conclusion is that correcting the gravity signal at measurement points by the effect of masses exceeding or lacking, according to the relative position of T and S, will strongly smooth the gravity signal along S, therefore making the observation density compatible with a continuous model, namely with a B.V.P.. As a matter of fact, using the theory of the first lecture, it is easy to see that if we want to get a 5 mGal signal on a 10 Km wavelenghts with measurements with $\sigma_\nu \sim 0,1$ mGal, it is enough to have one observation every $1 \div 1,5$ Km; but this is true only if we move on a flat land. In fact any height variation between two measurement points, which can easily amount to 100 m can produce a much stronger signal oscillation due to the pure monopole term $\frac{\mu}{r^2}$; with 100 m in fact we have ~ 30 mGals of variation of $\frac{\mu}{r^2}$, which would completely hide the "horizontal" high frequency variation of 5 mGals.

It is interesting to observe that to perform the corrections we must know at least some average height of the terrain (DTM), for instance in terms of orthometric heights, as well as the height of the observation points.

In this way in fact we come to define somehow the difference between T and S and it is through this difference that is implicitly defined S, which, we shouldn't forget, is one of our unknowns.

Summarizing the solution process goes as follows:

- we define differences $T - S$ through a digital terrain model (this implicitly defines what is S), we correct observations by a residual terrain correction and then we interpolate them on S,

- we solve a free boundary B.V.P. thus recovering the surface S and the gravity potential u on it,

- we restore the effect of the residual terrain correction and thus we find our solution u throughout the space.

What we want to underline here as conclusion of the paragraph is that this remove-restore procedure is not only a practical tool to perform computations, but also a very essential tool on a logical ground, without which we cannot consistently arrive to state our problem in terms of a B.V.P.

5 Errors from the DTM

We want to verify, in this paragraph, that realistic data in the form of a DTM can be provided, allowing to perform the operations described in § 2, without increasing too much the noise in g, that we have placed at the 0.1 mGal level; also we want to be sure that a fairly dense gravity material with this kind of accuracy will produce a good knowledge of the potential u, or, what amounts to be the same, of the geoid N.

Finally we have to verify that the restored potential will not affect too badly the estimate of the geoid.

All computations will be done here in planar approximation, because we are interested only in the order of magnitude of the errors and not in their precise calculation.

So, referring to Fig. 2 for symbols, we will need the following formulas

$$A_T(P) = \int dS_{Q0} \int_{\tilde{H}(Q_0)}^{\overline{H}(Q_0)} \mu \frac{z - H}{[r^2 + (z - H)^2]^{\frac{3}{2}}} dz$$

(5.1)

$$(H = H(P_0)), \quad r = |P_0 - Q_0|,$$

$$u_T(P) = \int dS_{Q0} \int_{\tilde{H}(Q_0)}^{\overline{H}(Q_0)} \mu \frac{1}{[r^2 + (z - H)^2]^{\frac{1}{2}}} dz$$

(5.2)

$$N(P) = \frac{u(P_0)}{\gamma} = \frac{1}{\gamma} \int dS_{Q0} \frac{g_z(Q_0)}{r} ,$$

(5.3)

where (5.1) and (5.2) express respectively the attraction and the potential due to residual topographic masses, i.e. those between T and S, while (5.3) is a purely planar approximation of Stokes formula, which however is suitable for our purposes.

Let us start now from A_T; we want to understand what is the influence of a noise in the knowledge of $\overline{H}(Q_0)$ on its computation, so that, if η is the error in $\overline{H}$ we can use the linearized formula

$$\delta A_T(P) = \int dS_{Q0} \mu \eta \frac{\Delta H}{[r^2 + \Delta H^2]^{\frac{3}{2}}}$$

(5.4)

$$(\Delta H = \overline{H}(Q_0) - H(P_0)) .$$

Here, following the geodetic tradition, it is preferable to split the computation of (5.4) into two parts, the inner zone and the outer zone. This is also conceptually correct because for the outer zone we can consider η as the error of the block averaged terrain model (e.g. with blocks of $100 \times 100 m^2$), while in the inner zone, practically inside a single block, we have to consider the true topographic error with respect to the height of the gravity station; in fact, as we know from experimental gravimetry, a large contribution to A_T comes from very close topographic masses.

So let us model the outer zone error as a Wiener measure $D_{\mu_\eta} = (\eta dS)$, such that the average error of a 100 m side block has a fixed amount

$$E\left\{\left[\frac{1}{S} \int \eta dS\right]^2\right\} = E\left\{\left[\frac{1}{S} \int d\mu_\eta\right]^2\right\} = \frac{\sigma_\eta^2}{S} = \sigma_{\overline{H}}^2 ;$$

(5.5)

we shall perform our computation under two hypotheses, $\sigma_H = 1m$ or $\sigma_H = 3m$.

Let us observe that if S has to have size $10^4 m^2$ and we choose a circular shape for it, the radius R of the circle has to be $R \simeq 56m$.

So we find

$$\sigma^2(\delta A_{Tf}) = \mu^2 \sigma_\eta^2 \int_{r>R} dS_Q \frac{\Delta H^2}{[r^2 + \Delta H^2]^3} \quad ; \qquad (5.6)$$

to make it simple, let us assume $\Delta H \sim \frac{r}{\sqrt{2}}$ in $(5.6)^5$, and we find an upper bound

$$\sigma(\delta A_{Tf}) \stackrel{<}{\sim} 0,02 \text{ mGal} \quad \text{for } \sigma_H = 1m$$

$$\sigma(\delta A_{if}) \stackrel{<}{\sim} 0,06 \text{ mGal} \quad \text{for } \sigma_H = 3m .$$

This shows that mean heights can very well be wrong by 3 m but not much more than that.

We come now to the inner zone, extending from P, to 56 m distant points.

We assume that within the inner zone, topography is directly observed, as it is common in practice; therefore we can claim that the r.m.s. of the height error should have an amplitude modulated by the distance of the point Q_0 from P_0.

This could correspond to a model

$$E\left\{ \left[\frac{1}{S} \int_S \eta dS \right]^2 \right\} = E\left\{ \left[\frac{1}{S} \int_S d\mu_\eta \right]^2 \right\} = \frac{\sigma_\eta^2}{S^2} \int_S \left(\frac{r}{R} \right)^2 dS = \frac{\sigma_\eta^2}{2S} = \sigma_H^2 .$$

$$(5.7)$$

Let us make the simple hypothesis $\sigma_H^2 = 0,5m^2$, which amounts to assuming that from our direct measurements we can determine the mean height of the block around P with an accuracy of 70 cm; then from (3.7) we get for the present case

$$\frac{\sigma_\eta^2}{S} = 1m^2 . \qquad (5.8)$$

---

[5] This is in fact the value that maximizes the integrand $\Delta H^2 [r^2 + \Delta H^2]^{-3} = f(\Delta H)$, so that $f(\frac{r}{\sqrt{2}}) = \frac{4}{27} r^{-4}$.

To avoid divergencies, related only to the linear approximation in (5.4), let us further assume that $\eta \equiv 0$ below 1 m of distance from P. Then we have

$$\sigma^2(\delta A_{TI}) = \mu^2 \sigma_\eta^2 \int_{\delta < r < R} dS_Q \frac{\Delta H^2}{[r^2 + \Delta H^2]}$$

$$(\delta = 1m , \quad R = 56m) \; ;$$

(5.9)

putting again $\Delta H \sim \frac{r}{\sqrt{2}}$, from (5.8), (5.9) we get

$$\sigma(\delta A_{TI}) \lesssim 0,05 \; mGal \; ;$$

(5.10)

this figure, in a flatter area, would become much smaller.

As a partial conclusion we can say that both $\sigma(\delta A_{Tf})$ and $\sigma(\delta A_{TI})$ are compatible with the hypothesis that the finally reduced observation G has an error in the range of 0,1 mGal.

We can pass now to evaluate the effect of the DTM errors in the restore term, which is derived in linear approximation from (5.2), as

$$\delta N = \frac{\delta u(P)}{\gamma} = \frac{\mu}{\gamma} \int_d S_Q \eta \frac{1}{[r^2 + \Delta H^2]} \; .$$

(5.11)

Fortunately we don't need in this case do distinguish between inner and outer zone, because, as we will prove, this term is in any way negligible even under pessimistic hypotheses.

So we perform the propagation from (5.11), just stopping the integration at 1 m distance from P_0; moreover we shall assume that η is such that

$$E\left\{\left[\frac{1}{S}\int_S \eta dS\right]^2\right\} = E\left\{\left[\frac{1}{S}\int d\mu_\eta\right]^2\right\} = \frac{\sigma_\eta^2}{S} = \sigma_{\overline{H}}^2 = 10^2 m^2 \; ,$$

so that $\frac{\sigma_\eta}{\sqrt{S}} = \sigma_{\overline{H}} \simeq 10m$, with $S = 10^4 m^2$.

In this case, with $\delta = 1m$,

$$E\left\{\delta N^2\right\} = E\left\{\frac{\delta u_T^2}{\gamma^2}\right\} = \frac{\mu^2 \sigma_\eta^2}{\gamma^2} \int_{r > \delta} dS_Q \frac{1}{[r^2 + \Delta H^2]} \; ;$$

(5.12)

now the integral (5.12) diverges for $r \to \infty$, but this is a typical drawback of the planar approximation and everybody can probably agree that if we limit (5.12) to 20.000 Km (half the diameter of the earth) we are likely to be not too much optimistic.

Moreover, to make it simpler, we use the inequality $\dfrac{1}{r^2} \leq \dfrac{1}{r^2 + \Delta H^2}$ in (5.12); so we find

$$E\{\delta N^2\} \leq \frac{\mu^2 \sigma_\eta^2 2\pi}{\gamma^2} \log \frac{D}{\delta} \quad,$$

$$(D = 2 \cdot 10^7 m \quad, \quad \delta = 1m)$$

i.e. with the numbers above specified

$$\sigma(\delta N) = \sqrt{E\{\delta N^2\}} \lesssim 0,16 \cdot 10^{-3} m \tag{5.13}$$

which is indeed negligible.

So our second partial conclusion is that the remove restore procedure is compatible with errors of the order of 0,1 mGal in gravity and of 1 cm in geoid.

The final question is now whether a 1 Km grid of gravity points with 0,1 mGal noise is sufficient to derive a geoid at the 1 cm error level.

We shall use (5.3), which is valid only for a plane boundary; yet the figure we shall obtain is so good that we believe that the result is indicating in any way a well acceptable situation.

In fact, from (5.3) rewritten as

$$\delta N = \frac{1}{\gamma} \int dS \frac{\delta g}{r} \quad, \tag{5.14}$$

we compute the error propagation by assuming that $\delta g dS = d\mu_g$ is a suitable Wiener measure, i.e. for the mean on a block S of $1 Km^2$ it has a variance of $0,1^2 mGal^2$,

$$E\left\{\left[\frac{1}{S} \int_S \delta G dS\right]^2\right\} = E\left\{\left[\frac{1}{S} \int d\mu_g\right]^2\right\} = \frac{\sigma_g^2}{S} = 0,1^2 mGal^2 \quad;$$

this implies that

$$\sigma_g = 0,1 mGal \times 10^3 m \quad. \tag{5.15}$$

Now we can take the variance of (5.14) but we have choose inferior and superior limits; we put $\delta = 1m$ for the lower bound and $D = 2 \times 10^7 m$ for the upper bound.

The result then is

$$\sigma(\delta N) = \sqrt{2\pi \log \frac{D}{\delta} \frac{\sigma g}{\gamma}} \simeq 10^{-3} m \quad . \tag{5.16}$$

The conclusion of this paragraph is then: with a DTM of $100m \times 100m$ mean heights, known better than a 3 m error, with a 1 Km net of gravity measurements, of the 0,1 mGal error level, we can (implicitly) define a smooth surface S, wandering not too far from measure points, such that we can assume that the gravity modulus is known on S, still with an error of the order of 0,1 mGal, but having considerably smoothed the horizontal variations due to residual topographic masses; the solution of a B.V.P. for S using g as a datum, should lead us to the knowledge of the geoid at an error level better than a centimeter and the addition (restore) of potential of residual topographic masses should lead us back to the (averaged) topographic surface with no loss of accuracy.

Auxilia mathematica

6 The Legendre transform

The Legendre transform is just a clever transform of coordinates, under which some free boundary B.V.P. attains a convenient form.

Remark 6.1: The Legendre transform as applied to geodetic B.V.P.'s is also known as the gravity space approach, for the very simple reason that in 1974, when the author conceived it, he was not aware that this was one further application for the very general method of Legendre transform already well known for instance in mechanics.

Let us then assume that we have a filed u in Ω and a B.V.P.

$$\begin{cases} F(U, \nabla u, u, x) = 0 & \text{in } \Omega \\ G(\nabla u, u, x) = 0 & \text{on } S \end{cases} \tag{6.1}$$

with

$$S = \partial \Omega = \{\underline{x} = \underline{x}(\alpha)\}$$

the boundary of Ω, and

$$U = [u_{ik}] = \left[\frac{\partial^2 u}{\partial x_i \partial x_k}\right] . \tag{6.2}$$

Let us further assume that the vector transform

$$\underline{p} = \nabla u(\underline{x}) = \underline{p}(\underline{x}) \tag{6.3}$$

is biunivocal on Ω and continuous there with its first derivatives, such that the inverse transform

$$\underline{x} = \underline{x}(\underline{p}) \tag{6.4}$$

exists and is unique for $\underline{x} \in \Omega$; moreover $\underline{x}(\underline{p})$ is also continuous with its first derivatives.

Then, we can observe that

$$\frac{\partial p}{\partial x} = U = U^+ = \left(\frac{\partial p}{\partial x}\right)^+ \tag{6.5}$$

as it is obvious from the fact that $\bar{p}$ is the gradient of some function u; therefore we have too

$$\frac{\partial x}{\partial p} = U^{-1} = [U^{-1}]^{+} = \left(\frac{\partial x}{\partial p}\right)^{+} . \qquad (6.6)$$

If Ω is simply connected, then its transform $\Lambda \equiv \left\{p = \nabla u(\underline{x}), \underline{x} \in \Omega\right\}$ is simply connected too and, due to (6.6), we expect that also the transformation (6.4) be of potential type.

Namely we expect that there is an adjoint potential ψ, such that

$$\underline{x} = \nabla_p \psi(\underline{p}) . \qquad (6.7)$$

Now we have to find the relation between Ψ and u, U. As for the second it is simply

$$\frac{\partial x}{\partial p} = \Psi = U^{-1} \qquad (6.8)$$

with

$$\Psi = [\Psi_{ik}] = \left[\frac{\partial^2 \Psi}{\partial p_i \partial p_k}\right] ; \qquad (6.9)$$

as for the relation with u, one finds

$$\frac{\partial u}{\partial p_i} = \frac{\partial u}{\partial x_k}\frac{\partial x_k}{\partial p_i} = p_k \frac{\partial^2 \Psi}{\partial p_i \partial p_k} =$$

$$= \frac{\partial}{\partial p_i}\left(p_k \frac{\partial \Psi}{\partial p_k}\right) - \delta_{ik}\frac{\partial \Psi}{\partial p_k} = \qquad (6.10)$$

$$= \frac{\partial}{\partial p_i}\left\{p_k \frac{\partial \Psi}{\partial p_k} - \Psi\right\} = \frac{\partial}{\partial p_i}\left\{p\frac{\partial \Psi}{\partial p} - \Psi\right\} .$$

From (6.10) one can see that, apart from an inessential constant always present in the definition of a potential, we can put

$$u = p\frac{\partial \Psi}{\partial p} - \Psi \qquad (6.11)$$

Accordingly the problem (6.1) is transformed into

$$\begin{cases} F(\Psi^{-1}, p, p\frac{\partial u}{\partial p} - \Psi, \nabla\Psi) = 0 & \text{in } \Gamma \\[2mm] G(\underline{p}, p\frac{\partial \Psi}{\partial p} - \Psi, \nabla\Psi) = 0 & \text{on } \partial\Gamma ; \end{cases} \qquad (6.12)$$

as it is obvious, when $\underline{p}(\alpha) = \nabla u[\underline{x}(\alpha)]$ is known on the unknown boundary S, as it is the case for the vector problem of Molodensky, we succeed in this way to transform it into a fixed boundary B.V.P.

7   The hodograph-Legendre transform

This is a partial Legendre transform, performed on one of the coordinates only.

This might be useful to treat B.V.P.'s where the boundary is so to say unknown only along one of the coordinates, as it might be the scalar problem of Molodensky, in its quasi-linear form.

Therefore let us assume that the coordinate vector $\underline{x} \equiv \{x_i\} \equiv \{x_\alpha, x_3, \alpha = 1, 2\}$ is transformed according to the law

$$\begin{cases} y_1 = x_1 \\ y_2 = x_2 \\ y_3 = \dfrac{\partial u}{\partial x_3} \end{cases} \tag{7.1}$$

where u is some potential satisfying a problem like (6.1).

Let us further assume that u is continuous with its first two derivatives and that the 3^d of (7.1) is invertible with respect to x_3, everywhere through Ω, providing the function

$$x_3 = x_3(y_1, y_2, y_3) \ . \tag{7.2}$$

We define an adjoint potential v through

$$v = x_3 \frac{\partial u}{\partial x_3} - u = x_3 y_3 - u \ ; \tag{7.3}$$

we have

$$\frac{\partial v}{\partial y_3} = x_3 + y_3 \frac{\partial x_3}{\partial y_3} - \frac{\partial u}{\partial y_3} \ . \tag{7.4}$$

On the other hand, from (7.1) we derive

$$\frac{\partial y}{\partial x} = \begin{bmatrix} 1 & 0 & 0 \\ 0 & 1 & 0 \\ u_{13} & u_{23} & u_{33} \end{bmatrix} \ , \tag{7.5}$$

so that

$$\frac{\partial x}{\partial y} = \left(\frac{\partial y}{\partial x}\right)^{-1} = \begin{bmatrix} 1 & 0 & 0 \\ 0 & 1 & 0 \\ -\dfrac{u_{13}}{u_{3}3} & -\dfrac{u_{23}}{u_{3}3} & \dfrac{1}{u_{23}} \end{bmatrix} . \tag{7.6}$$

Therefore we can compute

$$\frac{\partial u}{\partial y_3} = \frac{\partial u}{\partial x_i}\frac{\partial x_i}{\partial y_3} = \frac{\partial u}{\partial x_3}\frac{\partial x_3}{\partial y_3} = y_3\frac{\partial x_3}{\partial y_3} \tag{7.7}$$

which substituted in (7.4) gives us the fundamental relation

$$\frac{\partial v}{\partial y_3} = x_3 . \tag{7.8}$$

Analogously we have, from (7.3),

$$\frac{\partial v}{\partial y_1} = y_3\frac{\partial x_3}{\partial y_1} - \frac{\partial u}{\partial y_1} ; \tag{7.9}$$

but, using also the shape of $\frac{\partial x}{\partial y}$ in (7.6)

$$\frac{\partial u}{\partial y_1} = \frac{\partial u}{\partial x_1} + \frac{\partial u}{\partial x_3}\frac{\partial x_3}{\partial y_1} = \frac{\partial u}{\partial x_1} + y_3\frac{\partial x_3}{\partial y_1} ,$$

which substituted back in (7.9) provides

$$\frac{\partial v}{\partial y_1} = -\frac{\partial u}{\partial x_1} . \tag{7.10}$$

In the same way we prove that

$$\frac{\partial v}{\partial y_2} = -\frac{\partial u}{\partial x_2} . \tag{7.11}$$

Furthermore from (7.8) we see that we can write

$$\frac{\partial x}{\partial y} = \begin{bmatrix} 1 & 0 & 0 \\ 0 & 1 & 0 \\ v_{13} & v_{23} & v_{33} \end{bmatrix} , \tag{7.12}$$

which composed with (7.6) gives

$$\begin{cases} u_{33} = \dfrac{1}{v_{33}} \\[3mm] u_{13} = -\dfrac{v_{13}}{v_{33}} \\[3mm] u_{23} = -\dfrac{v_{23}}{v_{33}} \ . \end{cases} \tag{7.13}$$

Finally, for any two indexes $1 \leq \alpha, \beta \leq 2$, also recalling (7.10), (7.11), we can write

$$\frac{\partial^2 u}{\partial x_\alpha \partial x_\beta} = -\frac{\partial}{\partial x_\alpha} v_\beta = -\frac{\partial v_\beta}{\partial y_\alpha} - \frac{\partial v_\beta}{\partial y_3} u_{\alpha 3} = -v_{\beta\alpha} + \frac{v_{\beta 3} v_{\alpha 3}}{v_{33}} \ . \tag{7.14}$$

Therefore the equation (7.14) rewritten in the form

$$u = x_3 y_3 - v = y_3 \frac{\partial v}{\partial y_3} - v \ , \tag{7.15}$$

the equation

$$\frac{\partial u}{\partial x_3} = y_3 \ , \tag{7.16}$$

and the equations (7.10), (7.11), (7.13), (7.14) allow to express x_i, u, u_i, u_{ik} in terms of y_i, v, v_i, v_{ik} so that the transformation process is completed.

8   Holder spaces and Schander estimates for the Dirichlet problem

We remember first of all that $u \in C^k(\overline{\Omega})$ means that u is continuous on the closed set $(\overline{\Omega})$ together with its first k derivatives, so that the norm

$$|u|_k = \operatorname*{Sup}_{x \in \overline{\Omega}} \left(|u(x)| + |Du(x)| + ... + |D^k u(x)| \right) \tag{8.1}$$

results to be bounded.
Here we mean

$$Du(x) = [\partial_i u(x)], \quad D^2 u(x) = [\partial_{ik} u(x)]$$

etc.

Also a Holder norm of exponent $0 < \lambda \leq 1$ is defined as

$$|u|_\lambda = |u|_0 \operatorname*{Sup}_{x,y \in \overline{\Omega}} \frac{|u(x) - u(y)|}{|x - y|^\lambda} \quad ; \tag{8.2}$$

moreover for any integer k we can put

$$|u|_{k,\lambda} = |u|_k + |D^k u|_\lambda \quad , \tag{8.3}$$

the corresponding Holder space being usually indicated as $C^{k,\lambda}(\overline{\Omega})$. We arrange all these Holder spaces in a sequence

$$H_a = C^{k,\lambda} \, , \quad a = k + \lambda \quad ; \tag{8.4}$$

we note explicitly that with this definition $H_k \neq C^{k}$, for integer k. Let us now consider the regularity of surfaces in R^3, like $S = \partial\Omega$; we say that $S \in H_a$ if S can be split into overlapping patches $\{S_j\}$, $S = \bigcup_{j=1}^u S_j$, such that any point $x \in S$ is internal at least to one S_j, and there are n suitable reference systems such that S_j in the j-th system has equation

$$x_3 = f_j(x_1, x_2) \tag{8.5}$$

with (x_1, x_2) varying over the bounded set B_j corresponding to the projection of S_j on the (x_1, x_2) plane, and moreover

$$f_j \in H_a(B_j) \quad .$$

We say that a functions $g(x)$ defined on $S \in H_a$, is in $H_b(S)$, $b \leq a$, if

$$g[x_1, x_2, f_j(x_1, x_2)] \in H_b(B_j) \quad . \tag{8.6}$$

In what follows we will also need weighted Holder spaces. Given a closed set $\overline{\Omega}$ and calling $B_{x,\delta} \equiv \{y; \, |y - x| \leq \delta\}$ the ball of center x and radius δ, we define

$$\overline{\Omega}_\delta \equiv \{x \in \overline{\Omega}; \, B_{x,\delta} \subset \overline{\Omega}\} \quad .$$

Then we put, for a and b such that $a + b \geq 0$,

$$|u|_{a,\overline{\Omega}}^{(b)} = |u|_a^{(b)} = \operatorname*{Sup}_{\delta > 0} \delta^{a+b} |u|_{a,\overline{\Omega}_\delta} \quad ; \tag{8.7}$$

when this norm is finite for a function u, we say that $u \in H_a^{(b)}(\overline{\Omega})$.

Lemma 8.1: The norm (8.7) is equivalent to the norm

$$|\overline{u}|_a^{(b)} = \operatorname*{Sup}_{x \in \overline{\Omega}} d_x^{a+b}|u(x)| + \operatorname*{Sup}_{x,y \in \overline{\Omega}} d_{x,y}^{a+b} \frac{|u(x) - u(y)|}{|x - y|^a}$$

with

$$0 < a < 1 , \quad d_x = \operatorname{dist}(x, \partial\Omega) , \quad d_{x,y} = \operatorname{Sup} d_x, d_y .$$

Let us list some of the properties of weighted Holder spaces:

- $$H_a^{(-a)} = H_a \tag{8.8}$$

- $$|u|_a^{(b)} \leq c|u|_a^{(b^1)} , \quad b > b^1 \tag{8.9}$$

- $$|u|_{a^1}^{(b)} \leq c|u|_a^{(b)} , \quad a > a^1 > 0 \quad a^1 + b > 0 \tag{8.10}$$

- $$|u|_a^{(b)} \leq c \left(|u|_{a_1}^{(b_1)}\right)^t \left(|u|_{a_2}^{b_2}\right)^{1-t} \tag{8.11}$$

$$(a, b) = t(a_1, b_1) + (1 - t)(a_2, b_2) .$$

Moreover if $a^1 + b^1 = a + b$ and $a^1 < a$, the embedding of $H_a^{(b)}$ in $H_{a^1}^{(b^1)}$ is compact.

Now we consider the Dirichlet problem for the Laplace operator in a bounded domain Γ

$$\begin{cases} \Delta u = f & \text{in } \Gamma \\ u|_{\partial\Gamma} = g \end{cases} \tag{8.12}$$

and we ennunciate two theorems; the proof of the first can be found in any textbook on elliptic partial differential equations like [Miranda], [Gilbarg-Trudinger], while the second is in the second textbook only.

Theorem 8.1: Assume Γ is an H_a $(a = k + \lambda, 0 < \lambda < 1)$ domain, and $f \in H_{a-2}(\Gamma)$, $g \in H_a(\partial\Gamma)$, then (8.12) has one and only one solution in $H_a(\partial\Gamma)$ with

$$|u|_a \le C(|f|_{a-2,\Gamma} + |g|_{a,\partial\Gamma}) \quad . \tag{8.13}$$

We will need this classical estimate for $a = 2 + \lambda$.

Theorem 8.2: Let Γ be a bounded domain in H_c, $c \ge 1$ and let a, b be such that

$$0 < b \le a \quad , \quad a > 2 \quad , \quad b \le c \tag{8.14}$$

and assume that $f \in H_{a-2}^{(2-b)}(\Gamma)$ and $g \in H_b(\partial\Gamma)$, then (8.12) has one and only one solution in $H_a^{(-b)}$ satisfying the inequality

$$|u|_a^{(-b)} \le C \left(|f|_{a-2,\Gamma}^{(2-b)} + |g|_{b,\partial\Gamma} \right) \quad . \tag{8.15}$$

Again we will use this theorem for $a = 2 + \lambda$, $0 < \lambda < 1$.

9 On the local solvability of non-linear functional equations

Let $F(x) = y$ be a functional transformation between a Banach space X and another Banach space Y and assume we know that

$$y_0 = F(x_0) \quad ; \tag{9.1}$$

we want to address the question whether the equation

$$F(x) = y \tag{9.2}$$

has one and only one solution x in a neighborhood of x_0, when y is ranging in a convenient neighborhood of y_0.

Let us first of all recall the definition of Frechet derivative; we say that the linear bounded operator A, $X \to Y$, is the Frechet derivative of $F(x)$ at x_0 if

$$|F(x) - F(x_0) - A(x - x_0)|_Y = o(|x - x_0|_X) \quad . \tag{9.3}$$

In this case we say that

$$A = F'(x_0) \quad ; \tag{9.4}$$

we say that $F'(x)$ is continuous in x in a set $B \subset X$, if

$$|F'(x) - F'(z)| \to 0 \quad , \quad z \to x \quad , \quad z, x \in B \quad ; \tag{9.5}$$

the norm in (9.5) is the usual operator norm.

We aim now at proving the following well-known theorem of non linear analysis.

Theorem 9.1: Let us assume that $F(x)$ is Frechet differentiable is a neighborhood of x_0 and that $F'(x)$ is continuous there; furthermore assume that $k = F'(x_0)^{-1}$ is a bounded operator, $Y \to X$; then there are two numbers ρ, δ such that $\forall y \in B_{y_0,\delta}$ there is one and only one solution x of (9.2) i $B_{x_0,\rho}$.

In fact let us put

$$r_0(\xi) = F(x_0 + \xi) - F(x_0) - F'(x_0)\xi \ ;$$

indeed we have

$$r_0(o) = 0 \ , \quad r_0'(\xi) = F'(x_0 + \xi) - F'(x_0) \ , \tag{9.6}$$

$$\begin{cases} |r_0(\xi)| \le \epsilon(\xi)|\xi| \\ \\ \epsilon(\xi) \to 0 \quad \text{for } \epsilon \to 0 \end{cases} \tag{9.7}$$

$$|r_0'(\xi)| = \omega(\xi) \to 0 \quad \text{for } \epsilon \to 0 \ . \tag{9.8}$$

We now transform the equation (9.2) into

$$F(x_0 + \xi) - F(x_0) - F'(x_0)\xi + F'(x_0)\xi = y - y_0 = \eta$$

which can also be written as

$$\xi = K\eta + Kr_o(\xi) = T(\xi) \ ; \tag{9.9}$$

the equation (9.9) is in the form of a fixed point equation, to which we will apply the theorem for contractions.

First of all let us note that, setting $|K| = C$, $|K\eta| < C\delta$

$$\xi' = T(\xi) \Rightarrow |\xi'| \le C\delta + C\epsilon(\xi)|\xi| \ ; \tag{9.10}$$

so, on condition that $|\xi| < \bar{\rho}$ with suitable $\bar{\rho}$,

$$C\epsilon(\xi) \le q \le 1 \ ; \tag{9.11}$$

therefore $\forall \rho \le \bar{\rho}$, if we have

$$\delta \le \rho \frac{(1-q)}{C} \qquad (9.12)$$

we also have

$$|\xi'| \le \rho \ .$$

This means that $T(\xi) = K\xi - Kr_0(\xi)$ transforms the ball $B_{\theta,\rho}$ into itself if $\rho \le \bar{\rho}$ and $|\eta| \le \delta$, satisfying (9.11), (9.12). Moreover

$$|T(\xi) - T(\xi')| \le |kr_0(\xi) - kr_0(\xi')| \le C|r_0(\xi) - r_0(\xi')| \ ; \qquad (9.13)$$

on the other hand we can put

$$r_0(\xi) = r_0(\xi') + r_0'(\xi')(\xi - \xi') + \sigma(\xi, \xi')$$

with

$$\frac{|\sigma(\xi, \xi')|}{|\xi - \xi'|} = \zeta(\xi, \xi') \to 0 \ \text{ for } \xi \to \xi' \ , \qquad (9.14)$$

so that

$$|r_0(\xi) - r_0(\xi')| \le \{|r_0'(\xi')| + \zeta(\xi, \xi')|\xi - \xi'|\} \le \{\omega(\xi') + \zeta(\xi, \xi')\} |\xi - \xi'| \ . \qquad (9.15)$$

Considering (9.8) and (9.14) it is obvious that for $\rho \le \tilde{\rho}$, $\xi, \xi' \in B_{0,\rho}$ we can put

$$C \{\omega(\xi') + \zeta(\xi, \xi')\} \le q < 1 \ . \qquad (9.16)$$

Accordingly for $\rho \le$ min $(\bar{\rho}, \tilde{\rho})$; $T(\xi)$ transforms the ball $B_{0,\rho}$ into itself and it is a contraction on this set.

Therefore there is a unique fixed point, i.e. a unique solution of (9.9) which can be attained by a simple iteration scheme, starting from $\xi_0 = K\eta$ if $|\eta| \le \delta$ and δ satisfies (9.12); furthermore the solution ξ depends continuously on $K\eta$ in X, i.e. on $\eta \in Y$.

Corollary 9.1: Assume you have an equation of the form

$$u = h + GQ(u, u) \qquad (9.17)$$

where $u \in X$, $Q(u, v)$ is a bilinear operator from $X \to Y$ such that

$$|Q(u, v)|_Y \leq C|u|_X|v|_X \quad , \qquad (9.18)$$

G is a continuous linear operator $Y \to X$,

$$|G| \leq C \qquad (9.19)$$

and $h \in X$ is in $B_{0,\delta}$ for a suitably small δ; than (9.17) has one and only one solution in $C_{0,\delta}$.

In fact clearly for $h = 0$ we have $u = 0$; moreover

$$F(u) = u - GQ(u, v)$$

so that

$$F'(u)v = v - GQ(u, v) - GQ(v, u) \quad . \qquad (9.20)$$

Now clearly

$$|F'(u_1)v - F'(u_2)v|_X \leq |GQ(u_1 - u_2, v)|_X + |GQ(v, u_1 - u_2)| \leq 2C'^2|u_1 - u_2|_X|v|$$

so that

$$|F'(u_1) - F'(u_2)| \leq 2C'^2|u_1 - u_2|$$

and the Frachet differential $F'(u)$ is continuous everywhere. Furthermore

$$F'(0) = I$$

which is invertible with bounded inverse.

Then we can apply Theorem 9.1 and the Corollary 9.1 is proved.

This corollary, which is fairly particular being referred to an equation of the form (9.17), is however very useful in the context of the next Lecture.

The analysis of geodetic B.V.P.'s

10   The vector Molodensky's problem and the gravity space approach

The vector Molodensky's problem has been defined in Lecture II at the end of §1; to repeat, in the present formulation, it consists in determining the unknown gravity potential u and the earth's surface S, knowing that

$$\begin{cases} \Delta u = 0 & \text{in } \Omega \\ u|_S = u_0 & \text{on } S \\ \nabla u|_S = \underline{g} & \text{on } S \\ u = 0(r^{-1}) & r \to \infty \ . \end{cases} \tag{10.1}$$

In order of giving to the boundary conditions a more understandable form one can use $\alpha = (\Phi, \Lambda)$, the polar angles of the vector $-\underline{g}$, as parameters used to describe the unknown S in the form

$$\underline{r} = \underline{r}(\alpha) \ ; \tag{10.2}$$

accordingly one will write on S

$$\begin{cases} u[\underline{r}(\alpha)] = u_0(\alpha) \\ \\ \nabla u[\underline{r}(\alpha)] = \underline{g}(\alpha) \end{cases} \tag{10.3}$$

with $u(\alpha), g(\alpha)$ given functions.

This problem has been analyzed by $H\ddot{o}$rmander [[5]], using a hard inverse function theorem and by Sansó, using a Legendre transform, i.e. what is called the gravity space approach.

We follow here the second approach, which has yielded superior results. The first feature we immediately recognize in (10.1) is its lack of uniqueness. In fact if $\{\underline{r}(\alpha), u(\underline{x})\}$ is a solution of (10.1), (10.3) then so is

$$\underline{r}'(\alpha) = \underline{r}(\alpha) + \underline{t} \ , \quad u'(\underline{x}) = u(\underline{x} - \underline{t}) \tag{10.4}$$

for any constant (translation) vector $\underline{t}$.

This translation invariance is here related to the fact that we have eliminated the rotational part of the gravity field, which would cause indeed an impossibility of moving the field around, parallel to the equatorial plane, without changing the centrifugal terms.

To avoid non-uniqueness one has to introduce extra conditions to be imposed on u; these conditions should break the translational invariance. Traditionally one can impose three conditions which are well-known to be equivalent to placing the barycentre at the origin of the coordinates, namely we substitute the last of (10.1) with the more stringent condition

$$u \sim \frac{\eta}{r} + 0(r^{-3}) \text{ for } r \to \infty \qquad (10.5)$$

where η is physically known to equal the Newton's gravitational constant k, times the mass of the earth M.

One must consider that since M is not among the data of the problem, η is unknown too, so (10.5) has to be understood as: " *there is a constant η such that $u - \frac{\eta}{r} = o(r^{-3})$* ".

Since now we are imposing three more constraints, we should consider introducing more degrees of freedom in order to keep the balance between unknowns and equations.

A good choice in this sense has proven to be that one boundary condition, namely $u|_S = u_0(\alpha)$, be substituted with

$$u|_S = u_0(\alpha) - \underline{a} \cdot \underline{g}(\alpha) \quad ; \qquad (10.6)$$

in this way 3 more unknowns, the components of $\underline{a}$, have been introduced.

If we can reach a theorem of existence and uniqueness in the unknowns $u, S, \underline{a}$, with the addition of (10.5), (10.6) we can always say, about the original problem (10.1), that it admits one and only one solution, satisfying (10.5), if in the original solution we find $\underline{a} = 0$, which is an indirect way of imposing three conditions to the data.

Now the gravity space approach to (10.1), (10.5), (10.6) is just to apply the Legendre transform to this problem.

In this case the vector $\underline{p} = \nabla u(\underline{x})$, is conveniently called

$$\underline{g} = \nabla u(\underline{x}) \quad ; \qquad (10.7)$$

moreover, considering that here we have the purely gravitational potential, we expect $\underline{g} \to 0$ when $\underline{x} \to \infty$, so that the outer domain Ω with (unknown) boundary S, is mapped by (10.7) into the inner domain G with known boundary

$$\sum = \partial G \equiv \left\{ \underline{g} = \underline{g}(\alpha) \right\} \; ; \tag{10.8}$$

the origin of the gravity space $\underline{g} = 0$, has to lie inside G.
When we take (6.12) into account we see that $\psi = \psi(\underline{g})$ has to satisfy
the equation

$$\Delta u = TrU = Tr\psi' = 0 \quad \text{in } G \tag{10.9}$$

and also the boundary condition

$$\underline{g} \cdot \nabla \psi - \psi|_{\sum} = g \frac{\partial \psi}{\partial g} - \psi|_{\sum} = u_0(\alpha) - \underline{a} \cdot \underline{g}(\alpha) \; , \tag{10.10}$$

as it derives from (10.6). As we see (10.10) is an oblique derivative
problem, and if we assume that $\sum$ is a starshaped surface, i.e. that
when $\alpha \equiv (\psi, \Lambda)$ (polar angles of $-\underline{g}$) sweeps the unit sphere, $\underline{g}(\alpha)$
sweeps the surface $\sum$, in a one to one correspondence, with the function
$\underline{g}(\alpha)$ sufficiently regular (e.g. C^1), then $\sum$ should nowhere be tangent
to $\underline{g}$, and the oblique derivative problem is of regular type.
For instance we can expect it to satisfy the Fredholm alternative[6] [cf.
Miranda], while a singular oblique derivative problem does not.
Now we have to translate condition (10.5); since $g = 0(r^{-2})$, i.e. $r^{-1} = o(g^{\frac{1}{2}})$, we expect that

$$u(\underline{g}) = g \frac{\partial \psi}{\partial g} - \psi = \mu^{\frac{1}{2}} g^{\frac{1}{2}} + 0(g^{\frac{3}{2}}) \tag{10.11}$$

when $r \to \infty$, $g \to 0$; but how do we expect that ψ behaves close to
$g = 0$?
Since (10.11) is linear and since

$$\left(g \frac{\partial}{\partial g} - 1 \right) g^\lambda = (\lambda - 1) g^\lambda \; , \tag{10.12}$$

we see that

$$\psi = -2\mu^{\frac{1}{2}} g^{\frac{1}{2}} + 0(g^{\frac{3}{2}}) \; , \tag{10.13}$$

i.e. the lack of terms $0(g)$ in ψ seems to translate the condition of
uniqueness, namely that the origin is placed at the barycentre.

---

[6] In this context, this means that under conditions of uniqueness the existence
is also guaranteed.

As a matter of fact if ψ is a solution of our problem, than

$$S \equiv \left\{ \underline{x}(\alpha) = \nabla_g \psi[\underline{g}(\alpha)] \right\} \quad , \tag{10.14}$$

so that if we take $\overline{\psi} = \psi + \underline{b} \cdot \underline{g}$ ($\underline{b}$ constant) we find

$$\overline{S} = \left\{ \nabla_g \overline{\psi} \right\} = S + \underline{b} \quad ;$$

moreover

$$\overline{\psi} = \psi \Rightarrow Tr\overline{\psi}^{-1} = Tr\psi^{-1} = 0$$

and also, due to (10.12),

$$g\frac{\partial}{\partial g}\overline{\psi} - \overline{\psi} = g\frac{\partial \psi}{\partial g} - \psi = u(\underline{g}) \quad .$$

This shows that $\overline{\psi}$ is a solution of the same B.V.P. in gravity space as ψ but the two solutions in the geometry space give rise to two translated solutions $\overline{S} = S + \underline{b}$, $\overline{u}(\underline{x}) = u[\underline{g}(\underline{x} - \underline{b})]$, since now $\underline{g}(\underline{x})$ is derived from $\underline{x} = \nabla\overline{\psi} = \nabla\psi(\underline{g}) + \underline{b}$.

Of course in the family $\overline{\psi} = \psi + \underline{b} \cdot \underline{g}$ only one function satisfies (10.13), i.e. the condition

$$\nabla_g \left\{ \psi + 2\mu^{\frac{1}{2}} g^{\frac{1}{2}} \right\} |_{g=0} = 0 \quad . \tag{10.15}$$

To summarize we are looking for a solution of the problem

$$\begin{cases} Tr\psi^{-1} = 0 & \text{in } G \\ g\frac{\partial \psi}{\partial g} - \psi = u_0(\alpha) - \underline{a} \cdot \underline{g}(\alpha) & \text{on } \Sigma \\ \psi = -2\mu^{\frac{1}{2}} g^{\frac{1}{2}} + 0(g^{\frac{3}{2}}) & g \to 0 \quad ; \end{cases} \tag{10.16}$$

a solution of (10.16) is a couple $\psi, \underline{a}$ and once we have it we find a solution of the original problem as

$$\underline{x} = \nabla_g \psi(\underline{g}) \tag{10.17}$$

$$u(\underline{x}) = g\frac{\partial \psi(\underline{g})}{\partial g} - \psi(\underline{g}) \quad , \tag{10.18}$$

where $\underline{g} = \underline{g}(\underline{x})$ is derived by inverting (10.17) and then inserted into (10.18).

11   Several transformations of the basic problem

For our purposes it is convenient to make the following transformation of variables:

$$
\begin{cases}
\gamma_i = g^{\frac{1}{2}} g_i \\[2mm]
v = \frac{1}{2}\gamma^{-1}\left[\gamma\frac{\partial\psi}{\partial\gamma} - 2\psi - 2 \sim mu^{\frac{1}{2}}\gamma\right] \quad .
\end{cases}
\tag{11.1}
$$

Under the first of (11.1) the starshaped domain G is transformed into another starshaped domain Γ, containing the origin $\gamma = 0$.

The constant $\mu^{\sim\frac{1}{2}}$ is an approximation of $\mu^{\frac{1}{2}}$ which allows us to consider v a small quantity; in fact $\gamma\frac{\partial\psi}{\partial\gamma} - 2\psi = 2\mu^{\frac{1}{2}}\gamma + \gamma\frac{\partial\delta\psi}{\partial\gamma} - 2\delta\psi$ because we can always put

$$
\psi = -2\mu^{\frac{1}{2}}g^{\frac{1}{2}} + \delta\psi \quad ,
$$

the first term representing the purely spherical adjoint potential. As we already said, we cannot pretend to know μ, but we can always assume to know it approximately, leading to a convenient perturbation scheme for v.

To compute the matrix

$$
v = \left[\frac{\partial^2 v}{\partial\gamma_i\partial\gamma_k}\right]
\tag{11.2}
$$

is a lengthy, but not difficult, exercise, leading to the relation

$$
\psi = \left[\frac{\partial^2\psi}{\partial g_i\partial g_k}\right] = -\frac{\tilde{\mu}^{\frac{1}{2}}}{\gamma^3}\left(I - \frac{3}{2}P_\gamma\right) +
$$
$$
+\frac{v}{\gamma^3}\left(I - \frac{3}{2}P_\gamma\right) + \frac{2}{\gamma^2}\left(I - \frac{1}{2}P_\gamma\right)\int_0^\gamma v d\gamma \left(I - \frac{1}{2}P_\gamma\right)
\tag{11.3}
$$

where

$$
P_\gamma = \left[\frac{\gamma_i\gamma_j}{\gamma^2}\right] \quad .
$$

Now we can observe that from the identity

$$Tr\psi^{-1} = \frac{1}{det\psi}\left\{(Tr\psi)^2 - Tr\psi^2\right\}$$

we find that the field equation for the adjoint potential can be written

$$Tr\psi^2 - (Tr\psi)^2 = 0 \ . \tag{11.4}$$

Substituting (11.3) into (11.4) and simplifying one gets

$$2\int_0^\gamma TrV d\gamma \ = \tilde{\mu}^{-\frac{1}{2}}\{-2v\int_0^\gamma TrV d\gamma + 4\gamma[\left(Tr\left(I - \frac{3}{4}P_\gamma\right)\int_0^\gamma V d\gamma\right)^2 +$$

$$-Tr\left(\left(I - \frac{3}{4}P_\gamma\right)\int_0^\gamma V d\gamma\right)^2]\} \ ;$$

$$\tag{11.5}$$

finally differentiating along the radius γ we find

$$\Delta v \quad = \tilde{\mu}^{-\frac{1}{2}}Q(v,v)$$
$$Q(u,v) \quad = -Q_1(u,v) - Q_2(u,v) + Q_3(u,v) - Q_4(u,v) + Q_5(u,v) - Q_6(u$$
$$\tag{11.6}$$

where

$$Q_1(u,v) \ = u\Delta v$$

$$Q_2(u,v) \ = \frac{\partial u}{\partial \gamma}\int_0^\gamma (\Delta v)d\gamma$$

$$Q_3(u,v) \ = 2Tr\left(I - \frac{3}{4}P_\gamma\right)\int_0^\gamma V d\gamma \cdot Tr\left(I - \frac{3}{4}P_\gamma\right)\int_0^\gamma U d\gamma$$

$$Q_4(u,v) \ = 2Tr\left(I - \frac{3}{4}P_\gamma\right)\int_0^\gamma V d\gamma \left(I - \frac{3}{4}P_\gamma\right)\int_0^\gamma U d\gamma$$

$$Q_5(u,v) \ = 4\gamma Tr\left(I - \frac{3}{4}P_\gamma\right)U \cdot Tr\left(I - \frac{3}{4}P_\gamma\right)\int_0^\gamma V d\gamma$$

$$Q_6(u,v) \ = 4\gamma Tr\left(I - \frac{3}{4}P_\gamma\right)U\left(I - \frac{3}{4}P_\gamma\right)\int_0^\gamma V d\gamma \ .$$

$$\tag{11.7}$$

The equation (11.6), with the specification (5.7), is the new field equation holding in Γ.

As for the boundary condition we re-write the second (11.1) as

$$v = g^{-\frac{1}{2}}\left(g\frac{\partial\psi}{\partial g} - \psi - \tilde{\mu}^{\frac{1}{2}}g^{\frac{1}{2}}\right)$$

so recognizing that on $\partial\Gamma$

$$v|_{\partial\Gamma} = g^{-\frac{1}{2}}(\alpha)\left[u_0(\alpha) - \tilde{\mu}^{\frac{1}{2}}g^{\frac{1}{2}}(\alpha - \underline{a}\cdot g(\alpha))\right] = v_0(\alpha) - \underline{a}\cdot\underline{\gamma}(\alpha) \ . \quad (11.8)$$

Finally the condition at the origin becomes

$$\psi = -2\mu^{\frac{1}{2}}\gamma + 0(\gamma^3) \Rightarrow v = \left(\mu^{\frac{1}{2}} - \sim\mu^{\frac{1}{2}}\right) + 0(\gamma^2) \ , \quad (11.9)$$

thus showing that in v, as function of γ, the singularity at the origin has been eliminated; (11.9) can alternatively be written

$$\nabla\, v(0) = 0 \ . \quad (11.10)$$

Summarizing our problem has become to find $v, \underline{a}$ such that

$$\begin{cases} \Delta v = \tilde{\mu}^{-\frac{1}{2}}Q(v,v) & \text{in } \Gamma \\ v = v_0(\alpha) - \underline{a}\cdot\underline{\gamma}(\alpha) & \text{on } \Gamma \\ \nabla v(0) = 0 & \text{at } 0 \end{cases} \quad (11.11)$$

We now perform the last transformation of our problem; namely, assuming $\partial\Gamma$ to be at least of class $H_{1+\epsilon}$, we can be sure that there is a Green operator G, such that

$$\begin{cases} \Delta v = f \\ \\ v = v_0 \end{cases} \Rightarrow \quad v = G_\nu v_0 + Gf \quad (11.12)$$

with suitable regularity properties for v.

In particular $G_\nu v_0$ is just the harmonic function agreeing with $v_0(\alpha)$ on the boundary.

It follows that, since $\underline{a}\cdot\underline{\gamma}$ is harmonic in $\underline{\gamma}$, then

$$G_\nu\underline{a}\cdot\underline{\gamma}(\alpha) \equiv \underline{a}\cdot\underline{\gamma} \ . \quad (11.13)$$

Subsequently (11.11) can be transformed into

$$\begin{cases} v = G_\nu v_0 - \underline{a} \cdot \underline{\gamma} + \tilde{\mu}^{-\frac{1}{2}} GQ(v,v) \\ \nabla v(0) = 0 \end{cases} \tag{11.14}$$

Finally, let us call $h = G_\nu v_o$ and put

$$\nabla_0 f = \nabla f(0) \quad ;$$

then from (11.14), using the first equation in the second, we get

$$\underline{a} = \nabla_0 h + \tilde{\mu}^{-\frac{1}{2}} GQ(v,v)$$

which substituted back in the first, gives

$$v = h - \underline{\gamma} \cdot \nabla_0 h + \tilde{\mu}^{-\frac{1}{2}} [GQ(v,v) - \underline{\gamma} \cdot \nabla_0 GQ(v,v)] \quad . \tag{11.15}$$

This is finally an equation in the unknown v only, which is also of the form studied in the Corollary to Theorem 9.1.

12   Theorems of existence and uniqueness

The following Lemmas will be useful in this paragraph. The proofs are in some cases very simple, in other cases more technical; they can be in any way found in Sansó [[17]] and Sansó [[18]].

Lemma 12.1 If $u, v \in H_\lambda$, $0 < \lambda < 1$, $u,v \in H_\lambda$, i.e.

$$|uv|_\lambda \le C|u|_\lambda |v|_\lambda \quad . \tag{12.1}$$

Lemma 12.2: If Γ is starshaped, $\partial\Gamma$ is $H_{1+\lambda}$, $F \in H_\lambda(\Gamma)$ then

$$|\frac{1}{\gamma} \int_0^\gamma F(S,\alpha) ds|_\lambda \le C|F|_\lambda \quad . \tag{12.2}$$

Lemma 12.3: $\gamma, \frac{\gamma_i \gamma_j}{\gamma} \in H_\lambda$, $0 < \lambda \le 1$.

Lemma 12.4: If $u \in H_\lambda$, $v \in H_\lambda^{(1-2\lambda)}$ then

$$|uv|_\lambda^{(1-2\lambda)} \le C|u|_\lambda |u|_\lambda^{(1-2\lambda)} \quad . \tag{12.3}$$

Lemma 12.5: If $w \in H_\lambda^{(1-2\lambda)}$ then

$$|\int_0^\gamma w(s,\alpha) ds|_\lambda \le C|w|_\lambda^{(1-2\lambda)} \quad . \tag{12.4}$$

We are ready to prove the following theorems.

Theorem 12.1: If $\partial\Gamma$ is in $H_{2+\lambda}$, i.e. $\underline{\gamma} \in H_{2+\lambda}$, and $v_0(\alpha) \in H_{2+\lambda}$, then equation (11.15) has one and only one solution $v \in H_{2+\lambda}(\Gamma)$, on condition that, for a suitably small ρ,

$$|v_0|_{2+\lambda} \le \rho \ . \tag{12.5}$$

Before we get started, let us notice that

$$|\underline{\gamma} \cdot \nabla_0 f|_{2+\lambda} \le C|f|_1 \ , \tag{12.6}$$

so that $u - \underline{\gamma} \cdot \nabla_0 h$ is in a neighborhood of 0 when h is small in $H_{2+\lambda}$. Therefore in order to apply the Corollary to Theorem 9.1, we have only to ascertain that

$$|Q(u,v)|_\lambda \le C|u|_{2+\lambda}|v|_{2+\lambda} \tag{12.7}$$

because in force of the estimates (**??**), G is a bounded operator $H_\lambda \to H_{2+\lambda}$.

But to this aim it is just enough to use Lemmas D1, D2, D3 and the specific shape of $Q(u,v)$. So the theorem is immediately proved.

Theorem 12.2: Let's assume that $\partial\Gamma \in H_c$, and put $a = 2\lambda, b = 1 + 2\lambda, c = 1 + 2\lambda$, with λ small positive, $\lambda < \frac{1}{2}$; let us further assume that $h \in H_b(\partial\Gamma)$ and that it is suitably small

$$|h|_b < \rho \ ;$$

then equation (11.15) has one and only one solution $v \in H_a^{(-b)}(\Gamma)$. In fact for terms like $v\Delta v$ we can use a chain like

$$|v\Delta v|_{\lambda,\Gamma_\delta} \le C|v|_{\lambda,\Gamma_\delta}|v|_{2+\lambda,\Gamma_\delta} \le C|v|_{2+\lambda}^{-1-2\lambda} \cdot |v|_{2+\lambda}^{-1-2\lambda}\delta^{-1+\lambda}$$

entailing

$$|v\Delta v|_{\lambda,}^{1-2\lambda} = Sup\delta^{1-\lambda}|v\Delta v|_{\lambda,\Gamma_\delta} \le C \left(|v|_{2+\lambda}^{-1-2\lambda}\right)^2 \ .$$

For other terms like

$$\left(I - \frac{3}{4}P_\gamma\right)\int_0^\gamma V\,d\gamma$$

we can use majorizations like

$$\left| \left(I - \frac{3}{4} P_\gamma \right) \int_0^\gamma V \, d\gamma \right|_\lambda \leq C \left| \int_0^\gamma V \, d\gamma \right|_\lambda \leq C \cdot |V|_\lambda^{1-2\lambda} \leq C |v|_{2+\lambda}^{-1-2\lambda} \ ,$$

together with Lemma 12.1.

For terms like $Q_5(v,v)$, $Q_6(v,v)$ we can observe, for instance, that

$$|Q_5(u,v)|_\lambda^{1-2\lambda} \ \leq C|\gamma|_\lambda \cdot |Tr \left(I - \tfrac{3}{4} P_\gamma \right) \int_0^\gamma V \, d\gamma|_\lambda \cdot Tr| \left(I - \tfrac{3}{4} P_\gamma \right) |_\lambda \cdot |U|_\lambda^1$$

$$\leq C |v|_{2+\lambda}^{-1-2\lambda} \cdot |u|_{2+\lambda}^{-1-2\lambda} \ ,$$

and similarly for Q_6.

Therefore we come to state that $Q(v,v)$; $H_\lambda^{(-1-2\lambda)} \rightarrow H_\lambda^{(1-2\lambda)}$; however since G; $H_\lambda^{(1-2\lambda)} \rightarrow H_{2+\lambda}^{(-1-2\lambda)}$, we are exactly in the condition to apply the Corollary to Theorem 9.1, and the proof of this Theorem is completed.

Remark 12.1: Before closing the paragraph let us address the problem of the regularity of the final product of our analysis, namely the surface

$$S = \{\underline{x}(\alpha)\} = \left\{ \nabla_g \Psi[\underline{g}(\alpha)] \right\} \ . \tag{12.8}$$

For this purpose we go back to (11.3) and observe that we need to know the regularity of Ψ close to the boundary, i.e. far from the origin: also, if in the right hand side we substitute back $\gamma = g^{\frac{1}{2}}$, we immediately see the relation between Ψ and V in terms of g, and moreover, since

$$\frac{|\underline{g} - \underline{g}'|}{|\underline{\gamma} - \underline{\gamma}'|} = 0(1) \ ,$$

if we stay out of a neighborhood of the origin, the regularity in terms of $\underline{\gamma}$ or in terms of $\underline{g}$ is the same.

So in the hypotheses of Theorem 12.1 we need from (11.3) that $\Phi \in H_\lambda$, i.e. $\nabla \Psi \in H_{1+\lambda}$, so that, owing to (12.8), S is $H_{1+\lambda}$ too.

In the hypotheses of Theorem 12.2, recalling also Lemma 12.4, we find

$$|\Phi|_\lambda \leq C|v|_\lambda + b \left| \int_0^\gamma V \, d\gamma \right|_\lambda \leq C||_{\lambda+2}^{-11-2\lambda} + b|V|_\lambda^{1-2\lambda} \leq C^1 |v|_{-1-2\lambda}^{2+\lambda} \ ,$$

i.e. $\Phi \in H_{2+\lambda}$ again.

This implies again that $\nabla \Phi \in H_{1+\lambda}$, and $S \in H_{1+\lambda}$, what seems a reasonable result.

13 The quasi-linear scalar Molodensky problem

We present the material here more as an example of application of the hodograph-Legendre transform, than as an analysis of a true geodetic B.V.P.; yet, if treated in a rigorous ellipsoidal approximation, this problem becomes an extremely good approximation of the scalar Molodensky problem and as such it might help in understanding this important problem of geodesy.

The problem, mentioned in (3.10), will be now reformulated in a spherical approximation; namely, setting $\alpha = (\Phi, \lambda)$, to find the surface $S \equiv \{r = r(\alpha)\}$ and the potential u, such that

$$
\begin{cases}
\Delta u = 0 & r > r(\alpha) \\[2mm]
u[r(\alpha), \alpha] = u_0(\alpha) & r = r(\alpha) \\[2mm]
-\dfrac{\partial u}{\partial r}[r(\alpha), \alpha] = \gamma_0(\alpha) & r = r(\alpha) \\[2mm]
u = 0(r^{-1}) & r \to \infty \;.
\end{cases}
\tag{13.1}
$$

Due to the particular form of the B.V.P. it becomes natural to use the transformation described in § 2, Lecture III.

So we put

$$
\theta = \theta \quad \lambda = \lambda \quad \text{or } \alpha = \alpha
\tag{13.2}
$$

$$
\rho = \frac{\partial u}{\partial r} \quad r = \frac{\partial v}{\partial \rho}
\tag{13.3}
$$

$$
u + v = \rho r
\tag{13.4}
$$

$$
\nabla_\alpha u = -\nabla_a v
\tag{13.5}
$$

$$
u_{rr} = (v_{\rho\rho})^{-1}
\tag{13.6}
$$

$$
u_{\theta\theta} = -v_{\theta\theta} + \frac{v_{\theta\rho}^2}{v_{\rho\rho}}
\tag{13.7}
$$

$$u_{\lambda\lambda} = -v_{\lambda\lambda} + \frac{v_{\lambda\rho}^2}{v_{\rho\rho}} \ . \tag{13.8}$$

As a matter of fact we prefer to work here with the coordinate

$$\gamma = -\rho \tag{13.9}$$

because this will be always positive and we can interpret the coordinate system (γ, α) as a polar coordinate system in a suitable space.
As a consequence of (13.9) we must have

$$r = -\frac{\partial v}{\partial \gamma} \ , \tag{13.10}$$

while in all other formulas the signs remain unchanged; in particular

$$u + v = \rho r = -\gamma r = \gamma \frac{\partial v}{\partial \gamma} \ . \tag{13.11}$$

So, using all these transformations, we derive

$$\Delta u = \frac{1}{v_{\gamma\gamma}} + \frac{2\gamma}{v_\gamma} + \frac{1}{v_\gamma^2}\left\{ -v_{\theta\theta} + \frac{v_{\theta\gamma}^2}{v_{\gamma\gamma}} - \coth\theta v_\theta - \frac{1}{\sin^2\theta}v_{\lambda\lambda} + \frac{1}{\sin^2\theta}\frac{v_{\lambda\gamma}^2}{v_{\gamma\gamma}} \right\} = 0 \ .$$

Multiplying by $v_\gamma^2 v_{\gamma\gamma}$ and rearranging we get

$$v_\gamma^2 + 2\gamma v_\gamma v_{\gamma\gamma} + \left\{ -v_{\gamma\gamma}\Delta_\alpha v + v_{\gamma\theta}^2 + v_{\gamma\lambda}^2 \right\} = 0 \ , \tag{13.12}$$

where we have used the symbol Δ_α for the Laplace-Beltrami operator in $\alpha = (\theta, \lambda)$.
We can observe that (13.12) admits a purely spherical solution, corresponding to $\tilde{u} = \frac{\mu}{v}$, namely $\tilde{v} = -2\mu^{\frac{1}{2}}\gamma^{\frac{1}{2}}$.
With this in mind we can put

$$v = \tilde{v} + \varphi \tag{13.13}$$

with[7] $\tilde{v} = -2\tilde{\mu}^{\frac{1}{2}}\gamma^{\frac{1}{2}}$, so that φ can be considered as a perturbation.
Substituting into (13.12) we derive the new field equation

$$4\varphi_{\gamma\gamma} + \frac{2}{\gamma}\varphi_\gamma + \frac{1}{\gamma^2}\Delta_\alpha\varphi = 2\tilde{\mu}^{\frac{1}{2}}\gamma^{\frac{1}{2}}Q(\varphi, \varphi) \ , \tag{13.14}$$

---

[7]It has to be recalled once more that μ is not a datum, therefore we can assume to know only an approximate value $\tilde{\mu}$.

where

$$Q(\varphi, \varphi) = \varphi_\gamma^2 + 2\gamma\varphi_\gamma\varphi_{\gamma\gamma} - \varphi_{\gamma\gamma}(\Delta_\alpha\varphi) + \varphi_{\theta\gamma}^2 + \varphi_{\lambda\gamma}^2 \quad : \qquad (13.15)$$

together with (13.14) we must set up some boundary condition, which can be derived from (13.11), namely

$$\gamma\frac{\partial v}{\partial \gamma} - v = u_0(\alpha) \quad . \qquad (13.16)$$

Let me remark that (13.16) has to hold on the known boundary $\gamma = \gamma_0(\alpha)$, which is part of the data.

In analogy with the vector Molodensky problem we find it convenient to modify (13.16) and to put additional conditions on φ at the origin, namely

$$\gamma\frac{\partial \varphi}{\partial \gamma} - \varphi = u_0(\alpha) - \tilde{u}(\alpha) + \underline{a} \cdot \underline{\gamma} \qquad (13.17)$$

and

$$\varphi = \varphi_0\gamma^{\frac{1}{2}} + 0(\gamma^{\frac{3}{2}}) \quad \gamma \to 0 \quad . \qquad (13.18)$$

And, again exploiting the experience gained in § 2, we perform the change of variables

$$\begin{cases} w = 2\gamma^{-\frac{1}{2}}\left[\gamma\frac{\partial \varphi}{\partial \gamma} - \varphi\right] \\ \\ \xi_i = \gamma^{-\frac{1}{2}}\gamma_i \ \ or \ \ \xi = \gamma^{\frac{1}{2}} \ , \quad \alpha = \alpha \end{cases} \qquad (13.19)$$

However, in order to make it more digestible, we split it into two steps: at first we put

$$\begin{cases} \varphi = \xi\chi \\ \xi = \gamma^{\frac{1}{2}} \end{cases} \qquad (13.20)$$

and we get

$$\xi^{-1}\left[\chi_{\xi\xi} + \frac{2}{\xi}\chi_\xi + \frac{1}{\xi^2}\Delta_\alpha\chi\right] = 2\tilde{\mu}^{-\frac{1}{2}}\xi^{-1}Q(\chi, \chi) \qquad (13.21)$$

with

$$Q(\chi, \chi) = \tfrac{1}{4}\left(\chi_\xi + \tfrac{\chi}{\xi}\right)^2 + \tfrac{1}{4}\left(\chi_\xi + \tfrac{\chi}{\xi}\right)\left(\xi\chi_{\xi\xi} + \chi_\xi - \tfrac{\chi}{\xi}\right) +$$
$$- \tfrac{\xi^{-1}}{4}\left(\xi\chi_{\xi\xi} + \chi_\xi - \tfrac{\chi}{\xi}\right)\Delta_\alpha\chi + \tfrac{1}{4}\left(\chi_{\theta\xi} + \tfrac{\chi_\theta}{\xi}\right)^2 + \tfrac{1}{4}\left(\chi_{\lambda\xi} + \tfrac{\chi_\lambda}{\xi}\right)^2 \ . \tag{13.22}$$

Furthermore χ satisfies the boundary condition

$$\xi\chi_\xi - \xi = \frac{2}{3}\left[u_0(\alpha) - \tilde{u}(\alpha) + \xi\underline{a}\cdot\underline{\xi}\right] \tag{13.23}$$

and it has, at the origin, the behaviour

$$\chi = \chi_0 + 0(\xi^2)$$
$$(\chi_0 = \varphi_0 = constant) \ . \tag{13.24}$$

As a second step we put

$$w = \xi\chi_\xi - \chi \ ; \tag{13.25}$$

we then obtain, from

$$\chi = -w_0 + \xi \int_0^\xi \frac{w - w_0}{t^2}dt$$
$$(w_0 = -\chi_0) \tag{13.26}$$

$$\xi\chi_{\xi\xi} = w_\xi$$

$$\chi_\xi = \int_0^\xi \frac{w_\xi}{t}dt$$

$$\nabla_\alpha\chi = \xi \int_0^\xi \frac{\nabla_\alpha w}{t^2}dt \tag{13.27}$$

$$\nabla_\alpha\chi_\xi = \int_0^\xi \frac{\nabla_\alpha w_\xi}{t}dt$$

$$\Delta_\alpha\chi = \xi \int_0^\xi \frac{\Delta_\alpha w_\xi}{t^2}dt \ ,$$

the basic equation

$$\frac{w_\xi}{\xi} + \frac{2}{\xi}\int_0^\xi \frac{w_\xi}{t}dt + \frac{1}{\xi}\int_0^\xi \frac{\Delta_\alpha w}{t^2}dt = 2\tilde{\mu}^{-\frac{1}{2}}Q(w,w) \tag{13.28}$$

where

$$Q(w,w) = \frac{1}{4}\left(\int_0^\xi \frac{w_\xi}{t}dt - \frac{w_0}{\xi} + \int_0^\xi \frac{w-w_0}{t^2}dt\right)^2 +$$
$$+ \frac{1}{4}\left(w_\xi + \frac{w}{\xi}\right)\left(\int_0^\xi \frac{w_\xi}{t}dt + \int_0^\xi \frac{w-w_0}{t^2}dt - \frac{w_0}{\xi}\right) - \int_0^\xi \frac{\Delta_\alpha w}{t^2}dt +$$
$$+ \frac{1}{4}\left(\int_0^\xi \frac{w_{\theta\xi}}{t}dt + \int_0^\xi \frac{w_\theta}{t^2}dt\right)^2 + \frac{1}{4}\left(\int_0^\xi \frac{w_{\lambda\xi}}{t}dt + \int_0^\xi \frac{w_\lambda}{t^2}dt\right)^2 \; .$$

$$(13.29)$$

Multiplying (13.28) by ξ and taking $\frac{\partial}{\partial\xi}$ we finally obtain

$$w_{\xi\xi} + \frac{2}{\xi}w_\xi + \frac{\Delta_\alpha w}{\xi^2} = 2\tilde{\mu}^{-\frac{1}{2}}Q(w,w) + 2\tilde{\mu}^{-\frac{1}{2}}\xi\frac{\partial}{\partial\xi}Q(w,w) = B(w,w)$$

$$(13.30)$$

where $B(w,w)$ is a bilinear form in w, quite clearly bounded for instance from $H_{2+\lambda}$ into H_λ

$$B: \quad H_{2+\lambda} \otimes H_{2+\lambda} \to H_\lambda \; . \qquad (13.31)$$

The only difficulty in establishing (13.31) is the behaviour at the origin, which however is regular because of the chosen condition (13.24), which now becomes

$$w = w_0 + 0(\xi^2) \; . \qquad (13.32)$$

The equation (13.30) is now complemented by the B.V.P. of the Dirichlet type

$$w[\alpha, \xi_0(\alpha)] = \frac{2}{\xi_0(\alpha)}\left[u_0(\alpha) - \tilde{u}(\alpha) + \xi_0(\alpha)\underline{a}\cdot\underline{\xi}_0(\alpha)\right] \; . \qquad (13.33)$$

So we have at the end

$$\begin{cases} \Delta w = B(w,w) \\ w|_{\xi_0(\alpha)} = w_0(\alpha) + 2\underline{a}\cdot\underline{\xi}_0(\alpha) \\ w = w_0 + 0(\xi^2) \end{cases} \qquad (13.34)$$

which is exactly of the same type as the problem studied in the gravity space approach.

The following theorem of existence and uniqueness can then be maintained:

Theorem 13.1: Assume that $H_{2+\lambda}$ data $u_0(\alpha)$, $\gamma_0(\alpha)$ are given and that

$$w_0(\alpha) = \frac{2}{\gamma_0^{\frac{1}{2}}(\alpha)} \left[u_0(\alpha) - 2\tilde{\mu}^{\frac{1}{2}} \gamma_0^{\frac{1}{2}}(\alpha) \right] \qquad (13.35)$$

is conveniently small in $H_{2+\lambda}$, then there is one and only one w in $H_{2+\lambda}$ and a corresponding vector $\underline{a}$, satisfying (13.34).

Remark 13.1: From w, (13.20) and (13.26) we derive

$$\varphi = -w_0 \gamma^{\frac{1}{2}} + \gamma \int_0^{\gamma^{\frac{1}{2}}} \frac{w - w_0}{t^2} dt \qquad (13.36)$$

which is a $H_{2+\lambda}$ function too. So the surface

$$S \equiv \{ r = r(\alpha) \} \quad , \quad r(\alpha) = \frac{\partial}{\partial \gamma} \left[2\tilde{\mu}^{\frac{1}{2}} \gamma^{\frac{1}{2}} - \varphi \right] |_{\gamma = \gamma_0(\alpha)} \qquad (13.37)$$

turns out to be $H_{1+\lambda}$ as for the vector Molodensky problem.

Remark 13.2: An analogous of Theorem 12.2 should hold true in this case too.

We conclude the paragraph by remarking that an analogous theory could be developed for the case that γ is (the negative of) the derivative of u along the ellipsoidal normal, which would lead us much closer to the real situation.

The interested reader is invited to do it by itself.

References

[1] G.A. Backus *Application of a non-linear boundary value problem for Laplace's equation to gravity and geomagnetic intensity survey.* Quart J. Maech. and Appl. math., n. 21, part II, pp. 195-221, 1968.

[2] D. Gilbarg, L. Hörmander *Intermediate Schauder Estimates.* Archive for Rat. Mechanics and Analysis, Vol. 74, 1980.

[3] D. Gilbarg, N.S. Trudinger *Elliptic partial differential equations of second order.* Grundlagen der Math. Wiss., 224, Springfer-Verlag, 1983.

[4] P. Holota *The altimetry-gravimetry boundary value problem. I: Linearization, Friedrich's inequality.* Boll. Geod. Sc. Aff., XLII, pp. 14-32. *II: Weak Solution, V-ellipticity.* Boll. Geod. Sc. Aff., XLII, pp. 70-84, 1983.

[5] L. Hörmander *The boundary problems of physical geodesy.* Arch. Rat. Mech. An., 62, pp. 1-52, 1976.

[6] T. Krarup *Letters on Molodensky's problem.* IAG Special Study Group 4.31, unpublished, 1973.

[7] C. Miranda *Partial differential equations of elliptic type.* Springer-Verlag, 1970.

[8] Molodensky, Eremeev and Yurkina *Methods for study of the External Gravitational Field and Figure of the Earth.* Israel Program for Scientific Translation, Jerusalem, 1962.

[9] H. Moritz *Advanced physical geodesy.* H. Wichmann and Abacus Press, 1980.

[10] F. Sacerdote, F. Sansó *The scalar boundary value problem of physical geodesy.* Man. Geod., n. 11, pp. 15-28, 1986.

[11] F. Sacerdote, F. Sansó *New developments of boundary-value problems in physical geodesy.* proceedings of the IAG Symposia, IUGG XIX General Assembly, Vancouver, Tome II, pp. 369-390, 1987.

[12] F. Sacerdote, F. Sansó *On the analysis of the fixed-boundary gravimetric boundary-value problem.* II Hotine Marussi Symposium on mathematical geodesy, pp. 507-516, Pisa, 1989.

[13] F. Sacerdote, F. Sansó *The boundary-value problems of geodesy.* Acc. dei Lincei, n. 11, pp. 15-28, 1986.

[14] F. Sansó *The geodetic boundary-value problem in gravity space.* Atti Acc. Naz. Lincei, Cl. Sc. Fis. Mat. Nat., Memorie, S. VII, vol. XIV, pp. 41-97, 1977.

[15] F. Sansó *Recent advanced in the theory of the geodetic boundary value problem.* Rev. Geoph. Space Phys., 19, pp. 437-449, 1981.

[16] F. Sansó *Talk on the theoretical foundations of physical geodesy.* Geodetic theory and methodology, XIX General Assembly IUGG-IAG, Section IV, Vancouver, pp. 5-27, 1987.

[17] F. Sansó *The Wiener integral and the overdetermined boundary value problems of physical geodesy.* Man. Geod., 13, pp. 75-98, 1988.

[18] F. Sansó *On the aliasing problem with the harmonic analysis on the sphere. II Hotine-Marussi symposium on mathematical geodesy.* pp. 211-232, Pisa, 1989.

[19] F. Sansó *New estimates for the solution of Molodensky's problem.* Man. Geod., n. 14, pp. 68-76, 1989.

[20] F. Sansó *The long road from measurements to boundary value problem in physical geodesy.* Man. Geod., vol. 20, n. 5, pp. 326-344, 1995.

[21] F. Sansó, G. Sona *The theory of optimal linear estimation for continuous fields of measurements.* Man. Geod., Vol. 20, n. 3, 1995.

[22] K.J. Witsch *On a free boundary value problem of physical geodesy. I(Uniqueness),* Math. Meth. in the Appl. Sc., 7, 1985.

[23] K.J. Witsch *On a free boundary value problem of physical geodesy. II(Uniqueness),* Math. Meth. in the Appl. Sc., 8, 1986.

GBVP – Classical Solutions and Implementation

H. Sünkel

1 The Stokes solution

The determination of the geoid from gravity data may be considered as a classical problem of physical geodesy. Let us denote the actual gravity potential of the Earth by W, the normal gravity potential of a model ellipsoid by U, and the disturbing potential by T such that the tree potentials are related to each other by

$$W_P = U_P + T_P \qquad (1)$$

Then the geoid, as an equipotential surface of the real Earth's gravity field at mean sea level, can be defined in terms of

$$W = W_o = \text{const.} \qquad (2)$$

Accordingly, the model ellipsoid is characterized by a constant normal potential

$$U = U_o = \text{const.} \qquad (3)$$

If we assume that there are no masses outside the geoid, that both the Earth and the model ellipsoid have the same constant angular velocity, and a common zero gravity orgigin, then T is harmonic outside the geoid and fulfills the Laplace equation

$$\Delta T = 0. \qquad (4)$$

In a similar way we differentiate between actual gravity and normal gravity and introduce a gravity disturbance and a gravity anomaly:

$$
\begin{aligned}
g &= |\nabla W| & &\dots \text{ actual gravity} \\
\gamma &= |\nabla U| & &\dots \text{ normal gravity} \\
\delta g_P &= g_P - \gamma_P & &\dots \text{ gravity disturbance} \\
\Delta g_P &= g_P - \gamma_Q & &\dots \text{ gravity anomaly}
\end{aligned}
\qquad (5)
$$

Here P represents a point on the geoid and Q a corresponding point on the ellipsoid obtained by an orthogonal projection of P onto the ellipsoid.

By standard linearization we obtain

$$
\begin{aligned}
U_P &= U_Q + \left(\frac{\partial U}{\partial n}\right)_Q N &= U_Q - \gamma N \\
W_P &= U_P + T_P &= U_Q - \gamma N + T
\end{aligned}
\qquad (6)
$$

Considering $W_P = U_Q = W_o$ (geoid potential = ellipsoid potential) we obtain

$$T = \gamma N \qquad (7)$$

which leads to the famous Bruns formula:

$$N = \frac{T}{\gamma} \qquad (8)$$

The gravity disturbance δg is related to T by

$$\delta g = g_p - \gamma_P \doteq -\left(\frac{\partial W}{\partial n} - \frac{\partial U}{\partial n}\right) = -\frac{\partial T}{\partial n} \doteq -\frac{\partial T}{\partial h}. \qquad (9)$$

The gravity anomaly Δg is related to T as follows: With

$$\Delta g = g_P - \gamma_Q \qquad (10)$$

and the Taylor linearization of γ_P at the ellipsoidal point Q as Taylor point

$$\gamma_P = \gamma_Q + \frac{\partial \gamma}{\partial h} N \qquad (11)$$

and considering the expression for the gravity disturbance given above and Bruns formula, we obtain for the gravity anomaly

$$\Delta g = -\frac{\partial T}{\partial h} + \frac{1}{\gamma}\frac{\partial \gamma}{\partial h} T. \qquad (12)$$

This fundamental equation of physical geodesy represents a boundary condition which the anomalous potential T must satisfy at the geoid. Both the Laplace differential equation $\Delta T = 0$, valid outside the geoid, and this boundary condition are the key elements for the determination of the anomalous potential T, and therefore also for the geoid.

1.1 The Stokes operator

Spherical approximation:

Quantities of the anomalous gravity field are very small compared to the corresponding quantities of the actual or normal gravity field. Therefore, approximations may suffice if only moderate accuracies are required. Considering the earth's flattening of the order of $f \doteq 3 \cdot 10^{-3}$, a natural approximation is to disregard f in all equations relating quantities of the anomalous gravity field. Such an approximation introduces relative errors of the order of $3 \cdot 10^{-3}$.

In spherical approximation the boundary condition can be replaced by

$$\Delta g = -\frac{\partial T}{\partial r} - \frac{2T}{R} \qquad (13)$$

Harmonic representations:

The anomalous potential, being a harmonic function, may be represented in terms of a series of harmonic functions. Denoting by $T_n(\theta, \lambda)$ a Laplace surface harmonic of degree n, the anomalous potential is given by

$$T(r, \theta, \lambda) = \sum_{n=0}^{\infty} \left(\frac{R}{r}\right)^{n+1} T_n(\theta, \lambda) \qquad (14)$$

By differentiation with respect to r we obtain the gravity disturbance, and by considering the boundary condition we obtain the gravity anomaly:

$$\begin{aligned}
\delta g(r, \theta, \lambda) &= \frac{1}{r}\sum_{n=0}^{\infty}(n+1)\left(\frac{R}{r}\right)^{n+1} T_n(\theta, \lambda) \\
\Delta g(r, \theta, \lambda) &= \frac{1}{r}\sum_{n=0}^{\infty}(n-1)\left(\frac{R}{r}\right)^{n+1} T_n(\theta, \lambda)
\end{aligned} \qquad (15)$$

On the geoid (r=R) the anomalous potential, the gravity disturbance and the gravity anomaly become

$$T(R, \theta, \lambda) = \sum_{n=0}^{\infty} T_n(\theta, \lambda)$$

$$\delta g(R, \theta, \lambda) = \frac{1}{R} \sum_{n=0}^{\infty} (n+1) T_n(\theta, \lambda) \tag{16}$$

$$\Delta g(R, \theta, \lambda) = \frac{1}{R} \sum_{n=0}^{\infty} (n-1) T_n(\theta, \lambda)$$

Note: the gravity anomaly does not contain a first degree spherical harmonic !

By also expressing the gravity anomalies in terms of a series of Laplace surface harmonics

$$\Delta g(\theta, \lambda) = \sum_{n=0}^{\infty} \Delta g_n(\theta, \lambda) \tag{17}$$

and by comparing with the above series, a relation between the gravity anomaly and the anomalous potential in the spectral domain can be established:

$$\Delta g_n = \frac{n-1}{R} T_n \tag{18}$$

The gravity anomaly Laplace spherical harmonic Δg_n is given by the transformation

$$\Delta g_n = \frac{2n+1}{4\pi} \int\int_{\sigma} P_n(\cos\psi) \Delta g \, d\sigma \tag{19}$$

As a consequence the anomalous potential is then obtained by

$$T = \frac{R}{4\pi} \int\int_{\sigma} \left[\sum_{n=2}^{\infty} \frac{2n+1}{n-1} P_n(\cos\psi) \right] \Delta g \, d\sigma \tag{20}$$

The infinite series given in parentheses is the famous Stokes function $S(\psi)$:

$$S(\psi) = \sum_{n=2}^{\infty} \frac{2n+1}{n-1} P_n(\cos\psi) \tag{21}$$

The series also permits a closed expression as follows:

$$S(\psi) = \frac{1}{\sin(\psi/2)} - 6\sin(\psi/2) + 1 - 5\cos\psi - 3\cos\psi \ln\left[\sin(\psi/2) + \sin^2(\psi/2)\right] \tag{22}$$

Considering Bruns' equation we finally obtain the famous Stokes formula which is a linear mapping of the gravity anomaly to the geoid height N:

$$N = \frac{R}{4\pi G} \int\int_{\sigma} S(\psi) \Delta g \, d\sigma \tag{23}$$

In mathematical terms Stokes formula is a convolution on the sphere. This fact will be used in the subsequent sections in the context with spectral techniques.

Note that this Stokes formula holds only if the mass of the reference ellipsoid coincides with the mass of the earth, and if also the normal potential on the ellipsoid U_o coincides with the actual potential W_o on the geoid. However, this is never the case. Therefore, the Stokes equation has to be extended in order to account for those mass and potential differences.

Denoting the mass difference with δM and the potential difference with δW,

$$\delta M = M - M_o$$

$$\delta W = W_o - U_o, \tag{24}$$

the extended Stokes formula becomes

$$N = N_o + \frac{R}{4\pi G} \int \int_\sigma S(\psi) \Delta g d\sigma \tag{25}$$

with

$$N_o = \frac{k\delta M}{RG} - \frac{\delta W}{G} \quad . \tag{26}$$

Exterior potential:

The anomalous potential outside the geoid $r = R$ (in spherical approximation) can be computed by a spatial extension of the Stokes operator

$$S(r, \psi) = \frac{2R}{l} + \frac{R}{r} - 3\frac{Rl}{r^2} - \frac{R^2}{r^2} \cos \psi \left(5 + 3\ln \frac{r - R\cos\psi + l}{2r} \right) \tag{27}$$

yielding the anomalous potential and, by applying Bruns' formula, the separation between the geopotential surface $W = W_P$ passing through the computation point P and the corresponding spheropotential surface $U = W_P$:

$$
\begin{aligned}
T(r, \phi, \lambda) &= \frac{R}{4\pi} \int \int_\sigma S(r, \psi) \Delta g d\sigma \\
N(r, \phi, \lambda) &= \frac{R}{4\pi\gamma} \int \int_\sigma S(r, \psi) \Delta g d\sigma
\end{aligned}
\tag{28}
$$

1.2 Spherical and planar approximation

Spherical approximation:

The idea of spherical approximation is based on two concepts:

- Mapping of an ellipsoidal point with ellipsoidal (geodetic) coordinates onto a corresponding point on the sphere with numerically identical spherical ccordinates
- All first order terms and quantities of higher order are neglected

Planar approximation:

When local effects on anomalous gravity field quantities are being calculated, the spherical surface may be approximated by its tangent plane in the neighborhood of the computation point. Denoting by s the linear distance from the computation point, the leading term in Stokes' function becomes then

$$S(\psi) \doteq \frac{1}{\sin(\psi/2)} \doteq \frac{2}{\psi} \doteq \frac{2R}{s} \tag{29}$$

Then the contribution of this neighborhood $s < s_o$ (the so-called "inner zone") to the geoid height at a computation point P is obtained by

$$N_i = \frac{1}{2\pi G} \int_{\alpha=0}^{2\pi} \int_{s=0}^{s_o} \Delta g ds d\alpha \doteq \frac{s_o}{G} \Delta g_P \tag{30}$$

The relative error introduced by this approximation is of the order of 1% for $s = 10\,km$.

1.3   Ellipsoidal corrections

For high accuracy geoid determinations a spherical approximation can no longer be tolerated and ellipsoidal corrections must be applied. (Spherical approximation causes geoidal errors of 20 cm on a global average.) The derivation of these corrections is based on the following considerations:

1. A position on the ellipsoid is mapped one-to-one onto a corresponding position on a mean sphere.
2. The mean sphere ($\varepsilon = 0$) represents a "Taylor point" for a Taylor series of a function F defined on the ellipsoid ($\varepsilon > 0$)

Identifying the ellipsoidal coordinates φ, λ with spherical coordinates on the mean sphere it follows

$$F(\varphi, \lambda) = F_o(\varphi, \lambda) + \varepsilon F_1(\varphi, \lambda) + \varepsilon^2 F_2(\varphi, \lambda) + \dots \tag{31}$$

with $F_o(\varphi, \lambda)$ corresponding to $\varepsilon = 0$ (mean sphere). Due to the smallness of the flattening parameter ε it suffices to use spherical expressions for $F_i(\varphi, \lambda), i = 1, 2, \dots$.

Ellipsoidal correction for the geoid height:

The mapping of functions is now done in such a way that F_o on the mean sphere corresponds to F on the ellipsoid. Considering Bruns' formula

$$N = \frac{T}{\gamma} \tag{32}$$

with γ as the normal gravity on the ellipsoid and γ_o as the mean gravity

$$\gamma = \gamma_o \left[1 - \frac{1 - 3\sin^2\varphi}{4} \varepsilon \right] \tag{33}$$

we obtain on the ellipsoid:

$$N = \frac{T}{\gamma_o} \left[1 + \frac{1 - 3\sin^2\varphi}{4} \varepsilon \right] = N_o + N_1\varepsilon \tag{34}$$

with

$$N_o = \frac{T_o}{\gamma_o} = \frac{T}{\gamma_o} \qquad \text{(on the sphere)} \tag{35}$$

and therefore the first order correction term for the geoid height follows:

$$N_1 = N_o \left(\frac{1 - 3\sin^2\varphi}{4} \right) \tag{36}$$

Ellipsoidal correction for vertical deflections:

In a similar way we proceed with the derivation of the ellipsoidal correction terms for the vertical deflections:

$$\begin{aligned} \xi &= -\frac{\partial N}{\partial s_\varphi} \\[2mm] \eta &= -\frac{\partial N}{\partial s_\lambda} \end{aligned} \tag{37}$$

Here ds_φ and ds_λ denote the ellipsoidal line elements along a meridian and a parallel, respectively:

$$
\begin{aligned}
ds_\varphi &= \mu d\varphi \\
ds_\lambda &= \nu \cos\varphi d\lambda
\end{aligned}
\tag{38}
$$

With μ and ν as the principal radii of curvature we obtain

$$
\begin{aligned}
\frac{1}{\mu} &= \frac{1}{R}\left(1 + \frac{5 - 9\sin^2\varphi}{6}\varepsilon\right) \\
\frac{1}{\nu} &= \frac{1}{R}\left(1 - \frac{1 + 3\sin^2\varphi}{6}\varepsilon\right)
\end{aligned}
\tag{39}
$$

and with $N = N_o + N_1\varepsilon$ from the above expressions we obtain

$$
\begin{aligned}
\xi &= \xi_o + \xi_1\varepsilon \\
\eta &= \eta_o + \eta_1\varepsilon
\end{aligned}
\tag{40}
$$

with the zero order terms

$$
\begin{aligned}
\xi_o &= -\frac{1}{R}\frac{\partial N_o}{\partial\varphi} \\
\eta_o &= -\frac{1}{R\cos\varphi}\frac{\partial N_o}{\partial\lambda}
\end{aligned}
\tag{41}
$$

and the first order correction terms

$$
\begin{aligned}
\xi_1 &= \frac{13 - 27\sin^2\varphi}{12}\xi_o + \frac{3}{2}\sin\varphi\cos\varphi\frac{N_o}{R} \\
\eta_1 &= \frac{1 - 15\sin^2\varphi}{12}\eta_o
\end{aligned}
\tag{42}
$$

Ellipsoidal correction for the gravity anomaly:

The derivation of correction terms for the gravity anomaly is considerably more complicated due to the relation between gravity anomaly and the disturbing potential, given by the boundary condition on the ellipsoid:

$$
\Delta g = -\frac{\partial T}{\partial h} + \frac{1}{\gamma}\frac{\partial\gamma}{\partial h}T
\tag{43}
$$

After a tour de force we obtain

$$
\Delta g = \Delta g_o + \Delta g_1\varepsilon
\tag{44}
$$

with the spherical term

$$
\Delta g_o = \sum_{n=2}^{\infty}\sum_{m=0}^{n}\frac{n-1}{R}P_{nm}(\sin\varphi)\cdot(C_{nm}\cos m\lambda + S_{nm}\sin m\lambda)
\tag{45}
$$

and the first oder correction term

$$
\Delta g_1 = \frac{1}{R}\sum_{n=2}^{\infty}\sum_{m=0}^{n}P_{nm}(\sin\varphi)\cdot(G_{nm}\cos m\lambda + H_{nm}\sin m\lambda)
\tag{46}
$$

with G_{nm} and H_{nm} depending linearly on C_{ij} and S_{ij} (see [7]).

1.4 Stokes integration by 2D-FFT

In planar approximation the Stokes' integral, confined to an area E, can be expressed in local cartesian coordinates x, y by the following convolution:

$$
\begin{aligned}
N(x_P, y_P) &= \frac{1}{2\pi G} \int\!\!\int_E \frac{\Delta g(x, y)}{\sqrt{(x_P - x)^2 + (y_P - y)^2}} dx\, dy \\
&= \frac{1}{2\pi G} \Delta g * l_N
\end{aligned}
\tag{47}
$$

with the inverse distance as the planar Stokes' kernel

$$
l_N(x, y) = \frac{1}{\sqrt{x^2 + y^2}}
\tag{48}
$$

Considering the convolution theorem, such a two-dimensional convolution suggests itself for an application of fast Fourier techniques.

The 1-D discrete Fourier transform

Let a function $h(t)$ be sampled with a constant sampling interval Δt, carried over a finite interval T such that N samples are produced. Then the 1-D discrete Fourier transform $H(f)$, as a function of the frequency f, is given in a sampled form by

$$
H(m\Delta f) = \sum_{k=0}^{N-1} h(k\Delta t) e^{-i2\pi km/N} \Delta t
\tag{49}
$$

and its inverse, the function $h(t)$, as a function of t, in its sampled form, by

$$
h(k\Delta t) = \sum_{m=0}^{N-1} H(m\Delta f) e^{i2\pi km/N} \Delta f
\tag{50}
$$

Here the time and frequency period (T and F), the time and frequency spacing (Δt and Δf), and the time and frequency sample size (both N) are related as follows:

$$
T = \frac{1}{\Delta f} = N\Delta t \qquad\qquad F = \frac{1}{\Delta t} = N\Delta f
\tag{51}
$$

Due to the finite time period and the finite time sampling interval, the function $h(t)$ can only be recovered within a certain frequency band, spanning from a minimum frequency to a maximum frequency (corresponding to a maximum wavelength and a minimum wavelength, resp.). Aliasing and leakage prevent the recovery of frequencies outside this band.

The minimum frequency recovered is Δf. The maximum frequency recovered is given by $1/2\Delta t$; it is called the "Nyquist frequency".

The 2-D discrete Fourier transform

The two-dimensional discrete Fourier transform is a natural extension of the one-dimensional case. Let a function $h(x, y)$, with x and y being cartesian coordinates, be sampled with constant sampling intervals Δx and Δy, carried over finite intervals T_x and T_y with M samples in x - direction and N samples in y - direction.

Then the 2-D discrete Fourier transform $H(u, v)$, as a function of the two frequencies u and v, is given in a sampled form by

$$H(m\Delta u, n\Delta v) = \sum_{k=0}^{M-1} \sum_{l=0}^{N-1} h(k\Delta x, l\Delta y) \cdot e^{-i2\pi(km/M + ln/N)} \Delta x \Delta y \tag{52}$$

and its inverse, the function $h(x, y)$, as a function of x and y, in its sampled form, by

$$h(k\Delta x, l\Delta y) = \sum_{m=0}^{M-1} \sum_{n=0}^{N-1} H(m\Delta u, n\Delta v) \cdot e^{i2\pi(km/M + ln/N)} \Delta u \Delta v \tag{53}$$

Here the time and frequency periods (T_x, T_y and F_u, F_v), the time and frequency spacings ($\Delta x, \Delta y$ and $\Delta u, \Delta v$), and the time and frequency sample sizes (both M, N) are related as follows:

$$
\begin{aligned}
T_x &= \frac{1}{\Delta u} = M\Delta x & \qquad F_u &= \frac{1}{\Delta x} = M\Delta u \\
T_y &= \frac{1}{\Delta v} = N\Delta y & \qquad F_v &= \frac{1}{\Delta y} = N\Delta v
\end{aligned}
\tag{54}
$$

1-D convolution

A convolution of a function $g(t)$ with a function $h(t)$ produces another function $x(t)$ by the following integral:

$$s(t) = g(t) * h(t) = \int_{-\infty}^{\infty} g(\tau)h(t - \tau)d\tau \tag{55}$$

Its discretization is given by

$$s(k) = \sum_{m=0}^{N-1} g(m)h(k - m)\Delta t \tag{56}$$

2-D convolution

A convolution of a function $g(x, y)$ with a function $h(x, y)$ delivers another function $s(x, y)$ by the following integral:

$$
\begin{aligned}
s(x, y) &= g(x, y) * h(x, y) \\
&= \int_{-\infty}^{\infty} \int_{-\infty}^{\infty} g(\xi, \eta)h(x - \xi, y - \eta)d\xi d\eta
\end{aligned}
\tag{57}
$$

Its discretization is given by

$$s(k, l) = \sum_{m=0}^{M-1} \sum_{n=0}^{N-1} g(m, n)h(k - m, l - n)\Delta x \Delta y \tag{58}$$

Now, if point gravity anomalies are given on a cartesian grid, we may express the Stokes integral in planar approximation as follows:

$$N(x_k, y_l) = \frac{1}{2\pi G} \sum_{m=0}^{M} \sum_{n=0}^{N} \Delta g(x_m, y_n) \cdot l_N(x_k - x_m, y_l - y_n)\Delta x \Delta y \tag{59}$$

The singularity of the integral kernel l_N is accounted for by setting $l_N(0, 0) = 0$. The contribution at the computation point is provided by the cartesian analogue of the polar expression derived before:

$$\delta N(x_k, y_l) \approx \sqrt{\frac{\Delta x \Delta y}{\pi}} \frac{\Delta g(x_k, y_l)}{G} \tag{60}$$

Let us denote the discrete Fourier transform of the gridded gravity anomalies by $\Delta G(u, v)$:

$$\Delta G(u_m, v_n) = \mathbf{F}\{\Delta g(x_k, y_l)\}$$

$$= \sum_{k=0}^{M-1} \sum_{l=0}^{N-1} \Delta g(x_k, y_l) \cdot e^{-i2\pi(mk/M+nl/N)} \Delta x \Delta y \tag{61}$$

The Fourier transform of the planar Stokes' kernel $l_N(x, y)$ is denoted by $L_N(u, v)$ and given by

$$L_N(u_m, v_n) = \mathbf{F}\{l_N(x_k, y_l)\}$$

$$= \sum_{k=0}^{M-1} \sum_{l=0}^{N-1} l_N(x_k, y_l) \cdot e^{-i2\pi(mk/M+nl/N)} \Delta x \Delta y, \tag{62}$$

or alternatively, by the continuous Fourier transform of the planar Stokes' kernel l_N, called analytically-defined spectrum of the Stokes' kernel,

$$L_N(u, v) = \mathbf{F}\{l_N(x, y)\}$$

$$= \int\limits_{-\infty}^{\infty} \int\limits_{-\infty}^{\infty} l_N(x, y) \cdot e^{-i2\pi(ux+vy)} dx dy \tag{63}$$

$$= \frac{1}{\sqrt{u^2 + v^2}} = \frac{1}{q}$$

By considering the convolution theorem, the geoid height in planar approximation is given by

$$N(x, y) = \frac{1}{2\pi G} \mathbf{F}^{-1}\{\Delta G(u, v) L_N(u, v)\} \tag{64}$$

Note that by this convolution the gravity spectrum is devided by the radial frequency q. As a consequence, high frequencies of gravity are attenuated with increasing frequency. Therefore, the Stokes kernel can be considered as a low-pass filter and the Stokes' integration as a low-pass filtering process.

Spherical kernel

In order to avoid errors introduced by the planar approximation, the spherical form of the Stokes' integration can be used. Explicitly written, the spherical form of Stokes' integration is given by

$$N(\varphi_P, \lambda_P) = \frac{R}{4\pi G} \int\int\limits_{\sigma} S(\varphi_P, \lambda_P, \varphi, \lambda) \cdot \Delta g(\varphi, \lambda) \cos\varphi d\varphi d\lambda \tag{65}$$

and with gridded gravity anomalies by

$$N(\varphi_k, \lambda_l) = \frac{R}{4\pi G} \sum_{i=0}^{M-1} \sum_{j=0}^{N-1} S(\varphi_k, \lambda_l, \varphi_i, \lambda_j) \cdot \Delta g(\varphi_i, \lambda_j) \cos\varphi_i \Delta\varphi\Delta\lambda \tag{66}$$

This transformation can be turned into a proper convolution integral by applying a very smart trick invented by Strang van Hees (1990), [11]: in the computation of the spherical Stokes' kernel $S(\psi)$ he suggested to represent

$$\sin^2\frac{\psi}{2} = \sin^2\frac{\varphi_P - \varphi}{2} + \sin^2\frac{\lambda_P - \lambda}{2} \cos\varphi_P \cos\varphi \tag{67}$$

and to approximate

$$\cos\varphi_P \cos\varphi \approx \cos^2\bar\varphi - \sin^2\frac{\varphi_P - \varphi}{2} \tag{68}$$

such that we obtain

$$\sin^2 \frac{\psi}{2} \approx \sin^2 \frac{\varphi_P - \varphi}{2} + \sin^2 \frac{\lambda_P - \lambda}{2} \cdot (\cos^2 \bar{\varphi} - \sin^2 \frac{\varphi_P - \varphi}{2}) \tag{69}$$

With this approximation the discrete Stokes' transformation given above can be expressed as a proper convolution:

$$
\begin{aligned}
N(\varphi_k, \lambda_l) &= \frac{R}{4\pi G} \sum_{i=0}^{M-1} \sum_{j=0}^{N-1} S(\varphi_k - \varphi_i, \lambda_l - \lambda_j, \bar{\varphi}) \cdot \Delta g(\varphi_i, \lambda_j) \cos \varphi_i \Delta\varphi\Delta\lambda \\
&= \frac{R}{4\pi G} [\Delta g(\varphi_k, \lambda_l) \cos \varphi_k] * S(\varphi_k, \lambda_l, \bar{\varphi})
\end{aligned}
\tag{70}
$$

Again, the great efficiency of 2-D FFT can be exploited with great advantage to compute the geoid heights simultaneously on all grid points of the gravity anomaly grid.

There are two disadvantages of this type of approach:

- Approximation of the Stokes' kernel introduces errors in the computed geoid heights. (A multi-band approach suggested by Forsberg and Sideris (1992), [1] can miminize this error.)
- Large computer memory is absorbed by zero padding both in latitude and in longitude

1.5 Stokes integration by 1D-FFT

The approximation of the Stokes' kernel made above is due to the mean latitude $\bar{\varphi}$. As a consequence, the Stokes' kernel is accurate only along this mean latitude parallel. Based on this consideration Haagmans et al. (1993), [2] came up with the idea to use a 1-D FFT along each parallel and to sum up the results along all the parallels taking advantage of the addition theorem of the Fourier transform.

$$
\begin{aligned}
N(\varphi_k, \lambda_l) &= \frac{R}{4\pi G} \sum_{i=0}^{M-1} \left[\sum_{j=0}^{N-1} S(\varphi_k, \varphi_i, \lambda_l - \lambda_j, \bar{\varphi}) \cdot \Delta g(\varphi_i, \lambda_j) \cos \varphi_i \Delta\lambda \right] \Delta\varphi \\
&= \frac{R}{4\pi G} [\Delta g(\varphi_k, \lambda_l) \cos \varphi_k] * S(\varphi_k, \lambda_l, \bar{\varphi})
\end{aligned}
\tag{71}
$$

The expression in the bracket is a 1-D discrete convolution along a parallel and can therefore be treated by 1-D FFT. Therefore, the gridded geoid can be computed by

$$N(\varphi_k, \lambda_l) = \frac{R}{4\pi G} \mathbf{F}_1^{-1} \left\{ \sum_{i=0}^{M-1} \mathbf{F}_1 \{ S(\varphi_k, \varphi_i, \lambda_l) \} \cdot \mathbf{F}_1 \{ \Delta g(\varphi_k, \lambda_l) \cos \varphi_k \} \right\} \tag{72}$$

(Here $\mathbf{F}_1$ and $\mathbf{F}_1^{-1}$ represent the 1-D discrete Fourier transform operator.)
This approch has two main advantages:

- No approximation is made because the Stokes kernel is exact. If used on complete parallels, the result is identical to a direct pointwise numerical integration
- Compared to the 2-D FFT approach, the 1-D FFT technique needs much less computer memory because only one-dimensional arrays are processed each time.
- In terms of computation speed the 1-D FFT technique outperforms the direct summation method by a factor of about 200.

1.6 Stokes integration by FHT

As an alternative to the Fourier transform Hartley (1942), [3] has proposed a transformation which is based on real operations rather than on complex operations such as the Fourier transform. This transformation is denoted "Hartley transform". It represents the basis for the "fast Hartley transform" (FHT) as a counterpart to the fast Fourier transform (FFT) and outperforms the FFT by 30% - 50% in terms of computation speed.

The 1-D discrete Hartley transform

Let a function $h(t)$ be sampled with a constant sampling interval Δt, carried over a finite interval T such that N samples are produced. Then the 1-D discrete Hartley transform $H(f)$, as a function of the frequency f, is given in a sampled form by

$$H(m\Delta f) = \sum_{k=0}^{N-1} h(k\Delta t) cas\frac{2\pi km}{N}\Delta t \tag{73}$$

and its inverse, the function $h(t)$, as a function of t, in its sampled form, by

$$h(k\Delta t) = \sum_{m=0}^{N-1} H(m\Delta f) cas\frac{2\pi km}{N}\Delta f \tag{74}$$

Here we have used the following notation:

$$cas x = \cos x + \sin x \tag{75}$$

The 2-D discrete Hartley transform

The two-dimensional discrete Hartley transform is a natural extension of the one-dimensional case. Let a function $h(x,y)$, with x and y being cartesian coordinates, be sampled with constant sampling intervals Δx and Δy, carried over finite intervals T_x and T_y with M samples in x - direction and N samples in y - direction.

Then the 2-D discrete Hartley transform $H(u,v)$, as a function of the two frequencies u and v, is given in a sampled form by

$$H(m\Delta u, n\Delta v) = \sum_{k=0}^{M-1}\sum_{l=0}^{N-1} h(k\Delta x, l\Delta y) \cdot cas\frac{2\pi km}{M}cas\frac{2\pi ln}{N}\Delta x\Delta y \tag{76}$$

and its inverse, the function $h(x,y)$, as a function of x and y, in its sampled form, by

$$h(k\Delta x, l\Delta y) = \sum_{m=0}^{M-1}\sum_{n=0}^{N-1} H(m\Delta u, n\Delta v) \cdot cas\frac{2\pi km}{M}cas\frac{2\pi ln}{N}\Delta u\Delta v \tag{77}$$

The properties of the discrete Hartley transform are very similar to the discrete Fourier transform. For details the reader is recommended to study the relevent literature (e.g. Sideris (1994), [10]).

1.7 Practical examples and experience

For any spatial function such as the geoid we may distinguish between several spectral bands: a low, medium, and high frequency band corresponding to long, medium and short wavelength phenomena.

In the case of the gravity anomaly and the geoid the low freqency part is usually provided by a global geopotential model (GM) in terms of a harmonic series, complete up to a certain degree and order n_{max}. The medium frequency band is covered by a Stokes integration of residual gravity anomalies, and the high frequency band is covered by the effect of a high resolution digital terrain model (DTM). Thus the geoid height is split into three components:

$$N = N^{GM} + N^{Stokes} + N^{DTM} \tag{78}$$

Before applying the Stokes integration, the gravity anomalies must be reduced due to the geopotential model contribution and due to the DTM effect:

$$\Delta g = \Delta g^{FA} - \Delta g^{GM} - \Delta g^{DTM} \tag{79}$$

The contribution of the geopotential model to gravity and the geoid is given, in spherical approximation, by the following harmonic series:

$$
\begin{aligned}
\Delta g_P^{GM} &= G \sum_{n=2}^{n_{max}} (n-1) \sum_{m=0}^{n} P_{nm}(\sin \varphi_P) \cdot [C_{nm} \cos m\lambda_P + S_{nm} \sin m\lambda_P] \\
N_P^{GM} &= R \sum_{n=2}^{n_{max}} \sum_{m=0}^{n} P_{nm}(\sin \varphi_P) \cdot [C_{nm} \cos m\lambda_P + S_{nm} \sin m\lambda_P]
\end{aligned}
\tag{80}
$$

The Stokes formula is valid provided that there are no masses outside the geoid. Of course, this is not the case, and therefore, masses outside the geoid have to be removed or transferred numerically in order to make the geoid a boundary surface. The effect of such a mass reduction is applied on the gravity anomaly as input to the Stokes integration. Restoring the removed or shifted masses has an effect on the computed geoid height and must be considered as well. (It is usually called the indirect effect on the geoid.) This procedure has become a standard technique and is denoted "remove-restore technique".

Various methods to accomplish this task have been designed in the past such as

- Helmert's condensation reduction
- Isostatic reduction
- Residual terrain model reduction

Applying one of these reduction methods, the reduced gravity anomaly on the geoid is then given by

$$\Delta g = \Delta g_P + F + \delta A \tag{81}$$

Here δA represents the effect of the mass condensation or reduction on the gravity anomaly, and F represents the free-air reduction from the surface to the geoid.

Such a mass condensation or mass reduction changes the potential as well. Its effect on the potential is called indirect effect on the potential. Due to this change of potential, of course, the Stokes equation does not exactly deliver the geoid but rather a surface called co-geoid. This has two consequences:

1. Before applying the Stokes operator, the gravity anomalies must be transferred from the geoid to the co-geoid; this is accomplished by applying a small correction, the so-called indirect effect on gravity $\delta \Delta g$ which is actually a free-air reduction from the co-geoid to the geoid.
2. At the co-geoid supplied by the Stokes equation the indirect effect on the geoid δN must be applied by considering Bruns' formula.

$$
\begin{aligned}
\delta \Delta g &= -\frac{1}{\gamma} \frac{\partial \gamma}{\partial h} \delta T \\
\delta N &= \frac{\delta T}{\gamma}
\end{aligned}
\tag{82}
$$

In planar approximation the effect on the potential δT and on gravity δg for the Helmert condensation procedure, for example, is given by the following expressions:

$$
\begin{aligned}
\delta T_P &= -\pi k \rho h_P^2 - 2\pi k \rho \sum_{r=1}^{\infty} \frac{1}{(2r+1)!} \mathbf{L}^{2r-1} h^{2r+1} \\
\delta A_P &= 2\pi k \rho \sum_{r=1}^{\infty} \frac{1}{(2r)!} \mathbf{L}^{2r-1} (h - h_P)^{2r}
\end{aligned}
\tag{83}
$$

Here we have used the vertical derivative operator $\mathbf{L}$ which is defined as follows

$$
\mathbf{L} f = \frac{1}{2\pi} \int\!\!\int_{\sigma} \frac{f - f_P}{l^3} dx dy
\tag{84}
$$

where l is the planar distance between the integration point and the computation point P. Convergence is guaranteed for $h - h_P/l \le 1$.

2 The Molodensky solution

The problem of Stokes may be formulated as follows:

- Given the gravity potential $W = const.$ and gravity g at all points of the geoid.
- Determine the geoid.

The Problem of Molodensky is conceptionally similar to the Stokes problem:

- Given the gravity potential W and the gravity vector $\mathbf{g}$ at all points of the physical earth's surface S.
- Determine the surface S.

2.1 The Molodensky operator

In space both the gravity potential W and the gravity vector g are spatial functions, depending on three space coordinates. At the earth's surface S the gravity potential and the gravity vector are restricted to W_S and $\mathbf{g}_S$, respectively.

According to the solution of Dirichlet's boundary value problem the gravity potential W outside S can be uniquely determined if the gravity potential W_S is given on S. (Here we have extended the BVP for the centrifugal potential.) Now given W as a spatial function, we may determine the gravity vector $\mathbf{g}$ being the gradient of W, and therefore, the gravity vector can also be determined on the surface S.

As a consequence, $\mathbf{g}_S$ can be represented as a function of S and W_S :

$$
\mathbf{g}_S = F(S, W_S)
\tag{85}
$$

Compared to this direct problem, the Molodensky problem can be conceptionally formulated as an inverse problem:

$$
S = \Phi(W_S, \mathbf{g}_S)
\tag{86}
$$

Note that the Molodensky operator F is quite a complicated nonlinear operator which may be handled by a proper linearization.

Since W_S is given, we may formally consider $\mathbf{g}_S$ as a function of S only , and vice versa, we may express the surface S as a function of the gravity vector on the surface $\mathbf{g}_S$:

$$
\mathbf{g}_S = f(S) \qquad S = f^{-1}(\mathbf{g}_S)
\tag{87}
$$

We introduce an approximation S_o to the earth's surface as a "Taylor point"; the gravity vector at S_o is denoted $\mathbf{g}_o$. Then we obviously have

$$
\begin{aligned}
S &= S_o + \Delta S \\
\mathbf{g}_S &= \mathbf{g}_o + \Delta \mathbf{g}
\end{aligned}
\tag{88}
$$

with

$$
\mathbf{g}_o = f(S_o)
\tag{89}
$$

Then a Taylor series, terminated after the linear term, yields

$$
\mathbf{g}_o + \Delta \mathbf{g} = f(S_o + \Delta S) = f(S_o) + f'(S_o)\Delta S.
\tag{90}
$$

With

$$
\Delta \mathbf{g} = f'(S_o)\Delta S
\tag{91}
$$

we formally obtain the solution

$$
\Delta S = [f'(S_o)]^{-1}\Delta \mathbf{g} = M\Delta \mathbf{g}.
\tag{92}
$$

The elements introduced here have the following meaning:

$$
\begin{array}{lll}
S_o & \dots & \text{telluroid} \\
\Delta S & \dots & \text{height anomaly} \\
\mathbf{g}_o & \dots & \text{normal gravity on the telluroid} \\
\Delta \mathbf{g} & \dots & \text{gravity anomaly vector} \\
M & \dots & \text{linear Molodensky operator}
\end{array}
\tag{93}
$$

Newton's method may be employed to obtain approximations of higher order:

$$
S_{k+1} = S_k + [f'(S_k)]^{-1}[\mathbf{g}_S - f(S_k)], \qquad k = 0, 1, \dots
\tag{94}
$$

The telluroid is chosen such that the normal potential at the telluroid point Q coincides with the actual potential at the corresponding earth's surface point P, with Q located at the ellipsoidal normal passing through P. The ellipsoidal height h of the telluroid point Q is called "normal height" H^*, and the distance between P and Q is denoted "height anomaly" ζ:

$$
\begin{aligned}
U_Q &= W_P \\
\zeta &= h - H^*
\end{aligned}
\tag{95}
$$

In spherical approximation the boundary condition is identical to the one for the Stokes problem:

$$
\Delta g = -\frac{\partial T}{\partial r} - \frac{2T}{R}
\tag{96}
$$

Considering this boundary condition, the problem of the solution of the Laplace differential equation is denoted "simple linear Molodensky problem". Its solution provides the anomalous potential T in terms of a series

$$
T = \sum_{n=0}^{\infty} T_n.
\tag{97}
$$

The zero and first order terms ($n=0, 1$) are determined by

$$
T_n = \frac{R}{4\pi} \int\int_\sigma G_n S(\psi) d\sigma
\tag{98}
$$

with

$$
\begin{aligned}
G_o &= \Delta g \\
G_1 &= R^2 \int\!\!\int_\sigma \frac{h - h_P}{l_o^3} \kappa_o d\sigma
\end{aligned}
\tag{99}
$$

and

$$
l_o = 2R \sin \frac{\psi}{2}
\tag{100}
$$

and the auxiliary function

$$
\kappa_o = \frac{1}{2\pi} \left(\Delta g + \frac{3}{2} \frac{T_o}{R} \right)
\tag{101}
$$

Terms of higher order contain higher powers of $h - h_P$ and moreover, the tangent of the terrain inclination, the latter one making this series less suited for efficient numerical computations.

2.2 Solution by analytical continuation

An elegant solution of Molodensky's problem, which is based on the idea of analytical continuation, is due to Moritz (1969), [6]. This method turned out to be particularly well suited for efficient computations by fast Fourier techniques.

Let us consider a point Q at the telluroid and the equipotential surface $U = U_Q$ passing through Q (the so-called "point level"). Then a gravity anomaly Δg, given at the telluroid, may be analytically continued to that point level, yielding $\Delta g'$ such that

$$
\begin{aligned}
\Delta g &= \Delta g' + \left(\sum_{n=1}^{\infty} \frac{1}{n!} z^n \frac{\partial^n}{\partial z^n} \right) \Delta g' \\
&= \left(I + \sum_{n=1}^{\infty} z^n L_n \right) \Delta g'
\end{aligned}
\tag{102}
$$

with

$$
z = h - h_Q \,,
\tag{103}
$$

the identity operator I and the vertical differentiation operator L_n:

$$
L_n = \frac{1}{n!} \frac{\partial^n}{\partial z^n}
\tag{104}
$$

(The derivatives given above are vertical derivatives, and in spherical approximation, they are radial derivatives.)

Note that $\Delta g'$ referes to the level surface $U = U_Q$. Therefore, Stokes' formula can be applied to determine the anomalous potential T_Q :

$$
T = \frac{R}{4\pi} \int\!\!\int_\sigma \Delta g' \, S(\psi) d\sigma
\tag{105}
$$

By formal series inversion the gravity anomaly at point level $\Delta g'$ can be expressed by Δg through a series with terms determined by recursion:

$$
\Delta g' = \sum_{n=0}^{\infty} g_n
\tag{106}
$$

with

$$g_o = \Delta g$$

$$g_n = -\sum_{k=1}^{n} z^k L_k(g_{n-k}) \tag{107}$$

and the vertical differentiation operator

$$L_n = \frac{1}{n!} \underbrace{L_1 L_1 \cdots L_1}_{n} \tag{108}$$

given by

$$L(f) = \frac{R^2}{2\pi} \int\int_\sigma \frac{f - f_Q}{l_o^3} d\sigma \tag{109}$$

Considering Bruns' equation, the height anomaly is then provided by the following series:

$$\zeta = \frac{R}{4\pi G} \int\int_\sigma \Delta g\, S(\psi) + \sum_{n=1}^{\infty} \frac{R}{4\pi G} \int\int_\sigma g_n\, S(\psi) d\sigma \tag{110}$$

Restricting to $n = 1$ yields the so-called "gradient solution" for the height anomaly:

$$\zeta = \frac{R}{4\pi G} \int\int_\sigma \left[\Delta g - (h - h_Q)\frac{\partial \Delta g}{\partial h} \right] S(\psi) \tag{111}$$

Two-step continuation of gravity anomalies

Sideris (1987), [8] has suggested to determine the correction terms g_n by a two-step continuation procedure, such that the gravity anomalies Δg are analytically continued from the telluroid down to sea level, yielding a corresponding value Δg^o, followed by an upward continuation to the point level Δg^u which finally provides $\Delta g'$.

The great advantage of this approach is that, due to the zero height of the computation point at zero level, the resulting formulas for the computation of g_n do no longer depend on the height of the computation point, and therefore, FFT can be applied with great advantage.

At sea level the analytically continued gravity anomaly is then given by

$$\Delta g^o = \sum_{n=0}^{\infty} g_n^o \tag{112}$$

with

$$g_n^o = -\sum_{k=1}^{n} h^k L_k g_{n-k}^o, \qquad g_o^o = \Delta g \tag{113}$$

The upward continuation process to point level is then obtained by

$$\Delta g' = \Delta g^o + \Delta g^u \tag{114}$$

with

$$\Delta g^u = \sum_{n=0}^{\infty} g_n^u \tag{115}$$

and

$$g_n^u = \sum_{k=1}^n h_P^k L_k g_{n-k}^o, \qquad g_o^u = 0. \tag{116}$$

Consequently, the correction terms are computed as follows:

$$
\begin{aligned}
g_n &= \sum_{k=1}^n (h_P^k - h^k) L_k g_{n-k}^o \\
&= -\sum_{k=1}^n h^k L_k g_{n-k}^o + \sum_{k=1}^n h_P^k L_k g_{n-k}^o \\
&= g_n^o + g_n^u
\end{aligned}
\tag{117}
$$

In complete analogy we obtain the correction terms for the height anomaly, based on the Stokes operator which is applied to the correction terms g_n:

$$\zeta = \sum_{n=0}^\infty \zeta_n \tag{118}$$

with

$$\zeta_n = \zeta_n^o + \zeta_n^u \tag{119}$$

Height anomaly versus geoidal height

The difference between the height anomaly ζ and the geoid height N is equal to the difference between the orthometric height H and the normal height H^* :

$$\zeta - N = H - H^* \tag{120}$$

The orthometric height H and the normal height H^* are defined in terms of the geopotential number C, the mean value of the actual gravity along the plumbline between the geoid and the earth's surface $\bar{g}$, and the mean value of the normal gravity along the normal plumbline between the ellipsoid and the telluroid $\bar{\gamma}$:

$$H = \frac{C}{\bar{g}} \qquad H^* = \frac{C}{\bar{\gamma}} \tag{121}$$

Consequently the geoid height can be obtained from the height anomaly by

$$N = \zeta + \frac{\bar{g} - \bar{\gamma}}{\bar{\gamma}} H \tag{122}$$

Considering the definition of the Bouguer anomaly Δg_B, a good approximation of the difference between the height anomaly and the geoid height can be obtained (see Heiskanen and Moritz (1967), [4]):

$$(\zeta - N)_{[mm]} \approx -\Delta g_{B[mgal]} \cdot H_{[km]} \tag{123}$$

Since Δg_B is usually negative on the continents, the height anomaly is usually greater than the geoid height. Moreover, the height anomaly is correlated to the local topography. Therefore, the height anomalies mirror the topography.

2.3   Molodensky solution by FFT

When applying the operator L, the integration has theoretically to be extended over the whole sphere. However, due to the rapid decrease of the integral kernel with increasing distance from the computation point, it is sufficient to consider a rather limited area E centered at the computation point. In such a case planar approximation suffices.

The integral operator L in planar approximation is given by

$$L(f) = \frac{1}{2\pi} \int\int_E \frac{f - f_P}{l_o^3} dx dy \tag{124}$$

with

$$l_o = \sqrt{(x_P - x)^2 + (y_P - y)^2} \tag{125}$$

Introducing the auxiliary function

$$l_M = \frac{1}{\sqrt{x^2 + y^2}^3} = l_N^3 \tag{126}$$

and applying the convolution theorem, we obtain

$$L(f) = \frac{1}{2\pi} [f(x_P, y_P) * l_M(x_P, y_P) - f(x_P, y_P) \cdot L_M(0, 0)] \tag{127}$$

The spectrum of the integral kernel l_M can be obtained analytically by

$$\begin{aligned} L_M(u, v) &= \mathbf{F}\{l_M(x, y)\} \\ &= \int_{-\infty}^{\infty}\int_{-\infty}^{\infty} l_M(x, y) \cdot e^{-i2\pi(ux+vy)} dx dy \\ &= -2\pi\sqrt{u^2 + v^2} = -2\pi q \end{aligned} \tag{128}$$

Obviously the gravity gradient operator L acts as a filter which suppresses low frequencies and amplifies high frequencies.

Applying the convolution theorem, the correction terms g_n are given by

$$g_n = -\sum_{k=1}^{n} \frac{1}{k!} h^k \mathbf{F}^{-1}\left\{(-2\pi q)^k \cdot G_{n-k}^o\right\} + \sum_{k=1}^{n} \frac{1}{k!} h_P^k \mathbf{F}^{-1}\left\{(-2\pi q)^k \cdot G_{n-k}^o\right\} \tag{129}$$

In full analogy we obtain the correction terms for the height anomaly

$$\begin{aligned} \zeta_n &= \frac{1}{2\pi G}\mathbf{F}^{-1}\left\{\frac{1}{q} G_n^o\right\} + \frac{1}{2\pi G}\sum_{k=1}^{n} \frac{1}{k!} h_P^k \mathbf{F}^{-1}\left\{(-2\pi q)^k \frac{1}{q} G_{n-k}^o\right\} \\ &= \frac{1}{2\pi G}\mathbf{F}^{-1}\left\{\frac{1}{q} G_n^o\right\} - \frac{1}{G}\sum_{k=1}^{n} \frac{1}{k!} h_P^k \mathbf{F}^{-1}\left\{(-2\pi q)^{k-1} \frac{1}{q} G_{n-k}^o\right\} \end{aligned} \tag{130}$$

(Here G_n denotes the Fourier transform of the correction term g_n.)

2.4 Practical examples and experience

Very detailed numerical investigations have been performed by Sideris (1987), [9] and others for the area of North America in general and for Canada in particular. Molodensky correction terms up to degree 4 have been computed based on a coarse 5' × 5' grid and on a dense 1 km × 1 km grid.

In the following tables, which can be found in Sideris (1987), [9], results for the Molodensky correction terms $g_n, n = 1, ..., 4$ and for $\zeta_n, n = 1, ..., 4$, resp. are given for various continuation levels and for different gravity anomaly and corresponding DTM grids.

Grid: 5', results: [mgal]				
Level [m]	g_1	g_2	g_3	g_4
0	37.7	15.0	4.2	0.9
1670	7.4	0.7	0.1	0.0
2800	13.1	1.5	0.1	0.0

Table 1: Molodensky terms g_n - max. values

Grid: 5', results: [cm]				
Level [m]	ζ_1	ζ_2	ζ_3	ζ_4
0	48	10	2	0
1670	7	1	0	0
2800	21	2	0	0

Table 2: Molodensky terms ζ_n - max. values

Level: 1670 m, results: [mgal]				
Grid	g_1	g_2	g_3	g_4
5' × 5'	7.4	0.7	0.1	0.0
1 km × 1 km	38.3	18.2	7.8	1.3

Table 3: Molodensky terms g_n - max. values

Level: 1670 m, results: [cm]				
Grid	ζ_1	ζ_2	ζ_3	ζ_4
5' × 5'	7	1	0	0
1 km × 1 km	15	3	1	0

Table 4: Molodensky terms ζ_n - max. values

Similar results have been obtained by Kühtreiber (1990), [5] the vertical deflections in a high mountainous region in Austria.

The results suggest that even for a mountainous area the series do converge, and that a higher degree of smoothing reduces the maxima of the Molodensky correction terms. However, it remains questionable if the series are only "convergently beginning" and above all, if higher order terms are significant at all.

References

[1] Forsberg R. and Sideris M. (1992): *Geoid computations by the multi-band spherical FFT approach*. Proceedings of the First Continental Workshop on the Geoid in Europe (Eds.: P. Holota and M. Vermeer), Prague, pp. 335 - 347.

[2] Haagmans R., De Min E. and Von Gelderen M. (1993): *Fast evaluation of convolution integrals on the sphere using 1D FFT, and a comparison with existing methods for Stokes' integral*. Manuscripta Geodaetica, Vol. 18, pp. 227-241.

[3] Hartley R.V.L. (1942): *A more symmetrical Fourier analysis applied to transmission problems*. Proceedings of the I.R.E., pp. 144-150.

[4] Heiskanen W.A. and Moritz H. (1967): *Physical Geodesy*. Freeman, San Francisco.

[5] Kühtreiber N. (1990): *Untersuchungen zur gravimetrischen Bestimmung von Lotabweichungen im Hochgebirge nach Molodensky mittels Fast-Fourier-Transformation*. Disseration an der TU Graz, Graz.

[6] Moritz H. (1969): *Nonlinear solutions of the geodetic boundary value problem*. Rep. 126, Dep. of Geodetic Science, The Ohio State University, Columbus, Ohio.

[7] Moritz H. (1980): *Advanced Physical Geodesy*. Wichmannn, Karlsruhe.

[8] Sideris M. (1987): *On the application of spectral techniques to the gravimetric problem*. Invited paper, XIX IUGG Gen. Ass., Vancouver.

[9] Sideris M. (1987): *Spectral methods for the numerical solution of Molodensky's problem*. UCSE Report no. 20024, Dep. of Surveying Engineering, The University of Calgary, Calgary.

[10] Sideris M. (1994): *Geoid determination by FFT techniques*. Proceedings of the Int. School for the Determination and Use of the Geoid (Ed.: F. Sansò), Int. Geoid Service, Milano, pp. 165 - 229.

[11] Strang van Hees (1990): *Stokes formula using Fast Fourier Techniques*. Manuscripta Geodaetica, Vol. 15, pp. 235-239.

PART III: FROM THE GBVP TOWARDS
A 1 cm GEOID

Topographic Effects in Gravity Field Modelling for BVP

R. Forsberg and C.C. Tscherning

1. INTRODUCTION

The written form of these lecture notes is a modified version of the lecture notes "Terrain effects in geoid computations" by R.Forsberg, published in the lecture notes of the International School on the Determination and Use of the geoid, Milano, 1994. The text will therefore to a high degree refer to the treatment of topographic effects in relation to geoid computations. However, the solution of boundary value problems for the Earth's gravity field, including the determination of the (quasi-)geoid, is a part of the more general problem of modelling the gravity field. The main difference is that the data not necessarily will be associated with points at the surface of the earth, but may be points in space measured in an aircraft or a satellite. The problems associated with treating the topography, however, remains the same.

Geodetic gravity field modelling is, given some quantities - "observations" - of the earth's gravity field, what are other quantities - "predictions". The modelling procedures can be e.g. Stokes' or Vening-Meinesz' integrals, FFT methods, least-squares collocation, point-mass fitting, spherical or ellipsoidal harmonic expansions.

For geoid prediction the basic formula is the familiar Stokes' integral over the surface of the earth

$$N = \frac{R}{4\pi\gamma} \iint_{\sigma} \Delta g\ S(\psi)\ d\sigma \tag{1}$$

The topography in a mountainous area affects gravity field modelling in two ways:

1) A strong gravity signal is due to the gravitational attraction of the topographic masses itself, a signal which dominates at shorter wavelengths, and therefore information of topography can be used to smooth the gravity field prior to any modelling process, and

> *2) The topography implies that the basic observation data* - notably gravity anomalies *-are given on a non-level surface,* violating the basic requirements for Stokes' integral.

In the first case the gravity field may be smoothed by *terrain reductions*, in the second case *Molodensky* or *Helmert condensation* corrections are applied to offset the non-level surface. In the Molodensky theory the geoid is substituted by the quasi-geoid, and geoid undulations N by height anomalies ζ ("geoid heights at the surface of the topography"). Modified integral formulas of typical form

$$\zeta = \frac{R}{4\pi\gamma} \iint_{\sigma} (\Delta g + g_1)\ S(\psi)\ d\sigma \tag{2}$$

must be substituted instead of Stokes' formula, where g_1 is the first term in the Molodensky expansions, cf. Heiskanen and Moritz (1967, sec. 8) or Moritz (1980).

In these lecture notes we will review both of these "topographic" effects, starting with some basic concepts of the spatial "modern" gravity field description. The main emphasis will, however, be put on the definition and practical use of proper *terrain reductions*, which are dominant and usually far bigger than the non-level surface ("Molodensky-style") corrections. When "operational" methods such as least-squares collocation and point-mass modelling are used, these corrections are in principle irrelevant, since the spatial operational methods may directly take the varying height of the observation points into account.

The non-level surface corrections are of course irrelevant on the *ocean* as well, where the geoid and quasi-geoid are identical, and gravity observations refer to the geoid. However, the bathymetry generally has a strong "terrain effect" in the marine environment. We will in these notes by the term "terrain" in general include both topography and bathymetry, the bathymetry typically represented as negative terrain heights in the various program packages. Generally the influence of the bathymetry is comparable or even larger than the corresponding topography. In many cases the bathymetric effects are, however, neglected, either due to lacking data (high-resolution detailed bathymetric grids are not so common), or because the effects of bathymetry are considered marginal in improving e.g. a continental geoid prediction.

One of the impacts of local terrain reductions is to remove the correlation of free-air anomalies with height, and avoid the *aliasing* which might appear when gravity stations are systematically observed at different levels than the average topographic level, e.g. when gravity points have a tendency to be located in valleys in mountainous areas. Such aliasing errors can be very big and devastating for geoid prediction, but can be avoided through proper terrain reduction and gridding techniques. In the marine environment "non-random" gravity measurements are less likely to occur, as survey ship (or satellite altimeter) measurements are typically independent of bathymetry, and therefore terrain reductions as an "anti-aliasing filter" are of less importance.

We will in these notes focus on the continental terrain effects, marine follows the same principles except for changes of density etc. The next section provides a basic theory review, and in section 3 an overview of the practical computation principles is given.

2. BASICS OF THE GRAVITY FIELD AND TERRAIN REDUCTIONS

The field we have access to through measurements is the total gravity field, e.g. derivatives of the physical potential W. By using a normal ellipsoidal gravity field U, we construct the anomalous potential T = W-U, and obtain the usual quantities such as gravity anomalies Δg, height anomalies ζ, and deflections of the vertical (ξ,η) by the well-known linear operators

where γ is normal gravity. These quantities should in principle be viewed as *spatial expressions* - i.e. as functions of both latitude, longitude *and* radial distance or height. The basic linear functional relationships (3) may be viewed as the definitions of the corresponding spatial gravity field quantities. The anomalous potential $T(r,\phi,\lambda)$ is a spatial function, which outside the terrain masses will be harmonic, i.e. satisfy the Laplace equation

$$\Delta g = -\frac{\partial T}{\partial r} - \frac{2}{r} T$$

$$\zeta = \frac{T}{\gamma}$$

$$\xi = -\frac{1}{r\gamma}\frac{\partial T}{\partial r} \tag{3}$$

$$\eta = -\frac{1}{r\gamma\cos\phi}\frac{\partial T}{\partial r}$$

$$\Delta T = 0 \tag{4}$$

Inside the topography T = W-U is still well-defined, but will now fulfil the Poisson equation

$$\Delta T = -4\pi G\rho \tag{5}$$

where G is the gravitational constant and ρ the density.

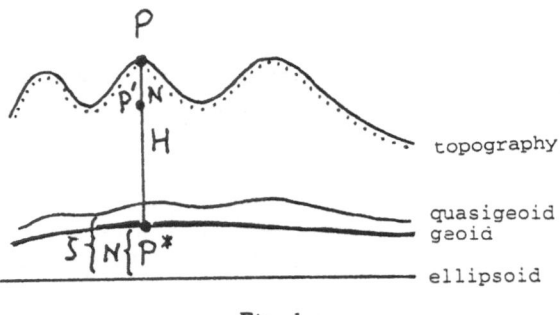

Fig. 1

For gravity anomalies the spatial concept means that the gravity anomaly refer to a point P at the earth surface, and not the corresponding geoid point P* (fig. 1). The gravity anomaly is defined as the difference between observed gravity at the surface, and normal gravity computed at the same height above the ellipsoid

$$\Delta g_P = g_P^{observed} - \gamma_{P\prime} \approx g_P^{observed} - \gamma_o + \frac{\partial\gamma}{\partial h} H \tag{6}$$

Here γ_o is the normal gravity on the ellipsoid, and the term $\partial\gamma/\partial h = 0.3086$ mgal/m is the normal free-air gradient. This "spatial" view of the definition of the gravity anomaly is fundamental for "modern" physical geodesy, but is often not stressed with sufficient clarity in introductory classes - here the free-air anomaly is frequently interpreted as the "gravity on the geoid, downward continued through free-air" (hence the name). However, actual gravity is not readily downward continued because the gradient $\partial g/\partial h$ can vary quite substantially (typically 5-10%) from the normal value 0.3086 mgal/m in rough topography.

2.1. Geoid, quasigeoid and harmonic continuation

Because the gravity field quantities are viewed as functions in space, the use of the term *height anomaly* is preferable to *geoid heights*. In these notes the term *geoid height* will be used in a somewhat loose fashion, referring either to the geoid or the quasigeoid, especially when talking about terrain effects. However, when geoid heights are needed at the cm-level, the difference between the concepts becomes quite significant, and it is important to be aware of the difference.

Rigorously the term *geoid* is reserved to one particular equipotential surface of the earth - the one approximating the global mean sea level. The geoid can be imagined as the surface which the water would stand at inside a tunnel connected with the ocean. This surface will on the continents in general be located inside the terrain masses, and thus not be an equipotential surface of a harmonic function (because T is not harmonic inside the topography). The geoid height N obeys the same formula as the height anomaly

$$N = \frac{T}{\gamma} \tag{7}$$

(Bruns' formula), but T in this case is the anomalous potential value at the geoid, *inside* the mass.

The *height anomalies* may be viewed as "geoid heights in space". The same formula as (7) holds for height anomalies, but now T is the potential value at the surface or aloft. Height anomalies evaluated at the topographic surface are known as the *quasigeoid*. Bruns' formula for the quasigeoid thus reads

$$\zeta(\phi,\lambda) = \frac{T\ [\phi,\lambda,r_{topo}(\phi,\lambda)]}{\zeta} \tag{8}$$

When the purpose of a geoid determination is e.g. to replace levelling with GPS, it is important to use the right kind of "geoid". Dependent on the national height system in use (orthometric heights or normal heights), the ellipsoidal height of a GPS point is expressed as either

$$h_{ellipsoisdal} = H^* + \zeta \tag{9}$$

where H* is the normal height, or

$$h_{ellipsoidal} = H + N \tag{10}$$

where H is the classical orthometric height. Therefore a national "geoid" in a country using normal heights should be a quasi-geoid rather than a geoid.

The classical dilemma in determining the "physical" geoid height N is that a knowledge of the density of the topographic masses above the geoid is required. The same problem occurs in the definition of the orthometric heights (heights along the plumbline from the geoid to the surface point). The basic observable in geodetic levelling is the geopotential number C (i.e., the potential difference W_o-W) at a surface point P, with practical formulas for orthometric and normal heights

$$\begin{aligned} H_{Helmert} &= \frac{C}{\bar{g}} \approx \frac{C}{g_P + 0.0424[mgal/m]H} \\ H^* &= \frac{C}{\bar{\gamma}} \approx \frac{C}{\gamma_o - 0.1543[mgal/m]H^*} \end{aligned} \tag{11}$$

The Helmert orthometric heights build on an assumption of a constant density of 2.67 g/cm³, using Bouguer plates (the Prey reduction) to estimate the mean gravity value inside the earth. For a detailed discussion of this see Heiskanen and Moritz (1967, sec. 4).

The point to be stressed here is that both heights (11) involve the potential at the surface point P (not at the geoid). It is relatively simple and straightforward at the approximation level of (11) to convert between geoid and quasigeoid: For a point

P at the surface of the topography (11) results in the simple formula

$$\zeta - N = H_P - H_P^* \approx - \frac{g_P - \gamma_o + 0.1967[mgal/m]H}{\gamma_o} H = - \frac{\Delta g}{\gamma_o}$$

(12)

where Δg_B is the Bouguer anomaly.

The conversion between geoid and quasigeoid is just a simple matter of applying a correction involving the Bouguer anomaly and the height. If gravity and heights are given on grids, this is straightforward, and - very important - if *Helmert* orthometric heights are used (what is nearly always done in practice), then the formula (12) may be considered as virtually exact. Trying to attempt a higher-order "downward continuation" of the quasigeoid is wrong in this case, because then corresponding higher order terms should also be applied to the computation of the mean gravity in the orthometric height formula (11).

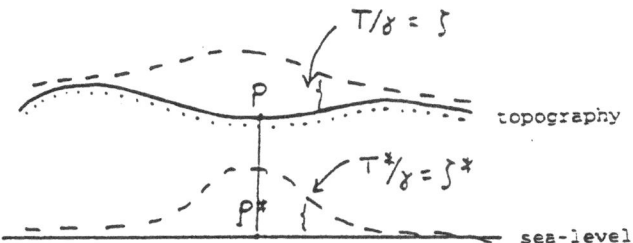

Fig. 2. Surface potential (T) and downward continued potential (T*)

In addition to the geoid height N and height anomaly ζ, another "geoid" term is also of relevance: the *height anomaly at sea level*, ζ^*. This quantity is directly related to the concept of *harmonic downward continuation*.

The harmonic downward continued anomalous potential T* is a potential which fulfils T* = T outside all mass, but is harmonic ($\Delta T^* = 0$) overall down to some level, e.g. the geoid level or an internal level (a Bjerhammar sphere). The existence of the T* solution inside the topography has been proven with the Runge-Krarup theorem, cf. Moritz (1980). The Runge-Krarup theorem says that it is always possible to find a T*-potential, harmonic down to any Bjerhammar sphere inside the earth, which approximates the outer potential T arbitrarily well.

The height anomaly at sea level ζ^* is corresponding to T*, and ζ^* (not N) will be the quantity obtained if e.g. least-squares collocation is rigorously spatially applied, and predictions are made at sea level. Downward continued and surface potential field values are related through the *Poisson integral*, which when written in planar approximation is of form

$$T_P = \frac{H_P}{2\pi} \int\int_{-\infty}^{\infty} \frac{T^*}{(x^2+y^2+H_P^2)^{3/2}} \, dxdy \qquad (13)$$

where H is the height of the computation point (Heiskanen and Moritz, 1967, sec. 6).

To first order the relationship between ζ and ζ^* is simply obtained by noting that

$$\zeta - \zeta^* \sim \frac{\partial \zeta}{\partial h} H = -\frac{\Delta g}{\gamma_o} H \qquad (14)$$

where the Δg is the free-air anomaly (opposed to the Bouguer anomaly in (12)). It should thus be stressed that N and ζ^* are very different quantities. In typical mountainous areas with, say, 1 km changes in height and corresponding gravity changes in the 100 mgal range, the differences in N, ζ and ζ^* will be at the 10 cm level.

2.2. Density anomalies and conventional gravity terrain reductions

The Helmert condensation or Molodensky theory approach is one aspect of the role of topography on geoid determination, the second aspect is the direct potential of the topography itself. Here the concept of *density anomaly* is important.

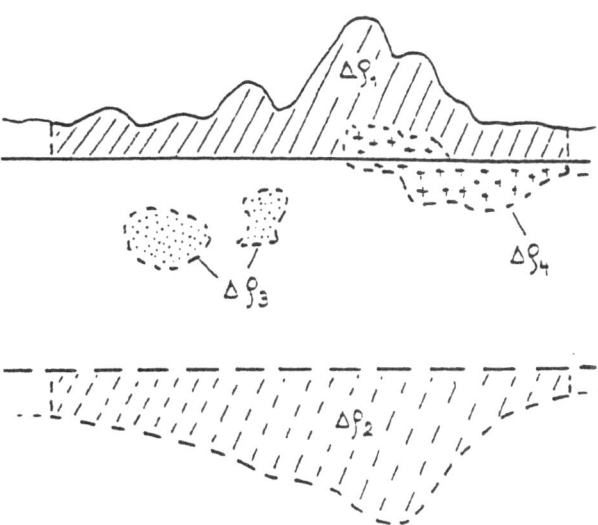

Fig. 3. Typical density anomalies ($\Delta\rho_1 \sim 2.67$, $\Delta\rho_2 \sim -0.4$, $\Delta\rho_{3-4} \sim \pm 0.1-0.3$)

We may view the anomalous potential both inside and outside the masses as being generated by an anomalous density distribution $\Delta\rho = \rho - \rho_{ref}$,

$$T(P) = G \iiint_{\Omega} \frac{\Delta\rho}{r_{PQ}} \, dV_Q \qquad (15)$$

where G is the gravitational constant, P the computation point (at, inside or outside the topography), and Ω the integration volume - in principle the whole earth.

The selection of a reference density function consistent with the ellipsoidal normal gravity field U is a complicated problem, but in *spherical approximation* (the approximation underlying (3)) the reference density distribution may be considered spherical, and in this case *any* reference density distribution with the correct total earth mass may be used - the potential of a spherical shell of constant density is equivalent to a point mass at the earth's center. It is therefore intuitively clear that the reference density distribution can be selected rather freely, and preferably in consistence with geophysical reality.

A schematic typical normal density model could be a crust of density ranging from 2.67 g/cm³ at the surface, smoothly increasing to 2.9 at the Moho interface around 30-35 km depth, with a density increase of around 0.4 across the Moho. With such a density model the major density anomalies would be the topography ($\Delta\rho \approx 2.67$), the ocean bathymetry ($\Delta\rho \approx -1.64$), and the isostatic compensation mass body ($\Delta\rho$

$\approx$ -0.4). Geological mass bodies inside the crust would typically have much lower density anomalies ($\Delta\rho \approx \pm0.1$-0.2, rarely larger).

In mountainous areas the topography itself produces a dominating signal in the anomalous gravity field, and the influence of the topography itself is removed by the *topographic* or *complete Bouguer* reduction

$$\Delta g_B = \Delta g - 2\pi G\rho H + c \qquad (16)$$

where c is an auxillary integral - "the classical terrain correction", given by an integral over the irregularities of the topographic mass body relative to a Bouguer plate passing through the computation point P. In planar coordinates and for constant density

$$c_P = G\rho \int\int_{-\infty}^{\infty} \int_{z=H_P}^{z=H(x,y)} \frac{z-H_P}{[(x_Q-x_P)^2+(y_Q-y_P)^2+(z_Q-H_P)^2]^{3/2}} \, dx_Q dy_Q dz_Q$$

$$(17)$$

The classical terrain correction c is typically one order of magnitude smaller than the dominant term $2\pi G\rho h$. Unfortunately, the terrain correction c also happens to be an approximation to the Molodensky term g_1, which often causes some confusion. Applying the terrain correction *alone* produces the socalled *Faye anomaly*, which is just as dependent on the local topographic height variations as the free-air anomaly itself (fig. 4).

The terrain correction c is an important quantity, and very dependent on the topography in the immediate surroundings of a gravity point. High-resolution digital terrain models are required for accurate computations in mountainous areas. Since the correction is always positive (a hill above the station level h_p and a valley below both yield the same sign of the contribution in (17)), insufficient resolution in the height models will yield terrain corrections which are systematically too low.

Another type of terrain-related density anomaly is the *isostatic compensation*. The isostatic compensation mass body is an idealized model for the earth lithosphere's general ability to provide a hydrostatic equilibirium of features of the crust. The Airy isostatic model assumes that all topography H is supported by a similar thickening of a crustal "root" below the normal crust of thickness T = αH, where the constant $\alpha = \rho/\Delta\rho$ is the root topography "magnification" factor.

The isostatic models serve as a useful model for providing smooth residual

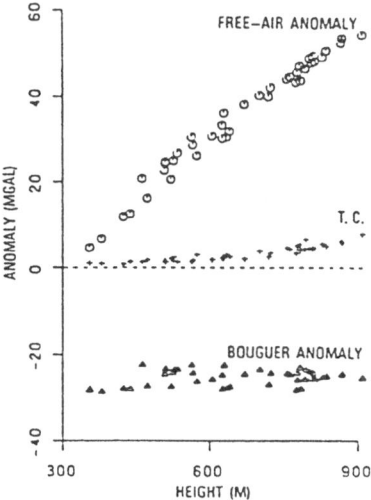

Fig. 4. Correlation of gravity with height in a local area in east Greenland

anomalies, but obviously the real earth does not follow such simple models. The strength of the earth's crust and dynamic forces may support topography without isostatic compensation (e.g. many islands), and the isostatic mass bodies might be deeper density anomalies in the upper mantle (e.g. mid-ocean ridges). However, in spite of the model oversimplifications, the practical applications of isostatic models do indeed produce the smoothest residual fields. For geoid modelling the Bouguer reduction must always be associated with some kind of isostatic or compensating reduction, because otherwise very large residual anomalies will occur. This is a because the Bouguer anomalies of the earth are systematically negative over the continents and positive over the oceans, a direct consequence of isostasy.

2.3. Residual density anomalies and RTM gravity anomalies

In practical geoid prediction schemes a global high-order spherical harmonic reference expansion is used as a reference field. Such a spherical harmonic expansion obviously includes the effects of the global topography. If a spherical reference field is used, i.e. gravity field modelling is applied on the residual

$$T' = T - T_{ref} \tag{18}$$

then the subtraction of a further topographic/isostatic effect may introduce long-wavelength effects into the residual potential. To avoid this a similar spherical harmonic expansion of global topographic/isostatic reference potential must be used, or - preferably - only the shorter wavelengths of the topographic/isostatic effect reference potential degree and order, then the topography will in principle be properly accounted for, but both too "short-wavelength" and "long-wavelength" reference surfaces may be used with advantage in practice. A short-wavelength surface is useful to minimize the magnitude of the RTM effects, and a more long-wavelength surface yields a better smoothing of the residual field. A mean height surface with resolution around 100 km will typically yield residual anomalies quite similar to isostatic anomalies in magnitude.

The topographic RTM density anomalies will make a "balanced set" of positive and negative density anomalies, representing areas where the topography is either above or below the reference topography. The effect of the RTM density anomalies will therefore in general cancel out in zones at larger distances from a computation point (say, e.g. a distance of 2-3 times the resolution of the mean height surface), which makes RTM reductions easy to work with in practice.

The RTM gravity terrain effect is in the planar approximation given by an integral of form

$$\Delta g_{RTM} = G\rho \int\limits_{-\infty}^{\infty}\int \int\limits_{z=h_{ref}(x,y)}^{z=h(x,y)} \frac{z-h_P}{[(x_Q-x_P)^2+(y_Q-y_P)^2+(z_Q-h_P)^2]^{3/2}} \, dx_Q dy_Q dz_Q$$

(19)

where h are the topographic heights, e.g. given by a digital terrain model. When the mean elevation surface is a sufficiently long-wavelength surface, the RTM reduction may be approximated by a Bouguer reduction to the reference level

$$\Delta g_{RTM} \sim 2\pi G\rho(h-h_{ref}) - c$$

(20)

This approximation again shows that the classical terrain correction is a key quantity, and it is therefore the basic "terrain" quantity, which should e.g. be stored in gravity data bases.

2.4. General remove-restore terrain reductions

For general gravity field modelling the terrain-reductions may be applied in a remove-restore fashion: Terrain effects are first removed from observations, then predictions are carried out, and finally the terrain effects are restored, symbolically written

$$L^c_{obs}(T) = L_{obs}(T) - L_{obs}(T_m)$$

$$\dots L_{obs} \rightarrow L_{pred} \dots \tag{21}$$

$$L_{pred}(T) = L_{pred}(T^c) + L_{pred}(T_m)$$

In the case of geoid computation from gravity, we would have the following observation and prediction functionals

$$L_{obs}(T) = L_{\Delta g}(T) = - \frac{\partial T}{\partial r} - \frac{2}{r} T$$

$$L_{pred}(T) = L_\zeta(T) = \frac{T}{\gamma} \tag{22}$$

The above scheme is only valid if the terrain reduction can be represented by a terrain potential T_m. This is the case for the complete Bouguer reduction, topographic/isostatic reduction and the RTM reduction, if 1) either a *fixed* area is taken into account (e.g. a square area between a given set of lat/lon limits), or 2) if - at least in principle - the corresponding density anomalies are taken into account globally. The terrain correction itself does *not* fit into this (the density anomalies associated with a c-computation are dependent on the actual computation point P), nor does terrain effects which are only computed inside a fixed spherical cap (e.g. the conventional computation of topographic/isostatic effects only out to a given radius, such as the Hayford 167-km zone).

Only RTM terrain effects may be computed in a spherical cap around the computation point, provided the cap is so big that the remote residual topography has a negligible effect, typically for a radius of computation 2-3 times the resolution of the mean height surface.

When the RTM reduction is used in a remove-restore fashion, special consideration is required for points under the reference topography level. If $h_p < h_{ref}$, and prisms are used to compute the residual density anomalies, i.e. "cutting away" mountains above the reference surface (density 2.67) and "filling" valleys below the surface (density -2.67), then the point P will end up *inside* the reference topography after the reduction, and T^c is *not* a harmonic function.

The non-harmonicity of the reduced potential below the reference height surface is a major theoretical problem with the RTM method when defined through a mean height surface. The problem may be circumvented through modifying the definition, representing the reference topography by a multipole masslayer expansion on the geoid, or it may be reformulated as a frequency domain method (Vermeer and Forsberg, 1992).

In the "conventional" RTM view a special correction is required after the terrain reduction computations, in order to change the value of a computed quantity $L(T^c)$ into the value $L(T^{c*})$, i.e. the value corresponding to the downward continued *outer* potential. Luckily this *harmonic correction* is quite straightforward if the reference topography is long-wavelength, so that the reference topography above the computation point P may be approximated by a Bouguer plate of density -ρ.

For downward continuation through a Bouguer plate, the harmonic correction for geoid and deflections of the vertical will be zero, whereas the harmonic correction for gravity will be

$$\Delta g_P^* - \Delta g_P = 4\pi G\rho(h_{ref} - h_P) \tag{23}$$

This follows from a gravity Prey reduction. If the approximation (20) is used to compute the RTM gravity anomaly the harmonic correction is automatically taken into account.

2.5. Direct or indirect use of terrain reductions

In the typical situation of geoid prediction from gravity anomalies using gridded data, the direct use of terrain reductions thus involves the following steps:

> 1. *Make terrain reduction $\Delta g^f = \Delta g - \Delta g_m$*
> 2. *Grid reduced gravity data $\Delta g^f \rightarrow \Delta g^{grid}$*
> 3. *Predict reduced geoid $\zeta^f = S(\Delta g^f)$*
> 4. *Restore terrain effects $\zeta = \zeta^f + \zeta_m$*

$S(\cdot)$ is the Stokes' operator, typically implemented by FFT methods. In addition to the above scheme a reference field like OSU91A would also be removed/restored. The advantage of the terrain remove/restore scheme is that the reduced gravity anomalies are smooth and with low variability, easy to grid, and further errors in the geoid computation step are mininized, especially the "Molodensky" errors due to the non-level surface the gravity observations refer to.

The above scheme is, however, in practice often modified to an "indirect" use of terrain reductions, so that terrain reductions are only applied in the gravity anomaly gridding process:

1. Terrain reduction $\Delta g^f = \Delta g - \Delta g_m$
2. Grid reduced gravity data $\Delta g^f \longrightarrow \Delta g^{f,grid}$
3. Restore free-air anomalies $\Delta g^{grid} = \Delta g^{f,grid} + \Delta g_m^{grid}$
4. Predict final geoid $\zeta = S(\Delta g)$

In this scheme the terrain reduction used is nearly always the complete Bouguer reduction

$$\Delta g_m = -2\pi G\rho h_P - c_P \qquad (24)$$

In the restore step (3.) only the simple Bouguer anomaly term ($2\pi G\rho h^{grid}$) is usually restored, resulting in a grid of *Faye* anomalies ($\Delta g+c$). This is done partly because it is difficult to compute average terrain corrections in a grid with sufficient accuracy, and partly because the Faye anomalies enter into the Helmert condensation approximation for the direct prediction of the geoid, cf. section 2.7.

The advantage of using the second "indirect" scheme is that large, national gravity grids can be directly utilized, and the use of terrain reductions is "hidden" in the generation of this grid. The drawback of the approach is that the full variablility of the gravity anomaliy field has to be handled in the geoid prediction, making the use of Molodensky corrections terms or high-order Helmert condensation terms more necessary.

2.6. Terrain reductions and Molodensky's theory

Molodensky's theory handles the problem of gravity anomalies referring to an non-level surface. It does not in any way remove the effects of topography, nor provide a smoothing of the gravity field allowing interpolation of sparse gravity measurements in mountainous areas. In the basic theory using harmonic continuation, the gravity field is found at a level surface by a sum of terms where

$$\Delta g^* = \sum_{n=0}^{\infty} g_n \,,$$

$$(25)$$

$$g_n = -\sum_{r=1}^{n} z^r L^r(g_{n-r}), \ g_o = \Delta g$$

z is the height difference to the level surface, and L the "upward continuation" operator, applicable to any surface function (not just a harmonic function) through the Poisson integral

$$L(T) = \frac{\partial T}{\partial z} = -\frac{T}{r} + \frac{r^2}{2\pi} \int\int_o \frac{T-T_P}{l_o^3} d\sigma \qquad (26)$$

At the level surface the usual Stokes operator may be applied to find the downward continued quasigeoid $\zeta^* = S(\Delta g^*)$, which is finally upward continued to the topographic surface by a similar sum

$$\zeta = \zeta^* + \sum_{n=1}^{\infty} \frac{1}{n!} z^n L^n \zeta^* \qquad (27)$$

For details see (Moritz, 1980) or Sideris (1987). To first order the Molodensky theory breaks down to a very simple scheme:

> 1. Predict vertical gravity gradient T_{zz} from Δg (e.g., by FFT)
> 2. Downward continue $\Delta g^* = \Delta g - T_{zz}h$
> 3. Apply Stokes operator $\zeta^* = S(\Delta g^*)$
> 4. Upward continue to surface by the vertical gradient
>   of the height anomaly (i.e., the gravity anomaly)
> $\zeta = \zeta^* + \Delta g^* z$

Obviously this scheme is much more stable if terrain-reduced quantities (Δg^c, T_{zz}^c etc.) are used. The whole Molodensky theory may in principle be applied to terrain-reduced data just as well as to the original free-air data, yielding much smaller Molodensky terms g_n^c. Molodensky theory and terrain reductions are therefore complementary, and both should in principle be applied for optimal results, cf. Forsberg and Sideris (1989).

The above first-order Molodensky scheme has been implemented in the GRAVSOFT FFT geoid prediction program GEOFOUR. In general the g_1^c is quite small when RTM anomalies are used, and may often be completely neglected. Without terrain reductions the g_1 integral becomes essentially an integral over the heights squared (similar to the terrain correction integral), and may often give large corrections (tens of mgal).

2.7. Helmert condensation

Helmert condensation is the "classical" solution to the problem of the non-level reference surface. In its conventional application it is a kind of mix of "Molodensky correction" and some elements of terrain reduction. The method is important because several major continental geoid models have been computed using the must be removed.

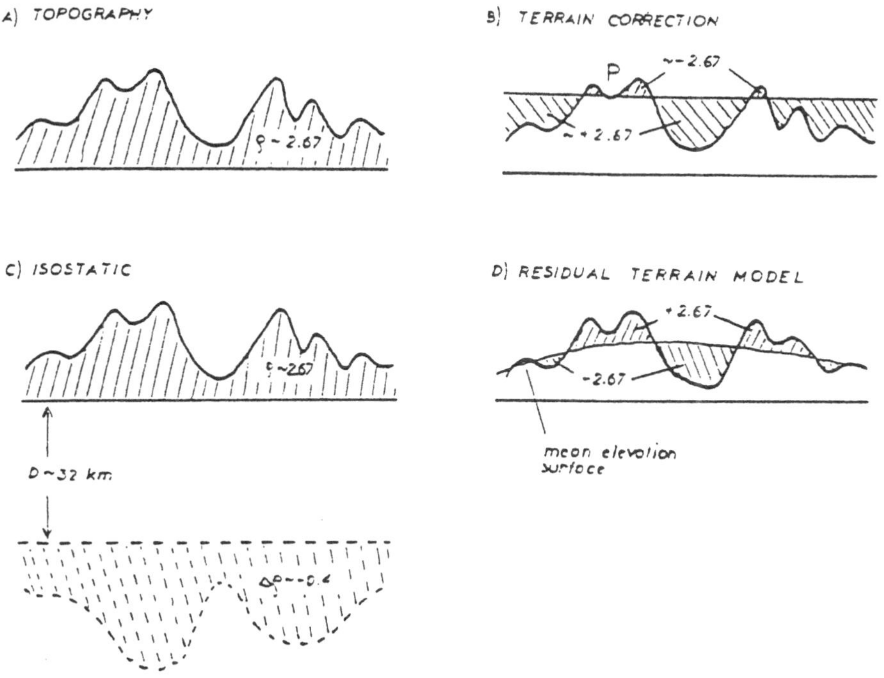

A) TOPOGRAPHY

B) TERRAIN CORRECTION

C) ISOSTATIC

D) RESIDUAL TERRAIN MODEL

Fig. 5. Examples of density distributions associated with different terrain effects

This may in practice be done using a *residual terrain model (RTM)*. In the RTM model a smooth *mean elevation surface* $h_{ref}(\phi,\lambda)$ is used to define an implicit reference density model ρ_{ref} which has crustal density (e.g., 2.67 g/cm³) up to the reference level h_{ref}. The reference surface could e.g. be defined as a surface corresponding to T_{ref} (e.g. a spherical harmonic expansion of global topography to same degree and order), but could as well be defined "freely" through a suitable filtering of local terrain heights.

If the "resolution" of the mean surface is corresponding to the spherical harmonic method, e.g. in the US (Milberts' Geoid-90) and Canada (e.g. GSD-solutions by Verenneau and Mainville). In Europe the joint European geoid model (currently in prep. by Denker) is computed as a quasigeoid.

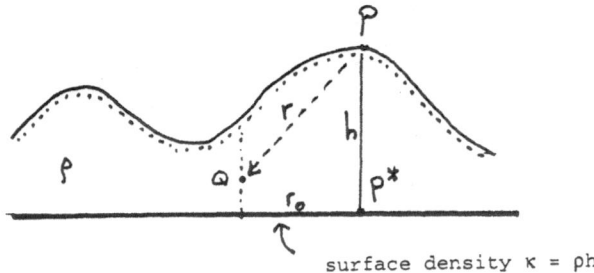

Fig. 6. Helmert condensation mass body

The Helmert condensation consists of shifting the topographic masses to a mass layer on the geoid of surface density $\kappa = \rho h$ (fig. 6). The "mass shift" is carried out with a downward continuation of the gravity anomalies in the following scheme:

1. *Remove complete Bouguer effect:* $\Delta g^B = \Delta g - (2\pi G\rho h - c)$
2. *Downward continue* $\Delta g^{B*} \approx \Delta g^B$ *(assume* T_{zz}^B *is zero)*
3. *Restore condensed topography* $\Delta g^* = \Delta g + 2\pi G\rho h$

The outcome of this process is the Faye anomaly ($\Delta g + c$), which is not smooth at all (the Faye
anomaly actually shows an even more perfect local correlation with height than the free-air anomalies), but it may still be viewed as the result of a terrain reduction of a density anomaly consisting of the topography (density ρ) and a masslayer on the geoid (density $-\kappa$), combined with a downward continuation assumption. This downward continuation assumption is reasonable because T_{zz}^B is much smoother than T_{zz} itself (there is currently in geodesy quite a debate on how exactly to interpret the Helmert reduction, and some theoretically inclined authors prefer not to apply any downward continuation assumption, making the method more theoretically correct, but less useful in practice).

In the above "classical" Helmert scheme the "terrain reduction" should be applied on the *full* gravity operator

$$\Delta g = - \frac{\partial T}{\partial r} - 2\frac{T}{r} \tag{28}$$

and not the just the gravity disturbance. So this leaves a contribution from the second term, called the *"indirect effect on gravity"*. To compute this small term (neglecting the downward continuation contribution) we need the Helmert potential terrain effect at sea level, i.e.

$$T_m(P^*) = G\rho \int\limits_{-\infty}^{\infty}\int\int\limits_{0}^{h} \frac{1}{r} dxdydz + G\rho h \int\limits_{-\infty}^{\infty}\int \frac{1}{r_o} dxdy \approx -\pi G\rho h_P^2 \tag{29}$$

$$r_o = \sqrt{(x-x_P)^2+(y-y_P)^2}, \quad r = \sqrt{r_o^2+z^2}$$

The approximative formula is obtained by a second-order expansion of r^{-1}. The "full" Helmert-condensed gravity anomaly at sea level Δg^{c*} may thus be converted to the quasigeoid at sea level ζ^{c*} by Stokes formula. By restoring the Helmert topography at sea-level the geoid undulation is now formally obtained (since the point P* after the restore correction will be located inside the mass), in other words

$$N = S(\Delta g + c - 2\frac{T_m}{\gamma r}) + \frac{T_m}{\gamma} \tag{30}$$

The last term is often called the *"first order indirect effect"*. The Helmert condensation method thus gives the geoid directly, but the theory itself is approximative. Higher-order expansions can be carried out (see e.g. Sideris, 1990), but if refined expansions are used for the downward continuation as well, consistency is lost with the conventional Helmert orthometric heights. It is therefore (in my view) preferable as far as possible to work with height anomalies, and only at the end of a geoid computation shift back to formal geoid undulations. However, the Helmert condensation method is very simple and straightforward to apply, and will in most cases yield a sufficient accuracy.

2.8. Reference fields in rough topography

All of the above theory is in practice applied relative to a reference field expansion by spherical harmonics, e.g. OSU91A

$$T_{ref}(r,\phi,\lambda) = \frac{GM}{r} \sum_{n=2}^{n_{max}} \sum_{m=0}^{n} (\frac{R}{r})^n [C_{nm}\cos m\lambda + S_{nm}\sin m\lambda] P_{nm}(\sin\phi)$$

$$(31)$$

Such an expansion is by its nature a *function in space*, and reference gravity at a point P should be evaluated at the correct elevation r = R+h_p. In practice reference effects are often computed in grids at a constant elevation. A 3-dimensional interpolation should therefore in principle be carried between at least reference two grids computed at different elevations (a "sandwich grid" interpolation). However, in many cases the effects of the removing a "sandwich" grid and restoring it in the final predictions is quite small, and can be neglected.

3. THE PRACTICAL COMPUTATION OF TERRAIN EFFECTS

Historically terrain corrections or -effects were computed using overlays on maps, subdivided in concentric circles and radial sectors. In each sector mean elevations were read from maps, and terrain corrections summed up using either tables or simple calculations based on the closed gravitational formulas for a cylindar, cf. Heiskanen and Moritz (sec. 3-1). The cylindrical overhead systems had names like "Hammer zones" or "Hayford zones". The Hammer system, used widely in geophysical prospecting, computed terrain corrections in a number of rings ranging in radius from zone B (2-16 m, 4 sectors) to zone K (15-22 km, 16 sectors). The Hayford system, used for regional work, had a similar system, in principle extending globally, but in practice extended only to zone "0_2" (167 km).

The Hayford terrain correction zone system often included a special correction - Bullard's term - taking into account the spherical earth, making a correction for a "spherical cap" rather than the usual planar Bouguer plate. It is very important when using "spherical" terrain-corrected Bouguer or isostatic anomalies, to be well-aware what has been done, e.g. to which distance have computations been carried out. Basically the full global "spherical" terrain correction is a quantity which is not usable at all: it is way too big, and stongly dependent on the *distance* out to which it is computed.

An example for illustration: consider a gravity point e.g. at the top of a large flat plateaux, with a negligible "planar" terrain correction (the edge of the plateaux is assumed to be far away), and no other topography on the earth. The complete topographic effect will in this case be -$2\pi G\rho h$. The spherical Bouguer correction corresponding to a spherical shell of thickness h is -$4\pi G\rho h$, so the global spherical terrain correction must be $2\pi G\rho h$.

The spherical "problems" can be completely avoided by using RTM-reductions, or by direct spherical computation of the complete topographic/isostatic attraction using e.g. prisms. The integration of terrain effects by prisms is nowadays the method of choice for space-domain computations, augmented by the much faster, but sometimes more approximative, FFT methods.

3.1 Terrain effect integration by prisms

The rectangular prism of constant density is a useful "building block" for numerical integrations of the basic terrain effects of form e.g. (17) or (19). Closed formulas for the gravitational potential and all derivatives exist, see e.g. Forsberg (1984), but the formulas are quite complicated. For a point P at the origin of the coordinate system, the gravity disturbance of a prism of density ρ in coordinate interval x_1-x_2, y_1-y_2, z_1-z_2 will be

$$\delta g_m = G\rho |x \log(y+r) + y \log(x+r) - z \arctan\frac{xy}{zr}|_{x_2,y_2,z_2}^{x_1,y_1,z_1}, \tag{32}$$

$$r = \sqrt{x^2+y^2+z^2}$$

The corresponding formula for the potential (and hence the geoid height) is even more complicated

$$T_m = G\rho | \; xy \log(z+r) + xz \log(y+r) + yz \log(x+r)$$
$$- \frac{x^2}{2} \arctan\frac{yz}{xr} - \frac{y^2}{2} \arctan\frac{xz}{yr} - \frac{z^2}{2} \arctan\frac{xy}{zr} \; |_{x_2,y_2,z_2}^{x_1,y_1,z_1} \tag{33}$$

It should be pointed out that in principle the computation of both (32) and (33) is required in order to rigorously compute the gravity anomaly, given by the functional

$$\Delta g = -\frac{\partial T}{\partial r} - \frac{2}{r}T \tag{34}$$

When computing topographic/isostatic effects the second term in (34) is called the *indirect effect on gravity* (again), and is often explained using the concept of gravity reduction from the geoid to the "cogeoid" (the cogeoid is corresponding to the terrain-reduced potential T-T_m). The "indirect" term may often be neglected in connection with RTM-reductions, where the associated geoid effects T_m are small (typically less than 1 m, corresponding to a 0.3 mgal "indirect effect").

The gravitational formulas for the rectangular prisms are slow and numerically unstable at large distances (they involve a number of small differences between large numbers, corresponding to the corners of the prisms). Therefore approximative formulas are useful at larger distances. Such formulas are based on an expansion of the prism field in spherical harmonics, which gives a surprisingly simple expansion of form (McMillan, 1958)

$$T_m = G\rho \; \Delta x \Delta y \Delta z \; [\frac{1}{r} + \frac{1}{24r^5}[(2\Delta x^2 - \Delta y^2 - \Delta z^2)x^2 + (-\Delta x^2 + 2\Delta \cdot$$

$$+(-\Delta x^2 - \Delta y^2 + 2\Delta z^2)z^2] + \frac{1}{288r^9}[\alpha x^4 + \beta y^4 + ...] + ...],$$

$$\Delta x = x_2 - x_1, \Delta y = y_2 - y_1, \Delta z = z_2 - z_1$$

(35)

This equation is easily differentiated for other gravity field quatitities. In the GRAVSOFT "TC"-program, such approximative formulas are automatically used in the prism integration when accuracy permits to obtain reasonable computation speeds. Approximative formulas are also used for the prism potential itself (corresponding to a mass-plane).

To further increase computation speed, and to allow use of less detailed, remote topography, it is practical to use a coarse/detailed grid system. This is fully implemented in TC. The detailed grid is used out to a minimum distance, and the coarse grid is used for the remainder of the topography. In a small innerzone of 3 x 3 grid points just around the computation point the topographic data are densified using a bicubic spline interpolation, so that a "finer" more smooth set of prisms is used to integrate the often large effects of the innerzone (fig. 7). This densification of the innerzone is essential to avoid a computation P being located at the edge of a prism, giving rise to artificial terrain effects from the "edges" of a prism.
Since gravity terrain effects are strongly dependent on the height of the computation point (through the $2\pi G\rho h$-term), a special precaution is necessary when the height of the computation point does not agree with the interpolated height from the DTM. Either the computation point can be "forced" to match the interpolated topography level, or the topography can be modified locally to give the "right" value at P. The modificiation in the TC program is done using a "smooth" correction in the innerzone (fig 8). The discrepancy between DTM and station heights will always

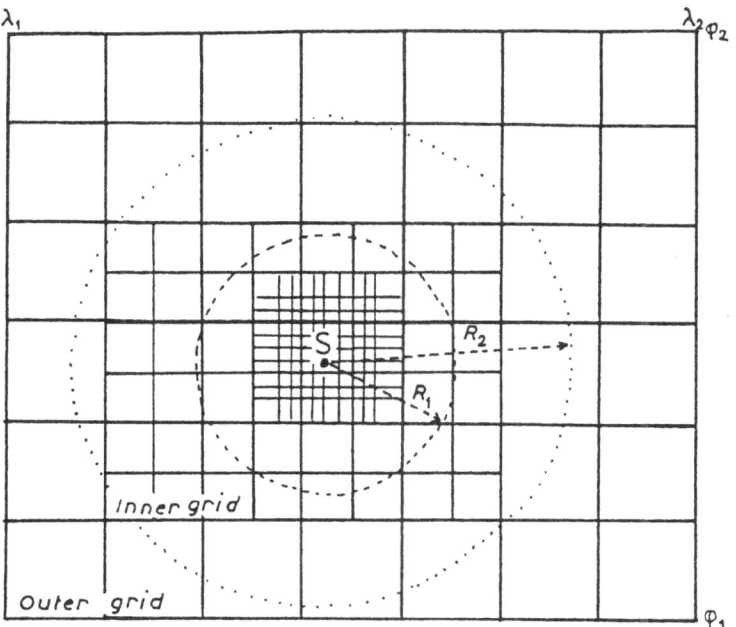

Fig. 7. Use of innerzone, detailed and coarse height grids in "TC"

be present, since the DTM's will hardly ever have sufficient resolution to represent all features in rugged topography.

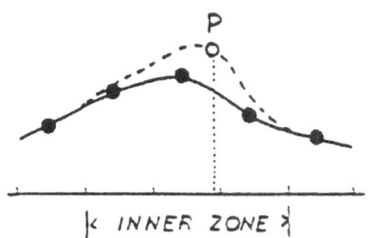

Fig. 8. Modification of terrain heights in innerzone

In the TC program the terrain grids to be used are assumed to be given as regular grids in either geographical coordinates or a UTM-zone. Grids are stored rowwise from north to south, and an additional grid must be provided by the user for the smooth mean height surface if RTM anomalies are to be computed. Mean height grids and smoothed reference surface grids (using a moving-average filter) may be computed by an auxillary program "TCGRID".

3.2. Terrain effects by Fourier transformation methods

Since digital heights or bathymetry is typically given in regular grids, Forurier transform methods is an obvious choice for speeding up computations of large sets of terrain effects, especially if a complete grid of terrain effects is required. The basis of most Fourier methods can be formulated in terms of *convolutions*, which can be evaluated very fast by Fast Fourier Transform methods.

A two-dimensional convolution integral is an integral over the x-y plane of form

$$g(x,y) = \int\int_{-\infty}^{\infty} f(x-x', y-y')\, h(x',y')\, dx'dy' = f*h(x,y) \quad (36)$$

In terms of signal analysis terminology f is the input function, h the convolution kernel, and g the output. Using the two-dimensional Fourier transform

$$\mathscr{F}(f) = F(k_x,k_y) = \int\int_{-\infty}^{\infty} f(x,y)e^{-i(k_x x + k_y y)} dxdy$$

$$\mathscr{F}^{-1}(F) = f(x,y) = \frac{1}{4\pi^2} \int\int_{-\infty}^{\infty} F(k_x,k_y)e^{i(k_x x + k_y y)} dk_x dk_y$$

$$(37)$$

the convolution f*h is readily obtained by the convolution theorem by a multiplication in the spectral domain

$$f*h = \mathscr{F}^{-1}[F \cdot G] \quad (38)$$

The convolution is in practice carried on gridded data over a final domain using FFT. Since FFT assumes the function grids to be periodic in x and y, the grids must be extended by a border zone of zeros (zero padding), or results discarded close to the grid edges. These topics are treated in more detail in Schwarz et al.(1990).

Most terrain effects are characterized by being unlinear integrals, so series expansions are typically required to make terrain computations by FFT. However, in many cases just one or two terms are sufficient. In the sequel I will illustrate how convolutions are obtained in a few examples.

Terrain corrections.

In the planar terrain correction integral

$$c_P = G\rho \int\limits_{-\infty}^{\infty} \int \int\limits_{h_P}^{h} \frac{z-h_P}{r^3} dx dy dz, \quad r = \sqrt{(x_P-x)^2+(y_P-y)^2+(h_P-z)^2} \quad (39)$$

the unlinear kernel may for small surface slopes be approximated by

$$\frac{1}{r^3} \sim \frac{1}{r_o^3}, \quad r_o = \sqrt{(x_P-x)^2+(y_P-y)^2} \quad (40)$$

This approximation is termed the *linear approximation* to the boundary value problem. In this approximation the z-integration of the integral yields

$$c_P = \frac{1}{2} G\rho \int \int\limits_{-\infty}^{\infty} \frac{(h-h_P)^2}{r_o^3} dx dy \quad (41)$$

This formula is not a convolution, but by writing the integrand in terms, a set of convolutions in h and h^2 is obtained

$$c_P = \frac{1}{2} G\rho [h^2 * f - 2h_P(h*f) + h_P^2 f_o], \quad f = (x^2+y^2)^{-\frac{3}{2}} \quad (42)$$

In this formula f_o is a singular integral

$$f_o = \int \int\limits_{-\infty}^{\infty} \frac{1}{r_o^3} dx dy \quad (43)$$

which, when evaluated with FFT over a finite domain, turns into a constant, readily obtained from the 0 (DC) value of the Fourier transform of the kernel. Similarly the singularity of r_o^{-3} at zero distance presents no problem in practice when discrete data are used, as the expression (42) will be independent of the actual value at zero of the kernel. For more details see Sideris (1985) or Forsberg (1985).

The computation of terrain corrections by FFT is of widespread use, and may be refined with higher order terms, see e.g. Harrison and Dickinson (1989). Based on terrain corrections RTM anomalies may be obtained by (20).

Isostatic, bathymetric or airborne terrain effects.

When gravity anomalies are wanted at a constant level above some topography or

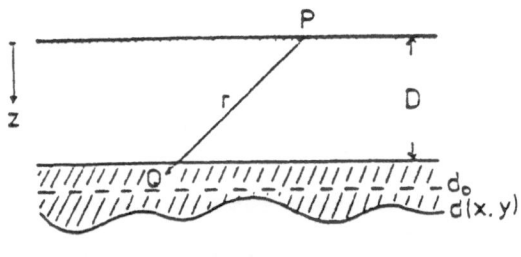

Fig. 9

bathymetry, a convolution series expansion of Fourier transforms is obtained, equivalent to the frequency domain Parker's formula. The general situation illustrated in fig. 9 can be used to compute isostatic effects, the effect of bathymetry on marine gravity, or - by changing the sign of d - to obtain gravity terrain effects in airborne gravimetry.

For the isostatic case, the gravity effect of the compensation mass body at sea level will be

$$g_m = G\Delta\rho \int\limits_{-\infty}^{\infty}\int \int\limits_{D}^{D+d} \frac{z}{r^3} dxdydz \tag{44}$$

The kernel may be evaluated in a power series around a suitable reference level d_o as

$$\frac{z}{r^3} = \frac{d_o}{r_o^3} + \frac{r_o - 3d_o}{r_o^5}(z - d_o) + \dots \quad , \quad r_o = \sqrt{(x - x_P)^2 + (y - y_P)^2 + d_o^2} \tag{45}$$

and the integration be carried out with respect to z, yielding to second order a set of convolutions in d and d^2

$$g_m = G\Delta\rho[d*f_1 + d^2*f_2]$$

$$f_1 = \frac{d_o}{r_o^3} + \frac{r_o^2 - 3d_o^2}{r_o^5}(D-d_o), \quad f_2 = \frac{r_o^2 - 3d_o^2}{2r_o^5} \tag{46}$$

A similar set of expansions may be obtained for geoid heights, using an expansion of the "geoid" kernel

$$\frac{1}{r} \approx \frac{1}{r_o} - \frac{d_o}{r_o^3}(z-d_o) \tag{47}$$

yielding to second order

$$\zeta_m = \frac{G\Delta\rho}{\gamma}[d*t_1 + d^2*t_2],$$

$$t_1 = \frac{1}{r_o} - \frac{d_o(D-d_o)}{r_o^3}, \quad t_2 = -\frac{d_o}{2r_o^3} \tag{48}$$

The above second-order formulas are efficiently implemented in FFT programs by obtaining simultanously the transforms of d and d^2 in one complex transform.

RTM geoid effects.

The above formulas for the geoid effects do not apply to RTM geoid effects evaluated at the surface of the topography. However, since RTM geoid effects are quite small, a simple condensation approximation will in most cases be sufficient, i.e.

$$\zeta_m = \frac{G\rho}{\gamma}\int_{-\infty}^{\infty}\int\int_{h_{ref}}^{h}\frac{1}{r}dxdydz \approx \frac{G\rho}{\gamma}(h-h_{ref})*\frac{1}{r_o} \tag{49}$$

Such geoid effects are often computed over large regions (opposed to gravity terrain corrections typically computed in local blocks), and spherical FFT methods may with advantage be used.

In the GRAVSOFT system the FFT terrain effect computations have been implemented in the planar "TCFOUR" program, or - to a less complete degree - in the spherical FFT program SPFOUR.

LITTERATURE & REFERENCES

Abd-Elmotaal, H. A.: On the calculation of the attraction of the topographic isostatic masses. In: Geodetic Theory Today, F.Sanso' (Ed.), pp. 372 - 380, IAG Symposia 114, 1995.

Arabelos, D. and C.C.Tscherning: Regional recovery of the gravity field from SGG and Gravity Vector data using collocation. J.Geophys. Res., Vol. 100, No. B11, pp. 22009-22015, 1995.

Arabelos, D., and I.N.Tziavos: The contribution of bathymetry to gravity field modelling. Pres. INSMAP 94, Hannover, Sep., 1994.

Arabelos, D., and I.N.Tziavos: Comparison of the global ETOPO5U model with local DTMs. Pres. INSMAP 94, Hannover, Sep., 1994.

Arabeols, D., & I.N.Tziavos: Gravity field improvement in the Mediterranean Sea by correcting bottom topography using Collocation. Presented XXI IUGG/IAG General Ass., G10, Boulder, July 1995.

Denker, H.: Hochaufloesende regionale Schwerefeldbestimmung mit gravimetrischen und topograpischen Daten. Wiss. Arb. Fachrichtung Vermessungswesen der Univ. Hannover, nr. 156, 1988.

Dronin, A.A. and S.V.Lebedev: Computation of terrain corrections on an electronic computer using terrain information in the form of contour lines (in russian). Izvestiya Vuz: Geodeziya i Aerofotos" yemka, No. 1, pp. 32-36, 1985.

Ferland, R. & J.A.R.Blais: Terrain Correction for Gravity measurements. Dept. of Geomatics Eng. Rep. 20009, University of calgary, 1984.

Forsberg, R.: A Study of Terrain Reductions, Density Anomalies and Geophysical Inversion Methods in Gravity Field Modelling. Reports of the Department of Geodetic Science and Surveying, No. 355, The Ohio State University, Columbus, Ohio, 1984.

Forsberg, R.: Gravity Field Terrain Effect Computations by FFT. Bulletin Geodesique, Vol. 59, pp. 342-360, 1985.

Forsberg, R.: Spectral Properties of the Gravity Field in the Nordic Countries. Boll. Geodesia e Sc. Aff., Vol. XLV, pp. 361-384, 1986.

Forsberg, R.: Terrain effects in Geoid computations. In: Lecture Notes, Int. school for the deter. and use of the geoid, Milano, Oct. 1994, pp.101-134, IGeS, Milano, 1994.

Forsberg,R. and M.Sideris: The role of topography in geodetic gravity field modelling. Proceedings Chapman Conf. on Progress in the Determination of the Earths gravity field, Ft. Lauderdale, Sept. 1988, pp. 85-88, American Geophysical Union, 1988.

Forsberg, R. and M.Sideris: On topographic effects in gravity field approximation. Festschrift to Torben Krarup, pp. 129-148, København, 1989.

Forsberg, R. and C.C.Tscherning: The use of Height Data in Gravity Field Approximation by Collocation. J.Geophys.Res., Vol. 86, No. B9, pp. 7843-7854, 1981.

Harrison,J.C. and M.Dickinson: Fourier transform methods in local gravity modelling. Bull. Geod., Vol. 63, pp. 1149-166, 1989.

Heiskanen, W.A. and H. Moritz: Physical Geodesy. W.H. Freeman & Co, San Francisco, 1967.

Jekeli, C.: The Downward Continuation to the Earth's surface of Truncated Spherical and Ellipsoidal Harmonic Series of the Gravity and Height Anomalies. Reports of the Department of Geodetic Science and Surveying No. 323, The Ohio State University, Columbus, 1981.

Knudsen, P. & O. Ba. Andersen: Ocean bottom topography from ERS-1 altimeter data. Earth Observation Quarterly, No. 51, pp. 16-18, 1996.

Mainville, A., M.Veronneau, R.Forsberg & M.Sideris: A comparison of geoid and quasigeoid modelling methods in rough topography. Pro. Geoid & Gravity, pp. 491-501, IAG Symp. 113, Springer Verlag, 1995.

Martinec, Z., C.Matyska, E.Grafarend & P.Vanicek:On Helmert's 2nd condensation Method. manus. geodaetica., Vol. 18, pp. 417-421, 1993.

McMillan, W.D.: The theory of potential. Theoretical mechanics vol. 2. Dover, New York, 1958.

McNutt, M.: Compensation of ocean topography: An application of the response function technique to the Surveyor area. J.Geophys. Res., Vol. 84, no. B13, pp. 7589-7598, 1979.

Moritz, H.: The use of terrain correction in solving Molodensky's problem. Dep. of Geodetic Science, Rep. no. 108, The Ohio State University, 1968./7854

Moritz, H.: Advanced Physical Geodesy. H.Wichmann Verlag, Karlsruhe, 1980.

Nagy, D.: The prism method for terrain corrections using digital computers. Pure applied Geophysics, Vol. 63, pp. 31-39, 1966.

Parker,R.L.: The theory of ideal bodies for gravity interpretation. Geophys.J.R.astr.Soc., Vol. 42, pp. 315-334, 1975.

Parker, R.L.: Improved Fourier terrain correction, Part I. Geophysics Vol. 60, pp. 1007 - 1017, 1995.

Parker, R.L.: Improved Fourier terrain correction, Part II. Geophysics Vol. 61, pp. 365 - 374, 1996.

Pellinen, L.P.: Accounting for topography in the calculation of quasigeoidal heights and plumb-line deflections from gravity anomalies. Bulletin Geodesique, Vol. 63, pp. 57-65, 1962.

Peng, M.: Topographic effects on gravity and gradiometry by the 3D FFT and FHT methods. Thesis, Dep. geomatics eng., Univ. of Calgary, 1994.

Peng, M., Y. Cai Li and M.G.Sideris: First results on the computation of terrain corrections by the 3D FFT method. Manuscr. geodaetica., Vol. 20, pp. 475-488, 1995

Rapp, R.H.: Degree variances of the Earth's potential, topography and its isostatic compensation. Bulletin Geodesique, Vol. 56, No. 2, pp. 84-94, 1982.

Rapp, R.H.: The decay of the spectrum of the gravitational potential and the topography for the Earth. Geophys. J. Int., Vol. 99, pp. 449-455, 1989.

Rapp, R.H., Y.M.Wang and N.K.Pavlis: The Ohio State 1991 Geopotential and Sea Surface Topography Harmonic Coefficient Models. Rep. of the Dep. of Geodetic Science and Surveying n. 410, The Ohio State University, Columbus, 1991.

Rummel, R., R.H.Rapp, H.Suenkel and C.C.Tscherning: Comparison of Global Topographic/Isostatic Models to the Earth's Observed Gravity Field. Reports of the Dep. of Geodetic Science and Surveying, No. 388, The Ohio State University, 1988.

Sanso', F., & G.Sona: Gravity reductions versus approximate B.V.P.S. In: Geodetic Theory today, F.Sanso' (Ed.), IAG Symposia no. 114, pp. 304-314, 1995.

Schwarz,K.P., M.G.Sideris and R.Forsberg: The use of FFT techniques in physical geodesy. Geophys. J. Int., Vol. 100, pp. 485-514, 1990.

Shaofeng, B. & W. Xiaoping: An improved prism integration for gravimetric terrain correction. Manusc. geodaetica, Vol. 20, pp. 515-518, 1995.

Sideris, M. G.: Computation of Gravimetric Terrain Corrections Using Fast Fourier Transform Techniques. Division of Surveying Engineering, The University of Calgary, publ. 20007, March 1984.

Sideris, M.: A Fast Fourier Transform method for computing terrain corrections. Manuscripta Geodaetica, Vol. 10, no. 1, pp. 66- 71, 1985.

Sideris, M.: Spectral Methods for the numerical solution of Molodensky's problem. UCSE rep. 20024, Dep. surveying Eng., Univ. of Calgary, 1987.

Sideris, M.: Rigorous gravimetric terrain modelling using Molodensky's operator. Manusc. geodaetica, Vol. 15, pp. 97-106, 1990.

Sjoeberg, L.: The total terrain effect in the modified Stokes' formula. Proc. Symp. Geoid & Gravity, pp. 616 - 623, IAG Symposia, Vol. 113, 1995.

Suenkel, H.: Digital Height and Density Model and its Use for Orthometric Height and Gravity Field Determination for Austria. Boll. di Geodesia e Sc. Aff., 1987.

Suenkel, H., N.Bartelme, H.Fuchs, M.Hanafy and W.D.Schuh: The Gravity field in Austria. Proceedings of the IAG Symposia, pp. 475-503, International Association of Geodesy, Paris, 1987.

Tscherning, C.C.: Gravity Prediction using Collocation and taking known mass density anomalies into account. Geophys. J.R. astr. Soc., Vol. 59, pp. 147 -153, 1979.

Tscherning, C.C.: Isotropic reproducing kernels for the inner of a sphere or spherical shell and their use as density covariance functions. Math. Geology, Vol. 28, pp. 161-168, 1996.

Tscherning, C.C., P.Knudsen and R.Forsberg: Description of the GRAVSOFT package. Geophysical Institute, University of Copenhagen, Technical Report, 1991, 2. Ed. 1992, 3. Ed. 1993, 4. ed, 1994.

Tscherning, C.C., R.Forsberg and P.Knudsen: The GRAVSOFT package for geoid determination. Proc. 1. Continental Workshop on the Geoid in Europe, Prague, May 1992, pp. 327-334, Prague, 1992.

Tscherning, C.C., R.Forsberg & P.Knudsen: First experiments with improvement of depth information using gravity anomalies in the Mediterranean Sea. GEOMED Report no. 4, Ed: Arabelos & Tziavos, pp. 133-148, Thesssaloniki, 1994.

Tscherning,C.C. and R.Forsberg: Harmonic continuation and gridding effects on geoid height prediction. Bulletin Geodesique, Vol. 66, pp. 41-53, 1992.

Tscherning, C.C. and H. Suenkel: A Method for the Construction of Spheroidal Mass Distributions consistent with the harmonic Part of the Earth's Gravity Potential. Manuscripta Geodaetica, Vol. 6, no. 2, pp. 131-156, 1981.

Tziavos,I.N., M.G.Sideris, R.Forsberg and K.P.Schwarz: The effects of the terrain on airborne gravity and gradiometry. J.Geophys. Res., Vol. 97, 99. 8843-8852, 1988.

Vermeer, M. and R.Forsberg: Filtered Terrain Effects: A Frequency Domain Approach to Terrain Effect Evaluation. Manuscripta Geodaetica, Vol. 17, pp. 215-226, 1992.

Wang, Y.M.: Downward continuation of the free air gravity anomalies to the ellipsoid using the gradient solution, Poisson's integral and terrain correction-numerical comparisons and the computations. Rep. Dep. geod. Sc. and Surv., no. 393, The Ohio State University, Columbus, 1988.

Wang, Y.M.: Comments on proper use of terrain correction for the computation of height anomalies. Manusc. geodaetica., Vol. 18, p. 53 - 57, 1993.

Wichiencharoen,C.: The indirect effects on the computation of geoid undulations. Rep. Dep. Geodetic Sc. and Surv., no. 336, The Ohio State University, Columbus, 1982.

Global Models for the 1cm Geoid – Present Status and Near Term Prospects

R.H. Rapp

1. Introduction

This discussion is designed to consider the calculation of geoid undulations from spherical harmonic representations of the Earth's gravitational potential. These potential coefficient models are called global models because once the potential coefficients are defined the geoid undulation can be computed at an arbitrary latitude and longitude. Of course the accuracy of the undulation estimate will vary from area to area.

The use of potential coefficient models for the calculation of geoid undulations has been carried out for many years. In the 1950's estimates of the Earth's gravitational potential from surface gravity data was carried out to degree 8 corresponding to a resolution of approximately 2500 km. Today one uses expansions to degree 360 routinely and soon we will be using expansions to much higher degrees.

The interest in geoid computations has been around for a long time. However in the past five to ten years there has become an increasing awareness of the need for an accurate determination of geoid undulations with a high accuracy on a variety of spatial and spectral scales that will be described later.

Why the renewed interest in global determinations of the geoid? There are two main reasons. First is the rapid growth of GPS positioning where the vertical component desired is the mean sea level height (orthometric height) or the normal height. To obtain such information it is necessary to compute absolute or relative geoid undulations or height anomalies. Second is the increasing interest in ocean circulation through the development of global circulation through the development of global circulation models from basic principles and data and the subsequent comparisons with circulation information derived from satellite altimeter data and geoid undulation or undulation slope data. So we may have a way of evaluating ocean circulation models if we have a good geoid, in a global system, or ultimately the merger of models to obtain a optimum combination of different estimation procedures.

There are of course many more specific purposes for which a global geoid model is needed. One such need relates to the estimation of a global vertical datum where the reference surface becomes the geoid whose position is defined in a consistent way through a high degree representation of the Earth's gravitational potential.

In the applications noted above one finds an increasing demand on the accuracy in which geoid undulations (or height anomalies) are needed. A few years ago a spatial accuracy of 1 m may have been considered adequate. Today there may be applications requiring undulation differences to a few cm. Or in the case of spectral components one may want undulations accurate to 1 cm at resolutions of 200 km. The estimation of such high accuracy geoid models is a challenge now being considered by numerous groups.

The discussion in these lectures is to review some basic procedures for estimating the Earth's gravitational potential and then review current developments and how the accuracy of such developments can be assessed. Details will be avoided so that the big picture can be more easily seen.

2. Preliminary Equations

We start from the representation of the Earth's gravitational potential, V, at a point P defined by geocentric distance r, geocentric co-latitude q, and longitude l:

$$V(r,\theta,\lambda) = \frac{GM}{r}\left[1 + \sum_{n=2}^{\infty}\left(\frac{a}{r}\right)^n \sum_{m=-n}^{n} C_{nm}Y_{nm}(\theta,\lambda)\right] \tag{1}$$

GM is the geocentric gravitational constant; a (usually the equatorial radius of an adopted mean-Earth ellipsoid) is the scaling factor associated with the fully normalized spherical geopotential coefficients, C_{nm}. In addition:

$$\begin{aligned}Y_{nm}(\theta,\lambda) &= P_{n|m|}(\cos\theta)\cos m\lambda \text{ if } m \geq 0\\ Y_{nm}(\theta,\lambda) &= P_{n|m|}(\cos\theta)\sin|m|\lambda \text{ if } m < 0\end{aligned} \tag{2}$$

The Pn|m| (cosq) values are the fully normalized associated Legendre functions of the first kind. Note that the summation in (1) on n is to ∞. In practice this is truncated to a finite degree that depends on the size of the mean anomaly (or undulation) cells used in the combination solution.

The estimation of the potential coefficients is important because these values (i.e. C_{nm}) can be used to determine geoid undulations and height anomalies as will be discussed in a later section. So our first line of discussion must relate to the procedure that are used to estimate the potential coefficients.

3. Potential Coefficient Estimation

The determination of the potential coefficients can be done in a variety of ways:

1) the analysis of close earth satellite tracking data to infer potential coefficient data. Such a solution leads to the satellite alone model.

2) the combination of the satellite model normal equations with normal equations generated from satellite altimeter data and surface gravity data. This leads to a combination model that is given to (today) degree 70. Degree 70 is a limit that primarily comes from computer time considerations.

3) the development of degree 360 models can be done in several ways. These include: a) the adjustment of the satellite model potential coefficient estimates with 30'x30' mean gravity anomalies and the use of quadrature procedures to obtain the adjusted 30' anomalies; b) the use of the satellite model with the 30' mean anomalies where the block diagonal normal equations of this data type are taken into account; c) a blended set of coefficients where the coefficients from degree 2 to 70 are taken from the degree 70 combination solution and the coefficients from degree 71 to 360 are taken from one of the 360 models.

In the following sections each of the above techniques is discussed.

3.1 The Satellite Model

The estimation of the potential coefficients from satellite tracking data has a long history in satellite geodesy. Today there are many more satellites, at different inclinations, and average height above the Earth, than ever before. In addition the tracking types have expanded significantly in the past 4 years. The traditional tracking of satellite laser ranging has been complemented by GPS tracking of the Topex/Poseidon satellite, of EUVE, and GPS/MET. More specifically the tracking data now available is: optical, laser, Doppler, GPS, and TDRSS. Altimeter data is also used but not in the estimation of the satellite alone model.

The estimation of the model is a complex task, partly caused by the need to estimate parameters other than just potential coefficient parameters. For example, the parameters of a tide model may be simultaneously estimated with potential coefficients as well as parameters that may be associated with a specific measurement type. Weighting of the data has to be carefully done so that data noise and data density is properly represented in the solution. One might consider large amounts of data that reflect redundant information about the Earth's gravitational signature because the data is acquired over many cycles with the same geographic coverage. Various groups have developed their own ways to handle this weighting procedure. The procedures used for the NASA/DMA project (Lemoine et al, 1996) follow the calibration methods used in previous model developments, such as GEM-T3 or JGM-2. The method used involves the estimation of models with data withheld from the solution and determining the changes in comparison with solutions that included the data. The process is time consuming because of the large number of data types being used in the model estimation.

The result of the data analysis will be the satellite model with a calibrated error covariance matrix. In the course of a major new field estimation there may be several satellite alone models that may be tested in the combination model development or by itself in orbit estimation and tracking data fits. Examples of satellite alone models have included GEM-T2S and GRIM4-S4.

3.2 The Low Degree Combination Model

Next we consider the procedure described as item 2 in the first part of section 3. We need to discuss how gravity anomalies and satellite altimeter data can be used, with the satellite model, to determine a set of potential coefficients to some maximum degree (e.g. 70) that is consistent with the computer resources available to the project.

We start with the introduction of the disturbing potential, $T(r,\theta,\lambda)$, which is the difference between the true gravity potential (W) at point P and the gravity potential (U) implied by a rotating equipotential ellipsoid of revolution. We have:

$$T(r,\theta,\lambda) = W(r,\theta,\lambda) - U(r,\theta,\lambda) \tag{3}$$

If W and U can be represented in spherical harmonic series, so can T:

$$T(r,\theta,\lambda) = \frac{GM}{r} \sum_{n=2}^{\infty} \left(\frac{a}{r}\right)^n \sum_{m=-n}^{n} C_{nm} Y_{nm}(\theta,\lambda) \tag{4}$$

The zero degree term in (4) has been set to zero assuming the equality of the actual mass of the Earth and the mass of the reference ellipsoid. In addition the even zonal coefficients in (4) represent the difference between the coefficients of the actual and normal gravitational potentials.

Gravity anomalies can be calculated from T using various relationships whose accuracy depends on the assumptions or approximations one wishes to make. For these discussions we will take relatively simple forms of the equations emphasizing the assumptions made on effects that are considered in practice. If point Q is on the telluroid, one can represent the radial component of the gravity anomaly as follows:

$$\Delta g(r,\theta,\lambda) = -\left(\frac{\partial T}{\partial r}\right)_Q - \frac{2}{r_Q} T_Q \tag{5}$$

The substitution of (4) into (5) yields:

$$\Delta g(r,\theta,\lambda) = \frac{GM}{r^2} \sum_{n=2}^{\infty} (n-1)\left(\frac{a}{r}\right)^n \sum_{m=-n}^{n} C_{nm} Y_{nm}(\theta) \tag{6}$$

As written, (6) represents a point gravity anomaly without concern for convergence effects, atmospheric effects, etc. Note that the summation to ∞ is replaced, in practice, by a summation to a finite degree M.

For combination solutions one may be given a mean gravity anomaly where the mean is usually defined as an area mean over a cell bordered by two meridians and two parallels. This may not be the best way (Jekeli, 1996) to define a mean value for coefficient estimation purposes but it is the way in which mean values are currently given. For the purposes of our discussion, in this section, we will consider the use of $1°\text{x}1°$ mean surface free-air anomalies (Δg). If one is to estimate the potential coefficients, from the anomaly data, to the same degree (M) that the satellite model has been given one must remove the information above degree M in the anomaly data. This removal can be done, in an approximate way, by the case of a preliminary high degree (360 or higher) potential coefficient model. Conceptually we express the corrected anomaly:

$$\overline{\Delta g}_c(r,\theta,\lambda) = \overline{\Delta g}(r,\theta,\lambda) - \overline{\Delta g}(M+1 \text{ to } \infty) \tag{7}$$

where ∞ is replaced by the maximum degree of the preliminary model. The bar over Δg indicates mean values. These mean values are computed using rigorous integration of the θ,λ terms in (6).

For some computations it is appropriate to reduce the anomalies from the telluroid to the surface of the ellipsoid. This can be done by applying a gradient correction term so that the residual anomaly on the ellipsoid is given by:

$$\overline{\Delta g}_c(\theta,\lambda) = \overline{\Delta g}_c(r,\theta,\lambda) - \frac{\partial \Delta g}{\partial h}(h+N) \qquad (8)$$

where h is the mean normal height of the cell for which the anomaly has been estimated. The calculation of the gradient term can be done from topography alone (with an assumption on the linear correlation of free air anomalies with elevation) or from the preliminary high degree potential coefficient model. One can then express the corrected mean anomaly with the potential coefficients that are being estimated:

$$\overline{\Delta g}_c(r_E,\theta,\lambda) = \frac{GM}{r^2} \sum_{n=2}^{M} (n-1) \left(\frac{a}{r_E}\right)^n \sum_{m=-n}^{n} C_{nm} Y_{nm}(\theta,\lambda) \qquad (9)$$

In practice the right hand side of (9) is calculated using integration over the θ,λ functions. In addition terms are used that convert a given anomaly to a radial component to be consistent with the assumptions leading to (5). Equation (9) forms the basis for the observation equation that will be used to form the normal equations of the surface gravity data that will be added to the normal equations of the satellite model as one determines the low degree combination model. In such least squares adjustment special consideration has to be given to the weights assigned to the anomalies so that anomalies of high accuracy in certain geographic regions do not dominate and distort the solutions.

Next we consider the use of satellite altimeter data where the distance to the ocean surface has been estimated. Knowing the ellipsoid radius one can determine the sea surface height h. Recognizing that the actual estimate of h will be impacted by errors in the environmental and geophysical corrections and errors in the radial component of the satellite orbit due to potential coefficient errors, unmodeled air drags, etc., one can write the idealized (or improved) sea surface height as

$$h_s = h_c + \Delta h_G + \Delta h_I \qquad (10)$$

where h_c is the computed sea surface height, Δh_G is the correction to h_c due to errors in the potential coefficients, and Δh_I is the correction associated with initial state errors, air drag, etc. Analytic expressions for Δh_G can be given as a function of changes in the potential coefficients used in the altimeter orbit determination process. A representation for Δh_I can contain terms that include a bias, a 1 cycle/revolutions effect with both time dependent and time independent amplitudes, and a two cycle/revolution effect with a time dependent amplitude. Other terms (e.g. resonant effects) can be added to the model for Δh_I. The quantities that appear in both Δh_G and Δh_I involve parameters that are estimated as part of the least squares adjustment process.

Now the sea surface height is the sum of the geoid undulation plus sea surface topography which can be represented by a spherical harmonic series whose coefficients are c_{nm}. We have

$$h_s = N_c + \Delta N_G + \Delta N_0 \zeta(c_{nm}) \qquad (11)$$

where N_c is the geoid undulation computed from the geopotential model that was used for the satellite orbit integration. ΔN_G are corrections to N_c based on corrections to the a priori potential coefficient model; ΔN_0 represents high frequency undulation terms and ζ is the sea surface topography representation. Equating (10) and (11) we can form the following observation equation:

$$\begin{aligned}\Delta h = {} & N_c + \Delta N_G(\Delta C_{nm}) + \Delta N_0(M+1 \, \text{to} \, \infty) \\ & + \zeta(c_{nm}) - h_c - \Delta h_G(\Delta C_{nm}) - \Delta h_I\end{aligned} \qquad (12)$$

Note the ∞ is replaced by the highest degree (e.g. 360) in the preliminary geopotential model. Equation (12) shows that the parameters associated with the altimeter observation equation are: corrections to the a priori potential coefficients to degree M; parameters of the sea surface topography representation which is usually a spherical harmonic expansion to degree 20 (approximately); parameters in the Δh_I term.

The altimeter observation is not taken for each altimeter measurement. Instead it can be taken for a value computed from a sequence (e.g. those taken over a 20 second interval) of measurements. Since ζ is not defined on land, estimates of the ζ coefficients based on ocean based altimeter data can lead to large (after adjustment) ζ values on land unless a priori estimates of the ζ coefficient magnitudes are used in the least squares adjustment.

We should note that the altimeter data used will probably come from two or more altimeter satellites. For example, today data is available from Geosat, Topex/Poseidon, and ERS-1 and ERS-2. Since this data most probably will cover different time periods several sets of sea surface topography coefficients will be estimated.

The combination solution can now be determined by combining the normal equations found from the satellite tracking data; the surface gravity anomalies and the altimeter sea surface height data.

Each set of normal equations will have numerous unknowns, not in common, and one set, the potential coefficient correction terms, in common. The simultaneous solution will yield the adjusted, low degree (to 70) potential coefficients and error covariance matrix of this combination model. Weighting considerations will continue to be a major concern, especially with respect to the use of repeat track altimeter data.

3.3 High Degree Model - Quadrature Approach

We now consider one approach to the estimation of a degree 360 (or possibly higher) geopotential model combining the satellite potential coefficient model and its error covariance matrix, with a global estimate of free-air anomalies properly corrected and reduced to the ellipsoid. The starting equation for this process is the following quadrature expression.

$$C_{nm} = \frac{1}{4\pi\gamma(n-1)} \iint_{\sigma} \Delta g(\theta,\lambda) Y_{nm}(\theta,\lambda) d\sigma \tag{13}$$

This equation is a spherical approximation. In practice, ellipsoidal harmonic expansions are used to reduce possible error, especially at the higher degrees, (Rapp and Pavlis, 1990). Equation (13) can be used as the basis for a least squares adjustment where the Δg values are regarded as observables and a satellite alone solution is given with its error covariance matrix.

Consider an a priori global anomaly (Δg) field of (e.g.) 30'x30' mean anomalies, which implies (through (13)) a set of potential coefficients C_{mn}^c. Let C_{mn}^U be the set of potential coefficients found from the satellite solution. The misclosure vector is then:

$$W = C_{nm}^0 - C_{nn}^c \tag{14}$$

Let B be the observation equation coefficients based on (13) and P_ℓ the weight matrix associated with the mean anomalies. Then the solution vector, V_X, is:

$$V_x = -\left(\left(B_\ell P_\ell^{-1} B_\ell^T\right)^{-1} + \sum_x\right)^{-1}\left(B_\ell P_\ell^{-1} B_\ell^T\right)W \tag{15}$$

The adjusted coefficients are then:

$$C_{nm}^a = C_{nm}^0 + V_x \qquad (16)$$

Recall that this parameter set contains the same coefficients (e.g. degree and order) as that of the earlier solution with the satellite tracking data. The solution of (15) is not, at the present time, the degree 360 solution.

An important part of this solution is the calculation of anomaly residuals (or corrections to the priori global 30'x30' (e.g.) anomaly field). We have:

$$V_\ell = P_\ell^{-1} B_\ell^T \sum V_x \qquad (17)$$

where the length of V_ℓ corresponds to the number of anomalies in the global field. For 30'x30' set, there are 180x2x360x2=259200 values. The adjusted anomalies are then:

$$\Delta g^a = \Delta g^0 + V_\ell \qquad (18)$$

These anomalies can then be used in equation (13) to obtain the expansion to the maximum degree consistent with the mean anomaly cell size. The appropriate way to evaluate (13) for these computations is one of continuing study. One approach (Rapp, 1989, eq. (24)) is the following discretized approximation to (13):

$$C_{nm} = \frac{1}{4\pi\gamma(n-1)q_n} \sum_{i=0}^{N-1} \sum_{j=0}^{2N-1} \Delta g^a \iint_{\sigma_{ij}} Y_{n|m|}(\theta,\lambda)d\sigma \qquad (19)$$

where q_n is a quadrature weight that depends on the area of the cell and some approximate rules for reducing sampling and noise error. The more precise formulation of the relationship involving the adjusted anomalies given on the surface of the ellipsoid and ellipsoidal harmonics is given by equation (50) in Rapp and Pavlis (1990). Using techniques taking advantage of the latitude and longitude symmetry the evaluation of (19) can be efficiently carried out. It is the result from equation (19) that yields the potential coefficients to degree 360 and even higher (e.g. 500) as recently described by Pavlis (1996)

The q_n values can be determined in a number of ways and the best approach has probably not yet been found. The q_n values used in the development of the OSU89B model and the higher degree part of the OSU91A model were based on a suggestion of Colombo

(1981). This led to the following:

$$\text{Type 2 (Colombo)} \quad q_n^i = \begin{cases} \left(\beta_n^i\right) & 0 \leq n \leq L/3 \\ \beta_n^i & L/3 < n \leq L \\ 1 & L < n \end{cases} \tag{20}$$

where $N = 180°/\Delta \lambda°$ where $\Delta \lambda$ is the longitude extent of the cell. β_n is the Pellinen smoothing operator and can be computed from:

$$\beta_n = \frac{1}{1 - \cos \psi_0} \frac{1}{2n+1} \left[P_{n-1}(\cos \psi_0) - P_{n+1}(\cos \psi_0) \right] \tag{21}$$

where ψ_0 is the semi-aperture of a spherical cap having the same area as the equiangular block at a specific latitude (i). We have:

$$\psi_0^i = \cos^{-1} \left[\frac{\Delta \lambda}{2\pi} (\cos \theta_{i+1} - \cos \theta_i) + 1 \right] \tag{22}$$

A disadvantage of these quadrature weights is the discontinuity that exists at N/3 and N. The impact of this is clearly reflected in the accuracy spectrum of models estimated with the procedure. Pavlis (1996) has suggested factors that avoid this undesirable behavior.

In the actual implementation of the above procedure numerous corrections are applied to the anomaly data, including the gradient reduction noted in (8). In addition the formulation is done through ellipsoidal harmonics although the final result is expressed in spherical harmonic coefficients completely compatible with those estimated from the satellite model estimation process.

The computation of the error covariance matrix is carried out through a two step process. The accuracy of those coefficients that are part of the satellite model is found from the usual normal matrix inversion needed for the evaluation of (12). For the other coefficients an error propagation can be done (e.g. Rapp and Pavlis (1990, eq. (51)). This procedure is not totally satisfactory since one assumes the anomalies are independently estimated. The procedure also does not give the error correlation between the estimated potential coefficients. This result represents the propagated anomaly error. Another error source is associated with the sampling error associated with the finite size of the cells in which the anomalies are given. Equations for sampling error were proposed by Jekeli in 1981 with no improvements made over the past 15 years.

The OSU89B potential coefficient model and the coefficients of the OSU91A model (Rapp, Wang, Pavlis, 1991) above degree 50 were estimated using the procedures described in this section.

3.4 High Degree Model – Block Diagonal Approach

The quadrature approach for the high degree solution is not a general least squares solution where a full degree 360 model is estimated because normal equations for the full model are not generated. A more rigorous solution would form observation equations and then normal equations when the full set of potential coefficients are being determined, the number of potential coefficient parameters would be 130321. Such a solution would be able to simultaneous adjust the data yielding a full degree 360 potential model and a full error covariance matrix. Currently this ideal process can not be carried out because of the extensive computational requirements.

Several authors have proposed non-optimum solutions that retain some of the aspects of the ideal solution and go beyond a quadrature type of solution. A review of some of the methods used is given by Pavlis et al (1996).

In the least squares approach to the estimation process dealing with gravity anomalies we can consider (9) as the relationship between the observables (Δg) and the parameters (C_{nm}). In the most general case the normal equations are formed with the parameters representing the potential coefficient solution to degree 360. If the data has special characteristics zero elements will occur in the normal equations making the inversion simple. The special characteristics are (Pavlis et al, 1996) (a) the data reside on a surface of revolution b) the grid is complete and the longitude increment is constant, c) the data weights are longitude independent, and d) the data weights are symmetric with respect to the equator.

In the non-optimal solution with real data only items a and b could be considered satisfied. In practice items c and d are not true but the actual cases will yield normal equations where off diagonal elements are small. Pavlis et al (ibid.) have carried out solutions where off diagonal terms are set to zero if they correspond to coefficients of different order. The resultant normal equations are then block diagonal. They can be combined with the normal equations of the satellite model. In this combination a special re-ordering of the unknown coefficients of the high degree model permits the efficient combination of the full satellite normals with the block-diagonal surface gravity normals. (Pavlis et al, ibid). Tests described by Pavlis et al(ibid) only incorporated the 30'x30' anomaly data in the solution with the satellite normals. The next step in the block diagonal approach would include normal equations formed from altimeter sea surface height information.

Finally it should be noted that the block diagonal approach requires care in its implementation and is not as straight forward in its evaluation as is the adjustment/quadrature procedure. Pavlis et al (1996) points out the need to use a high degree reference model calculating corrections to the reference model. In addition an off-diagonal "wing" component of the normal equations must be formed "which, for a given order in (≤70), correlates coefficients of degree ≤70 with those >70." Pavlis et al (ibid) also found that the "higher degree harmonics obtained from the block-diagonal technique appear to overestimate the power in the field". Later tests (Pavlis, private communication) indicate that this additional power can give better fits to independent observations.

3.5 Summary

In this section we have discussed the estimation of a satellite potential coefficient model, a low degree combination (with direct altimeter data and surface (1°x1°) anomaly data) solution and two procedures that yield degree 360 models. Each of the latter procedures have different assumptions that yield approximate solutions of the ideal case. The block diagonal procedure is conceptually closer to the true case than the quadrature procedure. In each method special care needs to be taken in the weighting of the data to achieve the best estimate of the potential coefficient model that can meet the needs of multiple types of users. This would span orbit determination to a geoid undulation determination for GPS applications and dynamic ocean topography determinations.

Finally one should note that there may be circumstances in which a blended model may be appropriate. In such a case the coefficients to degree M may be taken from the low degree combination solution to degree M and the coefficients from degree M+1 would be taken from the quadrature or block diagonal solution. This procedure was followed in the definition of the OSU91A model (Rapp, Wang, Pavlis, 1991).

4. Geoid Undulation Determination

An important aspect of this discussion relates to the use of geopotential models for the determination of geoid undulation and height anomalies. We first consider the definition and relationship between these quantities. We define the geoid to be an equipotential surface of the Earth's gravity field chosen to approximate the mean ocean surface, which is not an equipotential surface. The potential on the geoid is W_0. The reference surface is an equipotential surface of the biaxial reference ellipsoid. The potential on the surface of the ellipsoid is U_0. The separation between the geoid (P_0) and ellipsoid (Q_0) is the geoid undulation.

Next, consider the equipotential surface passing through point P on the surface of the Earth. The potential at this point is W_P and $C_P=W_0-W_P$ is the potential difference (geopotential number) between the geoid and the point P. We next define point Q located on the ellipsoid normal between P and Q_0 (on the ellipsoid) so that $U_0-U_P=C_P$. The locus of points Q forms the surface (non-equipotential) called the telluroid. The distance between P and Q is the height anomaly (z). These quantities are shown in Figure 4.1.

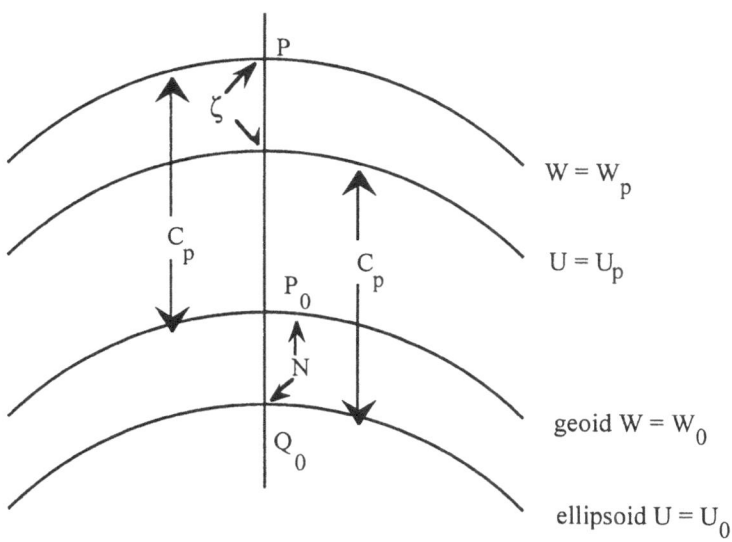

Figure 4.1 Relationship between geops and spherops

Based on the generalization of Brun's formula, one has (Heiskanen and Moritz, 1967, 2-178):

$$\zeta_p = \frac{T_p - (W_0 - U_0)}{\gamma_p} \tag{23}$$

where T_P is the disturbing potential at P. We let:

$$T_p = T_0 + T_p' \tag{24}$$

where T_0 is the zero degree term of the disturbing potential and T_p' is the disturbing potential excluding this term. We have:

$$T_0 \doteq \frac{GM - GM_0}{r} \tag{25}$$

where GM_0 is the geocentric gravitational constant of the reference ellipsoid and GM is the corresponding value of the Earth. Eq (23) can then be written:

$$\zeta_p = \zeta_0 + T_p' / \gamma_p \tag{26}$$

where

$$\zeta_0 = \frac{GM - GM_0}{r_p \gamma_p} - \frac{(W_0 - U_0)}{\gamma_p} \tag{27}$$

The value of ζ_0 can only be determined if estimates of GM and W_0 are known. This can be done to an increasing accuracy today with ζ_0 known to approximately 10 to 15 cm. In some applications ζ_0 is set to zero which implies that the values of ζ refer to an ellipsoid where $GM=GM_0$ and $W_0=U_0$.

To calculate geoid undulations we use the following (Heiskanen and Moritz, 1967, 8-100):

$$N = \zeta + \frac{\bar{g} - \bar{\gamma}}{\bar{\gamma}} H \tag{28}$$

where $\bar{g}$ and $\bar{\gamma}$ are average values of actual gravity and normal gravity, respectively, between the geoid and point P(for $\bar{g}$) and between the ellipsoid and the equipotential surface corresponding to U_P (for $\bar{\gamma}$). Heiskanen and Moritz (p. 327) show that:

$$\frac{\bar{g} - \bar{\gamma}}{\bar{\gamma}} H \approx \frac{\Delta g_B}{\bar{\gamma}} H \tag{29}$$

where Δg_B is the Bouguer anomaly. Then (28) becomes:

$$N = \zeta + \frac{\Delta g_B}{\bar{\gamma}} H \tag{30}$$

Using (26) and (4) we have:

$$N(r,\theta,\lambda) = \zeta_0 + \frac{GM}{\gamma_p r_p} \sum_{n=2}^{M} \left(\frac{a}{r}\right)^n \sum_{m=-n}^{n} C_{nm} Y_{nm}(\theta,\lambda)$$
$$+ \frac{\Delta g_B(\theta,\lambda)}{\bar{\gamma}} H(\theta,\lambda) \tag{31}$$

where the maximum degree of expansion is taken as M.

In the ocean areas, H=0 so that we can write (with ζ_0=0):

$$N(r,\theta,\lambda) = \frac{GM}{\gamma_p r_p} \sum_{n=2}^{M} \left(\frac{a}{r}\right)^n \sum_{m=-n}^{n} C_{nm} Y_{nm}(\theta,\lambda) \tag{32}$$

In a spherical approximation we can write (32) as:

$$N(\theta,\lambda) = R \sum_{n=2}^{M} \sum_{m=-n}^{n} C_{nm} Y_{nm}(\theta,\lambda) \tag{33}$$

when R is a mean Earth radius. For some applications one is interested in the geoid undulation by degree. At degree n, (32) can be written as:

$$N_n(r,\theta,\lambda) = \left(\frac{GM}{\gamma_p r_p}\right) \left(\frac{a}{r}\right)^n \sum_{m=-n}^{n} C_{nm} Y_{nm}(\theta,\lambda) \tag{34}$$

The geoid undulation degree variance (see Heiskanen and Moritz, p. 259) of the geoid undulation at degree n is:

$$C_{N,n} = \left(\frac{GM}{\gamma_p r_p}\right)^2 \left(\frac{a}{r}\right)^{2n} \sum_{m=-n}^{n} C_{nm}^2 \tag{35}$$

If e(c) represents the standard deviation of each potential coefficient, the undulation error degree variance is:

$$e^2(c_{N,n}) = \left(\frac{GM}{\gamma_p r_p}\right)^2 \sum_{m=-n}^{n} (e(c_{nm}))^2 \tag{36}$$

Generally, the coefficient in front of the summation is replaced by R^2. These undulation error degree variances are computed assuming the estimates of e(Cnm) are not correlated. Although the estimates are correlated (36) does give meaningful guidance to the undulation accuracy implied by a set of potential coefficients.

A more rigorous approach to the geoid undulation accuracy propagates the potential coefficient error covariance matrix into geoid undulation error. We could write:

$$\Sigma_N = G \Sigma_{Cnm} G^T \tag{37}$$

where Σ_{Cnm} is the coefficient error covariance matrix and G are the coefficients of Cnm in (34). For the combination solution described in 3.3, the low degree part of Σ_{Cnm} will be a full matrix while it will be a diagonal matrix for coefficients estimated through the quadrature procedure. Alternately one can use (37) to propagate the errors only up to some degree (e.g. 50 or 70).

The error degree approach to undulation error considerations (36) neglects the error correlation between the estimated coefficients. To understand the magnitude of this assumption one can calculate the undulation errors from (37) with the full error covariance matrix and then repeat the calculation using only the diagonal terms of the error covariance matrix. Such calculations were done for the GEM-T1 potential coefficient model, that is complete to degree 36, by Haagmans and van Gelderen (1991). They found differences reaching 50 cm (out of a total error of 110cm) in one region where, more typically the difference was on the order of 20% of the rigorous undulation error. The point here is that error correlations can play an important role in the determination of the commission error of a potential coefficient.

An important aspect of geoid accuracy is the estimate of the undulation difference between two points separated by a spherical distance ψ. If the square of the undulation accuracy is $C(\psi)$ then one can write:

$$C(\psi) = 2R^2 \sum_{n=2}^{M} e^2(C_{N,n})(1 - P_n(\psi)) \tag{38}$$

where P_n are Legendre polynomials of degree n.

So far our discussion has related to the computation of commission error, i.e. errors in the geoid undulation caused by the errors in the coefficients. We need to introduce the truncation error which is the error from the neglect of coefficients beyond degree M, the highest degree of the expansion being estimated. If c_n is a model for the anomaly degree variances of the Earth's gravitational field the truncation error for a point undulation would be:

$$m^2(N)_T = \left(\frac{R}{\gamma}\right)^2 \sum_{M+1}^{\infty} \frac{c_n}{(n-1)^2} \tag{39}$$

An approximation to this formula is:

$$m(N)_T \approx \frac{64}{M}(m) \tag{40}$$

If one has an expansion to degree 360, the root mean square truncation effect over the world is approximately ± 18 cm. There is also a truncation effect when calculating geoid undulation differences. Inferring from (36) we write.

$$C_T(\psi) = 2\left(\frac{R}{\gamma}\right)^2 \sum_{n=M+1}^{\infty} \frac{c_n}{(n-1)^2}(1 - P_n(\psi)) \tag{41}$$

The emphasis in this section has been on geoid undulation and geoid difference accuracies. Another important consideration is the accuracy of geoid slopes which is essentially equivalent to looking at deflection of the vertical accuracy. Such information on a global basis can be computed in a way analogous to geoid undulation accuracy as shown in Heiskanen and Moritz (1967, p. 260-261).

Using H/M (7-38) we write the mean square deflection commission error, to degree M, as:

$$\overline{\theta^2} = \frac{1}{\gamma^2} \sum_{n=2}^{M} \frac{n(n+1)}{(n-1)^2} \delta c_n \tag{42}$$

where c_n are the anomaly degree variances as used in (39). If we are interested in the error at degree n, we write.

$$\overline{\delta \theta_n^2} = \frac{1}{\gamma^2} \frac{n(n+1)}{(n-1)^2} \delta c_n \tag{43}$$

with the total commission error being:

$$\overline{\delta \theta_c^2} = \sum_{n=2}^{M} \delta \theta_n^2 \tag{44}$$

In (43) δc_n is the error anomaly degree variance and can be computed from the potential coefficient errors:

$$\delta c_n = \gamma^2 (n-1)^2 \sum_{m=-n}^{n} (e(c_{nm}))^2 \tag{45}$$

As with geoid undulation determination there will also be an omission error which represents the contribution of the neglected potential coefficients above degree M. This error would be:

$$\overline{\delta \theta_T^2} = \frac{1}{\gamma^2} \sum_{n=M+1}^{n} \frac{n(n+1)}{(n-1)^2} c_n \tag{46}$$

Given the error covariance matrix of the potential coefficients it is possible to calculated the geographic variation in deflection of the vertical error in a manner, similar to (37).

5. Geostrophic Velocities and Accuracy Using Satellite Altimeter Data and Geoid Undulation Models

Sea surface height determination from satellite altimeter data and knowledge of the geoid enables one to determine certain parts of the ocean circulation. Using geostrophic flow equations (e.g. Apel, 1987) one can calculate the north/south (v) and east/west (u) components of the "upper ocean geostrophic velocity." From (Tsaoussi and Koblinsky, 1994, p 24677):

$$v(\phi,\lambda) = \frac{g}{2R\omega\sin\phi\cos\phi} \frac{\partial\zeta(\phi,\lambda)}{\partial\lambda} \qquad (47)$$

$$u(\phi,\lambda) = \frac{-g}{2R\omega\sin\phi} \frac{\partial\zeta(\phi,\lambda)}{\partial\phi} \qquad (48)$$

where ζ is the dynamic ocean topography defined as the separation between the sea surface height and the geoid undulation. ω in (47) and (48) is the angular rotation velocity of the Earth. Although these equations break down near the equator other procedures can be used for velocity calculation. Considering (47) and (48) and the definition of ζ it is clear that the accuracy in the determination of the velocity depends , in part, on the accuracy of the geoid undulation slope in the north/south or east/west direction. A rigorous error propagation of errors in potential coefficients into velocity errors as a function of latitude and longitude has been described by Tsaoussi and Koblinsky (ibid)

6. Past and Current Geopotential Models

Considerable effort has been expended in the past 35 years in the determination of potential coefficient models from satellite tracking data and the combination with direct altimeter data and surface gravity data. A listing of some of these models to 1991 may be found in Rapp (1991). More recent reviews may be found in Vetter (1994), Nerem et al (1995) and Nerem (1995)

The accuracy of the geopotential models has increased and the highest degree of expansion has also increased in the past years as more data has become available. It is not the purpose of this discussion to review the past geopotential models. However one should briefly describe the current (May 1995) situation.

The JGM-1 and JGM-2 potential coefficient models were developed for use in the calculation of precise orbits for the Topex/Poseidon altimeter satellite. the models are complete to degree 70 (Nerem et al, 1994) and constitute the low degree combination model discussed in Section 3.2. A newer model that is a follow on to JGM-2 incorporating Topex GPS data and selected additional data is the JGM-3 potential coefficient model (Tapley et al, 1996). The use of the geopotential model has led to improved orbit determination for *Topex/Poseidon* and numerous other satellites. In addition the geoid undulations implied by JGM-3 are better, both in terms of actual values and accuracy issues, than JGM-2. We will discuss some accuracy issues in a later section.

Other low degree combination solutions have been developed in Europe with one of the most recent set of models being GRIM4-S4 and GRIM4-C4 (Reigber et al, 1996). the S4 (satellite model) is to degree 60 and the C4 (combination model) is complete to degree 72.

The development of the high degree expansions started in 1978 with a solution to degree 180 (OSU78). This was updated in 1981 with OSU81, also to degree 180. The GPM2 model (Wenzel), to degree 200 was produced in 1985. The first 360 model was in 1986 (OSU86). Tailored geopotential models, to 360, were produced in 1989 with the more general OSU model (OSU91A) released in 1991 (Rapp, Wang, Pavlis, 1991). The 1995 GFZ degree 360 model (GFZ95A) is described by Gruber et al (1996). Preliminary degree 360 (and higher) models developed as part of the joint GSFC/DMA gravity and geoid improvement project are described by Pavlis (1996).

7. Geoid Undulation Accuracy-Results from OSU91A and JGM-3

In this section we consider geoid undulation accuracy estimates from several geopotential models. We consider first the propagated potential coefficient error from the coefficients to degree 50 of the OSU91A model. This is shown in Figure 7.1. The high accuracy in the oceans is due to the use of Geosat altimeter data in the combination model. The geoid undulation error by degree and cumulatively, through error degree variances, is shown in Table 7.1.

An alternative error consideration for OSU91A set the coefficient standard errors equal to the magnitude of the coefficients above degree 260. In this case the ommission error, on a sphere whose mean radius is 6371 km, became 52 cm. The ommission error was estimated to be 24 cm, which led to an overall accuracy of 57 cm for a point geoid undulation calculation.

This single accuracy estimate is a geoid estimate with the accuracy expected to be better in some areas and poorer in others. Rapp (1992) made the following more specific estimates of the total undulation error:

ocean areas	26 cm
land area with good surface gravity data	38
land area with poor surface gravity data	56
land area with no surface gravity data	200

293

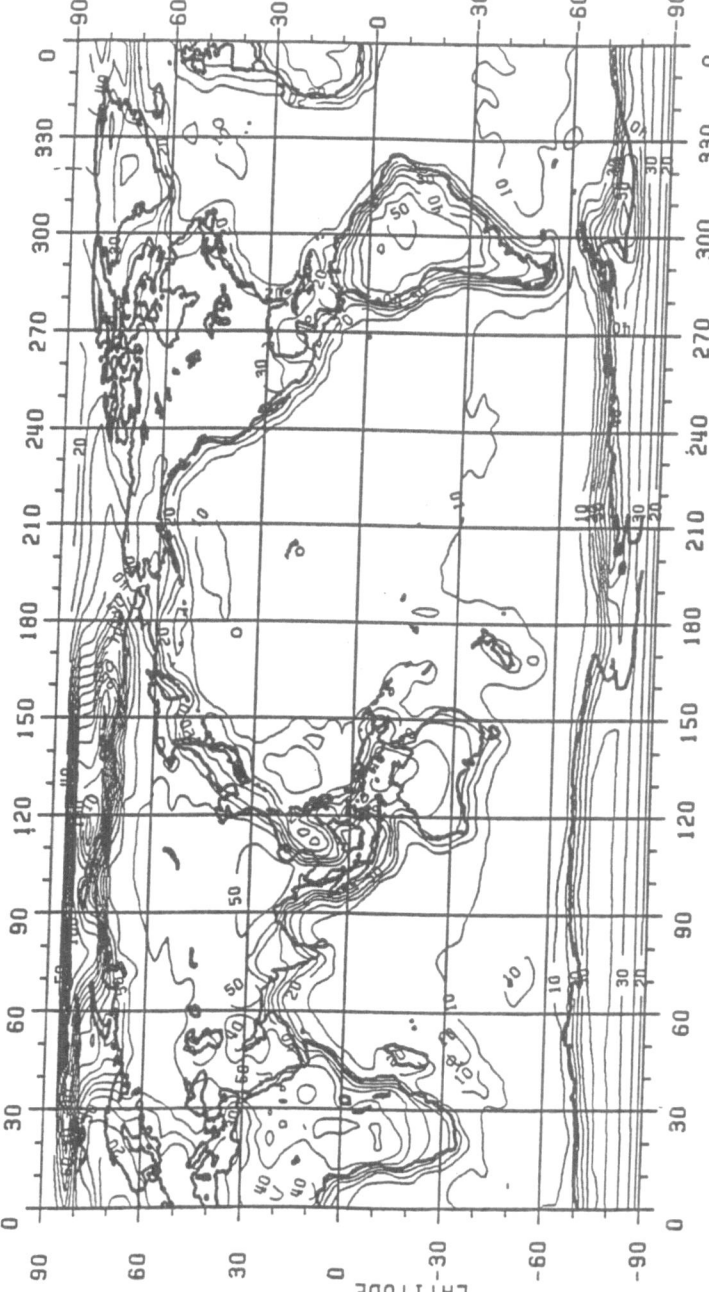

LONGITUDE

Figure 7.1   Geoid Undulation Accuracy Implied by Potential Coefficients to Degree 50 of OSU91A Model

Table 7.1
Geoid Undulation Commission Error for OSU91A Model

	OSU91A	
	By Degree	Cumulatively
2	0.2	0.2
6	1.3	2.2
10	2.4	5.0
20	3.6	10.6
30	4.3	16.8
50	3.0	24.8
75	3.7	32.3
100	3.2	36.5
180	2.2	43.2
360	1.3	48.7

Although OSU91A generally provides undulations with accuracies consistent with the above estimates a few locations have been found (e.g. in extreme latitudes (78°S)) where errors of 9m probably exist.

Undulation difference errors based on OSU91A are given in Table 7.2.

Table 7.2
Commission and Omission Error in the Computation of Geoid
Undulation Differences from the OSU91A Potential

Coefficient Model

Linear Separation	Commission Error	Truncation Error	Total Error
10km	9cm	10cm	14cm
20	18	18	26
30	27	23	35
40	34	25	43
50	41	26	49
70	52	25	58
90	60	23	64
100	62	22	66
200	72	23	65

The standard deviations of the potential coefficients of the JGM-3 potential coefficient model imply a geoid undulation commission error of ±54 cm, to degree 70. At lower degrees the cumulative is ±5 cm, on a global basis, considering an expansion to degree 14.

In addition to looking at the standard deviation by spherical harmonic degree, it is of interest to consider the accuracy in an ocean domain which is of interest for circulation studies. Such computations require the generation of functions that are orthonormal in the ocean domain of interest. Such a domain can be the usual oceans but it can also exclude areas outside specified latitude limits (e.g. 66°N and 66°S) and seas (e.g. Mediterranean Sea) that may not be suitable for global analysis procedures. Hwang (1991) developed the procedures now routinely used in our analysis. Other papers (Hwang, 1993, 1995, Wang and Rapp, 1994, and Rapp et al, 1996) have used the orthonormal system to understand geoid undulation accuracy estimates, on a spectral basis, in the ocean domain.

In Table 7.3 the geoid undulation accuracy for both JGM-2 and JGM-3, by degree and cumulatively, using spherical harmonic and orthonormal (for the ocean region) expansions, is given. One sees a clear improvement of JGM-3 over JGM-2. In addition one now sees undulation standard deviations below 1 cm at the lowest degrees.

The cumulative geoid undulation standard deviation, in the ocean regions, to degree 24 is 15.4 cm for JGM-2 and 12.5 cm for JGM-3. At degree 14 (which will be of interest to us later) the cumulative undulation error is 6.8 cm in the ON system.

From this discussion we see that we are (or may be) using potential coefficient information, at the longest wavelengths, that implies geoid undulation accuracies below 1 cm. But as additional degrees are considered the commission error increases significantly although the total geoid undulation error (commission plus truncation) generally will decrease. But it is clear that current solutions for the Earth's gravitational can not yield an accuracy of ±1 cm on all wavelengths. In fact one is very far from this goal.

8. The GSFC/DMA Joint Gravity and Geoid Improvements Project

Recognizing the need for improved determinations of the Earth's gravity field, DMA and NASA/GSFC, with the aid of several other groups, agreed in 1993 to develop a new potential coefficient model that would serve the many diverse needs of the scientific community. Initial thoughts on this project are given in Rapp and Nerem (1994). There were several drivers to the implementation of this project. The first part related to the need for an improved geopotential model for improved orbit determination, especially for altimeter satellites; geoid determination at the longer wavelengths where accuracies on the order of several cm are important; and two a global improvement of geoid undulation determination from a degree 360 model to an accuracy of ±50-100 cm could be expected.

Table 7.3
Geoid Undulation Accuracy by Degree and Cumulatively for
the JGM-2 and JGM-3 Potential Coefficient Models. Units are cm.

| | By Degree | | | | Cumulative | | | |
| | JGM-2 | | JGM-3 | | JGM-2 | | JGM-3 | |
Degree	SH	ON	SH	ON	SH	ON	SH	ON
2	0.1	0.5	0.0	0.3	0.1	0.5	0.0	0.3
3	0.5	0.8	0.1	0.4	0.5	0.9	0.2	0.5
4	0.5	1.1	0.2	0.6	0.7	1.4	0.3	0.8
5	1.1	1.5	0.4	0.8	1.3	2.1	0.4	1.2
6	1.1	1.8	0.5	1.0	1.7	2.7	0.7	1.6
7	1.9	2.3	0.7	1.3	2.6	3.6	1.0	2.0
8	2.0	2.6	1.0	1.5	3.2	4.4	1.4	2.6
9	2.6	3.0	1.3	1.9	4.1	5.3	1.9	3.2
10	2.6	3.2	1.5	2.1	4.9	6.2	2.4	3.8
11	3.2	3.4	2.0	2.4	5.8	7.0	3.1	4.5
12	3.1	3.5	2.2	2.7	6.6	7.9	3.8	5.3
13	3.6	3.7	2.6	2.9	7.5	8.7	4.6	6.0
14	3.5	3.8	2.7	3.1	8.3	9.5	5.4	6.8
15	3.8	3.9	3.1	3.3	9.1	10.3	6.2	7.5
16	3.7	3.9	3.1	3.4	9.8	11.0	6.9	8.2
17	3.9	4.0	3.3	3.5	10.6	11.7	7.7	8.9
18	3.9	4.1	3.4	3.6	11.3	12.4	8.4	9.6
19	4.2	4.2	3.7	3.7	12.0	13.1	9.2	10.3
20	4.3	4.1	3.8	3.6	13.6	13.7	9.9	10.9
21	4.6	4.1	4.0	3.6	12.8	14.4	10.7	11.5
22	4.8	3.8	4.2	3.3	14.4	14.8	11.5	12.0
23	5.0	3.4	4.4	3.0	15.3	15.2	12.3	12.3
24	5.2	2.2	4.6	1.9	16.1	15.4	13.1	12.5

The second major consideration related to the significant increase of terrestrial and altimeter implied gravity anomaly estimates over that put together in 1989 and used in the OSU91A model as well as for the low degree combination models of JGM-2, JGM-3, etc. The new data, in regions such as China, the former Soviet Union, South America, Greenland, was being collected by DMA directly and through its participation in other collection activities. In addition new satellite data, especially GPS tracking of several satellites, was becoming available and suggested an improvement of the potential coefficient estimation was possible.

One of the first steps in the project was the adoption of the following constants to be used in the project calculations: $a=6378136.3$ m, $GM=3986004.415 \times 10^8$ m^3 s^{-2}, $\omega=7292115 \times 10^{-11}$ rad s^{-1}. The tide free, second degree zonal coefficient of JGM-2 was also adopted: $C_{2,0}=-484.1654767 \times 10^{-6}$.

8.1 The JGP95E 5' Global Topographic Database

One of the initial steps in the joint project was the estimation of a 5' elevation file. The elevation file was needed for numerous reasons (e.g. terrain corrections, residual terrain modeling, calculation of topographic isostatic anomalies. Existing models (e.g. ET0P05U) were known to be deficient in some regions. The project considered elevations from a variety of sources including the DTED information of DMA, Terrain Base of NGDC, ice elevations from satellite altimeter data etc. Merged data sets were formed and subjected to spike tests to eliminate sudden jumps between adjacent cells. The elevations were classified in terms of one of six terrain types: These types and the percentage of the world's surface area for that type were as follows: (1) dry land below mean sea level (MSL), 0.08%; (2) lakes, 0.16%; (3) ice shelf, 0.20%, (4) ocean, 70.68%; (5) grounded glacier, 2.84%; (6) dry land above mean sea level (26.04%). The value and terrain type were assigned for the 9,331, 200 5' elevation values. The final elevation file is designated "The JGP95E 45' Global Topographic Database". The elevation file, whose length is 93,329.280 bytes is available for ftp distribution. For details contact N. Pavlis.

8.2 The Terrestrial Gravity Anomaly Files

The basic gravity anomaly data were the point gravity anomalies collected by the Defense Mapping Agency. The priority in this collection work was for land data as 30'x30' ocean region information was to be obtained from satellite altimeter data. The point gravity observations locations were transferred to a geocentric reference system, WGS84, so that cell values would be in a consistent system. The defined joint project need was for 30' x 30' mean free - air anomalies.

The 30' anomalies were estimated from the point data using the least squares collocation procedure. For most continental regions the computations of the free-air anomalies were carried out through terrain corrected Bouguer anomalies. In these calculations a degree 360 potential coefficient model (JGM-2/OSU91A) was used as the reference field. From this model spherical harmonic expansions of free-air anomalies (eq (9) and Bouguer anomalies (using elevation data) were determined. With the point data and the reference harmonic expansions, the following equations were used for the estimation of the mean anomaly and its accuracy (Kenyon and Pavlis, 1996):

$$\overline{\Delta g}_{30} = C_{\overline{\Delta g}\Delta g} \cdot (C_{\Delta g\Delta g} + V)^{-1} \cdot L + (\overline{\Delta g}_B(SH) - \overline{TC}$$
$$+ \Delta g(mean)) \tag{49}$$

$$M^2(\overline{\Delta g}_{30'}) = C_{\overline{\Delta g\Delta g}} - C_{\overline{\Delta g}\Delta g} \cdot \left(C_{\Delta g\Delta g} + V\right)^{-1} \cdot C_{\Delta g\overline{\Delta g}} \tag{50}$$

where:

$$\overline{\Delta g}_{30'} = C_{\overline{\Delta g}\Delta g} (C_{\Delta g\Delta g} + V)^{-1} L + (\overline{\Delta g}_B(SH) - \overline{TC} \tag{51}$$

$$\overline{\Delta g}_{30'} = \quad 30' \text{ mean Bouguer gravity anomaly}$$

$$L = \quad \Delta g_B - \Delta g_{BA}(SH) + TC - \Delta g(MEAN)$$

$$V = \quad \text{noise covariance matrix (diagonal) of point Bouguer gravity anomalies}$$

$$C_{\Delta g\Delta g} = \quad \text{signal covariance matrix of point Bouguer gravity anomalies}$$

$$C_{\overline{\Delta g}\Delta g} = \quad \text{signal cross convariance matrix between } 30' \text{ mean and point Bouguer anomalies}$$

$$TC = \quad \text{point terrain correction}$$

$$\Delta g_B(SH) = \quad \text{spherical harmonic Bouguer anomaly}$$

$$\Delta g_B(mean) = \quad \text{avearage of reduced poiont Bouguer anomalies over the computational area.}$$

$$\Delta g_B = \quad \text{point Bouguer anomaly}$$

$$M^2\overline{\Delta g}_{30'} = \quad \text{error variance of } 30' \text{ mean gravity anomaly}$$

$$C_{\overline{\Delta g\Delta g}} = \quad \text{signal covariance between } 30' \text{ mean gravity anomalies}$$

$$\overline{\Delta g}_B(SH), \overline{TC} = \quad \text{area-mean representations for the above defined quantities.}$$

The covariance values needed were calculated from the covariance function given by Forsberg (1987) using three parameters (D,T,C_0) that were determined through the analysis of the data in a local 1°x1° region. Note, through the L term in (49) that the point Bouguer anomalies were terrain corrected with a mean terrain correction restored after the residual anomaly has been predicted. The anomaly then available is a 30' mean surface free-air anomaly.

For some applications near coastlines and islands, where land and ship data was available, the 30' mean free-air anomaly was calculated from the point anomaly data without going through Bouguer anomalies. The least squares collocation equations were used with the covariance function parameters determined from the free-air anomaly residuals. In these computations a residual terrain model effect was removed from the point data and then restored to the final mean value. In predictions with both free-air and Bouguer anomalies the mean residual was removed from the data array and restored to the final mean as can be seen from eq. (49).

In addition to the 30' anomalies in land areas the project also required a set of $1°x1°$ anomalies in the ocean areas based on terrestrial measurements. These anomalies were to be used in the low degree combination solutions described in Section 3.2. The values used could not be based on altimeter derived anomalies because the terrestrial data was needed to separate the sea surface defined by altimeter data from the geoid. The $1°x1°$ data base was developed from DMA sources and from the data set used in the estimation of the OSU91A model. When $1°x1°$ in land areas were required they were formed from the 30'x30' values predicated through the collocation procedures.

Even though substantial progress was made in the collection of new gravity material in land areas, there were still areas of significant extent in which data was not available or very sparse. Such areas included "parts of the Amazon region in South America, Southwest Africa, Antarctica, and the Arctic region." For the calculation of the high degree models anomaly estimates for all 30' cells were needed. In order to determine such estimates the values were calculated from "a low degree satellite only model augmented with higher degree harmonic coefficients of the topographic-isostatic potential as implied by the Airy/Heiskanen isostatic compensation hypothesis (Pavlis and Rapp, 1990)." This fill in procedure had also been used in the development of the OSU91A model.

A summary of the anomaly counts by type discussed in this section, as of October 1995, is given in Table 8.1.

As noted earlier, the surface free air anomalies obtained as the result of the just described estimation process needed to be reduced (downward continued) to the ellipsoid for rigorous processing in the combination solutions. In the OSU91A development this process was carried out using g_1 terms calculated solely from an elevation grid. Alternative procedures were tested for the joint project including a downward continuation of the global 30' data set using the Poisson integral and the calculation of the anomaly gradient from a

Table 8.1
Type and Number of 30'x30' Terrestrial Anomalies
Used in Joint Project as of October 1995

TYPE	DMA Terr. 30'	DMA Altim. 30'	OSU Terr. 30'	OSU Terr. 60'	Topo/Iso Fill-in 30'
Number	72776	140701	4514	15291	25918
% of Area	29.7	65.1	0.5	1.4	3.3
RMS Δg (mgal)	34.3	25.5	36.6	31.9	28.1
RMS s (mgal)	5.6	1.7	17.4	38.5	36.0

degree 360 field. The latter two procedures have the advantage over the elevation based reduction in that they represent the actual gravity field behavior. For example, consider a topographically flat area with a gravity signature related to some geologic structure. The elevation based gradient correction would imply a zero gradient while the "real world" gradient would be non zero.

Tests have been carried out by Pavlis (1996, private communication) to see what gradient procedure would give the best results. Initial tests suggest the elevation based procedure gave the best results when tests were made in mountainous regions.

8.3 The Altimetry Derived Gravity Anomaly Files

Gravity anomalies in the ocean areas were derived from a 5'x5' grid of geoid undulations that were determined by DMA from the Geosat Geodetic Mission Data. The original grid file prepared by DMA was a sea surface height grid in the WGS84 system. These values were first converted to the reference system (ITRF91) used by the project. A dynamic ocean topography model, defined by a degree 20 spherical harmonic expansion, was developed by NASA and used to reduce the sea surface height values to geoid undulation values.

Least squares collocation was used to predict the anomalies from the undulation grid (Trimmer and Manning, 1996). The reference gravitational potential was defined as the JGM-2/OSU91A, the same as used for the terrestrial predictions. The Forsberg (1987) covariance functions were used with covariance function parameters tailored to the region of interest. The mean residual of the observation vector was removed from the original residual set, to center the data. This mean value was not converted to a correction to be restored to the predicted anomaly.

The Geosat satellite had a latitude range of ±72° so that there were ocean areas for which the geosat data was of no help in the anomaly estimation. However the gaps could be filled in by ERS-1 and ERS-2 data that extends to the latitude limits of ±82°. The anomalies derived from the ERS-1/2 data were provided to the project by KMS (Kort-og Matrikelstyrelsen) of Denmark and by NOAA (National Oceanic and Atmospheric Administration). Descriptions of the procedures used to estimate the gravity anomalies by these groups may be found in Andersen et al (1996a), Andersen et al (1996b) and Sandwell et al (1996).

Trimmer and Manning (1996) describe numerous tests to validate the accuracy of the altimeter derived gravity anomalies. Comparison of 27,610 30' anomalies from altimeter data with 30' anomalies derived from ship data showed a mean difference of 0.4 mgal and a standard deviation of ±2.3 mgal. As of October 1995 the number and source of the 30' anomalies made available to the project were: 141,082 (DMA, 7,559 (KMS ERS-1), 7890 (NOAA) for a total of 156,531 values. These numbers will most likely change in the final data set.

Here we should also note that special care was taken in several areas where a 30' cell could have land and ocean data. Examples would include islands areas such as Bermuda and the Hawaiian Islands and coastal cells. In such cases special predictions were made incorporating the 5'x5' grid from Geosat altimetry and from available ship and land gravity measurements. Without such a procedure an unrepresentative 30' estimate would be obtained.

The gravity anomalies from this altimeter data base refer to the geoid. To be consistent with the procedures used for the land data, the geoid referenced anomalies could be reduced to the ellipsoid using the gradients calculated from a degree 360 model.

8.4 The Merged 30' File

The land and altimeter derived anomalies were merged to create a single data set for use in the combination solutions. This set contains 259,200 values with information from the free-air predictions, Bouguer predictions, from separately supplied data in China, airborne measurements, altimeter derived anomalies (four sources), topographic (isostatic fill in values, plus split up of 1°x1° cells for which no 30' values were available.

8.5 The Preliminary Potential Coefficient Models

After the collection of the satellite tracking data, elevation data, gravity anomaly data, the efforts of the joint project have been directed to the determination of models based only on satellite tracking data, low degree combination models, least squares adjustment using quadrature solutions (OSU89B type) and block diagonal solutions. Numerous solutions have been run with different data sets, weighting factors, gradient reduction procedures, etc. In April 1996 the joint project released to the Evaluation Working Group which is led by Michael Sideris, four potential coefficient models that were the result of the joint project activities. These preliminary degree 360 models were designated EGM-X02, X03, X-04, and X-05 where EGM represents Earth Gravitational Model and the X indicates a preliminary model. In addition the project made available the merged model of JGM-3/OSU91A which could be used for comparison purposes with the newer models.

In addition to the EGM-X models other models have been developed and described in presentations by Lemoine et al (1996) and Pavlis et al (1996b). In the latter presentation Pavlis describes potential coefficient solutions that have been made to degree 500. Comparisons with GPS/leveling data and sea surface heights and along track slopes from Topex altimeter data based on a two year mean track from cycles 9 to 82 indicate the model contains information of use to about degree 460.

The GPS/leveling comparisons described by Pavlis et al (ibid) were for several lines including a north/south traverse in Europe a traverse along the south east coast of Australia, and set of stations in British Columbia. The standard deviation of the difference between the geoid undulation calculated from the GPS/leveling data and as computed from the potential coefficient model using (32) was determined from a solution designated V058 which was complete to degree 500. The results are shown in Table 8.2.

Table 8.2
Standard Deviation of GPS/Leveling Minus Model-Derived
UndulationDifferences With V058 model. Units are cm.
(from Pavlis et al (1996b))

	Europe	Australia	BC
Points	60	39	298
OSU91A	33	35	94
JGM3/91A	47	26	95
V058(280)			59
(360)	30	27	57
(460)	29	25	56
(500)			56

The results shown in Table 8.2 indicate improved results can be obtained by using the V058 model beyond degree 360 but not to degree 500. In addition, one clearly sees the improvement in the new models over OSU91A.

The Topex altimeter comparisons were made by computing a geoid undulation from the Topex sea surface height data using the POCM_4B dynamic topography model taken to degree 24 (Rapp, Zhang, Yi, 1996). Comparisons were made with the geoid undulation and along track slopes implied by the Topex data and the V058 geopotential model. The results (from Pavlis et al (1996b)) are given in Table 8.3.

Table 8.3
Standard Deviation at TOPEX/POCM_4B Derived
Geoid Undulation and V058 Model Values.
(from Pavlis et al (1996b))

Max Degree	Undulation Diferences	Along Track Slope Differences (*)
280	±31.2cm	2.19
360	28.0	1.97
460	27.0	1.87
500	27.0	1.88

From Table 8.3 one again sees that information above degree 360 has reduced the undulation and slope differences. Other tests, restricted to geographic regions of large gravity anomaly signature (such as the Aleutian Trench) show better agreement with observations when the potential coefficient model (V058) is used to degree 460 instead of 360.

8.6 Evaluation of Some Test Models at Ohio State

The preliminary potential coefficient models have been tested in two ways at Ohio State. The first set of tests have been made comparing the dynamic ocean topography implied by Topex sea surface height data and the geoid undulations implied by a potential coefficient model with the values of the POCM_4B model. The procedures used for such comparisons are described in Rapp, Zhang, and Yi (1996). Basically the Topex data is first referenced to JGM-3 orbits and the CSR 3.0 tide model. This data is used to form normal points of dynamic ocean topography by removing the geoid undulation value of the geopotential model being tested. This data is then used in a least squares adjustment to determine spherical harmonic coefficients to degree 24. In this determination a priori degree variances are used to help control the behavior of DOT estimates in land regions, and a data weighting procedure described in Wang and Rapp (1994, eq. (2-30)) that recognizes that points on the altimeter track become closer together as one goes from the equator to the latitude extremes of the data.

The DOT values of POCM_4B are given in a grid with a longitude spacing of 0.4° and a variable latitude spacing. This data is also used to determine a spherical harmonic expansion to degree 24.

Both expansions are converted to coefficients of an orthonormal expansion valid for the ocean domain (65°N to 66°S with the exclusion of the Black, Caspian, Mediterranean and Red Seas, and the Hudson Bay). The difference between the two sets was computed up to degree 14 as this degree is roughly the degree at which the undulation standard deviation of the JGM-3 geoid undulations (see Table 3) approaches the magnitude of the dynamic ocean topography estimates. The differences have been computed the JGM-3/V037 model, four test models and the PGS 6907 model (Pavlis, E et al 1996) augmented by the V037 model. PGS 6907 includes much of the data used in the test models but with improved weighting procedures. The comparisons are given in Table 8.4.

Table 8.4
Cumulative Difference Between Topex/Geoid and POCM_4B
Model Estimation of DOT Using Orthonormal
Coefficients to Degree 14. Units are cm.

Model	Standard Deviation
OSU91A	±15.5 cm
JGM-2/V037	14.2
JGM-3/V037	12.6
PGS6399/V058	12.1
V057	13.0
V058	13.0
HDM130	13.0
PGS6907/V037	10.9

This table shows the considerable reduction in the differences starting from the OSU91A model. Of the test models the blended model is the best of the four with the other three giving essentially the same results. A significant improvement is seen with the PGS 6907 solution. Rapp and Zhang (1996) describe the geographic regions in which significant improvement is seen in the use of PGS6907 over the other models.

The second type of test to which the geopotential models is the undulation difference tests from the comparison of GPS/leveling results with the geoid undulation from the model. In our comparisons we use a set of stations in the United States described by Milbert (1995). At these stations one has the ellipsoidal height determined from GPS analysis and Helmert orthometric heights in the NAVD88 system. This calculation was carried when N was evaluated using (32) with the radius r taken to the ellipsoid and also with (31) where the height anomaly is first computed and converted to the geoid undulation.

The evaluation of the geoid undulation through the height anomaly has using spherical harmonic expansions has been described by Rapp (1996) and tested using the OSU91A model. In this calculation a spherical harmonic expansion of $\Delta g_B H/ \bar{\gamma}$ is computed. In our calculations the 30'x30' mean elevation is computed from a degree 360 expansion of 30'x30' elevations formed from the JGP95E elevation model (Pavlis, 1995, private communication). Given a geopotential model 30'x30' mean free-air anomalies are computed using an integrated form of (6). This anomaly is converted to a Bouguer anomaly assuming a uniform crustal density and the 30' mean elevation. For each 30' cell the correction term $\Delta g_B H/ \bar{\gamma}$ is computed. If H is negative the term is set to zero. The global set of correction term values are then expanded into a degree 360 spherical harmonic expansion. Given a geopotential model one can first evaluate (32) and then the $\Delta g_B H/ \bar{\gamma}$ expansion at the station position. In the procedure described in Rapp (1996) a pseudo height anomaly is calculated on the ellipsoid using (32) and a correction term, represented by a spherical harmonic expansion, is then applied to calculate the height anomaly.

In the tests carried out for the joint project models and the Milbert data set, the full 1889 station complement and a thinned (960) station complement were used. The results of the thinned station set are more representative because clusters of stations were eliminated in the creation of the thinned data set. The results of the comparisons, with no height anomaly correction term applied and the "undulation" evaluated on the ellipsoid, and with proper undulation evaluated through the height anomaly are in Table 8.5.

First one sees the substantial improvement in all solutions tested over the results obtained with the OSU91A model. This is consistent with the tests shown in Table 8.2. It is especially interesting since the gravity coverage in the US was considered good when the OSU91A model was developed. The second point seen from Table 8.5 is the change between the solutions without and with the correction term. The mean difference is reduced by 14 cm. But the more important aspect is that the standard deviation of the differences is reduced from ±53 cm with no correction term applied to ±43 cm with the height anomaly correction term. This is an improvement of 19% in the undulation comparison. This difference is composed of 1) errors in the geopotential model; 2) errors in the ellipsoidal and orthometric heights; and 3) omission errors caused by the neglect of potential coefficients above degree 360.

Table 8.5
Comparisons of Geoid Undulation from Geopotential Models
to Undulation from Thinned (960) GPS Stations in the United
States With and Without the Height Anomaly
Correction Term. Units are cm

| Model | No Correction | | With Correction |
	Mean Difference	Standard Deviation	Mean Difference	Standard Deviation
OSU91A	- 98	±62	-84	±54
PGS6399/V058	-113	53.5	-99	42.6
V057	-112	53.4	-98	43.4
V058	-112	53.5	-98	43.4
HDM130	-112	53.4	-98	43.2

The two tests described in this section examine two different aspects of a geopotential model. One is its long wavelength accuracy through dynamic ocean topography comparisons and the second is the primarily short and medium wave length accuracy through undulation comparisons. Results seen with the preliminary models released to the evaluation group shows significant improvement over OSU91A. There is some indication that more improvement is possible with the existing data sets.

8.7 Formal Accuracy Estimates of the New Geopotential Models

The error analysis of the new models continues with final results not now available. At the long wavelength one will find geoid undulation accuracy estimates below JGM-3. Indications of the improvements at the higher frequencies have already been noted in the tests.

Using one of the new models the cumulative undulation to degree 10 is ±1.2 cm in contrast to ±4.9 cm for JGM-2 and ±2.4 cm for JGM-3. If the solution is taken to degree 70 the commission error has been estimated as ±18 cm in contrast to ±53 cm for JGM-3. More definite results await additional computations.

9. How Accurately Do We Know the Geoid Undulations From Potential Coefficient Models?

To answer this question we must consider what is of concern. Is one interested in the accuracy at a specific wavelength in the cumulative error to a specified degree or the total error considering all possible wavelengths (i.e. commission and omission errors)? Are you interested in the undulation error from a global point of view or are your interests in (e.g.) just the ocean region where the accuracy is better than in land areas?

In many applications the geopotential model is used with terrestrial gravity data to obtain more accurate results obtainable from just the model. How much improvement is possible with different gravity data sets and elevation variation of the region?

However for oceanographic applications where dynamic ocean topography is being determined the long wavelengths accuracies can be important. From Table 7.3 we see undulation accuracies, by degree that are smaller than 1 cm, globally to around degree for JGM-3. The joint project models will be better.

In the ideal case one might want geoid undulation accuracies of 1 to 2 cm to a resolution of 50 to 100 km. This is a difficult task and attempts to obtain such accuracy can only be done with the aid of a dedicated gravity field mapping mission.

And as we reach for such accuracies with such high resolution what are the theoretical problems that must be answered? What is the best way to do downward continuation of gravity anomalies to the ellipsoid? Or should the emphasis be on high degree solutions in which no downward continuation is needed? What is the appropriate way to estimate all smoothing factors for quadrature solution? Should anomalies be estimated in caps instead of rectangular cells to avoid aliasing effects?

10. Should We Be Considering Geoid Slope Accuracies?

For some applications the geoid slope is the important quantity as noted in Section . The slope propagates into velocity error which transports into transport error which is important in ocean circulation studies. Considering the JGM-3 potential coefficient model to degree 14, Tsaoussi (1995, private communication) has calculated the u and v velocity errors to be on the order of ±2 cm/s. As the potential coefficient expansion is taken to higher degree this error will increase and be larger than velocity errors expected from global circulation models. In such cases oceanography may give more accurate information than can be obtained from satellite altimeter data and geoid undulation data.

A proposal has been made by some oceanographers that an accurate geoid can be constructed using estimates of dynamic ocean topography and satellite altimeter data. Under what circumstances and accuracy considerations would the oceanographic approach be appropriate and possibly complementary to the gravimetric procedures for determining the geoid?

How accurately do we need geoid or geoid slope information to improve global circulation models? To answer this question models must be formulated that jointly consider oceanographic models and geoid undulation (magnitude or slope) information Wunsch (1996) has recently described approach to this problem.

11. Conclusions

The calculation of geoid undulation from geopotential models is a task that gets more complex as accuracy and resolution requirements become more stringent. Although we can talk about 1 cm geoid undulation information today at the longest wavelengths substantial work is needed to improve the current situation. The joint DMA/GSFC gravity model will yield improved geoid undulation information. But very significant improvements will require a gravity field mapping mission. The analysis of such data will pose some theoretical challenges to achieve a 1 cm or better accuracy.

Acknowledgment

The preparation of this paper was supported, in part, by the Defense Mapping Agency through Phillips Lab., Hanscom AFB, MA under contract F19628-94-K-0005, and through NASA's TOPEX Altimeter Research in Ocean Circulation Mission funded through the Jet Propulsion Laboratory under contract 958121. N.K. Pavlis, F. Lemoine and S. Kenyon provided material that greatly aided the preparation of this paper.

References

Andersen, O., et al. (1996) Comparisons of altimetric and shipborne gravity over ice-covered and ice-free polar seas. In: Global Gravity Field and Its Temporal Variations, eds. Rapp, Cazenave, Nerem, IAG Symposium 116, Springer Berlin Heidleberg New York.

Andersen, O., P. Knudsen, C.C. Tscherning (1996b) Global gravity field recovery from the dense ERS-1 geodetic mission altimetry. In: Global Gravity Field and Its Temporal Variations, eds. Rapp, Cazenave, Nerem, IAG Symposium 116, Springer Berlin Heidleberg New York

Apel, J. (1987) Principles of Ocean Physics, Academic Press, NY.

Forsberg, R., (1987) A new covariance model for inertia, gravimetry, and gradiometry, J. Geophy. Res, 42(B2):1305-1310.

Gruber, T., M. Anzenhofer, M. Rentsch (1996) The 1995 GFZ high resolution gravity model. In: Global Gravity Field and Its Temporal Variations, eds. Rapp, Cazenave, Nerem, Springer Berlin Heidleberg New York

Haagmans, RHN and M van Gelderen (1991) Error Variances-Covariance of GEM-T1: Their Characteristics and implications in geoid computation, J. Geophy. Res, 96(B12): 20,011-20,022.

Hwang, C. (1991) Orthogonal functions over the oceans and applications to the determination of orbit error, geoid and sea surface topography from satellite altimetry, Rpt. 414, Dept. of Geodetic Science and Surveying, The Ohio State University, Columbus, OH.

Hwang, C. (1993) Spectral analysis using orthonormal functions with a case study on the sea surface topography, Geophys, J. Int., 115:1148-1160.

Hwang, C. (1995) Orthonormal function approach for Geosat determinations of sea surface topography, Marine Geodesy, 18:245-271.

Jekeli, C. (1996) Methods to reduce aliasing in spherical harmonic analysis. In: Global Gravity Field and Its Temporal Variation , eds. Rapp, Cazenave, Nerem, Springer Berlin Heidleberg New York

Kenyon, S.C. and N.K. Pavlis (1996). The development of a global surface gravity data base to be used in the joint DMA/GSFC geopotential model. In: Global Gravity Field and Its Temporal Variations, eds. Rapp, Cazenave, Nerem, Springer Berlin Heidleberg New York

Lemoine, F., et al (1996) Latest results from the joint NASA GSFC and DMA gravity model project, (abstract) EOS, Vol. 77(17):541.

Milbert, D. (1995) Improvement of a high resolution geoid height model in the United States by GPS height on NAVD88 benchmarks, in New Geoid in the World, Bulletin d'Information 7., IGeS Bulletin, N. 4.

Nerem, R.S., et al (1994) Gravity model development for TOPEX/POSEIDON: Joint Gravity Models 1 and 2, J. Geophy. Res., Vol. 99, C 12, pp. 24,421-24,447.

Nerem, R.S., (1995) Terrestrial and planetary gravity fields, Reviews of Geophysics, supplement, pp. 469-476.

Nerem, R.S., C. Jekeli, W.M. Kaula (1995) Gravity field determination and Characteristics: Retrospective and Prospective, J. Geophys. Res., Vol. 100(B8): 15053-15074.

Nerem, R.S., et al (1996) Preliminary results from the joint GSFC/DMA gravity model project. In Global Gravity Field and Its Temporal Variations, eds. Rapp, Cazenave, Nerem, IAG Symposia 116 Springer Berlin Heidleberg New York

Pavlis, N.K. and R.H. Rapp (1990) The development of an isostatic gravitational model to degree 360 and its use in global gravity modeling, Geophys. J. Int. 100:369-378.

Pavlis, E. et al (1996) Earth gravity model improvement from GPS tracking of TOPEX/POSEIDON, GPS/MET and Explorer Platform (abstract), EOS, 77(17):S42.

Pavlis, N.K. (1996) Global geopotential solutions to Degree 500: Preliminary results, presentation at the 21st General Assembly, European Geophysical Society, Den Haag, The Netherlands.

Pavlis, N., J.C. Chan, F.J. Lerch (1996a) Alternative estimation technique for global high-degree gravity modeling, In: Global Gravity Filled and its Temporal Variation, Rapp, Cazenave, Nerem, eds., IAG Symposia 116, Springer Berlin Heidleberg NY.

Pavlis, N.K., J.C. Chan, R.H. Rapp, and D.E. Smith (1996b) Recent improvements in the high degree representation of the earth's gravitational potential (abstract), EOS, 77(17):S41.

Rapp, R.H. (1989) Combination of satellite, altimetric and terrestrial gravity data. In: Theory of Satellite Geodesy and Gravity Field Determination, Lecture Notes in Earth Science (25), F. Sansò and R. Rummel (eds.), Springer-Verlag, Berlin, Heidleberg, NY.

Rapp, R.H. (1991) The Earth's Gravity Field From Satellite Geodesy: A 30-Year Adventure. In proceedings of the workshop on Solid-Earth Mission ARISTOTLES, Anacapri, Italy, ESA SP-329.

Rapp, R.H. (1992) Computation and accuracy of global geoid undulation models, Sixth International Geodetic Symposium on Satellite Positioning, Columbus, OH.

Rapp, R.H., (1996) Use of potential coefficient models for geoid undulation determinatious using a spherical harmonic representation of the height anomaly/geoid undulation difference, submitted to the Journal of Geodesy, March.

Rapp, R.H. and N.K. Pavlis (1990) The Development and analysis of geopotential coefficient models to spherical harmonic degree 360, J. Geophy. Res, 95: 21,885-21,911.

Rapp, R.H., Y.M. Wang, N.K. Pavlis (1991) The Ohio State 1991 geopotential and sea surface topography harmonic coefficient models, Rpt. 410, Dept. of Geodetic Science and Surveying, The Ohio State University.

Rapp, R.H. and R.S. Nerem (1994) A joint GSFC/DMA project for improving the model of the Earth's gravitational field, Joint Symposium of the International Gravity Commission and the International Geoid Commissions, Graz, Austria.

Rapp, R.H. and C. Zhang (1996) Comparison and blending of dynamic ocean topography estimates from TOPEX data and global circulation models (abstract), EOS, 77(17):S76.

Rapp, R.H., C. Zhang, and Y. Yi, (1996) Analysis of dynamic ocean topography using TOPEX data and orthonomial functions, Journal of Geophy. Res. , 101(C10), 22, 583-22, 598.

Reigber, C. (1996) GRIM 4-S4/C4 Global gravity field model and recent extensions exploiting GFZ-1 and ERS-2/PRARE Tracking Data, EOS, 77(17): S42.

Sandwell, D., et al (1996) Marine gravity from satellite altimetry over ocean and sea ice. In: Global Gravity Field and Its Temporal Variations, eds. Rapp, Cazenave, Nerem, IAG Symposia 116, Springer Berlin Heidleberg.

Tapley, B.D. (1996) An improved earth gravity model (abstract), EOS, 77(17):542.

Tapley, B.D., et al (1996) The JGM-3 geopotential model, J. Geophy. Res, in press.

Trimmer, R.G. and D.M. Manning (1996) The altimetry derived gravity anomalies to be used in computing the joint DMA/GSFC earth gravity model. In: Global Gravity Field and Its Temporal Variations, eds. Rapp, Cazenaye, Nerem, IAG Symposia 116,.Springer Berlin, Heidleberg New York

Tsaoussi, L., and C. Koblinsky (1994) An error covariance model for sea surface topography and velocity derived from TOPEX/POSEIDON altimetry, J. Geophys. Res., 99(C2), 24, pp. 669-24,684.

Vetter, J. (1994) The evolution of earth gravitational models used in astrodynamics, Johns Hopkins APL Technical Digest, Vol. 15, No. 4, pp. 319-335.

Wang, Y.M. and R.H. Rapp (1994) Estimation of sea surface dynamic topography, ocean tides, and secular changes from TOPEX altimeter data, Rpt. 430, Dept. of Geodetic Science and Surveying, The Ohio State University, Columbus, OH.

Wunsch, C. (1996) Requirements on marine geoid accuracy for significant improvement in knowledge of the ocean circulation (abstract), EOS, 77(17):S40.

An Introduction to Airborne Gravimetry and its Boundary Value Problems

K.-P. Schwarz and Zuofa Li

Abstract

In this lecture, three major aspects of airborne gravimetry are presented:
- Introduction to system concepts
- Modelling of observables and estimation
- Boundary value problems.

After an introduction which overviews some of the historical development, the concepts of airborne gravimetry are briefly discussed. Three major approaches to the measurement of gravity from a moving platform are then outlined, namely scalar gravimetry, vector gravimetry, and gravity gradiometry. Scalar gravimetry is currently the most widely used, although test results are available for all three approaches. In the second part, the model equations for scalar and vector gravimetry are stated and their error models are derived. Special emphasis is given to scalar gravimetry in this part. Error sources are discussed in detail and some examples are used to illustrate major points. Insight gained in the process is used for the subsequent design of estimation procedures. It appears that at this stage of system development, the iterative use of time domain and spectral methods is the most appropriate way to extract the gravity signal from the noisy data. In the third part, three groups of geodetic boundary value problems (BVPs), related to airborne gravimetry, are introduced and solutions to these problems are presented. BVPs using airborne gravimetric data only are formulated first, and their solutions are given in terms of a Taylor expansion and a Fourier transform. BVPs combining airborne data with other data on the ground are considered next. Solutions to them are derived by way of planar harmonic analysis. Finally, the concept of multiresolution BVPs is introduced, and the solutions are developed based on discrete wavelet transforms. A brief assessment of current trends in airborne gravimetry concludes the course.

Part One: Introduction To System Concepts

In this part, a brief historical introduction to airborne gravimetry is given and some of its advantages with respect to land-based gravimetry are discussed. Then concepts of airborne gravimetry are outlined. After presenting the essential features of a gravimetric sensor, two major problems for the use of such sensors on moving platforms are identified. One is the orientation of the gravity sensor in space, the other, the separation of gravity from other sensor accelerations. After discussing possible solutions to these problems, three basic approaches to airborne gravimetry are identified. They are scalar gravimetry, vector gravimetry, and gravity gradiometry.

1   Introduction

Airborne gravimetry is not a new topic in geodesy. Proposals to implement such a technique go back to the late fifties, for instance Thompson (1959), and first flight experiments were done in the early sixties, Thompson and LaCoste (1960), Nettleton et al (1960). The major obstacle to a successful implementation of such systems at that time was the inaccuracy of the navigational data, especially velocity and acceleration, which are needed to obtain the desired precision. Although there were severe doubts at that time of ever achieving useful results with this method, see Hammer (1983), efforts to implement airborne gravimetry have continued. Why this persistence in the face of the apparently impossible?

The obvious answer is that airborne gravimetry at the appropriate level of accuracy is vastly superior in economy and efficiency to point-wise terrestrial methods. Geomagnetics is often quoted as an example. It was a marginal technique in geophysical exploration as long as it was ground-based. It developed into a major tool when is became airborne. As in gravimetry, it is not only the increased efficiency that brought this change about, but also the capability to survey remote areas, not easily accessible by land.

In addition, airborne gravimetry is a technique which lends itself to spectral analysis because it is essentially based on the difference of two time series of measurements. By defining the spectral band of interest in advance, it is possible to develop operational procedures which will optimally resolve this band. This is important because the spectral band required in geodesy and geotectonics on the one hand, and in geophysical exploration on the other, is quite different, see Schwarz et al (1991). By tuning operational procedures to the spectral band of interest, the estimated gravity profile will be of uniform accuracy and therefore well suited to be combined with gravity information from dedicated gravity satellite missions. They resolve the long wavelength features of the gravity field much better than terrestrial methods. This is especially important for many of the geodetic applications. When using terrestrial point gravity measurements, the resulting gravity field approximation will be non-uniform in accuracy, i.e. it will be extremely accurate at the measurement points and show interpolation errors which are a hundred times larger between the data points. To replace such a technique by one that has uniform accuracy over the whole range of interest, has numerous advantages.

Renewed interest in airborne gravimetry in the mid-eighties, led to improvements in scalar gravity system design, use of radar altimeters for vertical acceleration determination, and the selection of stable carriers to optimize operational conditions, see for instance LaCoste et al (1982), Hammer (1983), Brozena (1984), LaCoste (1988). The system concept used in this development stage was that of a precise accelerometer stabilized in vertical direction by a damped platform system. This concept will be called scalar gravimetry in the following because only the magnitude of the gravity disturbance vector is determined in this case. Systems of this type produced the first useful results, although the accuracy of the navigational data remained a concern.

In the late eighties and early nineties advances in GPS technology opened new ways to resolve the navigational problems, see for instance Schwarz (1987), Schwarz et al (1989), Brozena et al (1989), Kleusberg et al (1990), Wei et al (1991). The impact of this new technology led to two important developments. The first one was the perfection of existing scalar gravimeters to operational airborne gravity systems which could be used, on the one hand, for exploratory geophysical prospecting and, on the other hand, for large regional gravity surveys as required by geodesy and geotectonics. The second one was the development of new system concepts which made use of the full potential of existing inertial measuring units (IMU) for sensor stabilization and gravity vector determination.

In geophysical exploration, pioneering work was done by Carson Geophysics. Over the years this company has perfected their system by paying careful attention to the operational conditions under which a scalar gravimeter works best; for details see Gumert (1995). More recently, Harrison et al (1995) have developed a system for the same market and have made a careful analysis of the accuracy achievable. Wide-area airborne gravity surveys have been pioneered by the Naval Weapons Laboratory under the leadership of John Brozena. Their survey of Greenland, done in cooperation with the Danish National Survey and Cadastre (KMS) showed that airborne gravimetry could be used in a production mode. Key papers are Brozena (1992), Brozena and Peters (1994), Forsberg and Kenyon (1994). The recent airborne gravity survey of Switzerland, Klingele et al (1994) and Klingele et al (1995), and the ongoing surveys of Antarctica, see Brozena (1995) and Jones and Johnson (1995), show that this new technique is receiving wider and wider acceptance.

The use of INS in airborne gravimetry and the development of new system concepts was pioneered at the University of Calgary (U of C) and at the Inertial Technology Centre (ITC) in Moscow. The concept of a gravity vector system was developed at the U of C, see Knickmeyer (1990) and Schwarz et al (1991), and was implemented as a stable platform system in cooperation with Sander Geophysics Ltd. At the same time, the ITC equipped an existing Russian platform with a highly sensitive vertical accelerometer and tested it for both scalar and vector gravimetry. Airborne tests with this system were done in Canada in cooperation between the ITC, the U of C, and Canagrav Research Ltd. Results have been reported in Salychev et al (1994), Salychev and Schwarz (1995), and Salychev (1995). The use of a strapdown INS for vector gravimetry was jointly explored by the U of C and Canagrav Research Ltd. Its actual implementation was done by the U of C and first results have been reported in Wei and Schwarz (1996). Finally, the idea of a rotation invariant scalar gravimeter (RISC) which computes the magnitude of the gravity disturbance vector as the norm of its three vector components was developed in the early nineties and first published by Czompo in 1994. Results can be found in Czompo (1994) and Wei and Schwarz (1996).

As is apparent from this brief overview, airborne gravimetry is a rapidly developing field and much of the development is rather recent. Although scalar gravimetry can be considered as a production technique, efforts are being made to improve its accuracy and wavelength resolution from current RMS values of 2-6 mGal and half wavelengths of 5-10 km, to values of 1-2 mGal and 3-4 km. Once these accuracies can be reliably achieved for large regional surveys, many tasks in geodesy and geotectonics can be solved efficiently and with homogeneous accuracy. In the following chapters the concepts of airborne gravimetry will be outlined, major error sources discussed, the estimation problem described, and some boundary value problems associated with airborne gravimetry be formulated. A brief critical review of results achieved to date will conclude the lecture.

2 The concept of airborne gravimetry

Conceptually, **gravity sensors** used on the surface of the Earth or in its vicinity are highly sensitive accelerometers. They are, thus, realizations of Newton's second law of motion. In principle, an accelerometer consists of a proof mass, a weightless elastic suspension, a case and a scale attached to the housing, see Figure 2.1 for a simplified diagram of a possible realization. The proof mass m, considered as a point mass, is held in elastic suspension within the case. Assume that this can be done with three degrees of freedom, so that motion of the proof mass can be modelled in three-dimensional space. In an inertial frame of reference, the equation of motion for m can then be written as

$$m \frac{d^2 r_i}{dt^2} = m\ddot{r}_i = F_i + mG(r_i) \qquad (2.1a)$$

where r is the position vector from the origin of the inertial reference frame (i) to the centre of the proof mass m; $\ddot{r}$ is thus the acceleration of the proof mass with respect to an inertial frame of reference and is often called absolute acceleration; F is the force causing the elastic deformation of the suspension; $G(r)$ is the sum of the Newtonian attraction on m due to all bodies in the universe. Note that the assumption of an inertial reference frame considerably simplifies the formulas because inertial forces, due to frame rotation, are not considered. This case will be discussed in Chapter 3.

For an observer in the measurement frame of the accelerometer case, only the elastic deformation force is observable and can be measured on the internal scale. By dividing equation (2.1a) by m, one obtains

$$\ddot{\mathbf{r}}_i = \frac{\mathbf{F}_i}{m} + G(\mathbf{r}_i) \tag{2.1b}$$

Setting

$$\mathbf{f}_i = \frac{\mathbf{F}_i}{m}$$

and rearranging this formula with respect to $\mathbf{f}$, results in

$$\mathbf{f}_i = \ddot{\mathbf{r}}_i - G(\mathbf{r}_i) \tag{2.1c}$$

where $\mathbf{f}$, often called specific force, is the output of the spatial accelerometer. It has the unit of an acceleration because $\mathbf{f}$ is expressed as force per unit mass. The term on the left-hand side can be interpreted as representing the contact force exerted on the accelerometer by its support structure. The term on the right-hand side shows that $\mathbf{f}$ is the difference between absolute acceleration (i-frame) and net gravitational acceleration at m. Thus, a separation of $\ddot{\mathbf{r}}$ and $G(\mathbf{r})$ appears not to be possible on the basis of measurements $\mathbf{f}$.

However, experience shows that an accelerometer can be used as a relative gravimeter on the surface of the Earth. When it is set up at a point and aligned by levelling to the gravity vector at the point of support, the change of gravity relative to another point can be measured. How does this apparent discrepancy come about? The explanation can be given in terms of Newton's third law. By putting the support structure of the sensor on the surface of the Earth, gravity becomes a reaction force, counteracting the pull of the gravity field on the case. Thus, gravitational attraction and centrifugal force work as reaction forces and, thus, gravity is measured. Note, that this requires $\ddot{\mathbf{r}}$ to be known which is the case for a gravimeter stationary on the surface of the Earth if the rotation rate with respect to the i-frame is known.

In a similar way, when such sensors are mounted in an aircraft, the force needed to counteract the pull of the gravity field on the aircaft and thus on the support structure of the accelerometer is measured, together with aircraft motion and other inertial forces. Detailed formulas for this case will be given in Chapter 3. For the following conceptual discussion, the important point is that accelerometers, mounted in an aircraft, are sensitive to gravity in the specific sense discussed above. Thus, when the term 'sensitive to gravity' is used in the following, its exact physical meaning should be kept in mind. It should also be noted that for any type of free-fall motion, the output of an accelerometer will always be zero, even if it occurs in a gravity field, see Jekeli (1992).

In general, **accelerometers** are sensors which are only sensitive along one axis. One simple realization, the linear mass-spring system will be briefly discussed as an example. Its principle is shown in Figure 2.1. Typically, three such sensors would be mounted in an orthogonal triad to measure the specific force vector.

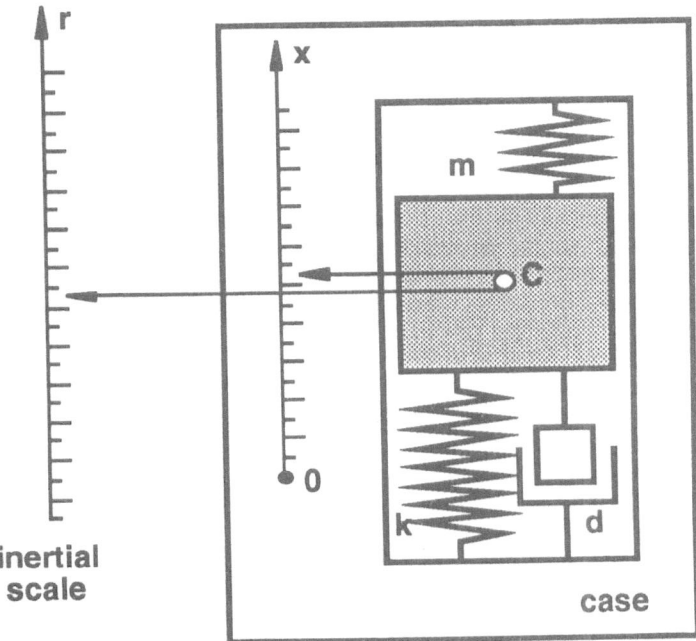

Figure 2.1 Principle of a linear mass-spring system

The essential components of such a device are a sensitive element of mass m constrained to move along its sensitive axis x, a restraining spring system, a damping device, a scale, and the housing. The restraining spring and the damping device have previously been called the elastic suspension. Point C is the centre of mass of the sensitive element. Point O indicates the equilibrium position of C when no external force is acting on the case along the sensitive axis. Point O is fixed to the case. The output indication of the instrument, supplied by displacement-sensitive transducers, is made proportional to the displacement x of the sensitive element relative to the case. The sensitive element is supported by a spring system which is attached to the case. The springs, assumed to be weightless, develop a restoring force proportional to the deflection of the sensitive element from its equilibrium position. Thus, if the spring has a stiffness coefficient of k > 0, then by Hooke's Law the restoring force is -kx. The sensitive element is also supported by a damper. Dampers are used to provide means of controlling the response of the instrument to dynamic inputs. The damper develops a viscous friction force $-c\dot{x}$, where c > 0 is the damping coefficient. In an actual instrument the spring and the damper take various forms.

The **specific force f** is proportional to the force needed to keep the proof mass of the accelerometer in equilibrium with respect to the sensor case. To obtain the differential equation which governs the dynamic behaviour of the accelerometer, Newton's second law is applied to the sensitive element. This is done by equating all forces acting on the element along the sensitive axis to its acceleration relative to inertial space times its mass. This equation can be solved for the displacement $x(C)$ from which f can be computed. It should be noted that the reading on the sensor scale x does not necessarily agree with the reading on a fictitious inertial scale r. The assumptions made in the definition of an operational inertial reference frame will usually result in biases between the two scales. At the current level of measurement accuracy, they will not affect the results.

Returning to the simple case of a stationary gravity measurement on the surface of the Earth, it should be noted that two important pieces of information are needed to obtain gravity reliably from such a measurement. The first is the knowledge of a direction in space, i.e. the direction of the gravity vector. In the stationary situation, this can be done by simply using a level bubble on the accelerometer box. In the more general case when the sensor is subject to motion, this is not so easy. The second is the knowledge that no rotation or acceleration with respect to the reference coordinate system is taking place. Again, this is relatively easy for the stationary case, but much more complicated when the sensor is subject to vehicle dynamics. Thus, in airborne gravimetry, the solution of the following two problems is fundamental:
- Sensor orientation (stabilization) in space under aircraft dynamics.
- Separation of gravitational and non-gravitational acceleration.

Figure 2.2 shows a number of different ways to answer the first question. Three of them use **inertial sensors for attitude stabilization**, one uses a GPS multi-antenna system. Figure 2.2a shows a damped two-axes platform system, as used for instance with the Lacoste-Romberg gravimeter. The platform, on which the gravity sensor is mounted, is mechanically stabilized by using accelerometers and gyros in a two-axes feedback loop. The damping period can be selected, typically at either 4 or 18 minutes, and is chosen to balance aircraft dynamics, feedback loop accuracy and gravity sensor resolution. This stabilization is still sensitive to the effects of horizontal acceleration on the gravity sensor output; although damped, this effect is not completely eliminated. A Schuler-tuned platform, see Figure 2.2b, with a damping period of 84 minutes eliminates this effect completely, at least in theory. In practice, it is limited by the long-term stability of its inertial sensors. As has recently been shown by Zhang (1995) and Zhang et al (1995), platform stabilization can be considerably improved by GPS velocity aiding. Electro-mechanical platform stabilization can be replaced by an analytical system which computes the transformation matrix between the body frame and the local-level frame at a high rate, using an inertial strapdown system, see Figure 2.2c. The inertial sensors used in these systems require a much greater bandwidth which usually has some adverse effects on accuracy. The major advantage in using strapdown systems is the availability of high rate digital data for post-mission analysis and the high linearity of drift effects for ring-laser gyros which are typically used in these systems. A system that has been proposed by Boedecker et al (1984) for attitude determination in airborne gravimetry is a GPS multi-antenna system, shown conceptually in Figure 2.2d. It determines aircraft attitude from coordinate changes of a rigid antenna array with respect to the GPS system of satellites. Currently, these systems do not achieve the accuracy and data rate of high-grade inertial systems, see e.g. El-Mowafy (1994) and Cohen (1994).

They also have the conceptual drawback that the sensor orientation is derived from an antenna array on the hull of the aircraft not on the sensor itself. Differential rotations between the multi-antenna system and the gravity sensor cannot be accounted for.

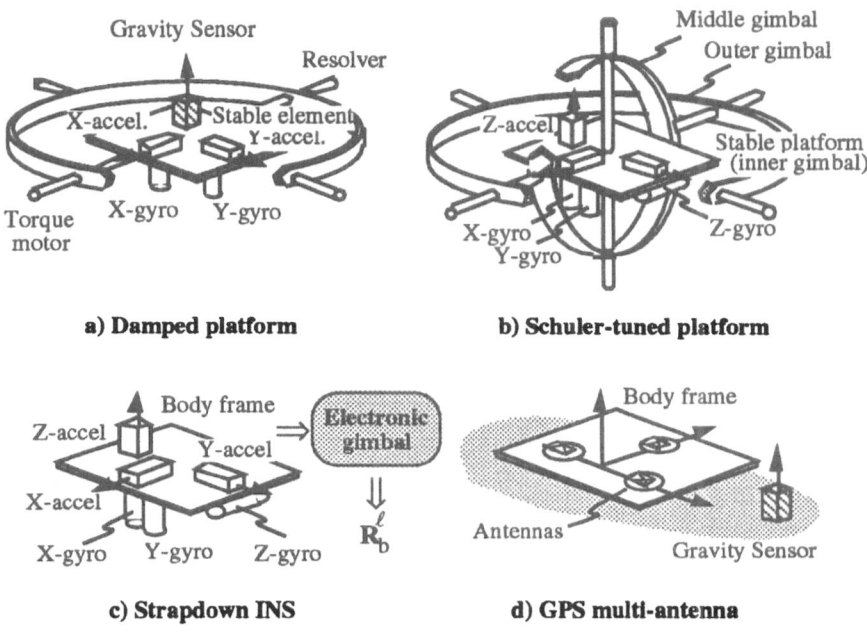

Figure 2.2 Attitude Stabilization for Airborne Gravimetry

To answer the second question, concerning the **separation of gravitational and non-gravitational acceleration,** two different approaches have been used. In the first, the acceleration output of two kinematic systems is differenced. One of the system outputs is sensitive to gravitational acceleration, the other one not. By differencing the two outputs in the same coordinate frame, common vehicle motion is eliminated and gravitation and the effect of system errors remains. The principle of the method is shown in Figure 2.3 where the time differencing of the two acceleration time series is shown along one axis. In this case, INS and GPS have been used. However, baro-altimeters and laser altimeters over water surfaces can be used instead of GPS, in cases where only vertical aircraft acceleration is required. The second approach to the separation problem is the elimination of vehicle motion by differencing the output of two accelerometers on a common base. If this base is rotationally stabilized, gravity gradient components can be obtained from the differenced readings. This is the principle underlying gravity gradiometry.

Data Streams

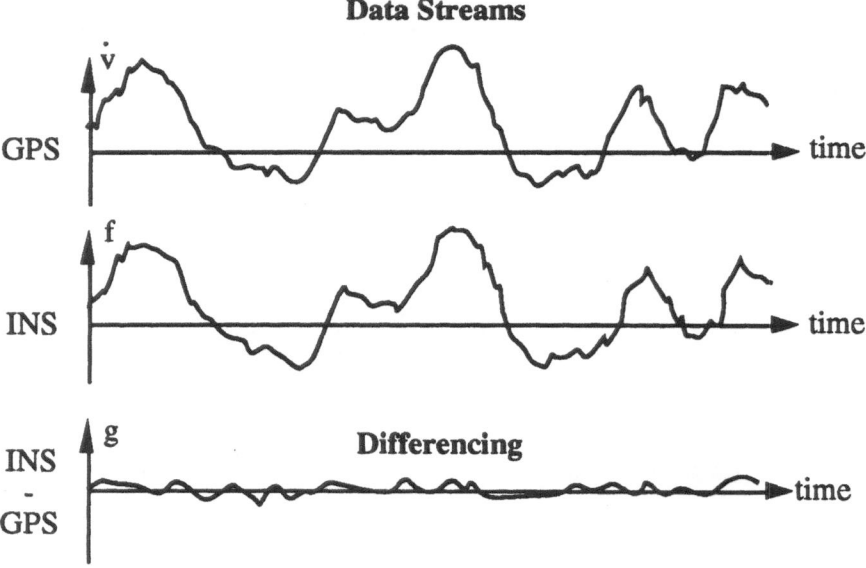

Figure 2.3   Principle of Airborne Gravimetry by INS/GPS

To give a conceptual framework to the following discussion, Figure 2.4 will be used. It shows the different **approaches to airborne gravimetry** that have either been implemented or proposed. They fall into three major groups:
- Scalar gravimetry
- Vector gravimetry
- Gravity Gradiometry.

In the rest of this chapter, each of these implementations will be briefly described. A mathematical description of scalar and vector gravimetry will be given in the next chapter.

In **scalar gravimetry** the magnitude of the anomalous gravity vector is determined. The basic idea can be implemented in either one of three ways. First, a precise vertical accelerometer can be mounted on a stabilized platform. Changes in the accelerometer readings, corrected for vertical aircraft acceleration, are then the desired gravity changes. The orientation problem is solved by platform stabilization, the separation problem by differencing two acceleration measurements, of which one is sensitive to gravity and the other is not. Second, a strapdown system can be used where the equivalent of a platform stabilization is done by computing the rotation matrix between the body frame of the vertical accelerometer and the local-level frame. In this case no separate accelerometer is used as gravity sensor, but the vertical accelerometer of the system is employed for this purpose. Again, its output has to be corrected for vertical aircraft acceleration. Finally, a triad of three orthogonal accelerometers can be used to obtain changes in the magnitude of gravity from changes in the difference between the specific force vector and the aircraft acceleration vector. In this case, the orientation problem is eliminated in theory and all three acceleration sensors contribute directly to the determination of gravity.

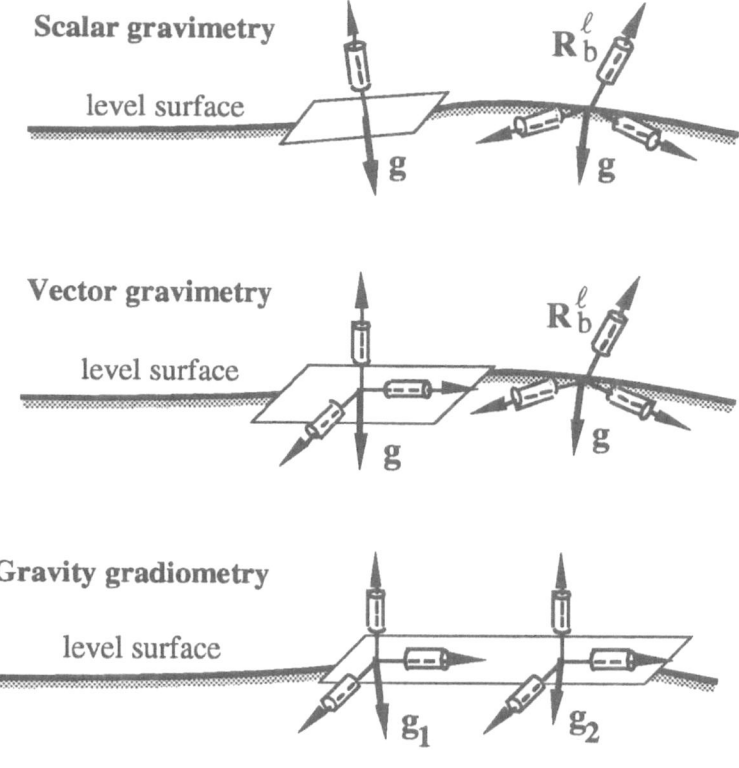

Figure 2.4   Approaches to Airborne Gravimetry

The main difference between the three methods is the way in which the orientation of the accelerometers with respect to the local vertical is established. In the first method, either one of the stabilizations shown in Figures 2.2a and 2.2b can be used. In principle, alternatives such as a GPS multi-antenna system or a star tracker, could also be applied. However, they are either not accurate enough or extremely expensive. In the second method, the stabilization shown in Figure 1c is employed. In the third method, the absolute orientation problem is replaced by a relative orientation problem. Instead of knowing the vertical orientation of one accelerometer in space, one has to know the relative orientation of two acceleration frames, namely that of the IMU local-level frame and the GPS local-level frame. Because no absolute orientation is needed in this case, the system has been called 'Rotation Invariant Scalar Gravimeter (RISG)'.

In **vector gravimetry**, the magnitude and direction of the anomalous gravity vector is determined. This can either be done by using a stable platform or a strapdown inertial system, see Figures 2.2b and 2.2c. In the first case, the accelerometers are essentially isolated from rotational motion and high resolution acceleration sensors with a relatively small measuring range can be used. Platform drifts are, however, a real problem because they affect the horizontal components directly and thus add quasi-systematic errors to the deflections of the vertical. Because the system works as a feedback mechanism, it is difficult to isolate these errors in post mission. In the second case, the accelerometer triad experiences full rotational motion and the error model becomes therefore much more complex. This is partially compensated for by the fact that a high-rate digital output is available and that it is therefore possible to carefully analyze data post mission. The orientation and separation problems are solved in essentially the same way as in scalar gravimetry.

In airborne **gravity gradiometry**, the second-order gradients of the anomalous gravity potential are determined along the flight trajectory. By integrating them using GPS velocity, first-order gradients and thus the anomalous gravity vector can be obtained. Conceptually, these systems can be viewed as an assembly of two carefully aligned accelerometer triads on a common stable platform. Thus, the orientation problem is again solved by platform stabilization, while the separation problem is solved by differencing sensor outputs on a common base. Although gravity gradiometers require a much higher relative sensor accuracy than the systems discussed above, they are currently the only systems that offer the promise of resolving the high-frequency gravity spectrum with high accuracy because they do not require independent kinematic acceleration measurements.

Current research in airborne gravimetry can be subdivided into two major areas: improvement of operational systems and development of new system concepts. In the first area existing gravimeters, such as the Lacoste-Romberg, the Bell, or the Bodenseewerke gravimeter, are applied to airborne gravity determination. Major problems to be overcome are the effects of insufficient platform stabilization, lag in the system output, and jumps in gravimeter readings. Companies and agencies currently involved in this area are the Naval Research Lab, Washington; Carson Services Inc, Austin; the Institute of Geodesy and Photogrammetry at the ETH, Zürich; the Institute of Physical and Astronomical Geodesy at the University FAF, Munich; Geodynamics Corporation, Santa Barbara; and Lamont-Doherty Geological Observatory, Palisades, N.Y.

Development work is currently concentrated in the area of airborne scalar and vector gravimetry, using some of the above concepts. This work is mainly done at universities and other research institutions, among them the Department of Geomatics Engineering at the University of Calgary; the Inertial Technology Scientific Center Ltd, Moscow; the Geoforschungszentrum in Potsdam; the Bayerische Akademie der Wissenschaften, Munich, and Sander Geophysics Ltd., Ottawa. Besides the work mentioned in the introduction, the use of superconducting technology in a scalar gravimeter is explored by the Geoforschungszentrum Potsdam and the use of a GPS multi-antenna system for sensor orientation is pursued by the Bayerische Akademie der Wissenschaften.

In addition to the two fundamental design problems of airborne gravimetry, major design changes are necessary when going from a stationary mode of operation to a dynamic one. The high accuracy of stationary gravimeters is due to the stable environment and the small measurement range. The first allows the averaging of measurements over time and thus results in improved accuracy. The second gives improved resolution. Airborne gravity meters have not only to function in high vibrational noise, which typically is a thousand times larger than the signal to be measured, but are also designed for operation under accelerations of 50 to 100 m/s^2. This severely limits the use of averaging techniques. Thus, resolution suffers and accuracy is affected. Measurement accuracies of better than 0.01 mGal (=0.0000001 m/s), which are typically achieved with stationary field instruments, cannot be expected from systems working under these dynamic conditions. The current objective to reliably achieve measurement accuracies of 1 mGal (=0.00001 m/s) is extremely difficult to attain. Why this is the case will be discussed in the next section.

Part Two: Modelling Of Observables And Estimation Methods

In this part, the equations for modelling inertial data are reviewed and their application to airborne gravimetry is discussed. These formulas are used to develop an error model for airborne gravimetry which is then analysed for the case that accurate position and velocity data from GPS is available. As will be seen, the interaction of attitude misalignment with horizontal acceleration, the errors of the accelerometer, the errors of determining acceleration from GPS, and the sychronization errors between GPS and INS are the dominant error sources to be considered. The error spectra of these four sources are discussed and an estimation strategy for minimizing them is outlined. A brief discussion of estimation methods currently used and their place within the general estimation strategy concludes this part.

3   The measurement model of airborne gravimetry

The principle of airborne gravimetry is based on Newton's equation of motion in the gravitational field of the Earth. In the local-level frame (l), it is of the form

$$\dot{\mathbf{v}}^l = \mathbf{f}^l - (2\Omega_{ie}^l + \Omega_{ie}^l)\mathbf{v}^l + \mathbf{g}^l \tag{3.1}$$

where $\dot{\mathbf{v}}^l$ is vehicle acceleration; $\mathbf{f}^l$ is the specific force measured by an inertial system; $\mathbf{v}^l$ is vehicle velocity, Ω_{ie}^l and Ω_{el}^l are skew-symmetric matrices of the angular velocities ω_{ie}^l and ω_{el}^l due to earth-rate and vehicle rate over the ellipsoid; and $\mathbf{g}^l$ is the gravity vector, i.e. the sum of gravitational and centripetal acceleration. For a more detailed discussion, see Schwarz and Wei (1990).

Gravity can be obtained from this equation by rewriting it in the following way

$$\mathbf{g}^l = \dot{\mathbf{v}}^l + (2\Omega_{ie}^l + \Omega_{el}^l)\mathbf{v}^l - \mathbf{f}^l \tag{3.2}$$

In this formula the first term on the right-hand side is the vehicle acceleration vector, the second term is the Coriolis acceleration vector, and the third term is the specific force vector.

The gravity vector $\mathbf{g}^l$ in equation (3.2) can be expressed as the sum of the normal gravity vector γ^l and the gravity disturbance vector $\delta\mathbf{g}^l$. Thus, the modelling problems connected with airborne gravimetry can be discussed in an elementary way using the equation

$$\delta\mathbf{g}^l = \dot{\mathbf{v}}^l - \mathbf{f}^l + (2\Omega_{ie}^l + \Omega_{el}^l)\mathbf{v}^l - \gamma^l \tag{3.3a}$$

Stricly speaking, this is the model of a local-level stable platform system, where the orientation of the vertical accelerometer or the accelerometer triad is obtained through an electro-mechanical feedback loop. Thus, all quantities are directly obtained in the l-frame. For a strapdown system, where accelerometer and gyro measurements are both in the body frame (b), the appropriate equations are

$$\delta\mathbf{g}^l = \dot{\mathbf{v}}^l - \mathbf{R}_b^l\mathbf{f}^b + (2\Omega_{ie}^l + \Omega_{el}^l)\mathbf{v}^l - \gamma^l \tag{3.3b}$$

where $\mathbf{R}_b^l$ is the transformation matrix which rotates the accelerometer measurements sensed in the body frame (b) to the local-level frame (l). This matrix thus replaces the stable platform mechanization in equation (3.3a). Equations (3.3) can be used for both vector and scalar gravimetry. In the latter case, only the last of the three equations in system (3.3) has to be considered. When written explicitly, the equation for scalar gravimetry is of the form

$$\delta g_U = \dot{v}_U - f_U - \left(\frac{v_E}{R_1 + h} + 2\omega_{ie}\cos\varphi\right)v_E - \frac{v_N^2}{R_2 + h} - \gamma_U \tag{3.4}$$

where the velocity vector is given as $\mathbf{v} = (v_E, v_N, v_U)$, the subscripts E,N,U stand for East, North, and Up in a local-level ellipsoidal frame, R_1 and R_2 are the prime vertical and the meridian radii of curvature.

Equations (3.3) give the gravity disturbance vector in terms of observed quantities, i.e. vehicle motion $\dot{\mathbf{v}}$, $\mathbf{v}$, $\mathbf{r}$, specific force $\mathbf{f}^b$, and integrated angular velocities $\mathbf{R}_b^l$. When using INS and GPS as measuring systems, the vehicle kinematic quantities $\dot{\mathbf{v}}^l$, v^l, $\mathbf{r}^l$ can be obtained using GPS carrier phase and phase rate measurements. The specific force vector $\mathbf{f}^b$ is measured by the three accelerometers of an inertial measuring unit (IMU) in case of vector gravimetry, and by a single vertical accelerometer in case of scalar gravimetry. To transform the specific force $\mathbf{f}^b$ to the l-frame, the transformation matrix $\mathbf{R}_b^l$ is needed. It is obtained by integrating the measured angular velocities between the body frame and the inertial frame ω_{ib}^b after correction for earth rotation ω_{ie}^l and vehicle rate ω_{el}^l. It should be noted that ω_{ie}^l and ω_{el}^l are computed from the known angular rate of the Earth and the position and velocity information from GPS.

Kinematic gravimetry, using a combination of GPS and INS, is different from kinematic positioning by GPS/INS integration. While in positioning the two data streams are integrated for optimal position accuracy, in gravimetry they are differenced to eliminate the common kinematic vehicle motion and extract the gravity disturbance vector δg which is a small quantity of acceleration type.

Gravity estimation is therefore much more affected by high frequency errors of INS and GPS and the resulting accuracy is determined by the error characteristics of both measurements. The resolution of the gravity spectrum depends on the signal to noise ratio in different frequency bands. Filter design for a band-limited gravity spectrum and specific aircraft dynamics is one of the critical problems in this application.

4   The error model of airborne gravimetry

The error model of airborne vector gravimetry can be obtained by linearizing equation (3.3) in the following way

$$d\delta g^l = F^l \varepsilon^l - R_b^l df^b + d\dot{v}^l + (2\Omega_{ie}^l + \Omega_{el}^l)dv^l - V^l(2d\omega_{ie}^l + d\omega_{el}^l) - d\gamma^l \quad (4.1)$$

where $d\delta g$ represents the errors in the gravity disturbance vector δg, ε^l represents attitude errors due to initial misalignment and gyro measurement noise, df^b represents accelerometer noise, $d\dot{v}^l$ represents the errors in the determination of vehicle acceleration $\dot{v}^l$, dv^l contains the velocity errors, $d\omega_{ie}^l$ and $d\omega_{el}^l$ are errors in angular velocity, dg^l are errors in the normal gravity computation, and E^l and V^l are skew-symmetric matrices containing the components of the specific force vector and the velocity vector, respectively.

Up to this point, it has been assumed that the two measurement system INS and GPS, are perfectly sychronized. In reality this can never be achieved and therefore a small synchronization error dT has to be added to the equation. Thus, we obtain

$$d\delta g^l = F^l \varepsilon^l - R_B^l df^b + d\dot{v}^l + (2\Omega_{ie}^l + \Omega_{el}^l)dv^l - V^l(2d\omega_{ie}^l + d\omega_{el}^l)$$
$$-d\gamma^l + \left(\dot{R}_b^l f^b + R_b^l \dot{f}^b\right)dT \quad (4.2)$$

With current GPS technology, post-mission position and velocity errors can be kept at standard deviations of about $\sigma_p = 0.3$ m and $\sigma_v = 0.05$ m/s, respectively. In that case, the errors generated by the last three terms in equation (4.2) are well below 0.5 mGal and can therefore be neglected in the following discussion. Thus, the significant errors affecting airborne gravimetry are given by

$$d\delta g^l = F^l \varepsilon^l - R_b^l df^b + d\dot{v}^l + \left(\dot{R}_b^l f^b + R_b^l \dot{f}^b\right)dT \quad (4.3a)$$

They are due to the effects of attitude errors ε^l, accelerometer errors df^b, errors in the determination of vehicle acceleration $d\dot{v}^l$ and synchronization errors dT. Thus, the first two terms are due to errors in the inertial system, the third due to errors in GPS, and the last due to timing errors between the systems. For scalar gravimetry, formula (4.3a) simplifies to

$$d\delta g = f_E \varepsilon_N - f_N \varepsilon_E - df_U + d\dot{v}_U + \left(\dot{R}_b^l f^b + R_b^l \dot{f}^b\right)dT \quad (4.3b)$$

Since an understanding of the errors affecting airborne gravimetry is essential for system design and filtering, each term in this equation will be briefly discussed.

Attitude error characteristics depend on the system used for attitude determination. Since inertial systems are currently the only ones that provide attitude stabilization with sufficient accuracy, the emphasis will be on these systems. For a brief analysis of the potential of GPS multi-antenna systems for sensor orientation, see Schwarz and Wei (1994).

Attitude errors of an INS can be approximated by the formula

$$\varepsilon^l = \varepsilon_o^l + \int_{t_1}^{t_2} R_b^l(t) d\omega_{ib}^b(t) dt \tag{4.4}$$

where the first term represents initial alignment errors, including anomalous gravity errors, while the second term represents integrated gyro errors which include non-orthogonality of the gyro axes, scale factor errors, linear and nonlinear gyro drift, correlated random errors, quantizer errors and fractal noise, see Savage (1978) and Li and Schwarz (1994) for details. Attitude errors are therefore time-dependent with a relatively complicated structure. A Schuler-type oscillation stemming from initial errors is overlaid by fast changing effects which are caused by the interaction of aircraft dynamics and constant or slow-changing sensor errors. In addition, correlated random errors can make a significant contribution to the error budget, while white noise effects are damped by the integration and are usually not critical. Some of these errors, such as non-orthogonality of the axes, can be modelled if the aircraft dynamics is known with sufficient accuracy.

The **gravity error due to the attitude** error term can be written in component form as

$$F^l \varepsilon^l = \begin{pmatrix} f_N \varepsilon_U - f_U \varepsilon_N \\ f_U \varepsilon_E - f_E \varepsilon_U \\ f_E \varepsilon_N - f_N \varepsilon_E \end{pmatrix} \tag{4.5}$$

where f_E, f_N, and f_U are the components of the specific force vector. Formula (4.5) shows clearly that attitude errors will affect gravity estimation only when specific force is present. Thus, constant velocity is the ideal operational environment because it eliminates specific force from the horizontal channels and, thus, the effect of attitude errors caused by horizontal acceleration. This is a major advantage of scalar gravimetry as compared to vector gravimetry. While the case of zero horizontal accelerations can at least be approximated, the case of zero specific force in the vertical cannot be obtained by operational procedures. The gravity field of the Earth will always account for about 10 m s^{-2} in the f_U-component.

In scalar gravimetry, a simple calculation shows that a misalignment error of 5 to 10 arcseconds, which is typical for high-grade inertial platforms and strapdown units, will generate a gravity error of 2.5 to 5 mgal under a horizontal acceleration of 1 m s^{-2}. If horizontal acceleration can be limited to a tenth of this value, the error will be reduced by the same ratio. Thus, the accuracy of airborne gravimetry can be influenced by the selection of a stable aircraft and a flying height where atmospheric perturbations are minimal. Since this requires a certain flying altitude, resolution of the high frequencies in the gravity spectrum will be affected, for details see Schwarz et al (1994). Because the perturbing horizontal accelerations are subject to rapid changes, the resulting gravity errors are fast changing and show essentially the spectral characteristics of the horizontal accelerations. Figure 4.1 illustrates this situation. Shown is the gravity error due to the interaction of measured horizontal acceleration at 3 km flying height and a misalignment of $\varepsilon_N=$

35" and ε_E= -45", respectively. The regularity of the error pattern in this case is caused by phugoid motion of the aircraft due to autopilot control.

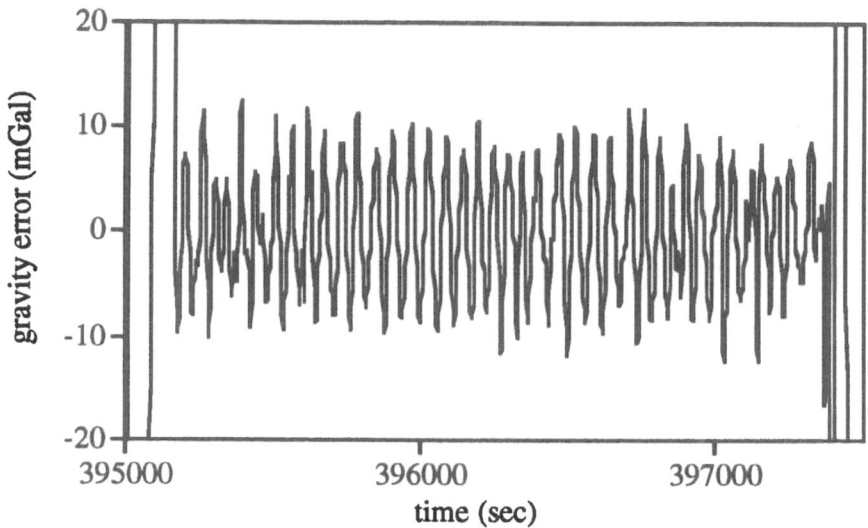

Figure 4.1   Gravity error due to an interaction of an alignment error
with horizontal acceleration

For vector gravimetry, the situation is more complex. As equation (4.5) shows, one misalignment component in each of the two horizontal channels is multiplied by f_U which is composed of a large nearly constant part due to gravity and a fast changing smaller part due to vertical aircraft acceleration. Thus, the contribution of the attitude errors ε_E and ε_N to the horizontal gravity errors (deflections of the vertical) will be much more pronounced than to the vertical error. Since specific force is almost constant in this case and attitude errors change only slowly with time, the induced gravity errors will be long-term in nature and will thus occupy a different part of the spectrum than in the scalar case. They will therefore overlap the gravity spectrum to a considerable degree. A detailed discussion is given in Wei and Schwarz (1994).

Accelerometer errors can be approximated by the formula

$$df = b + (s_1 + s_2 f)f + n \qquad (4.6)$$

where b is the accelerometer bias error, which changes slowly with time; s_1 and s_2 are the linear and quadratic scale factor errors which are usually a function of specific force; and n is sensor noise which usually has white noise and correlated components. Because gravity is obtained by differencing the INS and the GPS ouput and thus eliminating common vehicle motion, see Figure 2.2, any error in the vertical acceleration measurement will fully show up as a gravity error. This means that the bias error will appear as a low frequency error with considerable overlap into the gravity spectrum of interest, while the scale factor error will add an almost constant error term to the vertical channel and rapidly changing terms to the horizontal channels. Besides the actual sensor errors, vibrational noise and aircraft dynamics will have to be considered. If the INS and GPS sensors were in the same location on the aircraft, most of these errors would be eliminated by the differencing process. Since this is not the case, their difference will usually not tend to zero. The resulting errors will be again high frequency in nature. To illustrate the effect of aircraft vibration, Figure 4.2 should be consulted. It shows the 50 Hz output of an accelerometer in an aircraft, stationary on the tarmac, but with the engines switched on. The standard deviation which is about 1 mGal under lab conditions increases to about 25 000 mGal.

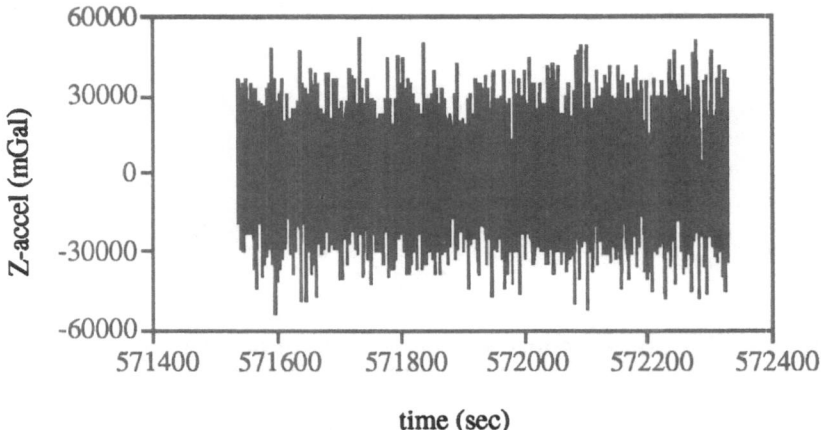

Figure 4.2   Effect of aircraft vibration on accelerometer measurements

Errors in the determination of vehicle acceleration are due to measurement errors in the GPS range and range-rate observables between GPS satellites and ground antennas. To derive acceleration from these observables, a differentiation process is necessary. This process amplifies the high frequency components in the measurements and thus measurement noise. By using standard double differencing techniques between satellites and ground receivers, quasi-systematic errors are eliminated or reduced to the noise level. Thus, orbit errors, clock errors, and most of the atmospheric errors can usually be neglected when double differenced carrier phase measurements are used for aircraft acceleration determination. The measurement model for double differenced carrier phase data has the following terms

$$\nabla\Delta\Phi = \nabla\Delta\rho + \lambda\nabla\Delta N + \nabla\Delta d_{trop} - \nabla\Delta d_{ion} + \nabla\Delta d_m + \nabla\Delta n \qquad (4.7)$$

where $\Delta\nabla$ is the symbol for double differencing between receivers and between satellites, Φ is the phase measurement, ρ is the receiver satellite distance, N is the ambiguity, d_{trop} and d_{ion} are the tropospheric and ionospheric effect, respectively, d_m is carrier phase multipath, and n is system noise. This indicates that residual atmospheric errors $\nabla\Delta d_{trop}$ and $\nabla\Delta d_{ion}$, multipath $\nabla\Delta d_m$, and measurement noise

$\nabla\Delta n$ are the most likely candidates for noise in the accelerations derived from $\nabla\Delta\rho$. The first two of these errors overlap the spectral band in which gravity has to be estimated. The third is a major factor for masking the gravity signal in the high frequencies.

Table 4.1 presents the dominant errors in different bandwidths as derived from static measurements, see Wei and Schwarz (1995) for details. These results show that the combined effect of low frequency errors due to orbit errors, atmospheric errors, and some multipath contributes a standard deviation of only 0.1 mGal to the total acceleration error. The medium frequency part is dominated by multipath errors and accounts for about 0.3 mGal of the total error. By far the largest portion of the error is in the high frequency part which is dominated by measurement noise. It amounts to a standard deviation of about 1 mGal. The subsequent analysis of airborne test data, see Wei and Schwarz (1995), indicates that these results are also valid for the aircraft environment, although GPS measurement noise tends to be higher. This is due to the fact that the bandwidth has to be opened up under dynamic measurement conditions which increases the noise.

Table 4.1   The effect of bandwidth on GPS acceleration determination

Bandwidth (Hz)	σ_a of SV 11 (mGal)	σ_a of SV 19 (mGal)	σ_a of SV 31 (mGal)
0.00005 - 0.005	± 0.09	± 0.09	± 0.09
0.005 - 0.01	± 0.28	± 0.30	± 0.24
0.01 - 0.02	± 1.15	± 1.23	± 0.89

Figure 4.3 shows this effect for a series of tests on a well-controlled dynamic base where the antenna position could be determined with an accuracy of about 0.01 mm. The resulting acceleration errors are increasing in parallel with the dynamics, for details see Zhang and Schwarz (1996). Thus, receiver noise is the critical component in the acceleration error budget while multipath and atmospheric errors are small and other errors are negligible. Receiver noise in the high frequencies masks the low-amplitude gravity signal in this part of the spectrum and is currently one of the limiting factors for short-wavelength resolution of the gravity field. Figure 4.4 illustrates this point. It shows the error spectrum of the GPS-derived acceleration versus the gravity spectrum at two different altitudes for a flying speed of 80 m/s. Both spectra were derived from actual data. Assuming that wavelength resolution is possible at a signal-to-noise ratio of 1, the minimum wavelength that can be resolved in this case is 4 km at a flying altitude of 0.5 km and 7 km at a flying altitude of 2.5 km. Better resolution can be achieved by flying at lower velocities.

As Figure 2.3 shows, airborne gravity determination is essentially a process of differencing two data streams. Although the computational algorithm may often hide this fact, synchronization of the two data streams is essential to achieving

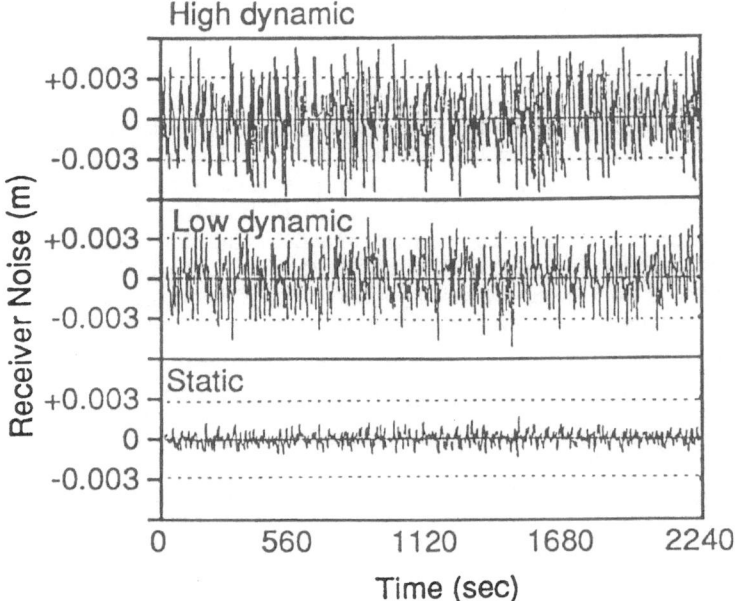

Figure 4.3   Dependance of GPS receiver noise on dynamics

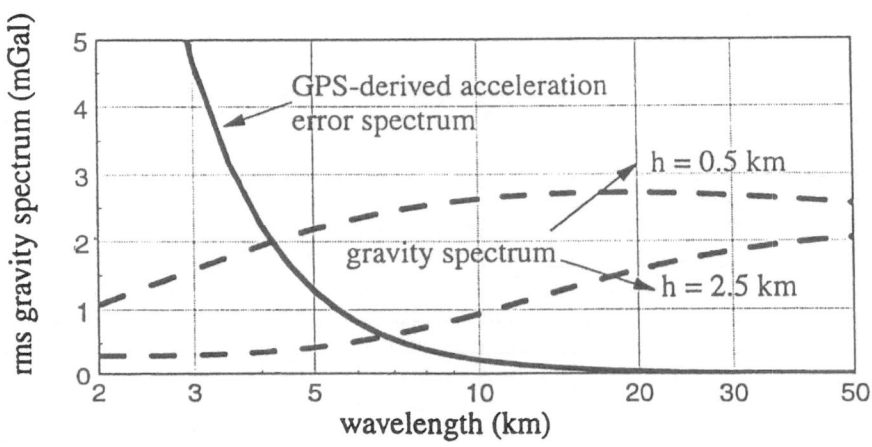

Figure 4.4   Influence of GPS error spectrum on minimum wavelength
resolution of gravity field at flight level.

accuracies at the 1 mGal level. The effect of **synchronization** errors can be expressed by

$$dg_{sync} = \left(\dot{R}_b^l f^b + R_b^l \dot{f}^b\right) dT \qquad (4.8)$$

Thus, the effect of these errors is strongly dependent on the dynamic environment in which measurements take place. Both translational and rotational motion contribute to the error. The size of the synchronization error alone is therefore not a measure for the resulting gravity error, but its interaction with aircraft dynamics. Therefore, a constant acceleration error is much more critical than a random error. Figure 4.5 shows the vertical gravity error generated by a 1 millisecond (ms) synchronization bias for the typical flight dynamics of a small airplane. It clearly shows that these errors are not negligible.

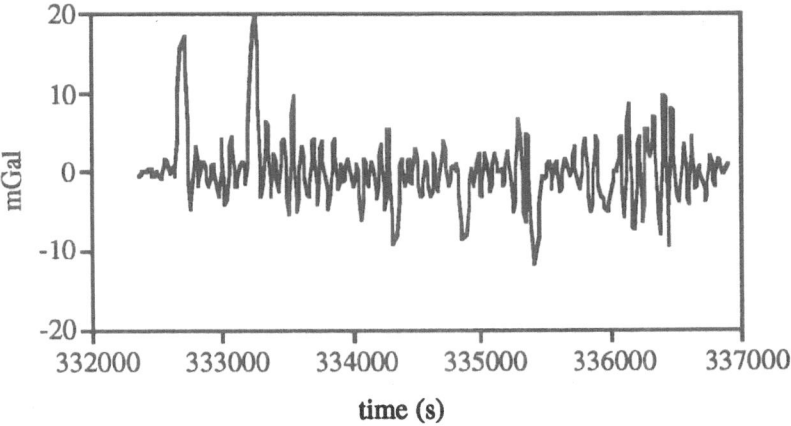

Figure 4.5 Effect of a constant synchronization error of 1 ms on the gravity signal

Synchronization errors can occur for a number of reasons, of which the following are the most important ones: System internal time delays, registration errors, and clock errors. The last two are usually summarized as time tagging errors. Each system has a delay between the epoch of the actual measurement and the epoch when this measurement is available as output. In general these time delays are constant and can either be directly measured or can be calibrated in post mission. There may, however, be a problem with the time definition of the output epoch which, in some cases, is simply given as 50 Hz without a precise definition of the epoch. Registration errors are related to the data logging process. Usually, data streams coming from different systems are logged by the computer and registered by the computer clock. Registration errors occur when the time registration process for any data stream is blocked by other processes with higher priority. These errors therefore vary randomly. They are typically at the level of 0.5 ms, but may reach several milliseconds in extreme cases. Usually these errors can be reduced to an appropriate level by moving the registration of the incoming data streams to a dedicated timing board. Clock errors may occur as drift, jump, or random noise. While random noise is sufficiently small, drifts and jumps affect the time tagging process. Drifts can usually be eliminated by resetting the clock regularly with a superior time signal. This can for instance be a hardware reset by the 1PPS signal of the GPS output. Jumps are fortunately rare and non cumulative. They can be largely eliminated by a good electronic design of the dedicated timing board.

5   The estimation problem

Gravity estimation from airborne measurements is a complicated procedure because noise in the INS-GPS differences
  • is extremely large (noise-to-signal ratios of 1000 and more)
  • has a similar spectral signature as the signal in the frequencies of interest
  • is highly dependent on aircraft dynamics.
To illustrate these points, Figures 5.1 to 5.7 will be briefly discussed. Figure 5.1 shows the difference between the vertical aircraft acceleration derived from GPS and the INS specific force in the vertical channel, corrected for normal gravity, Coriolis and tangential acceleration. The data rate is 2 Hz. The difference represents the **'raw' gravity signal** disturbed by noise. The gravity signal at flight level varies by about 120 mGal. The INS-GPS difference shown in Figure 5.1 varies by about 50 000 mGal. To extract a signal from such noise resembles finding the proverbial needle in the haystack.

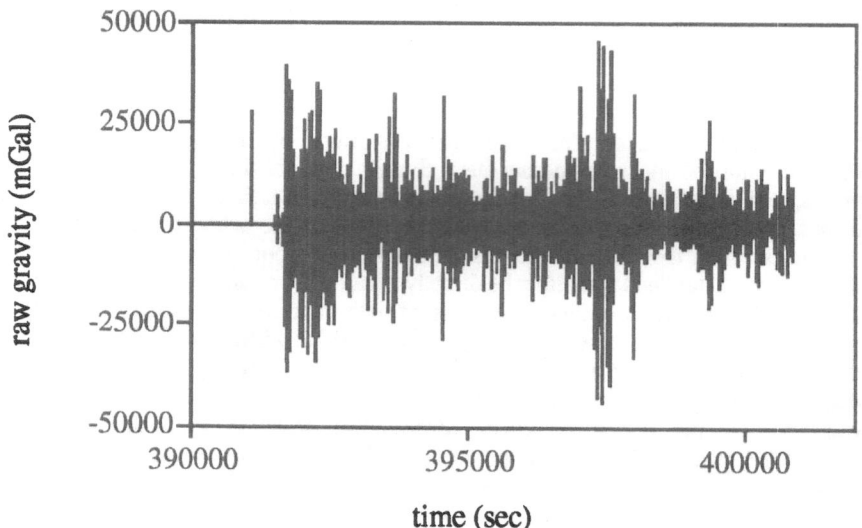

Figure 5.1   'Raw' gravity measurements (INS-GPS differences)

Fortunately, the discussion of error sources in the previous chapter provides some clues to the origin of this high noise. Much of it will be due to high frequency noise, caused by the effects of aircraft vibration on the INS and the amplification of GPS system noise when computing acceleration. That this is indeed the case can be seen from Figures 5.2 and 5.3 which show these two spectra. Amplitudes in the **vibration spectrum** go up to 3000 mGal and their cumulative effect would result in RMS values close to the ones shown in Figure 5.1. The spectrum of GPS acceleration errors is not quite as dramatic, but would also result in a cumulative RMS error of about 1000 mGal.

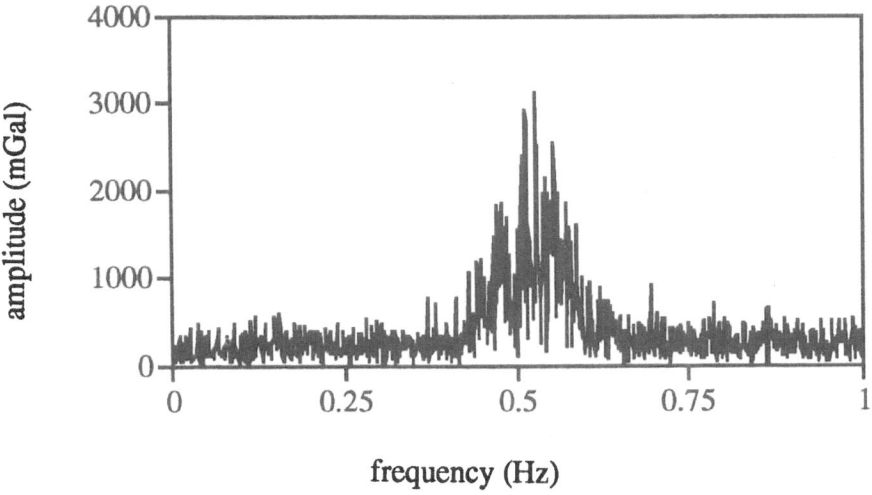

Figure 5.2 Spectrum of accelerometer error due to aircraft vibration (2 Hz)

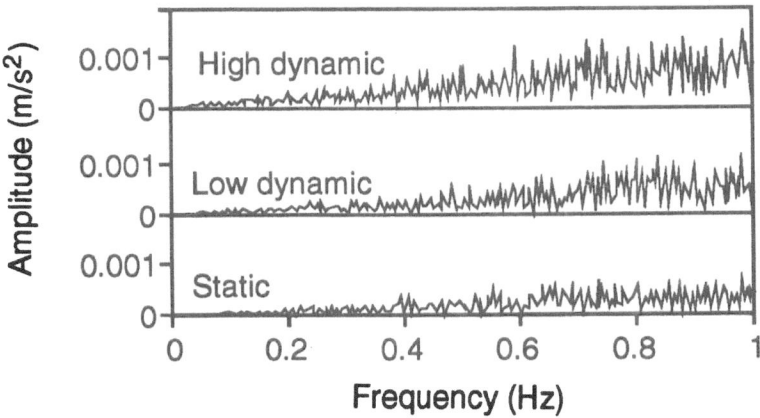

Figure 5.3 Spectrum of GPS acceleration error (2Hz)

The obvious way to eliminate these two noise effects is a **low-pass filter**. The problem with this approach is that part of the gravity information would be cut off at the same time. How serious would such a cut-off be? Figures 5.4a and 5.4b show the gravity spectrum for flat and mountainous terrain, as derived from actual ground gravity data. If the RMS acuracy of the estimated gravity signal is to be 1 mGal, the figures would indicate that a wavelength resolution of 10 km is required for flat terrain and of 4 km for mountainous terrain. For a small aircraft flying at 200 km/h, this would mean that the cut-off could be at 180 seconds in flat terrain and at 70 seconds in mountainous terrain, without causing a gravity error larger than 1 mGal (RMS). A careful analysis of signal-to-noise ratios (S/N ratio) in individual frequency bands shows, however, that wavelength below 5 km are very difficult to recover because their S/N ratio drops below 1. Thus, a 90 second FIR filter was chosen to eliminate high frequency noise. This should be sufficient for vibration and GPS error effects.

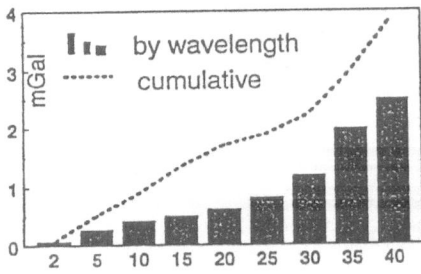
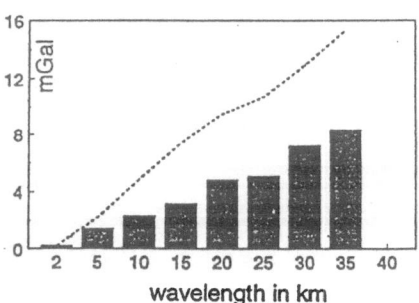

a) Flat Terrain   b) Mountainous Terrain

Figure 5.4   Anomalous gravity spectrum derived from land gravity data in a) flat terrain b) mountainous terrain (wavelength in km)

The result of applying the FIR filter to the original data (Fig. 5.1) is shown in Figure 5.5, where three flights over the ground same profile are presented. The variation of the filtered INS-GPS acceleration is now in the range expected for the gravity disturbances and shows clearly the structure of the **anomalous gravity field**. Note the mirror symmetry at time 397500 s which represents a 360 degree turn of the aircraft close to the largest gravity disturbance. When comparing the first flight over this disturbance at time 393000 s to the other two, it becomes obvious that the vertical accelerometer bias has not yet been eliminated. This means that the instrumental biases of the INS have to be determined in a separate procedure, using a GPS/INS Kalman filter optimized for this purpose. By estimating this bias separately and applying it to the data, Figure 5.6 is obtained. It shows an overlay of the three flights over the same ground profile and thus indicates repeatability of gravity estimation. It is at an RMS level of about 2 mGal. The RMS difference with respect to upward continued ground values is about 3 mGal (RMS). Figure 5.6 also shows that some of the very sharp peaks in the gravity profile are not adequately recovered. This is due to aircraft velocity which, for economical reasons, had to be higher than the optimal speed for recovering the short wavelengths of the gravity field. This problem can be easily overcome by either crusing at a lower velocity or by applying remove-restore techniques for the topography.

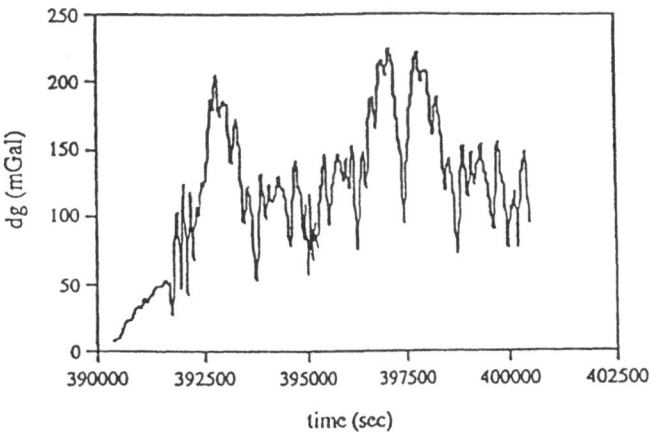

Figure 5.5 FIR filter (90s) applied to the 'raw' gravity data of Figure 5.1 for three flights over the same ground profile

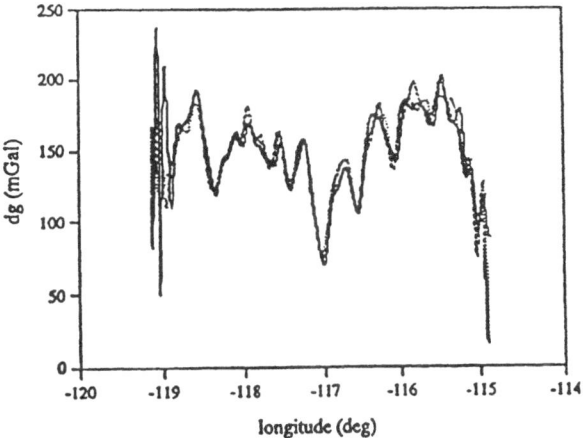

Figure 5.6 FIR filter (90s) after bias removal

Besides the errors discussed above, the interaction of **INS misalignment** errors with horizontal aircraft acceleration has to be considered. The spectrum of this error is shown in Figure 5.7. It predominantly reflects aircraft motion with a peak at about 55 s and and measurable effects to about 90 s. The peak represents the natural period of phugoid motion. It depends on aircraft properties and it cannot be assumed that this error will always be eliminated by a low-pass filter. It is therefore safer to estimate sensor misalignment via a Kalman filter and to correct the raw data for this effect using a smoothed form of the GPS-estimated horizontal accelerations in the process.

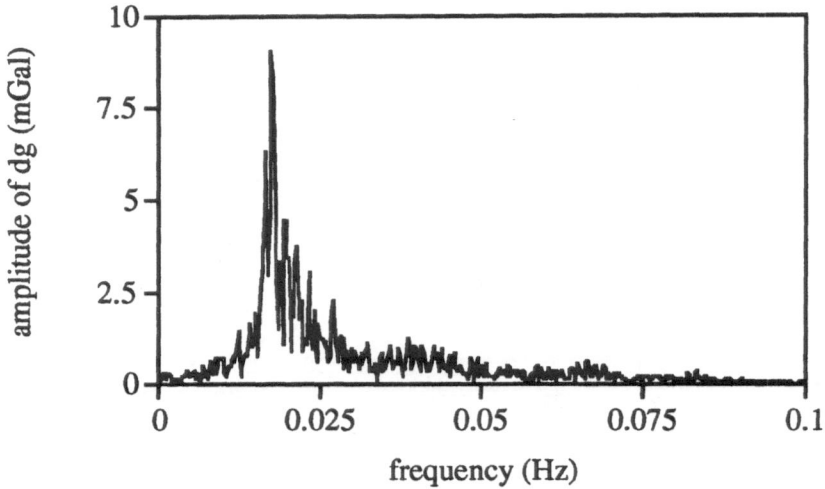

Figure 5.7   Spectrum of the gravity error induced by the effect of aircraft motion on misalignment

The rationale for an **estimation strategy** has therefore to take into account the high frequency noise of the 'raw' INS-GPS data, the effects of aircraft dynamics on the error budget, i.e. on misalignment errors and synchronization errors, and the remaining bias variation in the spectral range of interest for gravity estimation. This leads to an iterative procedure which is outlined in Figure 5.8. In the first step, long-term instrumental biases are estimated using an integrated INS/GPS Kalman filter for this purpose. In this step, gravity field variations are considered as noise. In the second step, the inertial data are corrected for these biases and a first estimate of the gravity disturbance is obtained by low-pass filtering the INS-GPS differences. These estimates are applied as corrections to the 'raw' specific force data and the procedure is repeated. The iteration is stopped when the RMS value of the difference between the current gravity disturbance estimate and the previous one is below a certain threshold value. In case of large INS biases where the linearity assumptions may not be met, it is safer to estimate corrections to the first bias estimate in subsequent iterations.

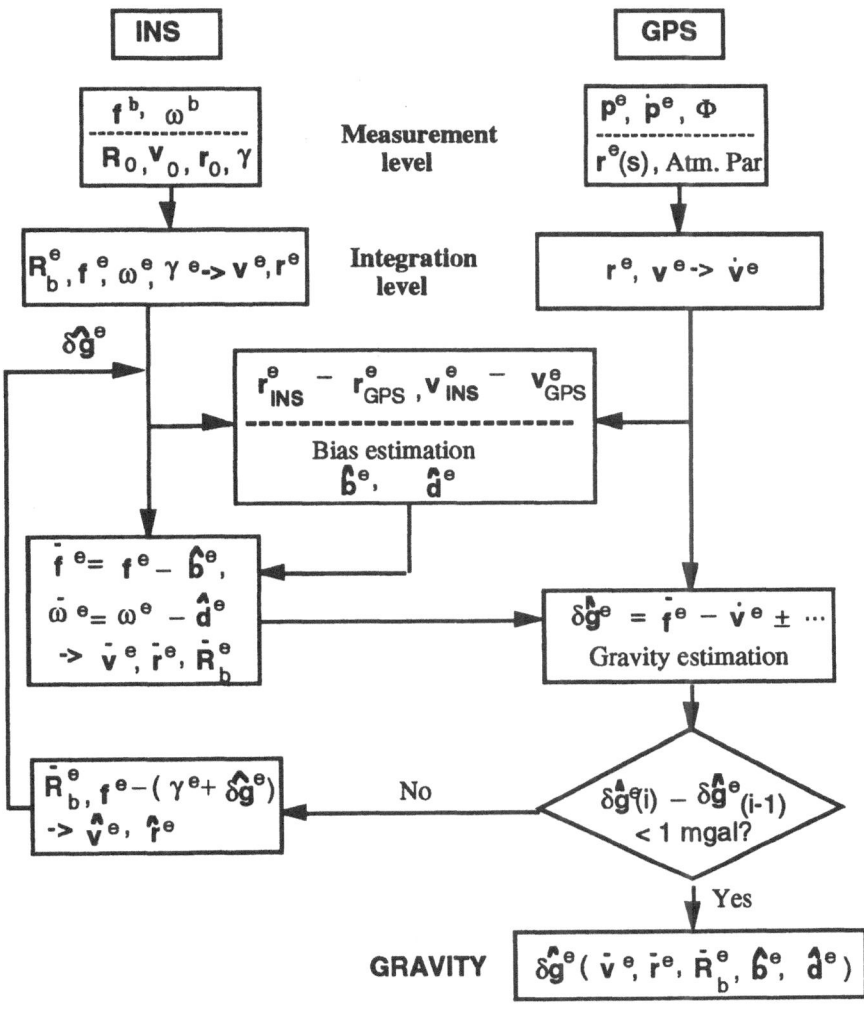

Figure 5.8   Gravity estimation strategy

Most of the methods proposed to date for airborne gravity estimation contain some aspects of the above scheme, but fall somewhat short of a full implementation. The method used in Wei and Schwarz (1996) comes very close to it, but is not iterative. Thus, gravity disturbance and biases are only estimated once, and their interaction is not yet considered. The wave algorithm, as proposed by Salychev and applied in Salychev and Schwarz (1995), uses an iterative way of estimating system biases and gravity disturbance. It also applies low-pass filtering to eliminate high-frequency noise, but uses a prescribed mathematical model to do this. This may lead to poorer approximation in areas with large gravity gradients, such as mountainous areas. Kalman filtering with a shaping filter for the gravity disturbance vector estimates the gravity disturbance and system biases simultaneously. This is an attractive alternative, as long as the a priori estimates of the covariance matrices are representative for the area in which the survey takes place. Since this is usually not the case - a gravity survey would not be necessary if the gravity field in the area is known - this method suffers most from the assumptions made. Wrong a priori assumptions in either the covariance information for the gravity disturbance vector or the sensor biases will affect gravity estimation adversely.

For a thorough discussion and comparison of these three methods, see Hammada (1996). He shows that all three of them are a form of low-pass filtering, but that the underlying mathematics makes for considerable differences in performance. While the approach shown in Figure 5.8 is essentially model independent, the two others are not. They take advantage of the proven reliability of estimation models, such as Kalman filtering or wave estimation, but may suffer in certain situations from the a priori assumptions going into that model. In the numerical example given in the above report, results of the model independent approach are significantly better than those of the model dependent approaches. In addition, the wave algorithm outperforms Kalman filtering in this case. Although such results must be taken with caution because they are based on only one example, they are an indicator that the estimation strategy is an important component of airborne gravity data processing.

Part Three: The Boundary Value Problems Of Airborne Gravimetry

The three approaches to airborne gravimetry mentioned earlier, result either in a scalar quantity, i.e the magnitude of the gravity disturbance vector, a vector quantity, i.e the complete gravity disturbance vector, or in a tensor of second-order gradients of the anomalous gravity potential. The determination of the gravity field by use of airborne gravimetric data at flying altitude leads to the investigation of boundary value problems (BVPs). For a general discussion of the state of the art in this field, see Heck (1996). Previous work in formulating BVPs for airborne gravity data has been done by Vassiliou (1986), Holota (1995) and Keller and Hirsch (1994). Work in this area is, however far from complete. In the following, three types of BVPs related to airborne gravimetry will be discussed with special emphasis on local and regional applications.

In chapter 6, the BVPs using only airborne gravimetric data are formulated, and the solutions are given using Taylor expansions and Fourier transform. The BVPs combining airborne data with other data on the ground are considered in chapter 7. Solutions are derived using the planar harmonic analysis method. The concept of multiresolution BVPs is introduced in chapter 8 and their solutions are developed based on discrete wavelet transforms.

6   BVPs using airborne gravimetric data only

For the sake of simplicity, the flight surface is assumed to be an equipotential surface of the gravity field. Only local or regional gravity fields will considered, which means a planar approximation can be used. A similar procedure can, however, be also employed for global gravity field determination, because only the gravity field between the flight surface and the Earth's surface has to be determined, while the gravity field outside the flight surface can be obtained using the conventional BVPs.

The BVPs using only airborne gravity data can be formulated as:

Given airborne gravity data at a flight surface Σ, the gravity field between the Earth's surface S and the flight surface Σ is to be determined.

Mathematically, these BVPs can formulated as follows:

$$\begin{cases} \Delta T(x,y,z) = 0 & \text{for } 0 \le z \le H \\ A_i T |_\Sigma = g_i \end{cases} \tag{6.1}$$
$$(i = 1, 2, ..., n)$$

where T is the disturbing potential; A_i is a linear functional relating T to the measurement g_i; n is the number of measurement types; (x, y, z) are the coordinates in a local coordinate system, the origin of which is in the centre of the area of interest, as shown in Figure 6.1.

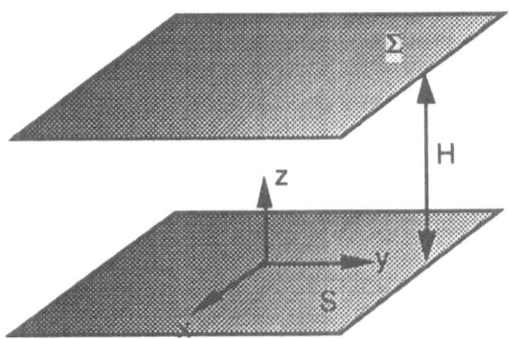

Figure 6.1   The coordinate system used for local gravity fields

Remark 6.1: This kind of BVP is different from the conventional ones in the sense that the boundary conditions are on the flight surface not on the geoid or telluroid.

Three special cases are of interest:

Case I: (Scalar BVP of Airborne Gravimetry)

$$\begin{cases} \Delta T(x,y,z) = 0 & \text{for } 0 \le z \le H \\ \dfrac{\partial T}{\partial z} \Big|_\Sigma = - \delta g^H \end{cases} \tag{6.2}$$

where δg_H is the gravity disturbance at flight level.

Case II: (Vector BVP of Airborne Gravimetry)

$$\begin{cases} \Delta T(x,y,z) = 0 \quad \text{for } 0 \le z \le H \\[2mm] \dfrac{\partial T}{\partial x}\Big|_\Sigma = \delta g_x^H \\[2mm] \dfrac{\partial T}{\partial y}\Big|_\Sigma = \delta g_y^H \\[2mm] \dfrac{\partial T}{\partial z}\Big|_\Sigma = -\delta g^H \end{cases} \qquad , \qquad (6.3)$$

where δg_x^H and δg_y^H are the two components of the gravity disturbance vector along the x-axis and the y-axis measured at flight level.

Case III: (BVP of Airborne Gradiometry)

$$\begin{cases} \Delta T(x,y,z) = 0, \quad \text{for } 0 \le z \le H \\[2mm] \dfrac{\partial^2 T}{\partial x \partial x}\Big|_\Sigma = g_{xx}^H, \quad \dfrac{\partial^2 T}{\partial x \partial y}\Big|_\Sigma = g_{xy}^H, \quad \dfrac{\partial^2 T}{\partial x \partial z}\Big|_\Sigma = g_{xz}^H \\[2mm] \dfrac{\partial^2 T}{\partial y \partial x}\Big|_\Sigma = g_{yx}^H, \quad \dfrac{\partial^2 T}{\partial y \partial y}\Big|_\Sigma = g_{yy}^H, \quad \dfrac{\partial^2 T}{\partial y \partial z}\Big|_\Sigma = g_{yz}^H \\[2mm] \dfrac{\partial^2 T}{\partial z \partial x}\Big|_\Sigma = g_{zx}^H, \quad \dfrac{\partial^2 T}{\partial z \partial y}\Big|_\Sigma = g_{zy}^H, \quad \dfrac{\partial^2 T}{\partial z \partial z}\Big|_\Sigma = g_{zz}^H \end{cases} \qquad , \qquad (6.4)$$

where $g_{xx}, g_{xy}, ..., g_{zz}$ are the nine components of the tensor of second derivatives of the disturbing potential, measured at flight level.

Remark 6.2: The BVP in Case I is a well-determined BVP, while the BVPs in both case II and case III are over-determined BVPs.

Remark 6.3: Only 5 out of 9 components in Case III are independent.

The solution to BVP (6.1) can be obtained using a Taylor series expansion:

$$T(x, y, z) = T(x, y, H) - \frac{\partial T}{\partial z}\Big|_{z=H} (H-z) + \frac{1}{2}\frac{\partial^2 T}{\partial z \partial z}\Big|_{z=H} (H-z)^2 + \dots$$

$$= \sum_{k=0}^{\infty} (-1)^k \frac{1}{k!} \frac{\partial^k T}{\partial z^k}\Big|_{z=H} (H-z)^k$$

$$(0 \le z \le H). \qquad (6.5)$$

The height anomaly ζ and the gravity disturbance δg^0 on the earth surface can be obtained as follows:

$$\zeta(x, y, 0) = \frac{T(x, y, 0)}{\gamma} = \frac{1}{\gamma}\sum_{k=0}^{\infty} (-1)^k \frac{1}{k!} \frac{\partial^k T}{\partial z^k}\Big|_{z=H} H^k \qquad (6.6)$$

and

$$\delta g^0 = -\frac{\partial T}{\partial z}\Big|_{z=0} = \sum_{k=1}^{\infty} (-1)^{k+1} \frac{1}{(k-1)!} \frac{\partial^k T}{\partial z^k}\Big|_{z=H} H^{k-1}. \qquad (6.7)$$

Because $z \ll R$ (the mean earth radius), only a few terms in (6.5) and (6.6) need to be computed. To estimate the truncation error, a simple mass model for the anomalous gravity field has been used. It is of the form

$$T = \frac{k\delta M}{R + z} \quad ,$$

where k is the gravitational constant and δM is a disturbing mass. For this model, we have

$$\frac{\partial^k T}{\partial z^k}\Big|_{z=H} = (-1)^{k-1} \frac{k! \delta g_M^0}{(R + H)^{k-1}} \quad ,$$

where $\delta g_M^H = -\frac{k\delta M}{(R + H)^2}$.

The truncation error for ζ and δg^0 can therefore be estimated using the following formulas

$$d\zeta = \frac{1}{\gamma} \sum_{k=M}^{\infty} (-1)^k \frac{1}{k!}(-1)^{k-1} \frac{k! \delta g_M^H}{(R + H)^{k-1}} H^k$$

$$= -\frac{\delta g_M^H H^M}{\gamma R (R + H)^{M-2}}$$

and

$$d\delta g^0 = \frac{1}{\gamma} \sum_{k=N}^{\infty} (-1)^k \frac{1}{k!}(-1)^{k+1} \frac{(k + 1)! \delta g_M^H}{(R + H)^k} H^k$$

$$\approx -(N + 1)(\frac{H}{R + H})^M \delta g_M^H.$$

Table 6.1 gives the truncation errors for ζ and δg^0 using N=2 and N=3 for the expansions and four different flight levels, i.e. H= 500 m, 1000 m, 5000 m and 10000 m. In this estimation, the gravity disturbance δg_M^H at flight level is assumed to be 100 mGal, R to be 6400 km and γ to be 1000000 mGal.

Table 6.1 Truncation errors of ζ and δg^0

H(m)	M	$|d\zeta|$(cm)	$|d\delta g^0|$(mGal)
500	2	0.0004	0.0000
1000	2	0.0016	0.0000
5000	2	0.0391	0.0002
10000	2	0.1563	0.0007
500	3	0.0000	0.0000
1000	3	0.0000	0.0000
5000	3	0.0000	0.0000
10000	3	0.0002	0.0000

Table 6.1 indicates that two terms in the expansions will usually give the required accuracy. However since high frequency parts of the gravity field signal have not been taken into account due to a simple model, higher order terms still might be needed, especially in mountainous areas.

Two well-known properties of the Fourier transform will be used in the following:

1) $\quad F(\dfrac{\partial^k f}{\partial z^k}) = (-2\pi q)^k F(f)$.

2) $\quad F(h(x,y) * g(x,y)) = F(h(x,y)) * F(g(x,y))$,

where the convolution, denoted by *, is defined as:

$$h(x,y) * g(x,y) = \int\limits_{-\infty}^{\infty} \int\limits_{-\infty}^{\infty} h(x_0, y_0) g(x_0 - x, y_0 - y) dx_0 dy_0.$$

Using the Fourier transform equation, (6.5) can be written as

$$F(T(x, y, z)) = F(T(x, y, H)) + 2\pi q F(T(x, y, H))(H - z) +$$
$$\frac{1}{2}(2\pi q)^2 F(T(x, y, H))(H - z)^2 + \dots \tag{6.8a}$$

or

$$T(x, y, z)) = F^{-1}(F(T(x, y, H)) + 2\pi q F(T(x, y, H))(H - z) +$$
$$\frac{1}{2}(2\pi q)^2 F(T(x, y, H))(H - z)^2 + \dots) \tag{6.8b}$$

where F is the 2D Fourier transform operator, which is defined as

$$F(u, v) = F(f(x,y)) = \int\limits_{-\infty}^{\infty} \int\limits_{-\infty}^{\infty} f(x,y) e^{-j(ux + vy)} dx dy,$$

F^{-1} is the inverse 2D Fourier transform. u and v are the spatial frequencies in the direction of x and y respectively, and $q = \sqrt{u^2 + v^2}$; j is the imaginary unit $(j = \sqrt{-1})$.

Remark 6.4: (6.8a) indicates that only the Fourier transform of the disturbing potential T at flight level needs to be computed in order to calculate $F(T(x, y, z))$. A similiar formula can be derived for ζ and δg^0.

To determine $F(T(x,y,H))$ using the data available, the following criterion will be used:

$$\int\limits_{\Sigma} (AT(x,y,H) - g(x,y,H))^T \int\limits_{\Sigma} C^{-1}(x - x', y - y')(AT(x',y',H) - g(x',y',H)) d\Sigma' d\Sigma = \text{min}$$

$$\tag{6.9a}$$

or

$$(AT(x,y,H) - g(x,y,H))^T * C^{-1} * (AT(x,y,H) - g(x,y,H)) = \text{min}, \tag{6.9b}$$

where $A = [A_1, A_2, \dots, A_n]^T$ and $g = [g_1, g_2, \dots, g_n]^T$. C is the measurement noise covariance matrix.

The minimization problem (6.9) is actually a variational problem, the solution of which leads to the following variational equation:

$$AdT(x, y, H)^T * C^{-1} * (AT(x, y, H) - g(x, y, H)) = 0 \qquad (6.10)$$

where dT is not an ordinary differential of T, but a first variation of T.

Applying the Fourier transform to (6.10) and assuming that $F(AT) = A(u,v)$ $F(T)$ leads to

$$A(u, v)^T F(dT) F(C^{-1})(A(u, v) F(T) - G) = 0 \qquad (6.11)$$

where G=F(g).

The function of dT in (6.11) is arbitrary. Therefore the following equation must hold:

$$A(u, v)^T F(C^{-1})(A(u, v) F(T) - G) = 0 \qquad (6.12)$$

which leads to

$$F(T) = (A(u, v)^T F(C^{-1}) A(u, v))^{-1} A(u, v) F(C^{-1}) G \cdot \qquad (6.13)$$

Formulas (6.8) and (6.13) provide the formula for the determination of the disturbance potential T(x,y,z) from measurements at flight level.

Applying these formulas to Case I first, $A(u, v) = 2\pi q$ and $G = F(\delta g^H)$. Therefore (6.13) becomes

$$F(T(x, y, H)) = \frac{F(\delta g^H)}{2\pi q} \qquad (6.14)$$

or

$$T(x, y, H) = \frac{1}{2\pi} \int_\Sigma \frac{\delta g^H}{\sqrt{(x - x')^2 + (y - y')^2}} dx' dy' \qquad (6.15)$$

i.e. the planar Stokes formulas is obtained.

In Case II, $A(u,v) = [j2\pi u \quad j2\pi u \quad 2\pi q]^T$, $G = [F(\delta g_x^H) \quad F(\delta g_y^H) \quad F(\delta g^H)]^T$.

Assuming the measurements to be uncorrelated, i.e.

$$F(C^{-1}) = \begin{bmatrix} C_1(u, v) & 0 & 0 \\ 0 & C_2(u, v) & 0 \\ 0 & 0 & C_3(u, v) \end{bmatrix} \cdot$$

and inserting A(u,v), G and $F(C^{-1})$, one obtains

$$F(T(x, y, H)) = \frac{j2\pi u C_1 F(\delta g_x^H) + j2\pi v C_2 F(\delta g_y^H) + 2\pi q C_3 F(\delta g^H)}{4\pi^2 (q^2 C_3 - u^2 C_1 - v^2 C_2)} \cdot \qquad (6.16)$$

Similiar formulas can be obtained for Case III if the uncorrelated measurements are assumed.

7 BVPs combining airborne gravity data with ground data

The determination of the gravity field between the flight surface and the Earth's surface using airborne gravity data only is a downward continuation process, i.e. it is inherently an unstable process. To stabilize it, the combination of airborne gravity data with ground data, e.g, terrestrial gravity data and altimetric data should be considered. This type of BVP has been investigated by Holota (1995) using a similar approach as Grafarend and Sanso (1984). In the following, BVPs for local and regional gravity field determination will be considered using both airborne and ground gravity data. Their solutions will be derived based on the planar Fourier analysis method.

BVPs combining airborne and ground data can be described as:

Given airborne gravity data at a flight surface Σ and gravity field related data on the Earth's surface S, the gravity field between the Earth surface S and the flight surface Σ is to be determined. Mathematically, this kind of BVP can be formulated as follows:

$$\begin{cases} \Delta T = 0 & \text{for } 0 \le z \le H \\ A_i T |_\Sigma = g_i & (i = 1, 2, ..., n) \\ B_j T |_S = f_j & (j = 1, 2, ..., m) \end{cases} \tag{7.1}$$

where A_i and B_j are linear functionals relating T to airborne data g_i and ground data f_j . n and m are the number of measurement types for airborne data and ground data respectively.

Some special cases are of interest:

Case IV: (Neuman Problem)

$$\begin{cases} \Delta T = 0 & \text{for } 0 \le z \le H \\ \dfrac{\partial T}{\partial z} |_\Sigma = \delta g^H \\ \dfrac{\partial T}{\partial z} |_S = \delta g^0 \end{cases} \tag{7.2}$$

Case V: (Mixed Neuman - Dirichllet Problem)

$$\begin{cases} \Delta T = 0 & \text{for } 0 \le z \le H \\ \dfrac{\partial T}{\partial z} |_\Sigma = \delta g^H \\ T |_S = T^0 \end{cases} \tag{7.3}$$

Case VI: (Mixed Gradiometry-Neuman Problem)

$$\begin{cases} \Delta T = 0 & \text{for } 0 \le z \le H \\ \dfrac{\partial^2 T}{\partial z \partial z} |_\Sigma = g_{zz}^H \\ \dfrac{\partial T}{\partial z} |_S = \delta g^0 \end{cases} \tag{7.4}$$

and

Case VI: (Mixed Gradiometry - Dirichllet Problem)

$$\left\{ \begin{array}{ll} \Delta T = 0 & \text{for } 0 \le z \le H \\ \dfrac{\partial^2 T}{\partial z \partial z}\big|_\Sigma = g_{zz}^H \\ T\big|_S = T^0 \end{array} \right. \tag{7.5}$$

Remark 7.1: Case IV and Case V are BVPs combining scalar airborne gravity data with ground gravity data and altimetric data respectively, while Case VI and Case VII are BVPs combining airborne vertical gradient data with ground gravity data and altimetric data, respectively.

To solve BVP of type (7.1), a planar harmonic expansion is used, see Bian and Zhang (1993) for details, i.e.

$$T(x,y,z) = \sum_{i=0}^{\infty}\sum_{j=0}^{\infty} e^{-z\omega_{ij}} [\cos(i\omega_1 x) \quad \sin(i\omega_1 x)] \begin{bmatrix} A_{ij} & B_{ij} \\ C_{ij} & D_{ij} \end{bmatrix} \begin{bmatrix} \cos(j\omega_2 y) \\ \sin(j\omega_2 y) \end{bmatrix} +$$
$$\sum_{i=0}^{\infty}\sum_{j=0}^{\infty} e^{z\omega_{ij}} [\cos(i\omega_1 x) \quad \sin(i\omega_1 x)] \begin{bmatrix} E_{ij} & F_{ij} \\ G_{ij} & H_{ij} \end{bmatrix} \begin{bmatrix} \cos(j\omega_2 y) \\ \sin(j\omega_2 y) \end{bmatrix} \tag{7.6}$$

where

$$\omega_{ij} = \sqrt{i^2 \omega_1^2 + j^2 \omega_2^2},$$

ω_1 and ω_2 are circular frequencies in the direction of the x or y-axis, respectively. The coefficients A_{ij}, B_{ij},..., H_{ij} are to be determined using the boundary conditions. It is easy to verify that $\Delta T = 0$ for $0 \le z \le H$.

To demonstrate how to determine these coefficients, only Case IV is considered here. The other cases can be treated in a similar way.

Using the boundary conditions in BVP (7.2), the following equations hold:

$$\delta g^H = -\frac{\partial T}{\partial z}\big|_{z=H} = \sum_{i=0}^{\infty}\sum_{j=0}^{\infty} \omega_{ij} e^{-H\omega_{ij}} [\cos(i\omega_1 x) \quad \sin(i\omega_1 x)] \begin{bmatrix} A_{ij} & B_{ij} \\ C_{ij} & D_{ij} \end{bmatrix} \begin{bmatrix} \cos(j\omega_2 y) \\ \sin(j\omega_2 y) \end{bmatrix} +$$
$$\sum_{i=0}^{\infty}\sum_{j=0}^{\infty} \omega_{ij} e^{H\omega_{ij}} [\cos(i\omega_1 x) \quad \sin(i\omega_1 x)] \begin{bmatrix} E_{ij} & F_{ij} \\ G_{ij} & H_{ij} \end{bmatrix} \begin{bmatrix} \cos(j\omega_2 y) \\ \sin(j\omega_2 y) \end{bmatrix} \tag{7.7}$$

and

$$\delta g^0 = -\frac{\partial T}{\partial z}\big|_{z=0} = \sum_{i=0}^{\infty}\sum_{j=0}^{\infty} \omega_{ij} [\cos(i\omega_1 x) \quad \sin(i\omega_1 x)] \begin{bmatrix} A_{ij} & B_{ij} \\ C_{ij} & D_{ij} \end{bmatrix} \begin{bmatrix} \cos(j\omega_2 y) \\ \sin(j\omega_2 y) \end{bmatrix} +$$
$$\sum_{i=0}^{\infty}\sum_{j=0}^{\infty} \omega_{ij} [\cos(i\omega_1 x) \quad \sin(i\omega_1 x)] \begin{bmatrix} E_{ij} & F_{ij} \\ G_{ij} & H_{ij} \end{bmatrix} \begin{bmatrix} \cos(j\omega_2 y) \\ \sin(j\omega_2 y) \end{bmatrix}. \tag{7.8}$$

On the other hand, the measurements δg^H and δg^0 can also be expanded as a planar harmonic series, i.e.

$$\delta g^H = \sum_{i=0}^{\infty}\sum_{j=0}^{\infty} [\cos(i\omega_1 x) \quad \sin(i\omega_1 x)] \begin{bmatrix} \alpha_{ij}^H & \beta_{ij}^H \\ \gamma_{ij}^H & \delta_{ij}^H \end{bmatrix} \begin{bmatrix} \cos(j\omega_2 y) \\ \sin(j\omega_2 y) \end{bmatrix} \tag{7.9}$$

and

$$\delta g^0 = \sum_{i=0}^{\infty} \sum_{j=0}^{\infty} [\cos(i\omega_1 x) \ \sin(i\omega_1 x)] \begin{bmatrix} \alpha_{ij}^0 & \beta_{ij}^0 \\ \gamma_{ij}^0 & \delta_{ij}^0 \end{bmatrix} \begin{bmatrix} \cos(j\omega_2 y) \\ \sin(j\omega_2 y) \end{bmatrix}. \tag{7.10}$$

Comparing (7.7) and (7.8) with (7.9) and (7.10) results in

$$\text{(I)} \quad \begin{cases} \omega_{ij}(e^{-H\omega_{ij}}A_{ij} - e^{H\omega_{ij}}E_{ij}) = \alpha_{ij}^H \\ \omega_{ij}(A_{ij} - E_{ij}) = \alpha_{ij}^0 \end{cases}, \tag{7.11}$$

$$\text{(II)} \quad \begin{cases} \omega_{ij}(e^{-H\omega_{ij}}B_{ij} - e^{H\omega_{ij}}F_{ij}) = \beta_{ij}^H \\ \omega_{ij}(B_{ij} - F_{ij}) = \beta_{ij}^0 \end{cases}, \tag{7.12}$$

$$\text{(III)} \quad \begin{cases} \omega_{ij}(e^{-H\omega_{ij}}C_{ij} - e^{H\omega_{ij}}G_{ij}) = \gamma_{ij}^H \\ \omega_{ij}(C_{ij} - G_{ij}) = \gamma_{ij}^0 \end{cases}, \tag{7.13}$$

and

$$\text{(IV)} \quad \begin{cases} \omega_{ij}(e^{-H\omega_{ij}}D_{ij} - e^{H\omega_{ij}}H_{ij}) = \delta_{ij}^H \\ \omega_{ij}(D_{ij} - H_{ij}) = \delta_{ij}^0 \end{cases}. \tag{7.14}$$

The solutions to the linear equations (7.11) to (7.14) are

$$\begin{cases} A_{ij} = \dfrac{1}{\omega_{ij}} \cdot \dfrac{\alpha_{ij}^H - e^{H\omega_{ij}}\alpha_{ij}^H}{e^{-H\omega_{ij}} - e^{H\omega_{ij}}} \\[3mm] E_{ij} = \dfrac{1}{\omega_{ij}} \cdot \dfrac{\alpha_{ij}^H - e^{-H\omega_{ij}}\alpha_{ij}^H}{e^{-H\omega_{ij}} - e^{H\omega_{ij}}} \end{cases}, \tag{7.15}$$

$$\begin{cases} B_{ij} = \dfrac{1}{\omega_{ij}} \cdot \dfrac{\beta_{ij}^H - e^{H\omega_{ij}}\beta_{ij}^H}{e^{-H\omega_{ij}} - e^{H\omega_{ij}}} \\[3mm] F_{ij} = \dfrac{1}{\omega_{ij}} \cdot \dfrac{\beta_{ij}^H - e^{-H\omega_{ij}}\beta_{ij}^H}{e^{-H\omega_{ij}} - e^{H\omega_{ij}}} \end{cases}, \tag{7.16}$$

$$\begin{cases} C_{ij} = \dfrac{1}{\omega_{ij}} \cdot \dfrac{\gamma_{ij}^H - e^{H\omega_{ij}}\gamma_{ij}^H}{e^{-H\omega_{ij}} - e^{H\omega_{ij}}} \\[3mm] G_{ij} = \dfrac{1}{\omega_{ij}} \cdot \dfrac{\gamma_{ij}^H - e^{-H\omega_{ij}}\gamma_{ij}^H}{e^{-H\omega_{ij}} - e^{H\omega_{ij}}} \end{cases}, \tag{7.17}$$

and

$$\begin{cases} D_{ij} = \dfrac{1}{\omega_{ij}} \cdot \dfrac{\delta_{ij}^H - e^{H\omega_{ij}}\delta_{ij}^H}{e^{-H\omega_{ij}} - e^{H\omega_{ij}}} \\[3mm] H_{ij} = \dfrac{1}{\omega_{ij}} \cdot \dfrac{\delta_{ij}^H - e^{-H\omega_{ij}}\delta_{ij}^H}{e^{-H\omega_{ij}} - e^{H\omega_{ij}}} \end{cases}. \tag{7.18}$$

The coefficients α_{ij}, β_{ij}, γ_{ij} and δ_{ij} can be determined using the measurements in the following way. Assume that δg^H and δg^0 are available in given areas $\Sigma'=\{(x,y,z)|\ -L_x \le x \le L_x,\ -L_y \le y \le L_y,\ z=H\}$ and $S'=\{(x,y,z)|\ -L_x \le x \le L_x,\ -L_y \le y \le L_y,\ z=0\}$, respectively, and are absolutely integrable. Then δg^H and δg^0 can be expanded to the whole plane periodically. In this case, $\omega_1 = \pi/L_x$ and $\omega_2 = \pi/L_y$.

Using the Fourier expansion, the coefficients α_{ij}, β_{ij}, γ_{ij} and δ_{ij} of δg^H and δg^0 can be computed by the following formula:

$$\begin{bmatrix} \alpha_{ij}^H & \beta_{ij}^H \\ \gamma_{ij}^H & \delta_{ij}^H \end{bmatrix} = \frac{1}{\varepsilon_i \varepsilon_j L_x L_y} \int_{-L_x}^{L_x} [\cos(i\omega_1 x) \ \ \sin(i\omega_1 x)] \otimes \int_{-L_y}^{L_y} \delta g^H(x,y) \begin{bmatrix} \cos(j\omega_2 y) \\ \sin(j\omega_2 y) \end{bmatrix} dxdy$$

(7.19)

and

$$\begin{bmatrix} \alpha_{ij}^0 & \beta_{ij}^0 \\ \gamma_{ij}^0 & \delta_{ij}^0 \end{bmatrix} = \frac{1}{\varepsilon_i \varepsilon_j L_x L_y} \int_{-L_x}^{L_x} [\cos(i\omega_1 x) \ \ \sin(i\omega_1 x)] \otimes \int_{-L_y}^{L_y} \delta g^0(x,y) \begin{bmatrix} \cos(j\omega_2 y) \\ \sin(j\omega_2 y) \end{bmatrix} dxdy$$

(7.20)

where

$$\varepsilon_k = \begin{cases} 2 & k=0 \\ 1 & k \ne 1 \end{cases} \quad ,$$

$\otimes$ is the Kronecker product.
Formulas (7.6), (7.15)-(7.18) and (7.19)-(7.20) provide all necessary formulas for the determination of the local or regional gravity field using scalar airborne gravity data and ground gravity data.

8   Multiresolution BVPs

In practice, the data used in gravity field modelling have different resolutions. For example, the resolution of data collected using ground, airborne, and satellite techniques are different. Therefore it is natural to consider the multiresolution BVPs.

Multiresolution BVPs can be described as follows:

Given gravity field related measurements at multiple scales (the measurements can be on the flight surface Σ or the Earth surface S or both), the gravity field between the earth surface S and the flight surface Σ is to be determined.

Mathematically, multiresolution BVPs can be formulated as:

$$\begin{cases} \Delta T = 0 & \text{for } 0 \le z \le H \\ A_i T|_{\Sigma_i} = g_i + v_i & i=0,1,..,N \end{cases} \quad ,$$

(8.1)

where g_i are measurements from $N+1$ sensors or measurements at different resolution levels i ($i=0, 1, ...N$). The scale N corresponds to the highest resolution, while the scale 0 corresponds to the lowest scale. Σ_i is a boundary surface on which multiresolution data g_i are given. A_i is a linear functional. v_i ($i=0, 1,...,N$) are stochastic processes whose first and second moments are assumed to be known, i.e.

$$E(v_i)=0, \quad E(v_i v_i^T) = R_i,$$
$$i=0, 1,...N.$$

Four special cases are of interest:

Case VIII: (Neuman multiresolution BVP)

$$\left\{ \begin{array}{ll} \Delta T = 0 & \text{for } 0 \leq z \leq H \\ \dfrac{\partial T_1}{\partial z} \big|_{\Sigma} = \delta g_1^H + v_1 & \\ \dfrac{\partial T_0}{\partial z} \big|_S = \delta g_0^0 + v_0 & \end{array} \right. , \tag{8.2}$$

Case IX: (Neuman - Dirichllet multiresolution BVP)

$$\left\{ \begin{array}{ll} \Delta T = 0 & \text{for } 0 \leq z \leq H \\ \dfrac{\partial T_1}{\partial z} \big|_{\Sigma} = \delta g_1^H + v_1 & \\ T_0 \big|_S = T_0^0 + v_0 & \end{array} \right. , \tag{8.3}$$

Case X: (Gradiometry-Neuman multiresolution BVP)

$$\left\{ \begin{array}{ll} \Delta T = 0 & \text{for } 0 \leq z \leq H \\ \dfrac{\partial^2 T_1}{\partial z \partial z} \big|_{\Sigma} = g_{zz1}^H + v_1 & \\ \dfrac{\partial T_0}{\partial z} \big|_S = \delta g_0^0 + v_0 & \end{array} \right. , \tag{8.4}$$

and

Case XI: (Gradiometry-Dirichllet multiresolution BVP)

$$\left\{ \begin{array}{ll} \Delta T = 0 & \text{for } 0 \leq z \leq H \\ \dfrac{\partial^2 T_1}{\partial z \partial z} \big|_{\Sigma} = g_{zz1}^H + v_1 & \\ T_0 \big|_S = T_0^0 + v_0 & \end{array} \right. . \tag{8.5}$$

Remark 8.1: Case VIII and Case VIIII correspond to BVPs combining fine-scale scalar airborne gravity data with coarse-scale ground gravity data and altimetric data, respectively, while Case X and Case XI correspond to BVPs combining fine-scale airborne vertical gradient data with coarse ground gravity data and altimetric data respectively.

To solve BVPs of type (8.1), discrete wavelet transforms are briefly introduced. Reference is made to Mallat (1989) and Beylkin et al (1991) for details.

For a given 1D sequence of a signals $f_{i+1}(n)$ at resolution level $i+1$, the lower resolution signal $f_i(n)$ can be derived by lowpass filtering with a half-band lowpass having impulse response $h(n)$ (note that larger i corresponds to higher resolution or scale and smaller i corresponds to lower resolution or scale). At the same time the added detail $d_i(n)$, also called wavelet coefficients, can be computed by using a highpass filter with impulse $g(n)$, i.e.

$$f_i(n) = \sum_k h(k - 2n)f_{i+1}(k) \qquad (8.6)$$

$$d_i(n) = \sum_k g(k - 2n)f_{i+1}(k) \qquad (8.7)$$

or

$$f_i = Hf_{i+1} \qquad (8.8)$$
$$d_i = Gf_{i+1} \qquad (8.9)$$

This process is referred to as the decomposition of the signal. The same decomposition procedure can be applied to a lower resolution signal until the lowest resolution of interest is reached.

Reversing this process, the synthesis form of a wavelet transform is obtained in which finer and finer representation via a coarse-to-fine scale recursion is achieved, i.e.

$$f_{i+1}(n) = \sum_k h(n - 2k)f_i(k) + \sum_k g(n - 2k)d_i(k) \qquad (8.10)$$

or

$$f_{i+1} = H^*f_i + G^*d_i \qquad (8.11)$$

where * indicates the conjugate operation.

This process is also referred to as the reconstruction of the signal. Figures 8.1 illustrate the block diagram for decomposition and reconstruction.

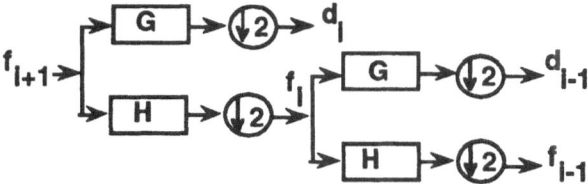

Figure 8.1a: Decomposition of 1D signal

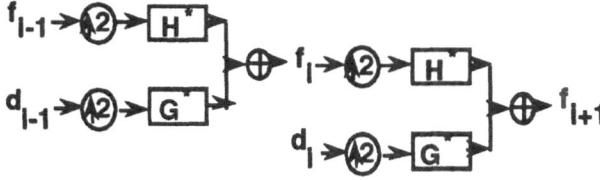

Figure 8.1b: Reconstruction of 1D signal

The simplest example of a discrete wavelet transforms is the Haar wavelet transform with

$$h(n) = \begin{cases} 1 & n = 0,1 \\ 0 & \text{otherwise} \end{cases}$$

and

$$g(n) = \begin{cases} 1 & n = 0 \\ -1 & n = 1 \\ 0 & \text{otherwise} \end{cases}$$

The 1D discrete wavelet transforms can be extended to 2D discrete wavelet transforms. In this case, the decomposition and reconstruction of a 2D signal take the following form:

$$f_i(n,m) = \sum_{k,l} h(k-2n)h(l-2m)f_{i+1}(k,l), \tag{8.12}$$

$$d_{i,1}(n,m) = \sum_{k,l} g(k-2n)h(l-2m)f_{i+1}(k,l), \tag{8.13}$$

$$d_{i,2}(n,m) = \sum_{k,l} h(k-2n)g(l-2m)f_{i+1}(k,l), \tag{8.14}$$

$$f_{i,3}(n,m) = \sum_{k,l} g(k-2n)g(l-2m)f_{i+1}(k,l), \tag{8.15}$$

or

$$d_{i,2} = (H \otimes G)\, f_{i+1}, \tag{8.16}$$
$$d_{i,2} = (H \otimes G)\, f_{i+1}, \tag{8.17}$$
$$d_{i,2} = (H \otimes G)\, f_{i+1}, \tag{8.18}$$
$$d_{i,3} = (G \otimes G)\, f_{i+1}. \tag{8.19}$$

and

$$\begin{aligned} f_{i+1}(n,m) = &\sum_{k,l} h(n-2k)h(m-2l)f_i(k,l) \\ &+ \sum_{k,l} g(n-2k)h(m-2l)d_{i,1}(k,l) \\ &+ \sum_{k,l} h(n-2k)g(m-2l)d_{i,2}(k,l) \\ &+ \sum_{k,l} g(n-2k)g(m-2l)d_{i,3}(k,l), \end{aligned} \tag{8.20}$$

or

$$f_{i+1} = (H^* \otimes H^*)\, f_i + (G^* \otimes H^*)d_{i,1} + (H^* \otimes G^*)d_{i,2} + (G^* \otimes G^*)d_{i,3} \tag{8.21}$$

where, f_i, $d_{i,1}$, $d_{i,2}$ and $d_{i,3}$ represent vectors formed by stacking the rows of matrices from 2D signals.

One of the attractive features of wavelet transforms for the analysis of signals is that they cannot only be computed recursively in scale, from fine to coarse, but can also be completely reconstructed from coarse to fine scale. Therefore different scales can be related to each other. Such a feature is very useful for the development in this chapter.

After introducing discrete wavelet transforms, we now concentrate on developing a method to solve the BVPs of type (8.1). To begin, some notations will be defined. $x_{i|i}$ denotes the estimate of x at scale i based on all measurements with resolution higher or equal to i. $x_{i|i+1}$ denotes the estimate of x at scale i based on all measurements with resolution higher or equal to i+1. A similar notation is used for other quantities. x_{si} denotes the estimate of x at scale i based on all available measurements.

The solution of such BVPs is now given in four steps:

Step 1: Estimation of $T_{N|N}$ and the corresponding error covariance of $T_{N|N}$ at the finest scale N:

$T_{N|N}$ can be computed using least-squares collocation, i.e.

$$T_{N|N} = C_{T_N t_N} (C_{t_N t_N} + C_{v_N v_N})^{-1} g_N \qquad (8.22)$$

with error covariance

$$C_{\varepsilon_{N|N}} = C_{T_N T_N} - C_{T_N t_N}(C_{t_N t_N} + C_{v_N v_N})^{-1} C_{t_N T_N}, \qquad (8.23)$$

where $t_N = A_N T$, $C_{t_N t_N}$ and $C_{v_N v_N}$ are the covariance of t_N and v_N, respectively, $C_{T_N t_N}$ is covariance between T_N and t_N, $C_{\varepsilon_{N|N}}$ is the error covariance of estimate $T_{N|N}$ and $C_{T_N T_N}$ the covariance of T_N.

It can also be computed using a numerical integral formula if the type of measurements at the finest scale is the same, i.e.

$$T_{N|N} = \Delta x \Delta y \sum_{i,j} K(i,j) g_N(i,j) \qquad (8.24)$$

with error covariance

$$C_{e_{N|N}} = (\Delta x \Delta j)^2 \sum_{i,j} K^2(i,j) C_{v_N v_N} \quad , \qquad (8.25)$$

where K is the kernel function and Δx and Δy are the grid increments along x and y.

Step 2: Estimation of $T_{N-1|N}$ from $T_{N|N}$:

$T_{N-1|N}$ can be computed using the discrete wavelet transforms described above, i.e.

$$T_{N-1|N} = (H^* \otimes H^*) T_{N|N} \qquad (8.26)$$

with error covariance

$$C_{\varepsilon_{N-1|N}} = (H^* \otimes H^*)^T C_{\varepsilon_{N|N}} (H^* \otimes H^*). \qquad (8.27)$$

At the same time, additional details can also be computed by

$$d_{N-1,1} = (G \otimes H) T_{N|N} \qquad (8.28)$$

$$d_{N-1,2} = (H \otimes G) T_{N|N} \qquad (8.29)$$

$$d_{N-1,3} = (G \otimes G) T_{N|N} \quad , \qquad (8.30)$$

which will be used in Step 4.

Step 3 : Measurement update:

$T_{N-1|N}$ can be updated using the measurements at scale N-1, i.e.

$$T_{N-1|N-1} = T_{N-1|N} + K_{N-1}(g_{N-1} - B_{N-1}T_{N-1|N})$$

$$K_{N-1} = C_{N-1}\overline{C}_{N-1|N-1}^{-1}$$

$$B_{N-1} = C_{N-1|N}C_{N|N}^{-1}$$

$$C_{N-1|N} = C_{t_{N-1}t_N} + C_{v_{N-1}v_N}$$

$$C_{N|N} = C_{t_N t_N} + C_{v_N v_N}$$

$$C_{N-1} = C_{T_{N-1}t_{N-1}} + C_{T_{N-1}t_N}C_{N|N}^{-1}C_{N-1|N}$$

$$\overline{C}_{t_{N-1|N-1}} = C_{N-1|N-1} - C_{N-1|N}C_{N|N}^{-1}C_{N-1|N}$$

$$C_{\varepsilon_{N-1|N-1}} = C_{\varepsilon_{N|N}} - C_{N-1}\overline{C}_{t_{N-1|N-1}}^{-1}C_{N-1}^T \tag{8.31}$$

Step 3 provides the mechanism of combining data from two different resolution levels for disturbance potential estimation. Step 2 and Step 3 are repeated until the lowest scale 0 is reached.

$\boxed{\text{Step 4}}$: Estimation of the signal based on all available data at scales 0 and N: Once the optimal estimate $T_{0|0}$ at scale 0 has been computed, the estimates at other scales (i=1,2,...,N) can be derived using the reconstruction procedure of the wavelet transforms, i.e.

$$T_{s(i+1)} = (H^* \otimes H^*) T_{si} + (G^* \otimes H^*)d_{i,1} + (H^* \otimes G^*)d_{i,2} + (G^* \otimes G^*)d_{i,3}$$
$$(i=0,1,...,N-1). \tag{8.32}$$

The main advantage of this approach is the estimation of gravity field signals at multiple resolutions.

Part Four: Assessment Of Current Trends In Airborne Gravimetry

Airborne gravimetry has either been applied or its use proposed for three different application areas:
- Continental gravity surveys for geotectonics,
- Regional gravity surveys for geoid determination,
- Local gravity surveys for oil and mineral exploration.

A brief discussion of each area will now be given, outlining the state of the art and the future potential in view of the error characteristics discussed in Part 2.

9   Application Areas

Airborne gravity surveys of continental extent usually aim at generating a first map of the gravity field with sufficient resolution to interpret features related to plate boundaries and intraplate deformation, such as large-scale flexures and rifts. The gravity system is often used as one component of an airborne multi-sensor system, containing aerogeomagnetic sensors, laser profilers, ice-penetrating radar and others. Characteristical of this application are long flight lines and wavelength resolution of the gravity field from about 20 km to 1000 kilometres with a standard deviation of about 5 mGal. Because of these requirements flying altitudes of several kilometres and flying speeds of 400 km/h are acceptable.

Pioneering work in this application area has been done by the Naval Weapons Laboratory under the leadership of John Brozena. The survey of Greenland, done in cooperation with the Danish National Survey and Cadastre (KMS), was the first example of a successful continental gravity survey of continental extent, see e.g. Brozena (1992). It showed many features of geophysical interest unknown at that time. A total of about 200 000 line kilometres were flown at a height of about 4 km in a large stable military aircraft at flight speeds between 370 and 450 km. Scalar gravity systems of the damped platform type were used in conjunction with differential GPS. The low pass filtering of the data (2-3 minutes) resulted in an RMS accuracy of 3-5 mGal as determined from cross-over points and of 5 mGal as determined from comparison with sparse upward continued data; for details see Brozena and Peters (1994), Forsberg and Kenyon (1994). Work of this type is continuing in the Arctic and in Antarctica and shows cross-over RMS values of about 2.5 mGal, see Brozena and Peters (1994).

In this application the resolution of the long wavelengths of the gravity field needs further investigation. Both the stability of DGPS over distances of up to 1000 km and the stability of gravity system biases over flight periods of several hours has to be checked in a controlled environment. The recent airborne gravity survey of Switzerland over an area with well-known gravity and topography will allow a detailed analysis of these questions and will also give further information on the consistency of cross-over RMS values with those derived from upward continued gravity, see Klingele et al (1995). The use of a Schuler-tuned inertial systems as the gravity sensor may be a major advantage in these applications.

Regional gravity surveys for geoid determination are an application area of considerable potential. Requirements with respect to short wavelength resolution are not as stringent as for explorations geophysics and flight lines are not as long as in some of the geotectonical applications. Typically, a regional geoid of 250-500 km extent with an accuracy of 0.2 ppm will require a minimum wavelength resolution of 5-15 km, depending on the terrain, for details see Schwarz and Li (1996). This can be achieved by flying with a speed of 300 km/h at altitudes of 2-3 km above ground.

Little practical work has been done in this application area, except for the survey of Switzerland, Klingele et al (1995), which could be used for this purpose. A recent test with a strapdown inertial system over part of the Rocky Mountains, reported in Wei and Schwarz (1996), shows that the required accuracy and wavelength resolution can be achieved. First results indicate that a relative geoid accuracy of 2-3 cm is possible over distances of 200 km. An interesting result of this test was that accelerometer biases were very stable over a total profile length of more than 700 km and a flight time of about three hours. Based on these initial results a large scale test to assess airborne gravimetry for geoid determination over the Rocky Mountains has just been completed.

In this application, downward continuation of gravity to the geoid is an important problem to be dealt with. In flat areas, with flying altitudes of 2-3 km and a required wavelength resolution of about 12-15 km, it appears not to be a critical problem. In mountainous areas, the application of remove-restore techniques for the topography may be sufficient to reduce downward continuation errors to a reasonable size. Stability of the gravity system over long periods of timestill needs further investigation. It appears that it is not as critical for a strapdown inertial system as for damped platform systems.

Local gravity surveys for oil and mineral exploration are the oldest application of airborne gravity. Surveys of this type aim at detecting density changes with lateral extensions between 1-10 km and magnitudes of the gravity signal below 0.5 mGal. Thus, the emphasis in this application is on short wavelength resolution of relatively small gravity signals. Long-term stability of the gravity sensor is usually not considered as a problem.

Although airborne gravity surveys of this type have been available for some time, very few detailed data analyses have been published and some of those published appear to be somewhat optimistic. As early as 1983 Hammer (1983) claimed a standard deviation of about 0.4 mGal for a "single airborne gravity measurement", based on actual data. He also pointed out the difficulty of detecting small features of significance to exploration geophysics from airborne data due to the attenuation effect. Today, after considerable further development has been done in this area, most companies will not claim results of better than about 1 mGal for the wavelenght range of interest. Even this accuracy is only achievable with slow aircraft velocities (50m/s or less) and low flying altitudes (100 m to 500 m). The first is necessary to obtain sufficient resolution for the GPS-derived accelerations, the second is needed to minimize the attenuation effect for short wavelength gravity signals with small amplitudes. The low flight altitude makes it very difficult to keep errors resulting from the interaction of attitude errors and flight dynamics at a reasonable level. For some recent results, see Gumert (1995),Harrison et al (1995), and Salychev and Schwarz (1995).

In this application, filtering periods of 20 seconds or less are needed with a slow flying aircraft to achieve the required resolution. In addition, the gravity signal to be detected is often not more than a few tenths of a mGal. To resolve such small, short-priodic signals with any reliability is extremely difficult both in terms of gravity sensor resolution and acceleration derived from differential GPS. Because the interaction of attitude errors and vehicle dynamics often generates gravity errors with longer wavelengths, low-pass filtering will only work, if the filtering period is longer than the period of phugoid motion. Similarly, optimized bandpass filters will only perform reliably if attitude biases of the gravity system are extremely small. Thus, long-term stability of the gravity sensor remains a problem. The use of aircraft other than fixed-wing airplanes, such as airships or helicopters, would allow longer filtering periods because of their slower speed and might alleviate some of these problems. Although this is the oldest area of application for scalar gravimetry, it appears that it contains some of the most challenging problems.

Acknowledgements: The authors would like to acknowledge the contribution of the research group working on these problems at The University of Calgary in clarifying the concepts of airborne gravimetry. Contributions have been referenced to specific persons, whenever possible. However, discussions are hard to acknowledge that way, although they are often instrumental in shaping new concepts. In this respect, special credit is due to Dr. Ming Wei, who has creatively influenced this work in many ways, to Drs. E. Knickmeyer, J. Czompo, J.Q. Zhang, all former Ph.D. students, and to Y. Hammada and Y. Li who currently work on these problems. All had their part in shaping the thinking and, thus, the presentation of these lecture notes.

References

Beylkin, G. , Coifman, R. and Rokhlin, V. (1991). Fast Wavelet Transforms and Numerical Algorithms I, Communication on Pure and Applied Mathematics, Vol. XLIV, pp.141-183.

Bian, S. and Zhang, K.(1993). The Planar Solution of the Geodetic Boundary Value Problem, Manuscripta Geodaetica, Vol.18, No.5.

Boedecker, G., F. Leismüller, T. Spohnholtz, J. Kuno, K.H. Neumayer (1994). Tests Towards Strapdown Airborne Gravimetry. Proc. IAG Symp. G4 Airborne Gravity Field Determination, XXI General Assembly of the IUGG, Boulder, Colorado, July 2-14, 1995, pp. 457-462 (Published by University of Calgary).

Brozena, J.M. (1984). A Preliminary Analysis of the NRL Airborne Gravimetry System. Geophysics, 49, 7, pp. 1060-1069.

Brozena, J.M. (1992). The Greenland Aerogeophysics Project: Airborne Gravity, Topographic and Magnetic Mapping of an Entire Continent. Proc. IAG Symposium G3: Determination of the Gravity Field by Space and Airborne Methods, Vienna, Aug. 1991, Springer Verlag, 1992.

Brozena, J.M. (1995). Integrated Aerogeophysical Measurements: Airborne Measurements in a Multi-Sensor Environment. Proc. IAG Symp. G4 Airborne Gravity Field Determination, XXI General Assembly of the IUGG, Boulder, Colorado, July 2-14, 1995, pp. 105-107 (Published by University of Calgary).

Brozena, J.M., Mader, G.L. and Peters, M.F. (1989). Interferometric Global Positioning System: Three-Dimensional Positioning Source for Airborne Gravimetry. J. Geophys. Res. 94, B9, pp. 12153-12162.

Brozena, J.M. and Peters, M.F.(1994). Airborne Gravity Measurement at NRL. Proceedings of the International Symposium on Kinematic Systems in Geodesy, Geomatics and Navigation (KIS 94), Banff, Canada, August 30 - September 2, pp. 495-506.

Cohen, C.E., Parkinson, B.W. and Mcnally, B.D. (1994). Flight Tests of Attitude Determination Using GPS Compared Against an Inertial Navigation Unit. Navigation, 41, 1, pp. 83-97.

Czompo, J. (1994). Airborne Scalar Gravimetry System Errors in the Spectral Domain. Ph.D. Thesis, UCGE Report No. 20067, Dept. of Geomatics Engineering, The University of Calgary, Calgary, Alberta.

El-Mowafy, A.M. (1994). Kinematic Attitude Determination from GPS. UCGE Report No. 20074, Department of Geomatics Engineering, The University of Calgary.

Forsberg, R. and Kenyon, S. (1994). Evaluation and Downward Cotinuation of Airborne Gravity Data - The Greenland Example. Proc. of the International Symposium on Kinematic Systems in Geodesy, Geomatics and Navigation, Banff, Canada, August 30 - September 2, pp. 531-538.

Grafarend, E. and Sanso,F. (1984). The Multibody Space-Time Geodetic Boundary Value Problem and the Honkasalo Term, Goephys.J.R, pp.255-275.

Gumert, W.R. (1995). Third Generation Aerogravity System.Proc. IAG Symp. G4 Airborne Gravity Field Determination, XXI General Assembly of the IUGG, Boulder, Colorado, July 2-14, 1995, pp. 153-161 (Published by University of Calgary).

Hammada, Y. (1996). A Comparison of Filtering Techniques for Airborne Gravimetry. UCGE Report No 20089, Department of Geomatics Engineering, The University of Calgary.

Hammer, S. (1983). Airborne Gravity is Here! Geophysics, 48, 2, pp. 213-223.

Harrison, J.C., J.D. Macqueen, A.C. Rauhut, J.Y. Cruz (1995). The LCT Airborne Gravity System. Proc. IAG Symp. G4 Airborne Gravity Field Determination, XXI General Assembly of the IUGG, Boulder, Colorado, July 2-14, 1995, pp. 163-168 (Published by University of Calgary).

Heck, B. (1996). Formulation and Linearization of Boundary Value Problems: From Observables to a Mathematical Model. Lecture Notes, Int. Summer School of Theoretical Geodesy, Como, Italy, May 27-June 7, 1996 (this volume).

Holota,P.(1995). Boundary and Initial Value Problems in Airborne Gravimetry, Proceedings of the Symposium on Airborne Gravimetry, XXI IUGG General Assembly, July 2-14, 1995, Boulder, Colorado.

Jekeli, C. (1992). Vector Gravimetry Using GPS in Free-fall and in an Earth-fixed Frame. Bulletin Geodesique, 66, 1, pp. 54-61.

Jones, P.C. and A.C. Johnson (1995). Airborne Gravity Survey in Southern Palmer Land, Antarctica. Proc. IAG Symp. G4 Airborne Gravity Field Determination, XXI General Assembly of the IUGG, Boulder, Colorado, July 2-14, 1995, pp. 117-123 (Published by University of Calgary).

Keller, W. and Hirsch, M. (1994). A Boundary Value Approach to Downward Continuation, Manuscripta Geodaetica, Vol.19, No.2, pp.101-118.

Kleusberg, A, D. Peyton and D. Wells (1990). Airborne Gravity and the Global Positioning System. Proc. of IEEE PLANS 1990, pp. 273-278.

Klingele, E., M. Halliday, M. Cocard, H.-G. Kahle (1995). Airborne Gravimetric Survey of Switzerland. Vermessung, Photogrammetrie, Kulturtechnik, 4, 1995, pp. 248-253.

Knickmeyer, E. (1990). Vector Gravimetry by a Combination of Inertial and GPS Satellite Measurements. Ph.D. Thesis, UCGE Report No. 20035, Dept. of Geomatics Engineering, The University of Calgary.

LaCoste, L.J.B., Ford, J., Bowles, R. and Archer, K. (1982). Gravity Measurements in an Airplane Using State-of-the-Art Navigation and Altimetry. Geophysics, 47, 5, pp. 832-838.

LaCoste, L.J.B. (1988). The Zero-Length Spring Gravity Meter. Geophysics, 53, pp. 20-21.

Li, Z. and Schwarz, K.P. (1994). Chaotic Behaviour in Geodetic Sensors and Fractal Characteristics of Sensor Noise. Geodetic Theory Today, Third Hotine-Marussi Symposium on Mathematical Geodesy, L'Aquila, Italy, May 29-June 3, 1994, IAG Symposium Series No. 114, Springer Verlag, pp. 246-258.

Mallat, S. G.(1989). A Theory for Multiresolution Signal Decomposition: the Wavelet Representation, IEEE Trans. PAMI 11, pp.674-693.

Nettleton, L.L., LaCoste, L.J.B. and Harrison, J.C. (1960). Tests of an Airborne Gravity Meter. Geophysics, 25, 1, pp. 181-202.

Salychev, O.S., Bykovsky, A.V., Voronov, V.V., Schwarz, K.P., Liu, Z., Wei, M., Panenka, J. (1994). Determination of Gravity and Deflections of the Vertical for Geophysical Applications Using the ITC-2 Platform. Proceedings of the International Symposium on Kinematic Systems in Geodesy, Geomatics and Navigation (KIS 94), Banff, Canada, August 30 - September 2, pp. 521-529.

Salychev, O.S. (1995). Inertial Surveying - ITC Ltd. Experience. Baumann MSTU Press, Moscow.

Salychev, O.S. and K.P. Schwarz (1995). Airborne Gravimetric Results Obtained with the ITC-2 Inertial Platform System. Proc. IAG Symp. G4 Airborne Gravity Field Determination, XXI General Assembly of the IUGG, Boulder, Colorado, July 2-14, 1995, pp. 125-141 (Published by University of Calgary).

Savage, P.G. (1978). Strapdown Sensors. AGARD Lecture Series No. 95, Strap-Down Inertial Systems.

Schwarz, K.P. (1987). Approaches to Kinematic Geodesy. In Contributions to Geodetic Theory and Methodology, presented to the XIX General Assembly of the IUGG, Vancouver, Can., August 9-22, 1987. Publication No 60006, Department of Surveying Engineering, The University of Calgary.

Schwarz, K.P. and Wei, M. (1990). A Framework for Modelling Kinematic Measurements in Gravity Field Applications. Bulletin Geodesique, 64, 331-346.

Schwarz, K.P., Cannon, M.E. and Wong, R.V.C. (1989). A Comparison of GPS Kinematic Models for the Determination of Position and Velocity along a Trajectory. Manuscripta Geodaetica, 14, pp. 345-353.

Schwarz, K.P., Colombo, O., Hein, G. and Knickmeyer, E.T. (1991). Requirements for Airborne Vector Gravimetry. Proceedings of the IAG Symposium G3 'From Mars to Greenland: Charting Gravity with Space and Airborne Instruments', Springer-Verlag, New York, pp.273-283.

Schwarz, K.P. and Wei, M. (1994). Some Unsolved Problems in Airborne Gravimetry. Proc. IAG Symposium No. 113 "Gravity and Geoid', Graz, Austria, Sept. 11-17, 1994. Published by Springer Verlag, pp. 131-150.

Schwarz, K.P., Li, Y. and Wei, M. (1994). The Spectral Window for Airborne Gravity and Geoid Determination. Proceedings of the International Symposium on Kinematic Systems in Geodesy, Geomatics and Navigation, Banff, Canada, August 30 - September 2, pp. 445-456.

Schwarz, K.P. and Y.C. Li (1996). What can Airborne Gravimetry Contribute to Geoid Determination? JGR, Vol. 101, B8, 17 873 - 17 881.

Thompson, L.G.D. (1959). Airborne Gravity Meter Test. J. Geophys. Res., 64, 488.

Thompson, L.G.D. and Lacoste, L.J.B. (1960). Aerial Gravity Measurements. J. Geophys. Res., 65, 1, 305-322.

Vassilou, A.A.(1986). Numerical Techniques for Processing Airborne Gradiometer Data, UCSE Report No.20017, Division of Surveying Engineering, The University of Calgary.

Wei, M., Ferguson, S.T. and Schwarz, K.P. (1991). Accuracy of GPS-Derived Acceleration from Moving Platform Tests. Proceedings of the IAG Symposium G3 'From Mars to Greenland: Charting Gravity with Space and Airborne Instruments', Springer-Verlag, New York, pp. 235-249.

Wei, M. and Schwarz, K.P. (1994). An Error Analysis of Airborne Vector Gravimetry. Proceedings of the International Symposium on Kinematic Systems in Geodesy, Geomatics and Navigation, Banff, Canada, August 30 - September 2, pp. 509-520.

Wei, M. and K.P. Schwarz (1995). Analysis of GPS-Derived Acceleration from Airborne Tests. Proc. IAG Symp. G4 Airborne Gravity Field Determination, XXI General Assembly of the IUGG, Boulder, Colorado, July 2-14, 1995, pp. 175-188 (Published by University of Calgary).

Wei, M. and K.P. Schwarz (1996). Flight Test Results from a Strapdown Airborne Gravity System. Submitted to Journal of Geodesy

Zhang, Q.J. (1995). Development of a GPS-Aided Inertial Platform for an Airborne Scalar Gravity System. Ph.D. Thesis, UCGE Report No. 20080, Dept. of Geomatics Engineering, The University of Calgary.

Zhang, Q.J., K.P. Schwarz, O.S. Salychev (1995). Accuracy of Inertial Platform Stabilization by GPS Velocity. Proc. IAG Symp. G4 Airborne Gravity Field Determination, XXI General Assembly of the IUGG, Boulder, Colorado, July 2-14, 1995, pp. 29-37 (Published by University of Calgary).

Zhang, Q.J. and K.P. Schwarz (1996). Estimating Double Difference GPS Multipath under Kinematic Conditions. Proc. of PLANS '96, Atlanta, Georgia, April 22-26, 1996, pp. 285-291.

Spherical Spectral Properties of the Earth's Gravitational Potential and its First and Second Derivatives

R. Rummel

1. Introduction

"But if, along with this interpretation, we also reject the fiction of a geoid, what to do with sea-topography?"
W. Baarda, 1995

Let me first discuss the role of the geodetic boundary value problem in the total framework of geodesy, the significance of the cm-geoid in this context and finally the contribution one could expect from satellite gradiometry.

More than one hundred years ago Bruns (1878) examined the state-of-art of geodesy at that time. Implicitly he based his considerations on a definition of geodesy as being the scientific discipline of the determination of the geometric shape of the earth and of the difference in gravity potential between all terrain points. In a brilliant analysis he came to the conclusion that with the five groups of measurable quantities, available at that time, (1) astronomical positioning, (2) triangulation, (3) trigonometric levelling, (4) geometric levelling and (5) gravimetry, geodesy could meet its goals, in theory. The approach he envisaged was that of the determination of a global polyhedron, encompassing the entire globe, and of the potential differences between all vertices of the polyhedron. No solution of a boundary value problem would be required. However, he also pointed out that the practical realization of this approach was impossible. The oceans were not accessible to geodetic measurements at all and measurement of vertical angles, which ought to provide strength in vertical direction to the determination of the shape of the earth, suffer from large uncertainties due to vertical, atmospheric refraction.

This explains why during the period to follow the formulation of the geodetic boundary value problem, its theoretical solution as well as its practical implementation received greatest attention. The reasoning was to replace so-to-say ellipsoidal heights, as derived from trigonometric levelling, by the combination of heights above sea level and the solution of the geodetic boundary value problem. This effort culminated in a solution of the boundary value problem by Molodenskii (Molodenskii et al., 1962) free of hypotheses. His view was: "A hypothesis, however modest its role, is inadmissible for such a purpose" (ibid, p. 1).

With the advent of space age the situation changed fundamentally. First, ocean areas can be bridged without problem and the geometric shape of the ocean surface accurately measured by satellite altimetry. Second, Bruns' polyhedron can now be determined more elegantly, conveniently and accurately by advanced space positioning techniques, such as satellite laser ranging, VLBI and in particular GPS; no vertical angles need to be measured. Has this development made the solution of the geodetic boundary value problem superfluous? Indeed modern space methods have removed the principal obstacles pointed out by Bruns. In this sense the aims of geodesy can be met now with or without the boundary value approach. However, firstly, nowadays the determination of the earth's gravity field in its exterior space, as a solution of the geodetic boundary value problem, is precondition for satellite orbit determination and computed orbit trajectories are in turn the input for gravity field modelling. Secondly, the solution of the geodetic boundary value problem is the only means to come to a global and unified geodetic system and it gives, in good geodetic tradition, an independent check of regional vertical height determination (expressed by the simplified formula $\Delta h = \Delta H + \Delta N$), provided ellipsoidal, orthometric and geoid heights are available with comparable precision. Thus, the solution of the geodetic boundary problem is still of fundamental importance, its role however shifted in the light of what geodetic space techniques added to geodesy.

The word "geoid" should be seen in this context as a label, at most as a visualisation, for we are not very much interested in the computation of the actual geometric figure of level surfaces. What one actually tries to determine is potential differences between points. Any non-zero potential difference implies a state of imbalance, which manifests itself in flux, whether it is the flow of water from A to B, erosion of land masses, ice flow or ocean circulation. And the term "one-centimeter geoid" has to be read as the objective of geodesy to determine potential differences between points accurate to $0.1 \ m^2/s^2$.

The solution of the problem to such a high degree of accuracy requires highest sophistication of the theory and also excellent global data. For several reasons terrestrial data alone will never suffice. Despite great progress in global data collection in recent time, large gaps remain and in many areas terrestrial data are inconsistent, sparse or inaccurate. Global, satellite based gravity models, on the other hand, do not have the necessary spatial resolution. Although the applied method of orbit perturbation analysis has been more and more refined and although tracking data to an increasing number of satellites has been incorporated into these models, one can show (Rummel, 1992), that this technique has reached intrinsically its limit. Thus a quantum step forward is expected to come either from the technique of "satellite-to-satellite tracking" or from satellite gradiometry. Only by satellites global and homogeneous data can be collected within short time and only these two concepts seem to be capable of attaining the required spatial refinement and accuracy. This conclusion has been drawn already in 1969 when leading earth scientists proposed the realisation of SST or satellite gradiometry in the famous Williamstown report (Solid-Earth and Ocean Physics, 1969). Many studies followed in which these concepts were elaborated, both from the technological and methodological side, and their effect on earth sciences was described. So far neither of these two concepts has been realized. Satellite gradiometry is included in these lectures because it is the concept that with highest probability is capable to get geodesy closer to the realization of the "one-cm-geoid".

2. Satellite Gradiometry and the Meissl Scheme

"In addition, due precautions have to be taken
to exclude the effect of such temporary
masses as wandering cattle, ..."
M. Hotine, 1969

Only with satellites it is possible to cover the entire earth densly with measurements of uniform quality within a few weeks time. However for gravity field determination any satellite method has one fundamental disadvantage. Due to the altitude of the orbit, the effect of individual mass inhomogeneities of the earth, such as mountain belts, subduction zones, ocean ridges, hot spots, convection cells or core/mantle topography, is strongly damped. Gravity gradiometry is the measurement of one, several or all second derivatives of the gravity potential (the gradient components of the three components of the gravity vector). The classical gradiometric device is the torsion balance. However its use is very limited, because torsion balance measurements are very much affected by local masses (see the citation above). It is very laborious to deduce representative information from them. Spaceborne gradiometry seems an ideal compromise: local mass effects are strongly damped, but on the other hand the differential measurement can optimally counteract signal attenuation due to the satellite's altitude.

It is the purpose of this chapter to illustrate this basic feature of satellite gradiometry but at the same time place satellite gradiometry into the broader context of gravity sensors and gravity quantities in general and their respective information content. For a discussion of gradiometry principles it is referred to (Rummel,

1986) and (Colombo, 1989). The broader context is offered by the so-called Meissl scheme. In 1971 Meissl formulated, what one could call the principles of a spectral theory of gravity quantities given on a sphere. They have been transformed in (Rummel, 1975; 1979) and (Rummel & van Gelderen, 1992; 1995) into a concise scheme - a kind of pocket guide - that should provide easy insight into the basic characteristics of any strategy of gravity field determination. In 1983 Svensson introduced pseudodifferential operators as a new tool for the solution of geodetic boundary value problems. One could see his work as a natural generalization of that of Meissl (1971), in particular it is not restricted to the spherical case only. Keller (1991) or (Keller & Hirsch, 1994) are strongly advocating Svensson's approach. Very likely the most rigorous mathematical analysis of satellite gradiometry is given by Schreiner (1994), see also (Freeden et al., 1994).

Let us introduce the disturbance potential T as the fundamental quantity for all further considerations. It is the unknown part of the earth's gravity potential. The surface of the earth is assumed to be a sphere of radius R, $\Omega(0,R)$. This is a severe assumption that puts aside a number of serious complications, such as the problem of downward continuation of satellite information to the actual topographic relief, see e.g. (Hotine, 1967). However we are interested here in a more qualitative analysis. The disturbance potential is assumed to be a harmonic function in the exterior of Ω and regular at infinity:

$$\Delta T = \nabla^2 T = 0 \qquad \text{in } \Omega^{\text{ext}}$$

and

$$\lim_{r \to \infty} T = 0 \ .$$

We denote the unit sphere $\omega(0,1)$, its infinitesimal surface element $d\omega = \sin\theta \, d\lambda \, d\theta$ and therefore $d\Omega = R^2 \sin\theta \, d\lambda \, d\theta$.

The expansion of T into a series of spherical harmonics is written as

$$T(P) = U_0 \sum_{n=0}^{\infty} \left[\frac{R}{r_p} \right]^{n+1} \sum_{m=0}^{n} \bar{P}_{nm}(\cos \theta_p) \, [\Delta \bar{C}_{nm} \cos m\lambda_p + \Delta \bar{S}_{nm} \sin m\lambda_p] \ , \qquad (2\text{-}1)$$

with   P ... evaluation point with spherical coordinates (θ, λ, r) and $r \geq R$,
        $U_0 = GM_0/R$ (gravitational constant times mass of the earth divided by its mean radius),
        $\bar{P}_{nm} (\cos \theta)$ fully normalized, associated Legendre polynomials, and
        $\Delta \bar{C}_{nm}$ and $\Delta \bar{S}_{nm}$ unknown (corrections to) spherical harmonic coefficients of degree n and order m.

Here a more concise notation[1] proves more convenient:

$$T(P) = U_0 \sum_{n=0}^{\infty} \left[\frac{R}{r_P} \right]^{n+1} \sum_{m=-n}^{+n} t_{nm} Y_{nm}(\theta,\lambda) \ , \qquad (2\text{-}2)$$

with surface spherical harmonics defined as

$$Y_{nm}(\theta,\lambda) = \overline{P}_{n|m|} (\cos \theta) \exp(\text{im } \lambda) \qquad (2\text{-}3)$$

and the series coefficients as

$$t_{nm} = \frac{1}{2} (\Delta \overline{C}_{nm} - i\Delta \overline{S}_{nm}) \qquad \text{for } m \geq 0$$

$$t_{n(-m)} = \frac{1}{2} (\Delta \overline{C}_{nm} + i\Delta \overline{S}_{nm}) \qquad \text{for } m < 0 \ . \qquad (2\text{-}4)$$

The surface spherical harmonics form an orthonormal system:

$$\langle Y_{nm}, Y_{kl} \rangle = \frac{1}{4\pi} \int\int_{\omega} Y_{nm}^*(\theta,\lambda) \ Y_{kl}(\theta,\lambda) \ d\omega = \delta_{nk} \ \delta_{ml} \qquad (2\text{-}5)$$

(with * complex conjugate). With the same notation it is

$$U_0 t_{nm} = \langle T, Y_{nm} \rangle_{r=R} \qquad (2\text{-}6)$$

and as short version of (2-1) and (2-2)

$$T = \sum_n \left[\frac{R}{r} \right]^{n+1} \sum_m \langle T, Y_{nm} \rangle_{r=R} \ Y_{nm} \equiv Yt \ . \qquad (2\text{-}7)$$

However the inner products in (2-6) and (2-7) require some assumptions on T. Being a geodesist I prefer mathematical definitions of the kind "T must be well behaved" or the function space must be "physically reasonable". More precisely the above inner product on Ω should be understood as $(\cdot,\cdot)_{L^2}$ in $L^2(\Omega)$. However the situation is somewhat trickier, because, especially when dealing with gradiometry, we have to make sure that even the second derivatives of T are square-integrable and that their series representation converges. For that purpose T on Ω

---

[1] One could consider to change to dimensionless quantities as is done in (Rummel, 1991) and introduce, in addition, in the sense of (Baarda, 1979; 1995) a fundamental datum point. To keep things in familiar context and notation this step has not been taken here.

has to be element of the Sobolev space $H^{3/2}$ (Ω). This is a Hilbert space of functions that fullfills the required smoothness criteria. This result is derived in (Schreiner, 1994).

The harmonicity is the most relevant property of the disturbance potential and as we discuss the space outside a mean earth sphere Ω it is usefull to express the Laplace operator in spherical coordinates (θ,λ,r):

$$\nabla^2 = \nabla_r^2 + \frac{1}{r^2} \nabla_\omega^2 \tag{2-8}$$

with the radial operator

$$\nabla_r^2 \equiv \frac{\partial^2}{\partial r^2} + \frac{2}{r} \frac{\partial}{\partial r} \tag{2-9}$$

and the so-called surface Laplace operator

$$\nabla_\omega^2 \equiv \sin^{-1}\theta \frac{\partial}{\partial\theta} (\sin\theta \frac{\partial}{\partial\theta}) + \sin^{-2}\theta \frac{\partial^2}{\partial\lambda^2} . \tag{2-10}$$

The surface spherical harmonics Y_{nm} (θ,λ) are simultaneously eigenfunctions of - ∇_ω^2 and of -i $\partial/\partial\lambda$:

$$-\nabla_\omega^2 Y_{nm} = n(n+1) Y_{nm} \tag{2-11}$$

and

$$-i\frac{\partial}{\partial\lambda}Y_{nm} = mY_{nm} \tag{2-12}$$

with eigenvalues n $(n + 1)$ and m, respectively.

Finally, with the (spatial) spherical harmonics defined as

$$\Psi_{nm} \equiv \left(\frac{R}{r}\right)^{n+1} Y_{nm} \qquad \text{and} \qquad r \geq R \tag{2-13}$$

we find

$$\nabla^2\Psi_{nm} = (\nabla_r^2 + \frac{1}{r^2}\nabla_\omega^2) \Psi_{nm}$$

$$= \nabla_r^2\Psi_{nm} - \frac{n(n+1)}{r^2}\Psi_{nm} = 0 . \tag{2-14}$$

Now we turn back to our objective to derive the Meissl-scheme. The operators ∇^2, ∇_r^2 and ∇_ω^2 are rotational invariant or isotropic. There exist many others. We consider three: the radial derivatives $\partial_r = {}^\partial/_{\partial r}$ and $\partial_{rr}{}^2 = {}^{\partial^2}/_{\partial r^2}$, and the so-called upward continuation operator C_h. The latter represents a mapping from the sphere Ω with radius R to a sphere Ω_r with radius $r = R + h$, where h may be the altitude of a satellite or of an aircraft.

We apply these operators to T, (eq. (2-1) or eq. (2-2)):

(A) $\quad C_h T(P) \quad = \quad U_0 \displaystyle\sum_{n=0}^{\infty} \sum_{m=-n}^{+n} t_{nm} C_h \Psi_{nm}(\theta,\lambda,r)\Big|_{r=R}$

$\qquad\qquad\qquad = \quad U_0 \displaystyle\sum_n \left[\frac{R}{r}\right]^{n+1} \sum_m t_{nm} Y_{nm}(\theta,\lambda) \quad \text{and } r=R+h>R \quad$ (2-15)

(B) $\quad \partial_r T(P)\big|_{r=R} \quad = \quad U_0 \displaystyle\sum_n \sum_m t_{nm} \partial_r \Psi_{nm}(\theta,\lambda,r)\big|_{r=R}$

$\qquad\qquad\qquad = \quad -U_0 \displaystyle\sum_n \frac{n+1}{R} \sum_m t_{nm} Y_{nm}(\theta,\lambda) \qquad \text{and } r=R \quad$ (2-16)

(C) $\quad \partial_{rr}^2 T(P)\big|_{r=R} \quad = \quad U_0 \displaystyle\sum_n \sum_m t_{nm} \partial_{rr}^2 \Psi_{nm}(\theta,\lambda,r)\big|_{r=R}$

$\qquad\qquad\qquad = \quad U_0 \displaystyle\sum_n \frac{(n+1)(n+2)}{R^2} \sum_m t_{nm} Y_{nm}(\theta,\lambda) \text{ and } r=R \quad$ (2-17)

and we find, for example, for an arbitrary combination of two:

(D) $\quad \partial_{rr}^2 C_h T(P) \quad = \quad U_0 \displaystyle\sum_n \sum_m t_{nm} \partial_{rr}^2 C_h \Psi_{nm}(\theta,\lambda,r)$

$\qquad\qquad\qquad = \quad U_0 \displaystyle\sum_n \frac{(n+1)(n+2)}{r^2} \left[\frac{R}{r}\right]^{n+1} \sum_m t_{nm} Y_{nm}(\theta,\lambda) \qquad$ (2–18a)

$\qquad\qquad\qquad = \quad \dfrac{U_0}{R^2} \displaystyle\sum_n (n+1)(n+2) \left[\frac{R}{r}\right]^{n+3} \sum_m t_{nm} Y_{nm}(\theta,\lambda) \; .$ (2–18b)

On purpose these formulas are derived in such a laborious manner. This should now facilitate their discussion.

The general pattern of the operation is as follows:
Application of the invariant operator S on T results in

$ST \quad = \quad U_0 \displaystyle\sum_n \sum_m t_{nm} S \Psi_{nm} \; (r=R)$

$\qquad = \quad U_0 \displaystyle\sum_n \lambda_n \sum_m t_{nm} \Psi_{nm} \; (r=R) \; .$ (2-19)

It allows three interpretations: First, application of S to T results in a modification of the spherical harmonic expansion. The modification is due to symbols λ_n that only depend on the degree n (not the order). Second, application of S to T is equivalent to application of S to Ψ_{nm} (r=R). This in turn represents an eigenvalue analysis of the operator S (see (Lanczos, 1961)):

$$S\Psi_{nm}\ (r=R)\ =\ \lambda_n\Psi_{nm}\ (r=R) \tag{2-20}$$

with λ_n the eigenvalues per degree (here: $(R/r)^{n+1}$, $-(n+1)/r$ and $(n+1)(n+2)/r^2$)) and Ψ_{nm} the complete orthonormal system of eigenfunctions of S on Ω. S can only be written in this form if it is self-adjoint (which corresponds to being symmetric in the terminology of eigenvalue analysis of matrices). From (2-20) one can derive an expansion of S:

$$
\begin{aligned}
S(P-Q) &= \sum_n \lambda_n \sum_m \Psi_{nm}(P;r=R)\Psi_{nm}(Q;r=R) \\
&= \sum_n \lambda_n\ (2n+1)P_n(P-Q)
\end{aligned}
\tag{2-21}
$$

and the determination of the eigenvalues

$$\lambda_n\ =\ \langle\langle S(P-Q),Y_{nm}(Q)\rangle\ ,Y_{nm}(P)\rangle\ . \tag{2-22}$$

Thereby P-Q denotes the distance of two points on a sphere, which is $\cos \psi_{PQ}$. Third, the invariant operators could be constructed employing ∇_ω^2, as done in (Svensson, 1983) with λ_n the "spherical symbols".
Now the basic Meissl scheme can be constructed, see TABLE 2.1.

It contains essentially all basic properties of gravity quantities and of the operators connecting them. Thereby we assume a local orthonormal triad $\underline{e}_i$, i=1,2,3 with $\underline{e}_{i=1} = \underline{e}_x$ pointing north, $\underline{e}_{i=2} = \underline{e}_y$ pointing east and $\underline{e}_{i=3} = \underline{e}_z$ pointing radially outwards (normal to the sphere). This makes $\partial_r T = \partial_z T = T_z$ and $\partial_{rr} T = \partial_{zz} T = T_{zz}$.

Discussion:
1. TABLE 2.1 shows T, $\partial_r T$ and $\partial_{rr}^2 T$ per spherical harmonic degree n on two horizontal levels, one time on the earth's (spherical) surface, one time at satellite altitude. It also contains eigenvalues (spherical symbols) that connect T, $\partial_r T$ and $\partial_{rr}^2 T$. The arrows indicate the direction in which they apply. The three upward continuation symbols connect the quantities on the earth's surface with the corresponding quantities at satellite altitude. Again the arrows show in which direction they apply.

Table 2.1: Meissl scheme: Eigenvalues (spherical symbols) per degree n, connecting the disturbance potential and its first and second radial derivatives, at the earth's surface and at altitude. The arrows indicate the direction for which the eigenvalues apply.

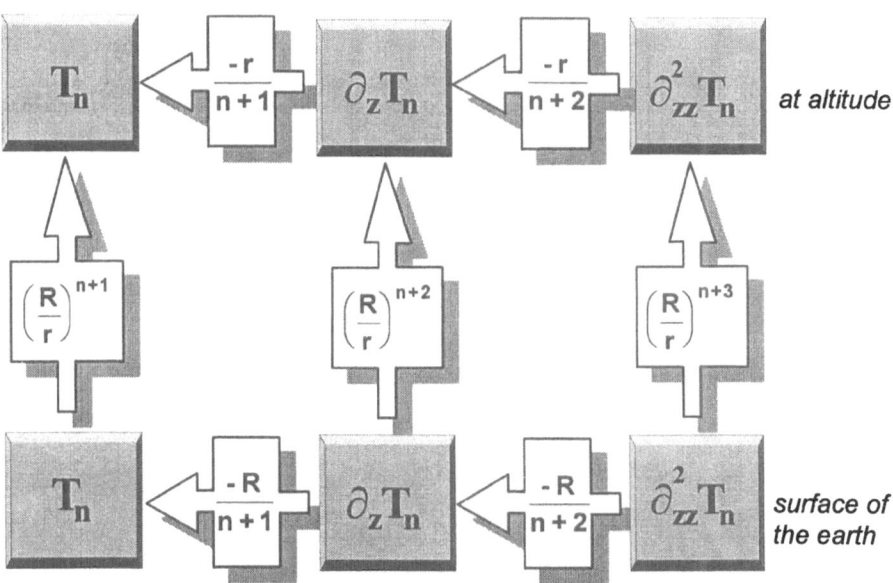

2. The scheme is a spectral or Fourier scheme. It applies per degree and order of the spherical harmonic expansion. The arrow direction corresponds to the direction of smoothing, which means the direction in which the coefficients of higher and higher degrees of the series expansion are attenuated. The eigenvalues (spherical symbols) corresponding to the opposite direction are just their inverses, i.e. $(^r/_R)^{n+1}$ instead of $(^R/_r)^{n+1}$ or $(n+1)/R$ instead of $R/(n+1)$. The opposite directions are unsmoothing directions in which the coefficients of higher and higher degrees are amplified.

Example 2.1: If the coefficients of the series expansion of T are t_{nm}, compare eq. (2-2), the coefficients of $\partial_r T$ are $-(n+1)/R\, t_{nm}$, compare eq. (2-16).

Degrees for which the eigenvalues can become zero are related to singularities and require special attention. See e.g. (Krarup, 1973) or (Sanso, 1981) or for the case of gradiometry (Schreiner, 1994; ch. 3.3).

3. In the space domain the smoothing direction corresponds to integrals. As all considered operators are invariant the kernel functions of the integral formulas are isotropic and can be written, according to eq. (2-21), as a Legendre series. The integral formulas are convolution integrals on the sphere (denoted by ".").

Example 2.2: Upward continuation or Poisson integral (Heiskanen & Moritz, 1967; 1-16)

spherical symbol:

$$\left[\frac{R}{r}\right]^{n+1}$$

integral kernel function:

$$C_h = \frac{R(r^2-R^2)}{l^3(P-Q)} = \sum_{n=0}^{\infty}\left[\frac{R}{r}\right]^{n+1}(2n+1)P_n(P-Q)$$

$$\left[l = \sqrt{(r^2+R^2-2Rr\,\cos\psi)}\right]$$

integral formula:

$$T(P) = \frac{1}{4\pi}\iint_\omega \frac{R(r_P^2-R^2)}{l^3(P-Q)}\,T(Q)\,d\omega_Q = C_h * T \; .$$

The unsmoothing direction corresponds to differential (or integro-differential) operators. The considered operators ∂_r, ∂_{rr}^2, ∇_r^2 and ∇_ω^2 are of this type, whereas C_h is the upward continuation operator of the example above, and therefore a smoothing operator.

In short: arrows correspond to smoothing and therefore to integration, the opposite direction to unsmoothing and therefore to differentiation.

4. The quadratic norm of T is finite:

$$\|T\|^2 = U_0^2 \sum_{n=0}^{\infty}\sum_{m=-n}^{+n} t_{nm}^2 < \infty \; . \tag{2-23}$$

If the quadratic norm of $\partial_{rr}^2 T$ is to be finite, too, it must hold

$$\| \partial_{rr}^2 T \|^2 = \frac{U_0^2}{R^4} \sum_{n=0}^{\infty} (n+1)^2 (n+2)^2 \sum_{m=-n}^{+n} t_{nm}^2 < \infty \ , \tag{2-24}$$

which implies, that the t_{nm} must decrease at least as

$$t_{nm}^2 < \frac{1}{n^4} \ .$$

This leads to the conclusion of Schreiner (1994) that T on Ω must belong to $H^{3/2}$. See also (Meissl, 1971).

As a general rule, once the norm of T is known, the norm of linear functionals of T, ST, follows from:

$$\| ST \|^2 = U_0^2 \sum_{n=0}^{\infty} \lambda_n^2 \sum_{m=-n}^{+n} t_{nm}^2 \ . \tag{2-25}$$

5. The scheme is commutative. The spherical symbols or eigenvalues connecting two arbitrary quantities of the scheme are unique, irrespective of the chosen path.

6. The scheme applies not only to the gravity field quantities themselves but as well to the observation noise contained in them. Thus it tells us in which directions noise is smoothed and in which it is amplified. More quantitatively, once a realistic error model of a measured quantity on Ω or Ω_r has been constructed, direct error propagation is possible with the scheme.

This aspect will be illuminated in chapter 3, but it should be clear that especially this aspect provides a basis for optimized experiment design.

7. All six functionals of the scheme can, on basis of their spectral signal content, be associated with one or more gravity quantities, such as geoid heights, surface layers, gravity anomalies, deflections of the vertical, torsion balance measurements, satellite velocities etc. This way the Meissl scheme can serve as an empirical pocketguide to physical geodesy. This association is shown in a second picture of the Meissl scheme, taken from (Rummel, 1979) where these other quantities are included at the respective place of the scheme. See table 2.2.

8. Already at this point it becomes clear that any satellite technique for gravity field determination is faced with severe error amplification (opposite to arrow in upward continuation direction). In theory this "downward continuation" may lead to a complete divergence of the solution and requires special attention. See chapter 4.

If we take the second radial derivative at satellite altitude as being representative of the technique of satellite gradiometry, the Meissl scheme shows that for $\partial_{rr}^2 T$ the effect of downward continuation is maximally counteracted by $r^2/(n+1)(n+2)$ $\approx r^2/n^2$. It is so-to-say

$$\left(\frac{R+h}{R} \right)^{n+1} \quad \text{against} \quad \frac{(R+h)^2}{n^2} \ .$$

Table 2.2: The six gravity functionals of the Meissl scheme can be associated qualitatively with geodetic quantities, such as geoid height, gravity anomaly, deflection of the vertical, surface density, torsion balance measurement, satellite gradiometry in a low earth orbiter (LEO), satellite to satellite tracking (SST) between satellites of the GPS and a LEO or between two LEO's or satellite altimetry. The "eigenvalu-es" connecting orbit acceleration, velocity and position corrections belong to perturbation theory (ω_0 orbit mean angular velocity, 2p running from zero to n).

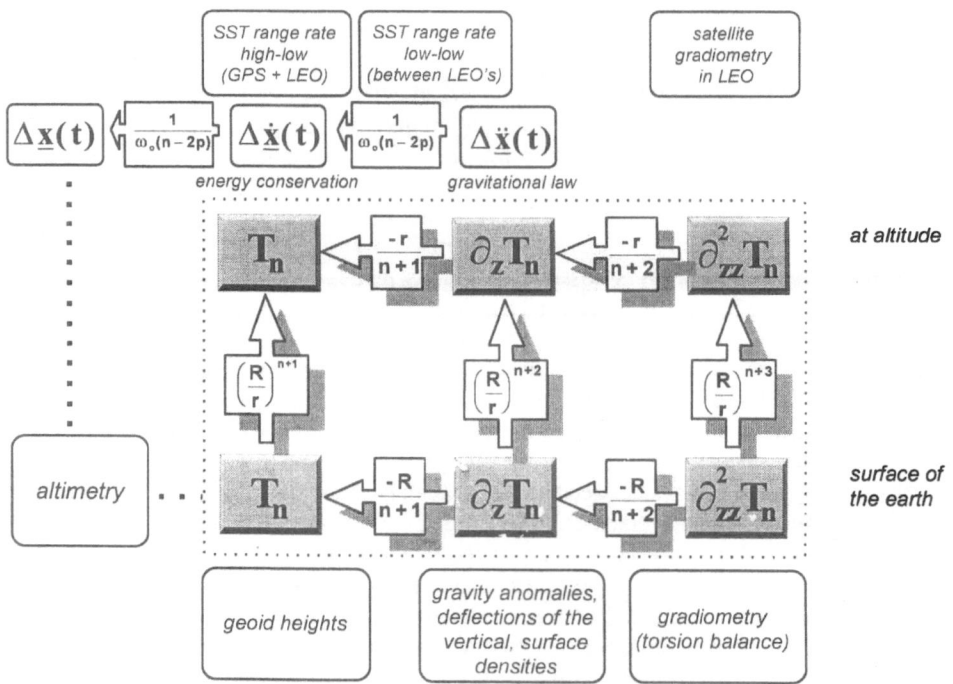

Extension to Non-self-adjoint operators

It is well-known from the discussion of geodetic boundary value problems that the inclusion of tangential components is not trivial. For the same reasons this is true for the considerations here. It is not obvious that deflections of the vertical should exhibit similar properties as e.g. $\partial_r T$ and it is also not self-explanatory that the other second derivative components of T should behave like $\partial_{rr}^2 T$. For the deflections of the vertical a first spectral theory is given by Groten & Moritz (1964). It received a more general foundation in (Meissl, 1971). An extension to second derivatives has been tried in (Rummel & van Gelderen, 1992) and is elaborated in (Freeden et al., 1994) and (Schreiner, 1994).

A principal feature is that a spectral theory of linear functionals of T that contains a derivative tangential to Ω, such as ∂_θ or ∂_λ, cannot be based on scalar spherical harmonics. The appropriate base functions are vectorial and tensorial spherical harmonics and also they can be defined such that a complete and orthonormal system of base functions is obtained.

One should be aware that vector and tensor spherical harmonics are important tools for the separation of certain types of partial differential equations, that occur e.g. in geophysical fluid dynamics, see e.g. (Backus, 1967) or (Phinney, 1973), theory of elasticity, see (Gurtin, 1971) or (Mochizuki, 1988), in electrodynamics, see (Jackson, 1925; 1975), or in general relativity, see (Zerilli, 1970) or (Thorne, 1980). See also the textbook by Jones (1985). In geodesy Hotine (1969) discussed vector and tensor spherical harmonics implicitly; Grafarend (1986) and Mikolaiski (1989) employed them for geodynamical problems. Unfortunately there seems neither an easy nor a unique access in particular to tensor spherical harmonics. As a result a variety of alternative representations exists, see again (Thorne, 1980). We shall not go into the derivation of vector and tensor harmonics, only use them. For a deeper discussion it is referred to the literature cited above.

Alternatively the same problem can be approached from the point of view of operator analysis. Then the difference to the previous case is that operators of tangential or mixed normal-tangential type (normal means normal to Ω here, i.e. radial) are not self-adjoint anymore. The extension of eigenvalue analysis of symmetric matrices to arbitrary rectangular $n \times m$ matrices A is singular value decomposition (SVD). Lanczos (1961) shows how to proceed: Complement A by its transpose A^T as to obtain once again a symmetric matrix, likewise complement an operator L by its adjoint L^t so as to form a self-adjoint operator. This leads to a bilinear identity of the form

$$\langle x_{nm}, L y_{nm} \rangle - \langle y_{nm}, L^t x_{nm} \rangle \equiv 0 , \tag{2-26}$$

with y_{nm} the orthonormal basefunctions of the domain space and x_{nm} those of the range space. However the operator case is more complex, because not only the operator needs to be considered but appropriate boundary conditions as well, (Lanczos, 1961; ch. 4).

We follow essentially (Meissl, 1971). It is $\nabla_\omega = \text{Grad} \equiv (-\partial_\theta, \sin^{-1}\theta\,\partial_\lambda)^T$ and $\nabla_\omega\cdot = \text{Div} \equiv (-\partial_\theta, \sin^{-1}\theta\,\partial_\lambda)$ surface gradient and surface divergence operator, respectively. For the surface Laplacian $\nabla_\omega^2 = \nabla_\omega \cdot \nabla_\omega = \text{Div Grad} = \text{Lap}$, see eq. (2-10).

Then it is

$$
\text{(E)} \qquad \nabla_\omega T \;=\; U_0 \sum_n \sum_m t_{nm} \nabla_\omega Y_{nm}\,(\theta,\lambda)
$$

$$
\;=\; U_0 \sum_n \sqrt{n(n+1)} \sum_m t_{nm} \underline{X}_{nm}\,(\theta,\lambda)
$$

$$(2\text{-}27)$$

and

$$
\text{(F)} \quad -\nabla_\omega \cdot (\nabla_\omega T) \;=\; -\nabla_\omega^2 T \;=\; U_0 \sum_n \sqrt{n(n+1)} \sum_m t_{nm} (-\nabla_\omega\cdot)\,\underline{X}_{nm}\,(\theta,\lambda)
$$

$$
\;=\; U_0 \sum_n n(n+1) \sum_m t_{nm} Y_{nm}\,(\theta,\lambda) \qquad\cdot (2\text{-}28)
$$

One observes that now two systems of base functions enter, the scalar spherical harmonics Y_{nm} and the vector spherical harmonics $\underline{X}_{nm}$ (denoted U_{nm} in (Meissl, 1971)). By Green's identities it can be shown that the $\underline{X}_{nm}$ form a complete and orthonormal system of eigenfunctions on ω for tangent, spherical vector fields. It is

$$
\langle \nabla_\omega f, \nabla_\omega g \rangle \;=\; -\,\langle f, \nabla_\omega^2 g \rangle \;=\; -\,\langle \nabla_\omega^2 f, g \rangle \;.
$$

$$(2\text{-}29)$$

With (2-27) and (2-28):

$$
\nabla_\omega Y_{nm} = \sqrt{n(n+1)}\;\underline{X}_{nm}
$$

$$(2\text{-}30)$$

and

$$
\sqrt{n(n+1)}(\nabla_\omega \cdot \underline{X}_{nm}) = \nabla_\omega^2 Y_{nm} = -n(n+1)Y_{nm}
$$

$$(2\text{-}31)$$

eq. (2-29) yields explicitly the orthonormality of $\underline{X}_{nm}$:

$$
\iint_\omega \underline{X}_{nm}^*(\theta,\lambda)\underline{X}_{kl}(\theta,\lambda)\,d\omega \;=\; \frac{1}{\sqrt{n(n+1)}}\frac{1}{\sqrt{k(k+1)}}\iint_\omega \nabla_\omega Y_{nm}^*(\theta,\lambda)\nabla_\omega Y_{kl}(\theta,\lambda)\,d\omega
$$

$$
\;=\; \frac{-1}{\sqrt{n(n+1)}}\frac{1}{\sqrt{k(k+1)}}\iint_\omega Y_{nm}^*(\theta,\lambda)\nabla_\omega^2 Y_{kl}(\theta,\lambda)\,d\omega
$$

$$
\;=\; 4\pi\,\delta_{nk}\,\delta_{ml}\;.
$$

$$(2\text{-}32)$$

See again (Meissl, 1971).

The extension of our approach from self-adjoint to non-self-adjoint and its analogy to matrix calculus are summarized in BOX 2.1

BOX 2.1   Eigenvalue Analysis

CASE A: self-adjoint (symmetric) operator

a	b	c
b	a	b
c	b	a

$\underline{u}$ $=$ λ $\underline{u}$ $\partial_r \Psi_{nm}(r=R) = -\dfrac{n+1}{R}\Psi_{nm}(r=R)$

$$S\,\underline{u} \;=\; \lambda\,\underline{u}$$

S self-adjoint (symmetric) operator
$\underline{u}$ eigenvector
$\underline{\Psi}_{nm}$ eigenfunction (compare eq. (2-19))
λ eigenvalue (e.g. $-(n+1)/R$)

CASE B: non-self-adjoint operator

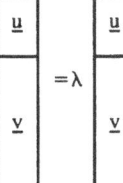

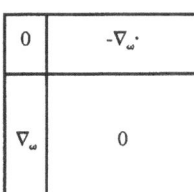

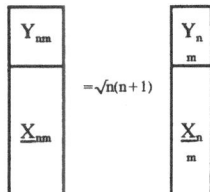

$A\,\underline{u} = \lambda\,\underline{v}$
$A^T\,\underline{v} = \lambda\,\underline{u}$

| $A^T A\,\underline{u} = \lambda^2\,\underline{u}$ | $-\nabla_\omega\!\cdot\nabla_\omega Y_{nm} = n(n+1)\,Y_{nm}$ |

Now we can complement our scheme with square-integrable, tangential vector functions to the sphere Ω of the form (2-27). We may write:

$$\underline{u}(\theta,\lambda) = U_0 \sum_n \sum_m u_{nm} \underline{X}_{nm}(\theta,\lambda) . \tag{2-33}$$

Since it is

$$u_{nm} = \sqrt{n(n+1)}\, t_{nm} \tag{2-34}$$

the squared norm of $\underline{u}$

$$\|\underline{u}\|^2 = U_0^2 \sum_n \sum_m u_{nm}^2 = U_0^2 \sum_n n(n+1) \sum_m t_{nm}^2 \tag{2-35}$$

can be compared with that of T, eq. (2-23). We also observe that $\underline{u}$ does not contain a zero-degree term. Thus given $\underline{u}$ the disturbance potential T can be reconstructed at most up to an arbitrary constant.

The N-S and E-W components of the deflection of the vertical ξ and η, respectively, are related to T as

$$\begin{pmatrix} \xi \\ \eta \end{pmatrix} = \begin{bmatrix} \dfrac{1}{U_0}\dfrac{\partial T}{\partial\theta} \\ \dfrac{-1}{U_0\sin\theta}\dfrac{\partial T}{\partial\lambda} \end{bmatrix} = \begin{bmatrix} -\dfrac{1}{\gamma_0}\dfrac{\partial T}{\partial x} \\ -\dfrac{1}{\gamma_0}\dfrac{\partial T}{\partial y} \end{bmatrix} = -\dfrac{1}{U_0}\,\underline{u}(\theta,\lambda) = -\dfrac{1}{U_0}\nabla_\omega T \tag{2-36}$$

(with $\partial T/\partial x = T_x$ and $\partial T/\partial y = T_y$ expressed in the local triad $\underline{e}_i$).

For the discussion of satellite gradiometry the approach followed above for tangential vectorial functions has to be extended to tangential and mixed normal-tangential tensorial functions. The most comprehensive analysis of this step is given in (Schreiner, 1994). Apart from the surface gradient ∇_ω Schreiner introduces a second tangential operator

$L_\omega \equiv (\sin^{-1}\theta\,\partial_\lambda, -\partial_\theta)$. The operators are denoted ∇^* and L^* in (Schreiner, ibid). They form the building blocks for the derivation of the tensor components, compare (Schreiner, ibid, ch. 2.2). We follow Regge & Wheeler (1957) and the more detailed derivation of Zerilli (1970). For our purpose, three kinds of tensor spherical harmonics are needed. We denote them $Z_{nm}^{(0)}$, $Z_{nm}^{(1)}$ and $Z_{nm}^{(2)}$. They suffice to represent the three kinds of tensor components $\{\Gamma_{zz}(=-\Gamma_{xx}-\Gamma_{yy})\}$, $\{\Gamma_{xz}, \Gamma_{yz}\}$, and $\{\Gamma_{xx}-\Gamma_{yy}, 2\Gamma_{xy}\}$. See (Rummel & van Gelderen, 1995). In the local triad $\underline{e}_i$ they are normal, mixed normal-tangential, and pure tangential, respectively.

$$Z^{(0)}\Psi_{nm} = \begin{bmatrix} 0 & 0 & 0 \\ 0 & 0 & 0 \\ 0 & 0 & \partial_{zz}^2 \end{bmatrix} \Psi_{nm} = \begin{bmatrix} 0 & 0 & 0 \\ 0 & 0 & 0 \\ 0 & 0 & \partial_{rr}^2 \end{bmatrix} \Psi_{nm} \equiv$$

$$\equiv \frac{(n+1)(n+2)}{R^2} Z_{nm}^{(0)} \tag{2-37}$$

$$Z^{(1)}\Psi_{nm} = \frac{1}{\sqrt{2}}\begin{bmatrix} 0 & 0 & \partial^2_{xz} \\ 0 & 0 & \partial^2_{yz} \\ * & * & 0 \end{bmatrix}\Psi_{nm} = \frac{1}{R^2\sqrt{2}}\begin{bmatrix} 0 & 0 & -r\partial^2_{\theta r}+\partial_\theta \\ 0 & 0 & \sin^{-1}\theta(r\partial^2_{\lambda r}-\partial_\theta) \\ * & * & 0 \end{bmatrix} \equiv$$

$$\equiv -\frac{(n+2)\sqrt{n(n+1)}}{R^2}\, Z^{(1)}_{nm} \qquad (2\text{-}38)$$

$$Z^{(2)}\Psi_{nm} = \frac{1}{\sqrt{2}}\begin{bmatrix} \partial^2_{xx}-\partial^2_{yy} & -2\partial^2_{xy} & 0 \\ * & \partial^2_{yy}-\partial^2_{xx} & 0 \\ 0 & 0 & 0 \end{bmatrix}\Psi_{nm} =$$

$$= \frac{1}{R^2\sqrt{2}}\begin{bmatrix} (\partial^2_{\theta\theta}-\cot\theta\partial_\theta-\sin^{-2}\theta\partial^2_{\lambda\lambda}) & 2\sin^{-1}\theta(\partial^2_{\theta\lambda}-\cot\theta\partial_\lambda) & 0 \\ * & -(\partial^2_{\theta\theta}-\cot\theta\partial_\theta-\sin^{-2}\theta\partial^2_{\lambda\lambda}) & 0 \\ 0 & 0 & 0 \end{bmatrix}\Psi_{nm} \equiv$$

$$= \frac{1}{R^2}\sqrt{\frac{(n+2)!}{(n-2)!}}\, Z^{(2)}_{nm} . \qquad (2\text{-}39)$$

(where Ψ_{nm} is taken at $r=R$).

Example 2.3: It should be understood that the Z are linear operators as before those in eqs. (2-15), (2-16), (2-17), (2-27) and (2-28). For example, it is with (2-39):

$$(G) \quad \Gamma^{(1)} \equiv Z^{(1)}T(P) = \frac{1}{\sqrt{2}}\begin{bmatrix} 0 & 0 & T_{xz} \\ 0 & 0 & T_{yz} \\ T_{xz} & T_{yz} & 0 \end{bmatrix} = U_0 \sum_n \sum_m t_{nm}Z^{(1)}\Psi_{nm}(\theta,\lambda,r) =$$

$$= U_0 \sum_n -\frac{(n+2)\sqrt{n(n+1)}}{R^2} \sum_m t_{nm}Z^{(1)}_{nm}(\theta,\lambda) . \qquad (2\text{-}40)$$

and

$$(H) \quad \Gamma^{(2)} \equiv Z^{(2)}T(P) = \frac{1}{\sqrt{2}}\begin{bmatrix} T_{xx}-T_{yy} & -2T_{xy} & 0 \\ -2T_{xy} & T_{yy}-T_{xx} & 0 \\ 0 & 0 & 0 \end{bmatrix} = U_0\sum_n \sum_m t_{nm}Z^{(2)}\Psi_{nm}(\theta,\lambda,r) =$$

$$= U_0 \sum_n \frac{\sqrt{(n+2)(n+1)n(n-1)}}{R^2} \sum_m t_{nm}Z^{(2)}_{nm}(\theta,\lambda) . \qquad (2\text{-}41)$$

Both, $\Gamma^{(1)}=Z^{(1)}T$ and $\Gamma^{(2)}=Z^{(2)}T$ are second-rank tensors. $Z^{(2)}T$ does not contain coefficients t_{nm} of degree zero and one.

The tensor spherical harmonics $Z_{nm}^{(0)}$, $Z_{nm}^{(1)}$ and $Z_{nm}^{(2)}$ are, apart from a scale factor, essentially identical to the pure-spin tensor harmonics of Thorne (1980) or the a_{nm}, b_{nm} and f_{nm} of Zerilli (1970). Again they form a complete and orthonormal system of base functions:

$$\int_\omega \int \sum_{i,j} \left[Z_{nm}^{(\alpha)}\right]_{i,j} \left[Z_{kl}^{(\beta)}\right]_{i,j} d\omega = 4\pi \, \delta_{nk}\delta_{ml}\delta_{\alpha\beta} \tag{2-42}$$

where α and $\beta = 0,1,2$ and $[\ ]_{ij}$ denote the tensor components. The orthogonality relationship together with expressions (2-40) and (2-41) open the way to determine the coefficients t_{nm} from the observable tensors $\Gamma^{(1)}$ or $\Gamma^{(2)}$. Also the norm of the latter can be calculated in analogy to eq. (2-35). Thus one can now introduce an extended Meissl scheme, TABLE 2.3, that contains the horizontal derivatives, too. All linear functionals of T, i.e. all geodetic, gravity related (linearized) observables, can now be classified according to a common scheme.

Discussion:

1. Spherical symbols or eigenvalues could be derived for tangential or mixed normal-tangential quantities, too. They are scalars, but they apply to vectorial or tensorial quantities. They do not apply to the individual component, e.g. to T_x or to T_{yz}.

2. These eigenvalues permit the extension of the Meissl scheme to gravity quantities not connected to T by a self-adjoint operator. Consequently all considerations discussed before for self-adjoint operators can now be extended to non-self-adjoint ones. The considerations include smoothing and unsmoothing behavior, construction of integral kernels, computation of norm, signal and noise propagation etc.

3. More specifically, the combination $\{\partial_x T, \partial_y T\}$ which corresponds to the deflections of the vertical (up to a constant) is on the same spectral level as $\partial_z T$, with eigenvalue $-\sqrt{n(n+1)}$ versus $-(n+1)$.

4. All gradiometric quantities, the components of the disturbance gravity tensor $\partial_{ij}^2 T$, can now be interpreted. The combinations $\Gamma^{(0)}=\partial_{zz}^2 T$, $\Gamma^{(1)}=\{\partial_{xz}^2 T, \partial_{yz}^2 T\}$ and $\Gamma^{(2)}=\{\partial_{xx}^2 T-\partial_{yy}^2 T, 2\partial_{xy}^2 T\}$ share the same spectral level with eigenvalues $(n+1)(n+2)$, $-(n+2)\sqrt{n(n+1)}$ and $\sqrt{(n+2)(n+1)n(n-1)}$, respectively.

The combination $\{\partial_{xx}^2 T-\partial_{yy}^2 T, 2\partial_{xy}^2 T\}$ resembles the observables of the classical torsion balance (see e.g. (Selényi, 1953) or (Rummel, 1986)). It is also the observable of some rotating gradiometer concepts although angular frequency modulated, compare (Pelka & DeBra, 1979).

5. The spectral behaviour of individual components can only be deduced in an approximate manner as discussed in (Groten & Moritz, 1964), (Sünkel, 1981), (Rummel & van Gelderen, 1992) or recently in (van Gelderen & Koop, 1996). We shall return to this point later.

Table 2.3: Extended Meissl scheme: The scheme of eigenvalues of Table 2.1 extended by the first and second tangential and normal-tangential derivatives.

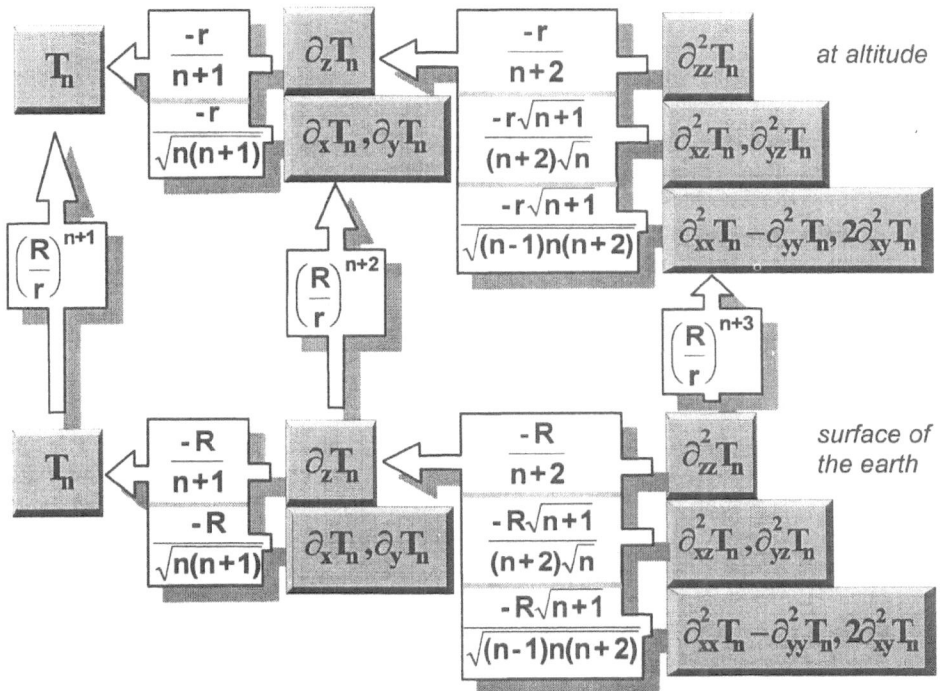

3. Spectral Analysis of Satellite Gradiometry

"If astronauts in orbit wish to detect the
gravitational field of the earth by measuring
the tide produced by the earth on a drop of water
they will find it desirable to use a
very large drop of water."
H.C. Ohanion, 1976

Considerations so far were dealing with the construction of a building that connects all gravity quantities. In the sequel we shall focus the discussion more on satellite gradiometry and in particular on the determination of T on the surface of a spherical earth or equally well the coefficients t_{nm} from satellite gradiometry. Thus we view the problem not as a boundary value problem but rather as its inverse, the determination of a boundary function from data in the exterior. The analysis shall still remain highly idealized, for we will consider, rather than individual gradiometry measurements sampled along the trajectory of a satellite, (1) one or several second order derivatives of T, $\partial_{ij}^2 T$, (2) given as a function and, (3) given on a sphere Ω_r in average satellite altitude h, where $r = R + h$. Nevertheless this analysis should exhibit all essential characteristics a satellite gradiometry mission could provide to geodesy.

What we have in mind is a signal/noise propagation analysis based upon the Meissl scheme. It should tell us where to rank satellite gradiometry among all other observation techniques, provide some hints on the experiment design and tell us whether one can expect to accomplish a unique and stable recovery of the disturbance potential on the earth sphere Ω. Taking on the one hand one of the well-known degree-variance models and on the other hand a degree variance model of the expected measurement and system noise a rather realistic error propagation analysis becomes feasible.

3.1 Signal Content of Satellite Gradiometry

Let us assume for simulation purposes the coefficients of one of the recent geopotential models, complete up to degree and order nmax, be available. The disturbance potential T can be determined using eq. (2-2). We assume a chosen type of second-order derivative $T_{ij} = \partial_{ij}^2 T$ on the "satellite altitude sphere" Ω_r be derived from it, using

$$T_{ij}(\theta,\lambda,r) = U_0 \sum_n \sum_m t_{nm} \, \partial_{ij}^2 \Psi_{nm}(\theta,\lambda,r) \tag{3-1}$$

or in matrix form, with t a vector containing all coefficients t_{nm}:

$$T_{ij} = Y_{(ij)}t \quad .$$

(3-2)

We still assume that i and j = 1,2,3 refer to a local orthonormal triad with $\underline{e}_{i=1} = \underline{e}_x$ pointing north, $\underline{e}_{i=2} = \underline{e}_y$ pointing east and $\underline{e}_{i=3} = \underline{e}_z$ pointing radially outwards. The explicit expressions for the components T_{ij} in terms of spherical coordinates are given in TABLE 3.1.

TABLE 3.1: T_i and T_{ij} in the local triad {N,E,radial}

$$T_i = \frac{\partial x^A}{\partial x^i} T_A$$

$$T_x = -\frac{1}{r}T_\theta$$

$$T_y = \frac{1}{r\sin\theta}T_\lambda$$

$$T_z = T_r$$

$$T_{ij} = \frac{\partial x^A}{\partial x^i}\frac{\partial x^B}{\partial x^j} T_{AB}$$

$$T_{xx} = \frac{1}{r}T_r + \frac{1}{r^2}T_{\theta\theta}$$

$$T_{xy} = \frac{1}{r^2\sin\theta}T_{\theta\lambda} + \frac{\cos\theta}{r^2\sin^2\theta}T_\lambda$$

$$T_{xz} = \frac{1}{r^2}T_\theta - \frac{1}{r}T_{r\theta}$$

$$T_{yy} = \frac{1}{r}T_r + \frac{1}{r^2\tan\theta}T_\theta + \frac{1}{r^2\sin^2\theta}T_{\lambda\lambda}$$

$$T_{yz} = \frac{1}{r\sin\theta}T_{r\lambda} - \frac{1}{r^2\sin\theta}T_\lambda$$

$$T_{zz} = T_{rr}$$

x^A ... "natural" coordinates {θ, λ, r}

Reference: (Reed, 1973; ch. 3), (Tscherning, 1976), (Koop, 1993)

FIGURE 3.1 shows global maps of T_{xx}, T_{xy}, T_{xz} (top), T_{yy}, T_{yz} (middle) and T_{zz} (below) computed at h = 400 km from model OSU91 by (Rapp et al., 1991) complete up to degree and order 360. The individual components are comparable to different illuminations of one and the same field. One observes that T_{zz} is particularly well focused, whereas T_{xy} is somewhat hazy.

Just to straighten our ideas one could think of three alternative ways to deal with the given $T_{ij}(\theta,\lambda,r)$.

(A) As any function given on a sphere each individual T_{ij} can be expanded into a series of surface spherical harmonics and get back a field representation of each

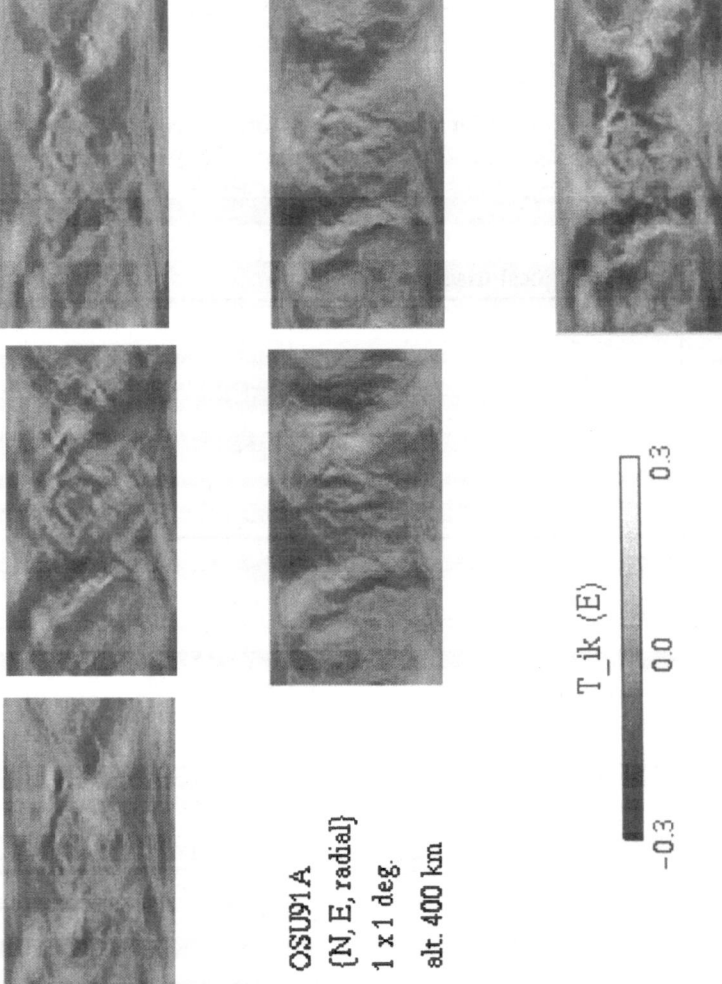

OSU91A
{N, E, radial}
1 x 1 deg.
alt. 400 km

T_ik (E)

-0.3   0.0   0.3

Figure 3.1: Global map of second derivatives of the disturbance potential (top row T_{xx}, T_{xy}, T_{xz}, middle row T_{yy} and T_{yz} and below T_{zz}) at 400km altitude (computed from the coefficient set OSU91 (Rapp & Pavlis, 1990) up to degree and order 360).

T_{ij} on Ω_r:

$$\tau_{nm}\{ij\} = \langle T_{ij}, Y_{nm} \rangle = \frac{1}{4\pi} \int_\omega \int T_{ij}(\theta,\lambda) Y_{nm}(\theta,\lambda) d\omega \tag{3-3}$$

and

$$T_{ij}(\theta,\lambda) = \sum_n \sum_m \tau_{nm}\{ij\} Y_{nm}(\theta,\lambda) \quad . \tag{3-4}$$

The signal degree variance for each component becomes

$$c_n\{ij\} = \sum_m \tau_{nm}^2\{ij\} \tag{3-5}$$

and the squared norm

$$\| T_{ij} \|^2 = \sum_n c_n\{ij\} \quad . \tag{3-6}$$

It is important to note that none of the coefficient sets, except $\tau_{nm}\{zz\}$, can be related to our fundamental unknowns, the coefficients t_{nm}, in a simple manner. From TABLE 2.1 or eq. (2-18) it follows that

$$\tau_{nm}\{zz\} = \frac{U_0}{R^2} (n+1)(n+2) \left[\frac{R}{r} \right]^{n+3} t_{nm} \quad . \tag{3-7}$$

For each other coefficient set $\tau_{nm}\{ij\}$ the connection becomes rather complicated and dependent on degree and order. See (Hotine, 1969) and (Reed, 1973; ch. 4). As an example the degree variances $c_n\{ij\}$ for all components taken at altitude h=200 km are shown in FIGURE 3.2.

(B) Alternatively eqs. (3-1) or (3-2) can be chosen as point of departure, in an attempt to solve for the coefficients t_{nm} by least squares. For an actual experiment analysis this case may become the most relevant one. It is discussed in some detail in (Rummel et al., 1993), advanced numerical aspects are treated in (Schuh, 1996). We only sketch this case briefly and for a rather idealized situation.

Let a specific component T_{ij} be given on a very dense equi-angular grid on Ω_r. Its prior error variance-covariance matrix be $Q_{(ij)}$ (and $E\{\epsilon_{(ij)}\}=0$). The grid values T_{ij} owe to be smoothed just enough to avoid any aliasing. The total number of grid points be M and the total number of coefficients t_{nm} N, with M>N. Then the least-squares solution becomes, compare eq. (3-2):

$$\hat{t} = (Y_{(ij)}^T Q_{(ij)}^{-1} Y_{(ij)})^{-1} Y_{(ij)}^T Q_{(ij)}^{-1} \tilde{T}^{(ij)} \tag{3-8}$$

$$\text{N·1} \qquad \text{N·M} \quad \text{M·M} \quad \text{M·N} \qquad \text{N·M} \quad \text{M·M} \quad \text{M·1}$$

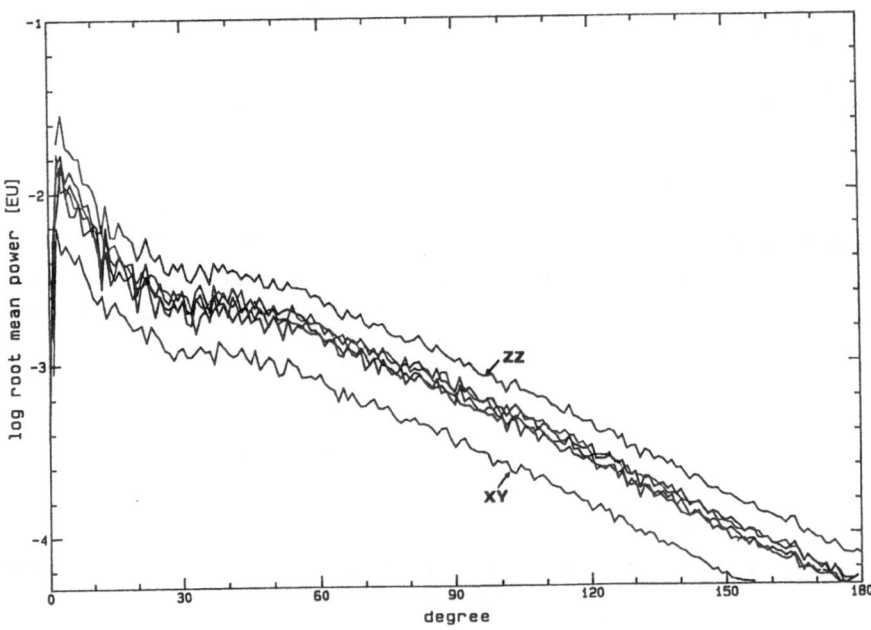

Figure 3.2: Average degree-order root-mean-square power (sqrt $(c_n\{ij\}/(2n+1))$) of the six gravity gradient components at altitude 200km. One observes the slightly higher power of component $\{zz\}$ and the slightly weaker power of component $\{xy\}$.

and the posteriori error variance-covariance matrix

$$\hat{Q}_{(nm)} = \sigma_0^2 \ (Y_{(ij)}^T \ Q_{(ij)}^{-1} \ Y_{(ij)} \ )^{-1} \tag{3-9}$$

$$\underset{N \cdot N}{} \qquad \underset{N \cdot M}{} \quad \underset{M \cdot M}{} \quad \underset{M \cdot N}{}$$

where σ_0^2 is the variance of unit-weight[2]. For $N = (nmax + 1)^2$ and $nmax \approx 200$ the size of the normal matrix becomes very large. However, as is well known, under some mild assumptions a block diagonal structure can be attained with the maximum sub-block of size $nmax \cdot nmax$, (Colombo, 1981), see also (Sneeuw, 1994) for a detailed analysis of this aspect.

However, one could equally well look for a system of base functions $\chi_{nm}(\theta, \lambda)$ such that

$$\langle \chi_{nm}, \partial_{ij}^2 Y_{kl} \rangle = \frac{1}{4\pi} \int \int_{\omega} \chi_{nm}(\theta, \lambda) \partial_{ij}^2 Y_{kl}^*(\theta, \lambda) d\omega = \delta_{nk} \ \delta_{ml} \quad . \tag{3-10}$$

In this case χ_{nm} and $\partial_{ij}^2 Y_{nm}$ are bi-orthonormal. Then it holds

$$U_0 t_{nm} \left(\frac{R}{r} \right)^{n+1} = \langle T_{ij}, \chi_{nm} \rangle \quad . \tag{3-11}$$

This approach is followed in (Brovelli & Sansò, 1990), (Albertella et al., 1994; 1995) for the component $\partial_{y'y'}^2 T = T_{y'y'}$, with y' perpendicular to an actual orbit of a sun-synchronous satellite. It may be very complicated to find the bi-orthonormal system, but it can always be derived numerically. Actually, the least-squares solution can be seen as a special finite method of derivation of χ_{nm}, for it is

$$(Y_{(ij)}^T Q_{(ij)}^{-1} Y_{(ij)})^{-1} Y_{(ij)}^T Q_{(ij)}^{-1} Y_{(ij)} = \langle X_{nm}, \partial_{ij}^2 Y_{kl} \rangle_Q = I \quad . \tag{3-12}$$

(I unit matrix).

Again for $\partial_{zz}^2 T$ the situation is straightforward, for in this case $\partial_{zz}^2 Y_{nm} = \partial_{rr}^2 Y_{nm} = \lambda_n Y_{nm}$ and therefore $\chi_{nm} = \lambda_n^{-1} Y_{nm}$.

(C) As long as we deal with either $\{T_{zz}\}$, $\{T_{xz}, T_{yz}\}$ or $\{T_{xx}-T_{yy}, 2T_{xy}\}$ a direct representation in terms of the respective, complete orthonormal system of tensor spherical harmonics is possible and as a consequence, a direct connection to the unknown t_{nm}, too. See previous chapter. This representation may even allow a qualitative judgement of the signal content of the individual components T_{ij}.

---

[2]In chapter 4 we will turn to a more concise notation.

Over the past thirty years models were determined for the average behavior of the gravity field. They were based upon coefficient sets of available satellite gravity models and on terrestrial gravity material, in particular mean gravity anomalies. A complete theoretical background for these models and their application is given in (Moritz, 1980). These models are expressed in terms of signal degree variances c_n (see eq. 3-5). [I would have preferred the notation c_n^2 to emphasize their meaning as quadratic measure but keep the generally accepted notation.] The particular gravity quantity, to which the degree variance refers is indicated in parenthesis, e.g. c_n {Δg} for gravity anomaly degree variances or c_{nm} {t_{nm}} for <u>dimensionless</u> disturbance potential degree variances. Three well-known degree-variance models are those by Kaula (1966), Tscherning & Rapp (1978) and Moritz modified by Jekeli (Moritz , 1976 und 1980), (Jekeli, 1978). They are listed in TABLE 3.2.

TABLE 3.2: Degree Variance Models

(I) $\quad c_n\{t_{nm}\} = s^{n+1} \dfrac{1{\cdot}6 \cdot 10^{-10}}{n^3} \qquad$ KAULA $\qquad\qquad$ (3-13)

here modified by the factor s^{n+1} with $s=(R_{BJE}/R)^2 < 1$
and R_{BJE} the radius of the Bjerhammar sphere, so as to secure convergence.

(II) $\quad c_n\{\Delta g\} = s^{n+2} \dfrac{A(n-1)}{(n-2)(n+B)} \qquad$ TSCHERNING–RAPP $\qquad$ (3-14)

for $n \geq 2$ and c_2 {Δg}$=7.6$ mGal^2; $s=0.999617$, $A=425.28$ mGal^2; $B=24$.

(III) $\quad c_n\{\Delta g\} = A_1 s_1^{n+2} \dfrac{n-1}{n+B_1} + A_2 s_2^{n+2} \dfrac{n-1}{(n+2)(n+B_2)}$ MORITZ–JEKELI $\quad$ (3-15)

$A_1 = 3.4050$ mGal^2, $A_2 = 140.03$ mGal^2, $s_1 = 0.998006$, $s_2 = 0.914232$, $B_1 = 1$ and $B_2 = 2$.

After these derivations we return to alternative (A). It was pointed out that there is no simple way to relate the coefficients $\tau_{nm}\{ij\}$ of an individual gradiometric component to the fundamental unknowns t_{nm}. However in the average sense, as a "signal power per degree" such a relationship can be established. This has first been done by Groten & Moritz (1964) for the case of deflections of the vertical. Based upon eqs. (3.17) similarly the average power of all gradiometric components has been derived in BOX 3-1.

The expected average signal content per $\{n,m\}$ is denoted

$$ave(c_{nm}) = (c)_{nm} = \frac{c_n}{2n+1} .$$ (3-16)

The signal degree variances of T_{zz}, $\{T_{xz}, T_{yz}\}$, $\{T_{xx}-T_{yy}, 2T_{xy}\}$ at altitude become, compare TABLE 2.3:

$$c_n\{zz;r\} = \left[\frac{U_0}{R^2}\right]^2 (n+2)(n+2)(n+1)(n+1)\left[\frac{R}{r}\right]^{2n+6} c_n\{t_{nm}\}$$ (3-17a)

$$c_n\{xz,yz;r\} = \left[\frac{U_0}{R^2}\right]^2 (n+2)(n+2)(n+1)n\left[\frac{R}{r}\right]^{2n+6} c_n\{t_{nm}\}$$ (3-17b)

$$c_n\{xx-yy,2xy;r\} = \left[\frac{U_0}{R^2}\right]^2 (n+2)(n+1)n(n-1)\left[\frac{R}{r}\right]^{2n+6} c_n\{t_{nm}\} .$$ (3-17c)

We observe that, in particular for higher degrees n, the degree variance spectra of the three gradiometric combinations are almost in coincidence. Now, employing one of the three degree variance models of TABLE 3.2, the average spectral behavior of satellite gradiometric quantities can be determined. We can compute

$c_n \{ij;r\}$ degree variances

$(c)_{nm} \{ij;r\}$ average signal size per spherical harmonic coefficient

$\|T_{ij}\|^2 = \sum_n c_n \{ij;r\}$ expected norm.

3.2 Noise Model

So far only a (deterministic) function model has been discussed. If a comparison were done of one of the gradiometric quantities with actual measurements discrepancies would be observed between measurement and model. In our case these discrepancies would be rather high because several idealizations were introduced for simplicity. However even if the most sophisticated model (no spherical approximation, no spherical boundary surface, actual orbits, sampling along orbit trajectories, effect of the attraction of sun, moon and planets a.s.o.) would be employed, a certain level of discrepancy would remain. These discrepancies have their origin in small, still remaining imperfections of the mathematical model and in small fluctuations of the measurement process itself. These latter effects could

---

BOX 3-1:   Average Signal Content (Power) per Degree
of Gradiometer Components

From eqs. (3.17) with "(.)" denoting average signal one finds:

$$(zz)^2 = c_n \{zz\} = (n+2)^2(n+1)^2 (U_0/R^2)^2 c_n\{t_{nm}\} \quad \text{(a)}$$
$$(xz)^2 + (yz)^2 = c_n\{xz,yz\} = (n+2)^2(n+1)n (U_0/R^2)^2 c_n\{t_{nm}\} \quad \text{(b)}$$
$$(xx-yy)^2 + (2xy)^2 = c_n\{xx-yy,2xy\} = (n+2)(n+1)n(n-1) (U_0/R^2)^2 c_n\{t_{nm}\} \quad \text{(c)}$$
$$(xx+yy)^2 = c_n\{zz\} = (n+2)^2(n+1)^2 (U_0/R^2)^2 c_n\{t_{nm}\} \quad \text{(d)}$$

We note that all spherical factors are approximately of the order: n^4. Then it holds on the average:

from (b)	$(xz)^2 \approx \frac{1}{2} (zz)^2$	(e)
from (b)	$(yz)^2 \approx \frac{1}{2} (zz)^2$	(f)
from (c)	$(xx-yy)^2 \approx \frac{1}{2} (zz)^2$	(g)
from (c)	$4(xy)^2 \approx \frac{1}{2} (zz)^2$	(h)

from (d) and (g):
$$(xx)^2 + 2(xx)(yy) + (yy)^2 + (xx)^2 - 2(xx)(yy) + (yy)^2 =$$
$$= 2[(xx)^2 + (yy)^2] = (zz)^2 + \frac{1}{2} (zz)^2 = 3/2(zz)^2 \quad \text{(i)}$$

from (i):	$(xx)^2 \approx \frac{3}{8} (zz)^2$	(j)
from (i):	$(yy)^2 \approx \frac{3}{8} (zz)^2$	(k)

Thus we find (with $\tau_n = \sum_m \tau_{nm}$):

$$c_n(zz) \approx n^4 (U_0/R^2)^2 c_n(t_{nm})$$
$$\tau_n (xz) \approx \frac{1}{2} c_n(zz) \quad \text{(e)}$$
$$\tau_n (yz) \approx \frac{1}{2} c_n(zz) \quad \text{(f)}$$
$$\tau_n (xx) \approx \frac{3}{8} c_n(zz) \quad \text{(j)}$$
$$\tau_n (yy) \approx \frac{3}{8} c_n(zz) \quad \text{(k)}$$
$$\tau_n (xy) \approx \frac{1}{8} c_n(zz) \quad \text{(h)}$$

---

be due to environmental, thermal, electronic, structural effects etc. inside the instrument. It is neither physically possible nor meaningful to try to extend the mathematical model such as to describe all these tiny influences. Rather, as the integral effect of all these contributions follows the laws of probability theory, their gross behavior is described by a stochastic model. We refer to (Baarda, 1967), see also (Rummel ,1984) or (Heck, 1987, ch. 4). The underlying thought is that of a large number of repetitions of an experiment and the choice of a probability function that describes the distribution of this ensemble of repetitions.

Translated to our case the situation is as follows. The gradiometer measurements are taken at a rate Δt along the orbit trajectory. Thus naturally the measurements are viewed as a time series spanning the entire mission duration. In this connection the concept of a repeat orbit is useful (Colombo 1984; Schrama, 1989). After a certain number N_0 of full orbit revolutions and - at the same time - after a certain number N_e of full revolutions of the earth under the (precessing) orbit plane the orbit trajectory closes in itself (bites its own tail). With each new repetition cycle a new time series starts. The repeat period could in our case be a few weeks or several months, depending on what density of the ground track pattern on the earth's surface or on Ω_r is desired. The orbit represents on the one hand a time series, on the other hand it is a line on Ω_r in the space domain. An orbit being repeat and polar generates a nicely regular and even pattern of measurements on Ω_r which is re-measured at regular intervals. For simplicity let us assume Ω_r to be covered by a sampling of equal density globally, i.e. with the same number of measurements per aqual area on Ω_r and the sampling being carried out repeatedly. In reality sampling is denser towards the poles.

In order to arrive at a spectral error model of sufficient simplicity the measurement process must be considered Gaussian, stationary and ergodic. Only if it is Gaussian (normally distributed) its first two statistical moments, mean value and variance, suffice for a complete description. Only if it is stationary (homogeneous and isotropic on the sphere) the error spectrum can be considered invariant and therefore representative over the mission period. Only if it is ergodic the stochastic characteristics can be deduced from a single realization in the time (or space) domain, instead of from its statistical ensemble. [Actually a process on a sphere cannot be simultaneously normal and ergodic. See (Yaglom, 1962) or (Moritz, 1980). But this fine point is considered beyond the context of this discussion.]

For the time being the gradiometer observations are not looked upon as a time series along the orbit, but rather a geographical function $\tilde{f}(\theta, \lambda)$ on the sphere Ω_r ("~" for stochastic). It is assumed that, with $\tilde{\varepsilon}$ the observation noise, it holds

$$E\{\tilde{\varepsilon}\} = 0 \qquad \text{which is} \qquad E\{\tilde{f}\} = f, \qquad (3\text{-}18)$$

with second-order moment $E\{\tilde{\varepsilon}(P), \tilde{\varepsilon}(Q)\}$ and P and Q two arbitrary points on Ω_r. $E\{\tilde{\varepsilon}(P), \tilde{\varepsilon}(Q)\}$ is supposed to have the following properties:

- It reflects the combined effect of measurement error and spatial sample point density. Sansó has investigated this aspect by measure theory. It is referred to (Rummel et al., 1993) and in particular to (Sansó & Rummel, 1994).
- In good approximation $E\{\tilde{\varepsilon}(P), \tilde{\varepsilon}(Q)\}$ be homogeneous and isotropic, i.e.

$$E\{\tilde{\varepsilon}(P), \tilde{\varepsilon}(Q)\} = D(P\text{-}Q) \qquad (3\text{-}19)$$

- There exists a small but finite spatial correlation between P and Q. Assuming no correlation (a Dirac function) would be in conflict with physical reality.

Moritz (1963) and Heiskanen & Moritz (1967; ch 7.7.) formulated an error theory for geodetic functions on a sphere. Since functions on a sphere are two-dimensional but also closed, i.e. periodic, their spectra must be discrete (line spectra) with two indices (degree n and order m, as $E\{\bar{\varepsilon}(P),\bar{\varepsilon}(Q)\}$ is a second-order moment, also its spectrum is to be of dimension two, i.e. of the form

$$\sigma^2_{nm,kl} = \frac{1}{4\pi} \int\int_\omega \frac{1}{4\pi} \int\int_{\omega'} Y_{nm}(P) \; E\{\bar{\varepsilon}(P),\bar{\varepsilon}(Q)\} \; Y_{kl}^*(Q) \; d\omega'_Q d\omega_P \qquad (3\text{-}20)$$

with $\sigma^2_{nm,kl}$ the spectral error co-variance. However for a homogeneous and isotropic process eq.(20) degenerates to a one-dimensional spectrum, (Papoulis,1965):

$$\sigma^2_{nm} = \frac{1}{4\pi} \int\int_\omega \frac{1}{4\pi} \int\int_{\omega'} Y_{nm}(P)D(P-Q) \; Y_{nm}^*(Q) \; d\omega'_Q d\omega_P \; . \qquad (3\text{-}21)$$

Moritz (1963) defines a so-called error constant S:

$$S = \int\int_{\Omega_r} D(P-Q) \; d\Omega_r = r^2 \int\int_\omega D(P-Q) \; d\omega \; , \qquad (3\text{-}22)$$

the volume under the error covariance function D(P-Q), as discussed by Strang van Hees (1986). It is a constant only for a stationary process. If, in addition, correlation length is very small, eq.(3-21) can approximately be written as:

$$\sigma^2_{nm} \approx \frac{1}{4\pi} \int\int_\omega D(P-Q) \; d\omega \cdot \frac{1}{4\pi} \int\int_{\omega'} Y_{nm}(Q)Y_{nm}^*(Q) \; d\omega'_Q = \frac{S}{4\pi r^2} \; . \qquad (3\text{-}23)$$

From the geometrical interpretation of S given above, it follows

$$S \approx \sigma^2 \Delta\Omega_r \qquad (3\text{-}24)$$

(variance σ^2 (in space domain) times equal area element $\Delta\Omega_r$ over which correlation disappears), or with $\Omega_r = 4\pi r^2$ and $M = \Omega_r/\Delta\Omega_r$ the number of samples on Ω_r (independent individual measurements or independent block averages):

$$\sigma^2_{nm} = \sigma^2 \frac{\Delta\Omega_r}{4\pi r^2} = \frac{\sigma^2}{M} \; . \qquad (3\text{-}25)$$

This result is identical to the spectral error model by Jekeli & Rapp (1980) and agrees for all practical purposes with Sansó's model (Rummel et al., 1993; eq. (2.78)).

It needs to be clarified, however, up to what maximum degree this equation holds. A useful guideline is, that the total number of estimable coefficients can be M at most, so that the maximum degree η becomes

$$\eta = \left| \sqrt{M} \right| - 1 \tag{3-26}$$

Since σ^2_{nm} is a constant, independent of degree and order (white noise), it is respresentative for any arbitrary coefficient of a particular degree, i.e.

$$(\sigma)^2_{nm} = \sigma^2_{nm} . \tag{3-27}$$

The error degree variance is simply

$$(\sigma)^2_n = (2n+1)\sigma^2_{nm} . \tag{3-28}$$

It should be added that more sophisticated error spectra are conceivable, in particular one could take into account the denser sampling towards the poles.

An alternative manner of derivation of eq. (3-24) is to express the isotropic error covariance function as

$$
\begin{aligned}
D(P-Q) &= \sum_{n=0}^{\eta} a_n (2n+1) P_n(P-Q) \\
&= \sum_n a_n \sum_m Y^*_{nm}(P) Y_{nm}(Q) .
\end{aligned}
\tag{3-29}
$$

Insertion in (3-21) yields

$$
\begin{aligned}
\sigma^2_{nm} &= \frac{1}{4\pi} \int\int_\omega \frac{1}{4\pi} \int\int_{\omega'} Y_{nm}(P) \left(\sum_{n'} a_{n'} \sum_{m'} Y^*_{n'm'}(P) Y_{n'm'}(Q) \right) Y^*_{kl}(Q) \, d\omega'_Q d\omega_P \\
&= a_n
\end{aligned}
\tag{3-30}
$$

and therefore

$$a_n = \sigma^2/M. \qquad\qquad \text{for } n \le \eta$$

For the <u>squared error norm</u> one arrives at the expected result

$$\| \epsilon \|^2 = \sum_n \sum_m \sigma^2_{nm} = \sum_n (2n+1)\sigma^2_{nm} = M\frac{\sigma^2}{M} = \sigma^2 . \tag{3-31}$$

The total error variance is evenly distributed over all spectral lines $\{n,m\}$, in the sense of <u>bandlimited white noise</u>. This error model can now be applied to the gradiometric quantities. The general expression reads, in analogy to eq. (3-21):

$$\sigma_{nm}^2 = \frac{1}{4\pi} \int\int \frac{1}{4\pi} \int\int Z_{nm}(P)D(P-Q)Z_{nm}^*(Q)\ d\omega'_Q d\omega_p \ . \tag{3-32}$$

For its evaluation $D(P-Q)$ could be written as Legendre tensor series. See (Freeden et al., 1994) and (Schreiner, 1994). This would be in analogy to eq. (3-29). Alternatively one may proceed as in eq. (3-23). If we can assume that the measurement of each gradiometer component is independent from all others and each is isotropic and homogeneous, the tensorial error variance-covariance function $D(P-Q)$ becomes diagonal. Let the error behavior (variance) also be the same for each component, then we may write, e.g. for $\{T_{xz}, T_{yz}\}$:

$$
\begin{aligned}
\sigma_{nm}^2\{xz,yz\} &\approx \frac{1}{4\pi} \int_\omega\int \sum_{i,j} D_{ij}(P-Q)d\omega \ \frac{1}{4\pi} \int_{\omega'}\int \sum_{ij} \left[Z_{nm}^{(1)}\right]_{ij} \left[Z_{nm}^{(1)}\right]_{ij} d\omega' \\
&= \frac{S}{4\pi r^2} \approx \frac{\sigma^2}{M} \ .
\end{aligned}
\tag{3-33}
$$

The two components together yield the same error spectrum as $\{zz\}$ alone. For the tensor $\{T_{xx}-T_{yy}, 2T_{xy}\}$ we assume $2T_{xy}$ derived from measuring (xy) and (yx). The corresponding error degree-order model becomes:

$$\sigma_{nm}^2\{xx-yy, xy+yx\} = \frac{2S}{4\pi r^2} \approx \frac{2\sigma^2}{M} \tag{3-34}$$

We observe that the combined gradiometric quantities yield an error level, which compares to that for T_{zz} as

$$\sigma_{nm}^2\{zz\} \div \sigma_{nm}^2\{xz,yz\} \div \sigma_{nm}^2\{xx-yy,xy+yx\} = 1 \div 1 \div 2 \tag{3-35}$$

[This is a correction to what was stated in (Rummel & van Gelderen, 1995).]

3.3 Spectral Estimation

Let us now turn to a situation, where a satellite gradiometry mission has actually been performed. Still under a very idealized situation we assume the following: a) all measurements are available on Ω_r in satellite altitude (supposing the orbit being perfectly polar; orbit uncertainty negligible; all observations being transfered from the actual orbit to Ω_r by interpolation);

b) the measurements are nicely distributed with equal area density, their total number being M (as deduced from sampling rate and mission duration);

c) the linearized gradiometric quantities are available:
$\Delta\tilde{\Gamma}^{(0)}=\{\partial^2_{zz}T\}$, $\Delta\tilde{\Gamma}^{(1)}=\{\partial^2_{xz}T,\partial^2_{yz}T\}$, $\Delta\tilde{\Gamma}^{(2)}=\{(\partial^2_{xx}T-\partial^2_{yy}T),\ 2\partial^2_{xy}T\}$,("~" denotes stochastic, i.e. measured)

d) all individual components are measured with the same quality, their variance being σ^2 .

Our objective would then be to determine the series coefficients of the disturbance potential T, $\hat{t}_{nm}$ ("^" for estimated) and a description of the error.

After spherical harmonic analysis (expansion) on Ω_r of $\Delta\tilde{\Gamma}^{(0)}$, $\Delta\tilde{\Gamma}^{(1)}$ and $\Delta\tilde{\Gamma}^{(2)}$ three coefficient sets complete up to degree and order η are available. The linear spectral model is (compare eqs. (2-40) to (2-41) or (Rummel & van Gelderen, 1992)):

$$\tilde{z}^{(0)}_{nm} = (n+1)(n+2)\left[\frac{R}{r}\right]^{n+3}\Gamma_0 t_{nm}+\tilde{\epsilon}^{(0)}_{nm} \tag{3-36a}$$

$$\tilde{z}^{(1)}_{nm} = -(n+2)\sqrt{n(n+1)}\left[\frac{R}{r}\right]^{n+3}\Gamma_0 t_{nm}+\tilde{\epsilon}^{(1)}_{nm} \tag{3-36b}$$

$$\tilde{z}^{(2)}_{nm} = \sqrt{(n+2)(n+1)n(n-1)}\left[\frac{R}{r}\right]^{n+1}\Gamma_0 t_{nm}+\tilde{\epsilon}^{(2)}_{nm} \tag{3-36c}$$

with $\Gamma_0=U_0/R^2$ and $\tilde{\epsilon}_{nm}^{(0)}$, $\tilde{\epsilon}_{nm}^{(1)}$ and $\tilde{\epsilon}_{nm}^{(2)}$ the unknown error contribution in $\tilde{z}_{nm}^{(0)}$, $\tilde{z}_{nm}^{(1)}$ and $\tilde{z}_{nm}^{(2)}$.

Even prior to any experiment we can employ one of the available signal degree variance models, see Table 3.2, and determine $c_n\{zz;r\}$, $c_n\{xz,yz;r\}$, and $c_n\{xx-yy,2xy;r\}$, eqs. (3-17). Eq. (3-16) gives the corresponding average coefficient variance per degree $(c)_{nm}$. On the other hand, eqs. (3-34) to (3-35) yield the corresponding errordegree-variance models. Thus the signal spectrum can be compared with that of the noise, as is done in Figure 3-3. Signal and noise spectrum intersect at the so-called <u>degree-of-resolution</u> η_{res}. We observe:

for $n<\eta_{res}$	$\Rightarrow$	$(c)_{nm}>\sigma_{nm}^2$
		signal dominates noise
for $n>\eta_{res}$	$\Rightarrow$	$(c)_{nm}<\sigma_{nm}^2$
		noise dominates signal

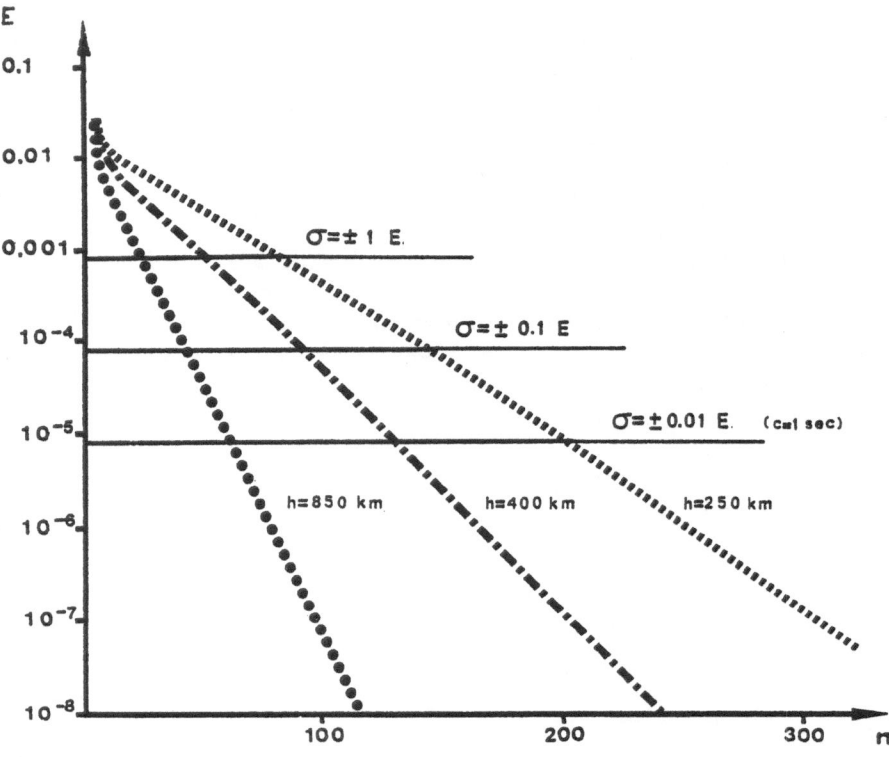

Figure 3.3: Comparison of signal and noise spectra: Displayed is the average root-mean-square power per spherical harmonic degree and order for T_{zz} at altitudes 250km, 400km and 850km and the error spectra for std.devs. of 1E, 10^{-1}E and 10^{-2}E (and correlation length of 1s).

Based on this comparison a so-called ideal low-pass filter can be constructed. Its spectral form is

$$h_n = \begin{cases} 1 & n \leq \eta_{res} \\ 0 & n > \eta_{res} \end{cases} \qquad (3\text{-}37)$$

Then the filtered coefficients become

$$\hat{z}_{nm} = h_n \bar{z}_{nm} \;. \qquad (3\text{-}38)$$

In the space-domain, on Ω_r, the filter equation becomes

$$H(P-Q) = \sum_n h_n(2n+1)P_n(P-Q) \ . \tag{3-39}$$

and eq. (3-38) takes the form

$$\hat{f}(P) = H * \tilde{f} = \frac{1}{4\pi} \int\int_\omega H(P-Q)\tilde{f}(Q)d\omega_Q \ . \tag{3-40}$$

The esimated coefficients $\hat{t}_{nm}$ are derived directly from the $\hat{z}_{nm}$ by means of the Meissl-scheme. The error variance of $\hat{z}_{nm}$ consists of two terms:

$$\sigma_C^2(\hat{z}_{nm}) = \sum_{n=2}^{\eta_{res}} \sigma_n^2(\tilde{z}_{nm}) \qquad \text{commission error} \tag{3-41}$$

and

$$\sigma_O^2(\hat{z}_{nm}) = \sum_{n=\eta_{res}+1}^{\infty} c_n(z_{nm}) \qquad \text{omission error} \tag{3-42}$$

and therefore

$$\sigma^2(\hat{z}_{nm}) = \sigma_C^2(\hat{z}_{nm}) + \sigma_O^2(\hat{z}_{nm}) \ . \tag{3-43}$$

The two error contributions are shown in Figure (3-3).

Despite its name, the ideal low-pass filter is not optimal. It does not produce the least error. Therefore we try to determine a linear filter with coefficients h_n that is optimal in this sense. With "M" an averaging operator per spherical harmonic degree, we solve

$$M\{(z_{nm} - h_n\tilde{z}_{nm})^2\} = \min \ .$$

It is

$$M\{(z_{nm} - h_n\tilde{z}_{nm})^2\} = M\{z_{nm}^2\} - 2h_nM\{z_{nm}\tilde{z}_{nm}\} + h_n^2M\{\tilde{z}_{nm}^2\} \tag{3-44}$$

Assuming spectral independence (orthogonality) of signal and noise, it is

$$M\{z_{nm}\tilde{z}_{nm}\} = M\{z_{nm}^2\} = c_n\{z_{nm}\} \tag{3-45}$$

and

$$M\{\tilde{z}_{nm}^2\} = \tilde{c}_n\{z_{nm}\} = c_n\{z_{nm}\} + \sigma_n^2(\tilde{z}_{nm}) \tag{3-46}$$

and the minimum is attained for

$$h_n = \frac{c_n\{z_{nm}\}}{c_n\{z_{nm}\} + \sigma_n^2} . \tag{3-47}$$

This is the well-known <u>Wiener-Kolmogorov filter</u>.
The estimate becomes

$$\hat{z}_{nm} = h_n \tilde{z}_{nm} = \frac{c_n}{c_n + \sigma_n^2} \tilde{z}_{nm} . \tag{3-48}$$

and eqs. (3-39) and (3-40) apply as before. The error contributions become now from eq.(3-44) merged into one expression:

$$\begin{aligned}
\sigma^2(\hat{z}_{nm}) &= c_n - c_n\frac{c_n}{c_n+\sigma_n^2} \\
&= c_n(1-h_n) .
\end{aligned} \tag{3-49}$$

The Wiener-Kolmogorov filter achieves an optimal balance between signal and noise. Its coefficients h_n can be determined, e.g. for simulations, prior to any experiment. In an actual experiment the synthetic estimate for $\tilde{c}_n$, eq.(3-46), should be replaced by the actual "signal + noise" variance per degree:

$$\tilde{c}_n\{zz;r\} = \sum_m \left(\tilde{z}_{nm}^{(0)}\right)^2 \tag{3-50a}$$

$$\tilde{c}_n\{xz,yz;r\} = \sum_m \left(\tilde{z}_{nm}^{(1)}\right)^2 \tag{3-50b}$$

$$\tilde{c}_n\{xx-yy,2xy;r\} = \sum_m \left(\tilde{z}_{nm}^{(2)}\right)^2 \tag{3-50c}$$

4. Regularization

"For a long time mathematicians felt that
ill-posed problems cannot describe
real phenomena and objects."
A.N. Tikhonov and V.Ya. Arsenin, 1977

It was explained in chapter 2, a principal drawback of any gravity field model-ling by satellites is the continuation of the attenuated gravity signal at altitude down to the surface of the earth. As the Meissl scheme shows, compare Table 2-1, it implies amplification of the spectral content by $(r/R)^{n+1}$ with $r = R+h > R$. Only if the gravity field signal at altitude meets certain strict smoothness criteria, convergence of the downward continued information is ensured. The gravity field itself and its derivatives fulfill these criteria intrinsically. However the measure-ment noise contained in the data would certainly violate them, were not some counter measures taken. These methods of stabilization are called regularization. They are supposed to maximally supress the high frequency noise without too much of signal loss. The most common and simplest technique of regularization is series truncation at some sensibly chosen spherical harmonic degree n. Twenty years ago, for gravity potential models based solely on satellite data, this trunca-tion degree used to be about 20. Nowadays, because of the enormous progress in data collection and because of the availability of powerful computers, degrees 30 to 50 and higher are possible. In the case of satellite gradiometry, where one aims at gravity modelling complete up to degree and order 200 or 300, refined regula-rization deserves serious attention.

Again we confine ourselves to the fundamentals. We introduce the linear model of gravity field recovery from satellite gradiometry in its plainest form and compare its solution by classical least-squares adjustment with that by least-squares collocation and ridge-type estimation. For the presentation we follow (Gerstl & Rummel, 1981) and in particular (Xu, 1992). See also (Schwarz, 1971), where regularization was first applied to the problem at hand, (Bouman, 1993), (Xu & Rummel, 1994a) and (Xu & Rummel, 1994b).

Linear Model: Suppose by some preprocessing (e.g. moving averaging or block averaging) our model world is confined to some finite but high spherical harmo-nic degree η. In the space domain our model be

$$\Gamma = L\ T \tag{4-1}$$

with Γ the observable gradiometric quantities, either {zz}, {xz,yz}, or {xx-yy,2xy}, linearized with respect to a chosen reference gravity field and Γ given on Ω_r, i.e. at satellite altitude; T is the unknown disturbance potential on Ω, referring to the same reference field; L is the linear operator connecting T and Γ. The latter contains the upward continuation operator from Ω to Ω_r, as well as the respective differential operator. With the theory of chapter 2, see in particular Box 2-1, L can symbolically be written as (by singular value decomposition (SVD)):

$$L = Z\Lambda Y^* \qquad \& \qquad L^* = Y\Lambda Z^* . \qquad (4\text{-}2)$$

Thereby Z contains as columns the complete set of surface tensor spherical harmonics up to $n=\eta$, in the same manner Y the surface spherical harmonics, and Λ, as diagonal matrix, the spherical symbols (or singular values) of differentiation and upward continuation. Orthogonality conditions, eqs. (2-5) and (2-42), become:

$$Z^*Z = ZZ^* = I \qquad (4\text{-}3)$$

and

$$Y^*Y = YY^* = I \qquad (4\text{-}4)$$

(I unit matrix, the dimensions may be different in (4-3) and (4-4)). Then (4-1) can be Fourier-transformed by inserting (4-2) into (4-1). We obtain

$$\Gamma = Z\Lambda Y^*T$$

$$\Rightarrow Z^*\Gamma = Z^*Z\Lambda Y^*T$$

$$\Rightarrow z = \Lambda t , \qquad (4\text{-}5)$$

where z is the vector of coefficients, either $z_{nm}^{(0)}$, $z_{nm}^{(1)}$, or $z_{nm}^{(2)}$ and t that of the coefficients t_{nm}. There hold the spherical harmonic synthesis and analysis formulae:

$$z = Z^*\Gamma \qquad \Leftrightarrow \qquad \Gamma = Zz \qquad (4\text{-}6,a\text{-}b)$$

and

$$t = Y^*T \qquad \Leftrightarrow \qquad T = Yt . \qquad (4\text{-}7,a\text{-}b)$$

Least-Squares Solution: We turn now to a system of observation equations: With eq. (4-1) the observation model becomes

$$\tilde{\Gamma} = LT + \tilde{E} \qquad (4\text{-}8)$$

with $\tilde{E}$ the vector of stochastic residuals. Furthermore the first and second statistical moments be:

$$E(\tilde{\Gamma}) = LT \qquad \text{and} \qquad D(\tilde{\Gamma}) = Q = \sigma_{nm}^2 ZZ^* \tag{4-9}$$

(compare eqs. (3-20) and (3-22), $\sigma_{nm}^2 = (\sigma)_{nm}^2 = \text{const.}$) and the a priori signal variance-covariance model of T

$$C_T = YC_{nm}Y^* \tag{4-10}$$

with C_{nm} diagonal and $(c)_{nm} = c_n/(2n+1)$ its diagonal elements, compare eq. (3-16). The $(c)_{nm}$ represent the prior average signal knowledge, translated into a signal variance per coefficient.

An appropriate model is chosen e.g. from Table 3.2. Classical least-squares adjustment yields for $M \geq N$:

$$\hat{T} = (L^*Q^{-1}L)^{-1}L^*Q^{-1}\tilde{\Gamma}$$

$$= (\sigma_{nm}^{-2}Y\Lambda Z^*ZZ^*Z\Lambda Y^*)^{-1}\sigma_{nm}^{-2}Y\Lambda Z^*ZZ^*\tilde{\Gamma}$$

$$= Y(\Lambda\Lambda)^{-}\Lambda Z^*\tilde{\Gamma} . \tag{4-11}$$

For the $N \cdot 1$ vector of estimated coefficients $\hat{t}$ one finds

$$\hat{t} = \Lambda^{-1}\tilde{z}$$

or

$$\hat{t}_{nm} = \frac{\tilde{z}_{nm}}{\lambda_n} = \frac{z_{nm}}{\lambda_n} + \frac{\tilde{\epsilon}_{nm}}{\lambda_n} . \tag{4-12a,b}$$

With $\lambda_n \approx \Gamma_0 \left[\dfrac{R}{r}\right]^{n+3} n^2$ it is

$$\hat{t}_{nm} \approx n^{-2} \left[\frac{r}{R}\right]^{n+3}(z_{nm}+\tilde{\epsilon}_{nm})/\Gamma_0 . \tag{4-13}$$

We observe a serious error amplification for higher degrees, where $(r/R)^{n+3}$ clearly dominates n^{-2}. This fact makes classical least-squares adjustment unsuitable for the solution of unstable problems. The a posteriori error variance-covariance becomes

$$\hat{Q} = \sigma_{nm}^2 Y \Lambda^{-2} Y^*$$

(4-14)

and the error variance coefficients

$$\hat{q}_{nm} = \frac{\sigma_{nm}^2}{\lambda_n^2} .$$

(4-15)

We should thereby keep in mind, that σ_{nm}^2 is a constant for all degrees and orders up to $n=\eta$.

Least-Squares Collocation or Regularization: The two methods least-squares collocation (LSC) and Tikhonov regularization are formally identical, compare (Moritz, 1980). For simplicity we assume that the disturbance potential T refers to the best, already available gravity reference model. Then LSC or regularization is built on the least-squares/minimum-norm condition

$$\| \tilde{\Gamma} - LT \|_{Q^{-1}}^2 + \| T \|_{C^{-1}}^2 = \min.$$

(4-16)

T can now be viewed as an element of an infinite dimensional Hilbert space with its series expansion extending up to infinity $(M < N)$. In this case classical least-squares adjustment fails. The strategy is to select among all possible solutions of T the one with minimum norm with respect to C_T. This is the principle of LSC, see (Krarup, 1969) or (Moritz, 1980). LSC is mostly used to determine an optimal representation of T based upon a finite and discrete set of given gravity functionals. Also for all recent satellite based geopotential model computations regularization is applied, so as to secure maximum spectral resolution, see e.g. (Lerch, 1989).

Another very illustrative interpretation is the following: To the given M-vector of measurements $\tilde{\Gamma}$ additional N (N may be infinite) pseudo-observations $\tilde{T}$ are added, e.g. as pseudo-observed coefficients $\tilde{t}_{nm}$. They get the value zero and possess the error variance-covariance matrix C_T (or C_{nm}).

From the minimum condition (4-16) the following formal solution is derived:

$$\hat{T} = (L^* Q^{-1} L + C_T^{-1})^{-1} L^* Q^{-1} \tilde{\Gamma}$$

$$= C_T L^* (LC_T L^* + Q)^{-1} \tilde{\Gamma} ;$$

(4-17a,b)

the latter form is that of LSC with $LC_T L^* = C_\Gamma$ and $C_T L^* = C_{T\Gamma}$. By SVD we find

$$\hat{T} = Y C_{nm} \Lambda (\Lambda C_{nm} \Lambda + \sigma_{nm}^2 I)^{-1} Z^* \tilde{\Gamma}$$

(4-18a)

and

$$\hat{t} = C_{nm}\Lambda(\Lambda C_{nm}\Lambda + \sigma_{nm}^2 I)^{-1}\tilde{z} \ . \tag{4-18b}$$

The estimated coefficients become

$$\hat{t}_{nm} = \frac{\lambda_n^2(c)_{nm}}{\lambda_n^2(c)_{nm} + \sigma_{nm}^2}\tilde{z}_{nm} = h_n\tilde{z}_{nm} \ . \tag{4-19}$$

The spectral filter h_n takes care of the optimal balance between pronounciation of signal and suppression of noise. Comparison with chapter 3 shows that the filter is identical to the Wiener-Kolmogorov filter already derived in eq. (3-47), when we observe that $c_n = (2n+1)(c)_{nm}$ and $\sigma_n^2 = (2n+1)\sigma_{nm}^2$ and e.g. $c_n\{z_{nm}\} = \lambda_n^2 c_n\{t_{nm}\}$.

The a posteriori variance-covariance matrix is now

$$\begin{aligned}\hat{Q} &= C_T - C_T L^*(LC_T L^* + Q)^{-1}LC_T \\ &= (L^*Q^{-1}L + C_T^{-1})^{-1} \ .\end{aligned} \tag{4-20a,b}$$

It yields as error variance per coefficient t_{nm}:

$$\hat{q}(I)_{nm} = (c)_{nm}\frac{\sigma_{nm}^2}{\lambda_n^2(c)_{nm} + \sigma_{nm}^2} = (c)_{nm}(1 - h_n) \ . \tag{4-21}$$

For low degrees $(1-h_n)$ is a small quantity, for high degrees it converges towards 1 and $\hat{q}(I)_{nm} \to (c)_{nm}$.

Very often in the literature a regularization factor α is introduced to balance the second term of eq. (4-16) with respect to the first one. This can also be achieved implicitly by incorporating α into σ_{nm}^2, i.e. assuming σ_{nm}^2 is multiplied by this constant.

Biased Estimation: Under certain idealized assumptions least-squares collocation can be considered an unbiased estimator. The argumentation is given in (Moritz, 1980). Considering the actual practical situation it is legitimate to view it as biased. Then LSC can be viewed a ridge-type estimator. This point is stressed in (Xu, 1992). Then the error considerations given above are incomplete and one has to investigate what the effect of a bias is.

Ultimately this problem can only be fully addressed if the unbiased result is available too. Still one can look into the mechanism in which a bias influences an unstable estimation process.

The effect of the bias on the estimate $\hat{T}$ is derived by inserting its expectation $E(\tilde{T}) = LT$, eq. (4-9), into eq. (4-17a).

This yields

$$\hat{T} = (L^*Q^{-1}L+C_T^{-1})^{-1}L^*Q^{-1}LT$$

$$= (L^*Q^{-1}L+C_T^{-1})^{-1}[L^*Q^{-1}L+C_T^{-1}-C_T^{-1}]T$$

$$= T-(L^*Q^{-1}L+C_T^{-1})^{-1}C_T^{-1}T \ . \tag{4-22}$$

Thus the bias in $\hat{T}$, Bia $(\hat{T})$, is :

$$\text{Bia}(\hat{T}) = \hat{T}-T$$

$$= -(L^*Q^{-1}L+C_T^{-1})^{-1}C_T^{-1}T$$

$$= -Y(\sigma_{nm}^{-2}\Lambda\Lambda+C_{nm}^{-1})^{-1}C_{nm}^{-1}Y^*T \ , \tag{4-23a}$$

or

$$\text{Bia}(\hat{t}) = -(\sigma_{nm}^{-2}\Lambda\Lambda+C_{nm}^{-1})^{-1}C_{nm}^{-1}t \ . \tag{4-23b}$$

This results in a bias per coefficient:

$$\text{Bia} \ (\hat{t}_{nm}) = -\frac{\sigma_{nm}^2}{\lambda_n^2(c)_{nm}+\sigma_{nm}^2}t_{nm} = -[1-h_n]t_{nm} \ . \tag{4-24}$$

Again we observe that the bias increases with increasing degree, reaching almost 100% for $n > \eta_{res}$.

The mean square error (MSE) consists now of two additional, bias related terms, compare (Xu, 1992):

$$\text{MSE} \ (\hat{T}) = \hat{Q}+\text{Bia}(T)\text{Bia}(T)^*$$

$$= (L^*Q^{-1}L+C_T^{-1})^{-1}$$

$$-(L^*Q^{-1}L+C_T^{-1})^{-1}C_T^{-1}(L^*Q^{-1}L+C_T^{-1})^{-1}$$

$$+(L^*Q^{-1}L+C_T^{-1})^{-1}C_T^{-1}[TT^*]C_T^{-1}(L^*Q^{-1}L+C_T^{-1})^{-1} \ . \tag{4-25}$$

The first term is $\hat{Q}$ as derived for LSC, eq. (4-20). The second term is negative. It results in an accuracy gain, a gain resulting from the use of biased estimation. The third term reflects the accuracy loss due to the Bia($\hat{T}$). By SVD these latter two terms become

$$\Delta\text{MSE(II)} = -Y(\sigma_{nm}^{-2}\Lambda\Lambda+C_{nm}^{-1})^{-1}C_{nm}^{-1}(\sigma_{nm}^{-2}\Lambda\Lambda+C_{nm}^{-1})^{-1}Y^* \tag{4-26}$$

or

$$\hat{q}(II)_{nm} = -(c)_{nm}[1-h_n]^2 \tag{4-27}$$

and

$$\Delta MSE(III) = Y(\sigma_{nm}^{-2}\Lambda\Lambda + C_{nm}^{-1})^{-1}C_{nm}^{-1}Y^*[TT^*]YC_{nm}^{-1}\cdot$$
$$\cdot(\sigma_{nm}^{-2}\Lambda\Lambda + C_{nm}^{-1})^{-1}Y^* \ . \tag{4-28}$$

Obviously it is $Y^*[TT^*]Y = [tt^*]$ and this leads to the spectral decomposition

$$\hat{q}(III)_{nmkl} = (1-h_n)[t_{nm}t_{kl}](1-h_k) \ . \tag{4-29}$$

One sees that covariance (correlation) terms appear, depending on $\{n,m\}$ as well as on $\{k,l\}$. The sum $\hat{q}(II)_{nm} + \hat{q}(III)_{nm}$ represents a measure of the deviation of the square of the actual coefficients t_{nm}^2 from their prior value $(c)_{nm}$:

$$\hat{q}(II)_{nm} + \hat{q}(III)_{nm} = (c)_{nm}(1-h_n)^2 \frac{t_{nm}^2 - (c)_{nm}}{(c)_{nm}} \tag{4-30}$$

$$\text{for } n=k \text{ and } m=l$$

and

$$\hat{q}(III)_{nmkl} = (1-h_n)t_{nm}t_{kl}(1-k_k) \tag{4-31}$$

$$\text{for } n \neq k \text{ or } m \neq l \ .$$

We can draw the following conclusions:

-    Regularization or LSC leads always to a smoothed estimate, i.e. $\|\hat{T}\|^2 < \|T\|^2$. The smoothing increases with increasing degree n, because $\lim_{n\to\infty} h_n = 0$.

-    If the actual coefficients t_{nm} decrease faster than the degree-order model variances $(c)_{nm}$, i.e. if $(c)_{nm} > t_{nm}^2$, then the error measure of LSC, $\hat{q}(I)_{nm}$, is conservative. If $(c)_{nm} < t_{nm}^2$ the actual error is underestimated.

References

Albertella, A., F. Migliaccio, F. Sansò (1994). "Application of the Concept of Biortho-gonal Series to a Simulation of a Gradiometric Mission", 3rd Hotine-Marussi Symposium on Mathematical Geodesy, L'Aquila.

Albertella, A., F. Migliaccio, F. Sansò (1995). "Global Gravity Field Recovery by Use of STEP Observations" in: "Gravity and Geoid" (eds. H. Sünkel & I. Marson), IAG-Symposia 113, 111-116, Springer, Heidelberg.

Baarda, W. (1967). "Statistical Concepts in Geodesy", Netherlands Geodetic Commission, New Series, 2, 4, Delft.

Baarda, W. (1979). "A Connection Between Geometric and Gravimetric Geodesy", Netherlands Geodetic Commission, New Series, 6, 4, Delft.

Baarda, W. (1995). "Linking up Spatial Models in Geodesy - Extended S-Transformations", Netherlands Geodetic Commission, New Series, 41, Delft.

Backus, G.E. (1967). "Converting Vector and Tensor Equations to Scalar Equations in Spherical Coordinates", Geophys. J. R. Astr. Soc., 13, 71-101.

Bouman, J. (1993). "The Normal Matrix in Gravity Field Determination with Satellite Methods", Delft University of Technology.

Brovelli, M., F. Sansò (1990). "Gradiometry: The Study of the V_{yy} Component in the BVP Approach", manuscripta geodaetica, 15, 4, 240-248.

Bruns, H. (1878). "Die Figur der Erde", Publication des Königl. Preussischen Geodätischen Instituts, Berlin.

Colombo, O.L. (1981). "Numerical Methods for Harmonic Analysis on the Sphere", Dept. Geodetic Science and Surveying, 310, The Ohio State University, Columbus.

Colombo, O.L. (1984). "Altimetry, Orbits and Tides", Report EG&G Washington Analytical Services Inc., NASA Technical Memorandum 86180.

Colombo, O.L. (1989). "Advanced Techniques for High-Resolution Mapping of the Gravitational Field" in: Theory of Satellite Geodesy and Gravity Field Determination (eds. F. Sansò & R. Rummel) Lecture Notes in Earth Sciences, 25, 335-369, Springer, Heidelberg.

Freeden, W., T. Gervens, M. Schreiner (1994). "Tensor Spherical Harmonics and Tensor Spherical Splines", manuscripta geodaetica, 19, 70-100.

van Gelderen, M., R. Koop (1996). "An Assessment of the Accuracy of Some Error Prediction Methods for Satellite Gradiometry", journal of geodesy (in print).

Gerstl, M., R. Rummel (1981). "Stability Investigations of Various Representations of the Gravity Field", Rev. Geoph. Space Phys., 19, 3, 415-420.

Grafarend, E. (1986). "Three-Dimensional Deformation Analysis: Global Vector Spherical Harmonic and Local Finite Element Representation", Tectonophysics, 130, 337-359.

Groten, E., H. Moritz (1964). "On the Accuracy of Geoid Heights and Deflections of the Vertical", Institute of Geodesy, Photogrammetry and Cartography, 38, The Ohio State University, Columbus.

Gurtin, M.E. (1971). "Theory of Elasticity", Handbuch der Physik, Bd. VI, 2nd ed., Springer, Berlin.

Heck, B. (1987). "Rechenverfahren und Auswertemodelle der Landesvermessung - Klassische und moderne Methoden", Wichmann, Karlsruhe.

Heiskanen, W.A., H. Moritz (1967). "Physical Geodesy", Freeman, San Francisco.

Hotine, M. (1967). "Downward Continuation of the Gravitational Potential", in: Hotine, M. (1991). "Differential Geodesy" 143-148, Springer, Berlin.

Hotine, M. (1969). "Mathematical Geodesy", U.S. Department of Commerce, Washington D.C..

Jackson, J.D. (1925; 1975). "Classical Electrodynamics", John Wiley, New York.

Jekeli, Ch. (1978). "An Investigation of Two Models for the Degree Variances of Global Covariance Functions", Dept. Geodetic Science, 275, The Ohio State University, Columbus.

Jekeli, Ch., R.H. Rapp (1980). "Accuracy of the Determination of Mean Anomalies and Mean Geoid Undulations from a Satellite Gravity Field Mapping Mission", Dept. Geodetic Science and Surveying, 307, The Ohio State University, Columbus.

Jones, M.N. (1985). "Spherical Harmonics and Tensors for Classical Field Theory", Wiley, New York.

Kaula, W.M. (1966). "Theory of Satellite Geodesy", Blaisdell Publ., Waltham Ma..

Keller, W. (1991). "Behandlung eines überbestimmten Gradiometrie-Randwertproblems mittels Pseudodifferentialoperatoren", ZfV, 116, 66-73.

Keller, W., M. Hirsch (1994). "A Boundary Value Approach to Downward Continuation", manuscripta geodaetica, 19, 2, 101-118.

Koop, R. (1993). "Global Gravity Field Modelling Using Satellite Gravity Gradiometry", Netherlands Geodetic Commission, New Series, 38, Delft.

Krarup, T. (1969). "A Contribution to the Mathematical Foundation of Physical Geodesy", Danish Geodetic Institute, 44, Copenhagen.

Krarup, T. (1973). "Letters on Molodensky's Problem", Communication to the members of IAG Special Study Group 4.31.

Lanczos, C. (1961). "Linear Differential Operators", Van Nostrand, London.

Lerch, F.J. (1989). "Optimum Data Weighting and Error Calibration for Estimation of Gravitational Parameters", NASA Technical Memorandum 100737.

Meissl, P. (1971). "A Study of Covariance Functions Related to the Earth's Disturbing Potential", Dept. Geodetic Science, 151, The Ohio State University, Columbus.

Mikolaiski, H.-W. (1989). "Synthetische Modelle zur Polbewegung eines deformierbaren Körpers", Deutsche Geodätische Kommission, C-354, München.

Mochizuki, E. (1988). "Spherical Harmonic Development of an Elastic Tensor", Geophys. J. Int. 93, 521-526.

Molodenskii, M.S., V.F. Eremeev, M.I. Yurkina (1962). "Methods for Study of the Ex ternal Gravitational Field and Figure of the Earth", Israel Program for Scientific Translations, Jerusalem.

Moritz, H. (1963). "Interpolation and Prediction of Point Gravity Anomalies", Publ. Isost. Int. Assoc. Geod., 40, Helsinki.

Moritz, H. (1976). "Covariance Functions in Least Squares Collocation", Dept. Geodetic Science, 240, The Ohio State University, Columbus.

Moritz, H. (1980). "Advanced Physical Geodesy", Wichmann, Karlsruhe.

Ohanian, H.C. (1976). "Gravitation and Spacetime", Norton, New York.

Papoulis, A. (1965). "Probability, Random Variables and Stochastic Processes", McGraw-Hill, New York.

Pelka, E.J., D.B. DeBra (1979). "The Effects of Relative Instrument Orientation upon Gravity Gradiometer System Performance", J. Guidance and Control, 2, 1, 18-24.

Phinney, R.A. (1973). "Representation of the Elastic-Gravitational Excitation of a Spherical Earth Model by Generalized Spherical Harmonics", Geophys. J. R. Astr. Soc., 34,4-51-487.

Rapp, R.H., Y.M. Wang, N.K. Pavlis (1991). "The Ohio State 1991 Geopotential and Sea Surface Topography Harmonic Coefficient Models", Dept. Geod. Science & Surveying, 410, The Ohio State University, Columbus.

Reed, G.B. (1973). "Application of Kinematical Geodesy for Determining the Short Wave Length Components of the Gravity Field by Satellite Gradiometry", Department of Geodetic Science, 201, The Ohio State University, Columbus.

Regge, T., J.A. Wheeler (1957). "Stability of a Schwarzschild Singularity", Phys. Rev., 108, 1063-1069.

Rummel, R. (1975). "Downward Continuation of Gravity Information from Satellite Gradiometry in Local Areas", Dept. Geodetic Science, 221, The Ohio State University, Columbus.

Rummel, R. (1979). "Determination of the Short-Wavelength Components of the Gravity Field from Satellite-to-Satellite Tracking or Satellite Gradiometry", manuscripta geodaetica, 4, 107-148.

Rummel, R. (1984). "From the Observational Model to Gravity Parameter Estimation", in: Proc. Beijing Intern. Summer School on Local Gravity Field Approximation, (ed: K.P. Schwarz), Beijing.

Rummel, R. (1986). "Satellite Gradiometry" ,in: Lecture Notes in Earth Sciences, 7 Ma thematical and Numerical Techniques in Physical Geodesy, (ed. H. Sünkel), 318-163, Springer, Berlin.

Rummel, R. (1991). "Fysische Geodesy II (Physical Geodesy 2)", Lecture Notes, Delft University of Technology, Delft.

Rummel, R. (1992). "On the Principles and Prospects of Gravity Field Determination by Satellite Methods" in: Geodesy and Physics of the Earth (eds. H. Montag & Ch. Reigber), IAG-Symposia 112, 67-70, Springer, Heidelberg.

Rummel, R., M. van Gelderen (1992). "Spectral Analysis of the Full Gravity Tensor", Geophys. J. Int., 111, 1, 159-169.

Rummel, R., M. van Gelderen (1995). "Meissl Scheme - Spectral Characteristics of Physical Geodesy", manuscripta geodaetica, 20, 379-385.

Rummel, R., F. Sansò, M. van Gelderen, M. Brovelli, R. Koop, F. Migliaccio, E.J.O. Schrama, F. Sacerdote (1993). "Spherical Harmonic Analysis of Satellite Gradiometry", Netherlands Geodetic Commission, New Series, 39, Delft.

Sansò, F. (1981). "Recent Advances in the Theory of the Geodetic Boundary Value Problem", Rev. Geoph. Space Physics, 19, 3, 437-449.

Sansò, F., R. Rummel (1994). "A Discussion on the Correct Way of Representing White Noise on the Sphere and its Propagation from Estimated Quantities", IAG Section IV Bulletin 1, 3-16.

Schrama, E.J.O. (1989). "The Role of Orbit Errors in Processing of Satellite Altimeter Data", Netherlands Geodetic Commission, New Series, 33, Delft.

Schreiner, M. (1994). "Tensor Spherical Harmonics and Their Application to Satellite Gradiometry", Doktorarbeit, Fachbereich Mathematik, Universität Kaiserslautern.

Schuh, W.-D. (1996). "Tailored Numerical Solution Strategies for the Global Determination of the Earth's Gravity Field", Habilitationsschrift, Technische Universität Graz.

Schwarz, K.-P. (1971). "Numerische Untersuchungen zur Schwerefortsetzung", Deutsche Geodätische Kommission, C-171, München.

Selényi, P. (editor), (1953). "Roland Eötvös gesammelte Arbeiten", Akademiai Kiado, Budapest.

Sneeuw, N. (1994). "Global Spherical Harmonic Analysis by Least-Squares and Numerical Quadrature Methods in Historical Perspective", Geophys. J. Int., 118, 707-716.

Solid-Earth and Ocean Physics (1969). Report of a study at Williamstown, Mass., NASA.

Strang van Hees, G. (1986). "Precision of the Geoid, Computed from Terrestrial Gravity Measurements", manuscripta geodaetica, 11, 1-14.

Sünkel, H. (1981). "Feasibility Studies for the Prediction of the Gravity Disturbance Vector in High Altitudes", Dept. Geodetic Science, 311, The Ohio State University, Columbus.

Svensson, S.L. (1983). "Pseudodifferential Operators - a New Approach to the Boundary Value Problems of Physical Geodesy", manuscripta geodaetica, 8, 1-40.

Thorne, K.S. (1980). "Multipole Expansions of Gravitational Radiation", Rev. Mod. Phys., 52, 2-I, 299-339.

Tikhonov , A.N., V.Y. Arsenin (1977). "Solutions of Ill-Posed Problems", John Wiley & Sons, New York.

Tscherning, C.C. (1976). "Comparison of the Second-Order Derivatives of the Normal Potential Based on the Representation by a Legendre Series", manuscripta geodaetica, 2, 71-92.

Tscherning, C.C., R.H. Rapp (1978). "Closed Covariance Expressions for Gravity Anomalies, Geoid Undulations, and Deflections of the Vertical Implied by Anomaly Degree Variance Models", Dept. Geodetic Science, 208, The Ohio State University, Columbus.

Xu, P.L. (1992). "The Value of Minimum Norm Estimation of Geopotential Fields", Geophys. J. Int., 110, 321-332.

Xu, P.L., R. Rummel (1994a). "Generalized Ridge Regression with Applications in Determination of Potential Fields", manuscripta geodaetica, 20, 1, 8-20.

Xu, P.L., R. Rummel (1994b). "A Simulation Study of Smoothness Methods in Recovery of Regional Gravity Fields", Geophys. J. Int., 117, 472-486.

Yaglom, A.M. (1962). "Stationary Random Functions", Dover, New York.

Zerilli, F.J. (1970). "Tensor Harmonics in Canonical Form for Gravitational Radiation and other Applications", J. Math. Phys., 11, 7, 2203-2208.

Satellite Altimetry, Ocean Dynamics and the Marine Geoid

E.J.O. Schrama

1. Introduction

When the organizers of this summer school asked us to organize these lectures they suggested as a subject the possibility of determining a marine geoid to within 1 centimeter. In this sense this subject is closely related to the title of this summer school which is after all on geodetic boundary value problems. There are many different ways in which one could solve such problems and various examples are given in this volume. Our strategy is to try to solve the problem by careful consideration of satellite altimetry allowing to accurately map the global mean sea surface (MSS) in a geometrical sense.

The word "geometrical" is chosen deliberately to distinguish from other techniques such as the application of an "inverse Stokes type of operator" on geoid heights to yield block averaged altimeter gravity anomalies as was done for the OSU91a model developed by Rapp et al. (1991b). In their case the actual application goes via least-squares collocation although this isn't really relevant for explaining the problem. To yield an eventual gravity model Rapp et al. (1991b) combine the altimeter anomalies with ship and terrestrial anomalies and other satellite gravity information in order to compute a set of spherical harmonic coefficients complete up to degree and order 360.

Here we want to avoid such schemes and intent to exploit geometry and sampling of the MSS which are the most attractive properties of satellite altimetry. After mapping the oceans for several months modern Earth orbiting altimeters can obtain an accuracy of better than 3 cm and resolution of less than 30 km which is utterly unattainable by any terrestrial technique. We will further specify this "3 cm level" which is made up of a variety of effects but also caused by the infrastructure required for altimetry consisting of a precise orbit determination scheme and an altimeter data processing procedure in section 2. The purpose of the data processing scheme is to correct for several well known geophysical and instrumental effects that are clearly visible in the data.

While discussing the altimeter data processing scheme special attention goes to the subject of tides discussed in section 3. To introduce the subject we will start with a straightforward derivation of the tide generating potential whose gradient is responsible for tidal forces causing the solid-Earth and the oceans to deform. This topic requires a discussion on a theory developed in the beginning of this century by Love (1927) explaining the Earth's geometric deformation as well as an induced potential as a result of astronomical forcing. The subject turns out to be more difficult than expected because there is also a contribution at the zero-frequency leading to a permanent deformation of the Earth affecting our definition of coordinate systems.

More difficult is the subject of ocean tides whose dynamics are described
by the Laplace tidal equations derived in appendix A. The Laplace tidal equations show that the deep oceans themselves respond at frequencies identical
to the astronomical frequencies. For the long period ocean tides a direct relation exists with Love's theory. However for the diurnal and semi-diurnal tides
it is required to compute amplitude and phase of the ocean tide depending
on the geographical location. Shallow seas are an exception; in this case the
combination of a dedicated sea tide storm surge finite element model may be
required to adequately deal with the problem of modeling tides.

Finally we see that tides are removed and that altimeter data can be
used to observe a MSS. This surface is by definition equal to the marine
geoid in case the MSS is free of tides, various non-tidal variations and also
additional topography caused by permanent ocean currents. To understand
this constant and time variable part of the MSS it is necessary to introduce
dynamic oceanography, see also section 4. However we will avoid presenting
a complete course so that we constrain ourselves to a number of essential
assumptions and definitions taken from physical oceanography. To introduce
the concept of the mean dynamic topography we will start with the equations
of motions in oceanography. A method for measuring the dynamic topography
by means of hydrographic density measurements is derived and comparisons
between the oceanographic and altimetric estimates of the mean dynamic
topography are discussed.

Yet we don't want to forget the time variable part of the dynamic ocean
topography which can be accurately monitored by means of altimetry. We finish the section on ocean dynamics by showing some of the temporal variations
in this topography observed by means of contemporary altimeter systems, see
also table 1.1.

Table 1.1. Satellite altimeter systems

Mission	Period	Organisation(s)	Height [km]	Inclination [degree]
GEOS-3	1975-1978	NASA	840	115
SEASAT	1978	NASA	800	108
GEOSAT	1985-1990	US Navy	800	108
ERS-1	1991-1996	ESA	800	98.5
TOPEX/Poseidon	1992-	NASA/CNES	1330	66
ERS-2	1995-	ESA	800	98.5

Keywords: Satellite altimetry, tides, sea-surface topography, ocean currents,
gravity field, geoid.

2. Satellite altimetry

2.1 Altimetric measurements

The measurement principle is, conceptually seen, an observation of the short-est distance from the radar antenna phase center to the sea surface beneath the spacecraft (s/c). Contrary to popular belief there is no such thing as the perfect pulse-radar, instead modern altimeters are based upon a frequency modulation (FM) technique where a linear chirp signal with a frequency range of 300 MHz is modulated on a 13 GHz carrier, see also figure 2.1. (The car-rier and modulation frequencies are just mentioned as examples and differ somewhat from the actual frequencies used for the TOPEX Ku-band altime-ter.) After receiving the chirp signal it is multiplied by the transmitted signal which allows us to derive the frequency difference being a measure of dis-tance. Certain ambiguities may occur which are in general avoided by chosing a proper modulation scheme and minimizing the satellite altitude variations with respect to the sea surface. The difference signal labeled "T-R" in fig-ure 2.1 is then provided to a Fast Fourier Transform processor returning the raw waveform samples. From this figure it is obvious that the inverse Fourier

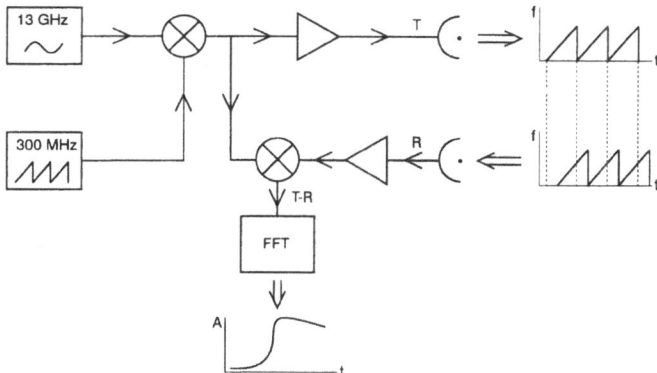

Fig. 2.1. Altimeter schematics based upon linear frequency modulation. Figure provided by courtesy of A. Smits.

transform of the "T-R" signal is equivalent to a phase (or distance) mea-surement of two saw-tooth signals and that the FFT processor will simply return a histogram of observed ranges. These radar waveform samples should be seen as the response of a sea surface equipped with wind waves to a short (but not infinitely short) radar pulse.

Normally, far too many raw waveform samples are generated so that statistical filtering methods based upon alpha-beta trackers or Kalman filters are applied to smooth and compress the data stream. This procedure is executed by the s/c and is optionally controlled from ground for certain altimeter systems. For TOPEX/Poseidon (see the JGR oceans special issues of December 1994 and 1995) one obtains 10 Hz ocean-mode waveform data which include range estimates. For ERS-1 and ERS-2 there are two programmable modes, one for flat ocean surfaces and another for rugged ice surfaces. The altimeter ice-mode is designed around the philosophy of measuring a wider domain of distances with decreased accuracies. This is accomplished by reducing the saw-tooth frequency range and relaxation of leading edge alignment criteria during the statistical processing of the raw waveform samples.

The linear FM radar technique describe above has the main advantage that power-hungry pulse radar methods may be avoided and that low-power solid-state electronics may be applied reducing the cost of implementing the radar altimeter. Clearly the radar waveform data are nothing more than a distribution of the reflected ranges in defined time slots. The typical shape of the radar waveforms is of course determined by the properties of the illuminated surface as a reflector, the antenna divergence and the off-nadir pointing angle of the altimeter. This illuminated sea surface, or radar footprint, is between 2 and 10 km in diameter depending on the state of the sea surface.

2.2 Radar correction algorithms

Some of the principles on which the corrections are based are introduced although it is not our intention to discuss in full detail all algorithms. Nevertheless these corrections are *essential for obtaining high precision measurements* and determine to a certain point the accuracy that can be achieved. The most frequently applied corrections concern the altimeter EM-bias, ionospheric and tropospheric delays, and the inverse barometer effect.

2.2.1 EM-Bias. The sea surface is a well defined reflector causing the radar waveform samples to be determined by surface wind waves in the radar footprint. A measure for the roughness of the surface is the so-called significant wave height (SWH) parameter, see for instance Hayne et al. (1994). The SWH varies between the lowest values of approximately 0.5 meters to up to 20 meters in the extremes with a global average of about 2 to 3 meters, see also Rummel (1993). Very low SWH values are usually indicative of ice reflections that must be removed from the radar signal. Extreme SWH values are usually indicative of extreme wind waves where the validity of the EM-bias correction is a problem. This correction is the result of the asymmetric shape of the sea surface since more radar signal is reflected from the troughs than the crests. It inherently leads to an electromagnetic bias or EM-bias since the measured surface will appear somewhat lower than the real surface. In reality the value of the EM-bias must be estimated from the SWH parameter prior to using the altimeter data. More sophisticated algorithms for the EM-bias correction incorporate knowledge about the wind speed (U) at the sea surface. Here it is interesting that U is estimated from other characteristics of the radar waveform samples, see also Brown (1977). Some remarks:

- Typically the EM-bias correction is of the order of -1% to -4% of the SWH with an uncertainty of the order of 2 cm (one sigma).
- The EM-bias correction should not be confused with the sea-state bias correction since the latter also includes the waveform tracker biases, cf. Gaspar et al. (1994). An example is the 4 parameter model (BM4) developed by Gaspar et al. (1994) where the sea state bias correction (SSB) is computed as:

$$\text{SSB}_m = \text{SWH} \left[a_0 + a_1 U + a_2 U^2 + a_3 \text{SWH} \right]$$

where $a_0 = -0.019$, $a_1 = -0.0037$, $a_2 = 0.00014$ and $a_3 = 0.0027$.

- The EM-bias of the altimeter is seen as one of the major limitations of altimetry not really allowing instantaneous height readings better than 2 cm (one sigma) over the oceans.

2.2.2 Media effects. The radar signal itself has to travel twice through the ionosphere and troposphere where the refractive index differs from 1. For frequencies in the electromagnetic spectrum near 13 GHz the ionospheric correction is usually less than 30 cm so that most altimeter missions have to rely on the presence of global ionospheric models to compute the correction. The accuracy of the global ionospheric models is of the order of 90% meaning that approximately 3 cm noise is introduced by this way of correcting the data. With the advent of the TOPEX dual-frequency altimeter measurements became available allowing the removal of the first-order ionospheric delay because of the dispersive nature of the ionosphere. Nevertheless, along-track filtering is necessary to reduce the inherent noise of the dual-frequency measurements. After this filtering process, the ionospheric error is typically less than 1 cm for the TOPEX altimeter.

Unfortunately dispersion of radar signals doesn't occur in case of the tropospheric signal delay. Usually two effects are distinguished here: the dry tropospheric correction describes the delay of the radar signal caused by the presence of dry air and the wet equivalent is related to the presence of water vapor. The latter should not be confused with water droplets in the atmosphere which are known to scatter the radar signal and in the extreme case of rain cells to even completely block the return signal.

The dry tropospheric correction amounts to approximately 2.3 meters and mainly depends upon air pressure for which meteorologic models are employed. As a rule of thumb meteorological models provide air pressure to within 3 mbar although they sometimes miss the extremes encountered in tropical cyclones. The 3 mbar is therefore a global number which should be compared to the nominal value of 1013.6 mbar of the standard atmosphere. This means that the relative accuracy of the dry tropospheric correction is no more than 0.3% which translates to a distance of 7 millimeters.

The wet tropospheric correction to the radar signal is much more of a problem. The nominal delay is roughly 30 cm and worst of all, if meteorologic models were used to compute a correction then 60% is an optimistic estimate for the relative accuracy of the correction. In practice this means that more than 5 cm noise easily remains thereby introducing one of the biggest difficulties in designing an accurate altimeter system. The remedy is to install a so-called water vapor radiometer (WVR) on the spacecraft to measure the brightness temperatures of the Earth at two or three lines in the electromagnetic spectrum near the water vapor absorption line at 22 GHz. Some altimeter systems, such as GEOSAT (1985–1990), did not carry a WVR and meteorological models or external oblique spaceborne radiometric data (e.g. SSM/I or NIMBUS data) had to be used to provide a correction. With the aid of WVR data on TOPEX/Poseidon and both ERS systems, the wet tropospheric correction can usually be measured to within 2 cm.

2.2.3 Inverse Barometer Correction. Apart from its role in computing the dry tropospheric range correction, air pressure will affect the sea level which responds as an inverse barometer. In this case we will see that there is a linear relation of about -1 cm per mbar; the minus sign tells that the sea level is depressed by 1 cm when air pressure is increased by 1 mbar, hence explaining its name: the inverse barometer effect. The practical way of dealing with the problem is to employ meteorological models such as provided by the ECMWF so that the inverse barometer correction itself may be computed to within 3 cm or so. Nevertheless some remarks should be made: a) theoretically the inverse barometer correction is more complicated than a simple linear response; the reason is that the barometric effect is forcing the sea surface so that the pressure gradient, Coriolis force and frictional forces should be taken into account, b) it turns out that the inverse barometric response is not very effective on time scales less than a day or so, c) on the 1 mbar level tidal signals exist in the atmosphere and one should find out whether the ocean tide model is in agreement with the pressure models being applied, d) in the tropics the natural variation in air pressure is small compared to other regions on Earth and statistical analysis of altimeter data, ie. comparison of air pressure variations against height variations of the sea surface in the tropics has shown that the -1 cm per mbar response is not always valid, cf. Fu and Pihos (1994).

2.2.4 Altimeter timing bias. The effect of the altimeter timing bias is as straightforward as multiplying the vertical speed of the s/c above the sea surface by the value of the timing bias. Consequently in order to obtain acceptable values of less than 1 cm it is required to get the time tags to within 500 μsec. However it turns out that there are no fail-safe engineering solutions to circumvent such problems other than to calibrate the s/c clock before launch and to continuously monitor the same clock during flight via a communications channel with its own internal delays. A practical way of solving the altimeter timing bias problem was suggested by Marsh and Williamson (1982) and is based upon estimating a characteristic lemniscate function that will show up in the altimeter profile. Nevertheless ERS-1 still exhibits time tag variations at the 1.1 msec level according to Scharroo (private communications) which corresponds to 2.5 cm mostly at two cycles per revolution. Provided that the s/c contains a GPS flight receiver, a preferable approach would be to rely on external GPS timing control which is normally better than 50 nsec.

2.3 Precise orbit determination

Any altimeter satellite places extreme requirements on the quality of orbital ephemeris where the goal is to compute the position of the center of mass of the spacecraft to within 3 centimeters. This task turns out to be a very difficult geodetic problem that did not achieve the desired radial orbit accuracy for many years. Initially the radial position error of the SEASAT and GEOS-3 altimeter satellites was typically 1.5 meters, manifesting itself as tracks in altimetric surfaces which are clearly identified as 1 cycle per revolution orbit errors. It turned out that these radial orbit excursions were mainly caused by the limited accuracy of existing gravity models as is described in Schrama (1989).

However altimetry as a technique was not useless because of poor orbits. Considerable effort went into the design of efficient processing techniques to eliminate the radial orbit effect from the data using empirical methods. Collinear track differences are insensitive to most gravity modeling errors and rather efficient adjustment techniques enable the removal of radial errors between overlapping altimeter tracks, see also Schrama (1989). Several papers have shown that such processing schemes result in realistic estimates of the oceanic mesoscale variability, which normally doesn't exceed the level of approximately 20 to 30 centimeters, see also Chelton et al. (1983).

Other processing schemes are based on a minimization of cross-over differences which are obtained as the sea surface height difference measured by the altimeter at intersecting ground tracks (cross-over points). In an attempt to reduce the orbit error, linear trend functions may be estimated from relatively short and intersecting orbits. Another possibility is to represent the radial orbit error as a Fourier series for a continuous altimeter arc spanning a full repeat cycle. A summary on the efficiency of such minimization procedures is discussed in more detail in Schrama (1992), who also mentions the null-space problem of cross-over adjustments. In fact the problem is rather similar to determining absolute heights from leveling networks where the measurements are always provided as height differences between two stations. The mathematical solution is to apply at least one constraint, known as a datum, which will determine the height offset of the network. However in the case of cross-over minimizations the datum problem is ill-posed and fully depends on the assumption of orbit error trend functions. Due to datum defects only a partial recovery of the orbit error function is feasible which will obscure the long-wavelength behavior of the sea surface modeled from altimeter data.

There are several reasons for not applying collinear or cross-over adjustment techniques in contemporary precision orbit determination schemes. A first reason is that we have seen significant advances in modeling the Earth's gravitational field. The older gravity models, such as GEM10b, cf. Lerch et al. (1981), were simply not adequate in describing the rich spectrum of perturbations of an orbiter at 800 km height such as ERS-1, GEOSAT and SEASAT. The Joint Gravity Model 2, (JGM-2, also named after the late James G. Marsh who was one of the early pioneers in precision orbit determination and satellite altimetry), is now complete to degree and order 70, cf. Nerem et al. (1994b). A second reason is in the design of nominal orbits being used. For instance, TOPEX/Poseidon flies at an altitude of 1330 km which inherently dampens out gravity modeling errors. Finally state of the art tracking systems such as DORIS (a French Doppler tracking system on TOPEX/Poseidon) and even spaceborne applications of GPS, cf. Bertiger et al. (1994) provide dramatic improvements. The result is that the orbit of TOPEX/Poseidon can be modeled to less than 2.5 cm rms which has completely revolutionized the processing strategy, and more importantly, our understanding of the oceans after interpretation of the altimeter data.

Nevertheless there are still a number of open problems that could stimulate future research. Efforts to further reduce any orbit errors are likely to continue hopefully reducing the effect to the sub-centimeter level. Another possibility are autonomous orbit determination systems that allow real time broadcast of sea level measurements thereby increasing the applicability of altimeter data in real time models such as shallow water models for predicting tides and storm surges.

3. Tides

3.1 Tide generating potential

It was suggested in Newton's Principia (1687) that the difference between the gravitational attraction of the Moon (and the Sun) at the Earth's surface and the Earth's center are responsible for tides, see also figure 3.1. According to

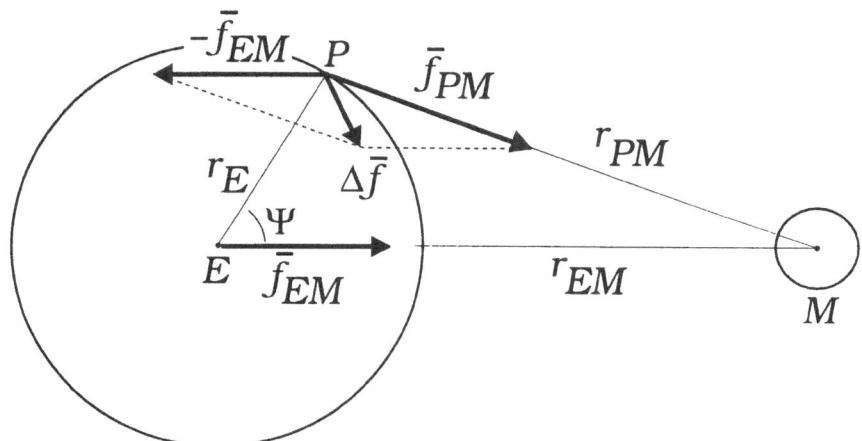

Fig. 3.1. The lunar gravitational force is separated into two components, namely $\bar{f}_{EM}$ and $\bar{f}_{PM}$ whose difference is responsible for the tidal force $\Delta \bar{f}$. Knowledge of the Earth's radius r_E, the Earth-Moon distance r_{EM} and the angle ψ is required to be able to compute a tide generating potential U whose gradient ∇U corresponds to the vector $\Delta \bar{f}$. Figure provided by courtesy of A. Smits.

this definition of the astronomical tides the corresponding tidal acceleration $\Delta \bar{f}$ becomes:

$$\Delta \bar{f} = \bar{f}_{PM} - \bar{f}_{EM} \tag{3.1}$$

whereby $\bar{f}_{PM}$ and $\bar{f}_{EM}$ are caused by the gravitational attraction of the Moon M. Implementation of eq. (3.1) is as straightforward as computing the lunar ephemeris and evaluating Newton's gravitational law, but is in practice not applied because it is more convenient to use a tide generating potential U whose gradient ∇U corresponds to $\Delta \bar{f}$ in eq. (3.1). To derive U we start with a Taylor series of $U = \mu_M / r$ developed at point E in figure 3.1 where μ_M is the Moon's gravitational constant and r the radius of a vector originating at point M. The first-order approximation of this Taylor series is:

$$\Delta \bar{f} = \frac{\mu_M}{r_{EM}^3} \begin{pmatrix} 2 & 0 & 0 \\ 0 & -1 & 0 \\ 0 & 0 & -1 \end{pmatrix} \begin{pmatrix} x \\ y \\ z \end{pmatrix} \tag{3.2}$$

where the vector $(x, y, z)^T$ is originating at point E and whereby x is running from E to M. We compute the potential U by evaluation of the work integral:

$$U^a = \int_{s=0}^{r_E} (\Delta \overline{f}, \overline{n}) \, ds \qquad (3.3)$$

under the assumption that it is evaluated at an idealized spherical surface with radius r_E. In our case $\overline{n}$ dictates the direction in which we have to integrate. Keeping in mind the situation depicted in figure 3.1 a logical choice is:

$$\overline{n} = \begin{pmatrix} \cos\psi \\ \sin\psi \\ 0 \end{pmatrix} \quad \text{and} \quad \begin{pmatrix} x \\ y \\ z \end{pmatrix} = \begin{pmatrix} s\cos\psi \\ s\sin\psi \\ 0 \end{pmatrix} \qquad (3.4)$$

so that $(\Delta \overline{f}, \overline{n})$ becomes:

$$(\Delta \overline{f}, \overline{n}) = \frac{\mu_M}{r_{EM}^3} \begin{pmatrix} 2s\cos\psi \\ -s\sin\psi \\ 0 \end{pmatrix} \cdot \begin{pmatrix} \cos\psi \\ \sin\psi \\ 0 \end{pmatrix} = \frac{s\mu_M}{r_{EM}^3} \left\{ 3\cos^2\psi - 1 \right\}$$

It follows that:

$$U^a = \int_{s=0}^{r_E} \frac{s\mu_M}{r_{EM}^3} \left\{ 3\cos^2\psi - 1 \right\} \, ds = \frac{\mu_M r_E^2}{r_{EM}^3} P_2(\cos\psi) \qquad (3.5)$$

which is the first term in the Taylor series where $P_2(\cos\psi)$ is the Legendre function of degree 2. But there are more terms, essentially because eq. (3.5) is of first-order. Another example is:

$$\Delta \overline{f_i} = \frac{\partial^3 U}{\partial x_i \partial x_j \partial x_k} \frac{\Delta x_j \Delta x_k}{3!} \qquad (3.6)$$

where $U = \mu/r$ for $i, j, k = 1, \cdots, 3$. Without any further proof we mention that the second term in the series derived from eq. (3.6) becomes:

$$U^a(n = 3) = \frac{\mu_M r_E^3}{r_{EM}^4} P_3(\cos\psi) \qquad (3.7)$$

By induction one can show that:

$$U^a = \frac{\mu_M}{r_{EM}} \sum_{n=2}^{\infty} \left(\frac{r_E}{r_{EM}} \right)^n P_n(\cos\psi) \qquad (3.8)$$

represents the entire series describing the tide generating potential. In the case of the Earth-Moon system $r_E = \frac{1}{60} r_{EM}$ so that rapid convergence of eq. (3.8) is ensured. Therefore in practice it doesn't make sense to continue the summation in eq. (3.8) beyond $n = 3$.

3.1.1 Some remarks. At the moment we can draw the following conclusions from eq. (3.8):

- The displacements of the equipotential surfaces due to the tide generating potential are for the Moon: 36.2 cm for $n = 2$ and 6.03 mm for $n = 3$ and for the Sun: 16.5 cm for $n = 2$ and 7.03 μm for $n = 3$.
- The $P_2(\cos\psi)$ term in the equation (3.8) resembles an ellipsoid with its main bulge pointing towards the astronomical body causing the tide. This is the main tidal effect which is, if caused by the Moon, at least 60 times larger than the $n = 3$ term in equation (3.8).
- The Sun and Moon are the largest contributors, tidal effects of other bodies in the solar system can be ignored.
- U^a is unrelated to the Earth's gravity field or the acceleration experienced by the Earth orbiting the Sun. Unfortunately there exist many confusing popular science explanations on this subject.
- The result of equation (3.8) is that astronomical tides seem to occur at a rate of 2 highs and 2 lows per day. The reason is of course due to Earth rotation since the Moon and Sun only move by respectively $\approx 13°$ and $\approx 1°$ per day compared to the 359.02° per day caused by the Earth's spin rate.
- Astronomical tides are too simple to explain what is really going on in nature, more on this issue will be explained later on in these notes.

3.1.2 Darwin symbols and Doodson numbers. Since equation (3.8) mainly depends on the astronomical positions of Sun and Moon it is not really suitable for applications where the tidal potential is required. A more practical approach was developed by Darwin (1883) who invented the harmonic method of tidal analysis and prediction and since then it is customary to assign "letter-digit combinations", also known as Darwin symbols, to certain main lines in a so-called spectrum of tidal lines. The M_2 symbol is a typical example; it symbolizes the most energetic tide caused by the Moon at a twice-daily frequency. Later Doodson (1921) calculated an extensive table of spectral lines which can be linked to the original Darwin symbols. With the advent of computers in the seventies, Cartwright and Edden (1973) and earlier Cartwright and Tayler (1971) (hereafter CTE), computed new tables to verify the earlier work of Doodson (1921). The tidal lines in these tables are identified by means of so-called Doodson numbers D which are "computed" in the following way:

$$D = k_1(5 + k_2)(5 + k_3).(5 + k_4)(5 + k_5)(5 + k_6) \tag{3.9}$$

where $k_1, ..., k_6$ is an array of small integers, corresponding with the description shown in table 3.1. In principle there exist infinitely many Doodson number although in practice only a few hundred lines are significant. To facilitate the discussion about this phenomenon it is common-practice to divide the table in several parts: a) All tidal lines with equal k_1, which happens to be the same as the order m in spherical harmonics (more on this issue in section 3.1.3), are said to form a *species*. The species indicated with $m = 0, 1, 2$ correspond to respectively long period, daily and twice-daily effects. b) All tidal lines with equal k_1 and k_2 terms are said to form *groups*. c) All lines with equal k_1, k_2 and k_3 terms are said to form *constituents*. In reality it is not necessary to go any further than the constituent level so that a year worth of tide gauge data can be used to define the amplitude and phase of a constituent.

3.1.3 Tidal harmonic coefficients. We computed a table of tidal harmonic coefficients under the assumption that accurate planetary ephemeris in the form of DE200/LE200, see also Standish (1990), are provided and that the Doodson numbers are prescribed rather than that they are selected by filtering techniques as in CTE. We recall that the tide generating potential U can be written in the following form:

$$U^a = \frac{\mu_M}{R_{em}} \sum_{n=2,3} \left(\frac{R_e}{R_{em}} \right)^n P_n(\cos \psi) \tag{3.10}$$

The first step in realizing the conversion of equation (3.10) is to apply the addition theorem on the $P_n(\cos \psi)$ functions so that:

$$P_n(\cos \psi) = \sum_{m=0}^{n} \sum_{a=0}^{1} \frac{1}{2n+1} Y_{nma}(\theta_m, \lambda_m) Y_{nma}(\theta_p, \lambda_p) \tag{3.11}$$

where Y_{nma} are normalized surface harmonics, θ_m, λ_m co-latitude and eastern longitude of the sub-lunar or sub-solar point, and θ_p, λ_p the corresponding coordinates of the point at which the tidal potential is computed. Substitution of equation (3.11) into (3.10) results in the following formulation:

$$U^a = \sum_{n=2,3} \sum_{m=0}^{n} \sum_{a=0}^{1} \frac{\mu_m (R_e/R_{em})^n}{(2n+1)R_{em}} Y_{nma}(\theta_m, \lambda_m) Y_{nma}(\theta_p, \lambda_p) \tag{3.12}$$

where:

$$Y_{nm0}(\theta, \lambda) = \cos(m\lambda) \overline{P}_{nm}(\cos \theta) \tag{3.13}$$

$$Y_{nm1}(\theta, \lambda) = \sin(m\lambda) \overline{P}_{nm}(\cos \theta) \tag{3.14}$$

with the overbar indicating that we are using normalized associated Legendre functions:

$$\overline{P}_{nm}(\cos \theta) = \left[(2 - \delta_{0m})(2n + 1) \frac{(n - m)!}{(n + m)!} \right]^{1/2} P_{nm}(\cos \theta) \tag{3.15}$$

Eq. (3.12) should now be related to the CTE equation for the tide generating potential:

$$U^a = g \sum_{n=2}^{3} \sum_{m=0}^{n} c_{nm}(\lambda_p, t) f_{nm} P_{nm}(\cos\theta_p) \qquad (3.16)$$

where $g = \mu/R_e^2$ and for $(n+m)$ even:

$$c_{nm}(\lambda_p, t) = \sum_v H^{(v)} \times [\cos(X_v)\cos(m\lambda_p) - \sin(X_v)\sin(m\lambda_p)] \qquad (3.17)$$

while for $(n+m)$ odd:

$$c_{nm}(\lambda_p, t) = \sum_v H^{(v)} \times [\sin(X_v)\cos(m\lambda_p) + \cos(X_v)\sin(m\lambda_p)] \qquad (3.18)$$

where it is assumed that:

$$f_{nm} = (2\pi N_{nm})^{-1/2} (-1)^m \qquad (3.19)$$

and:

$$N_{nm} = \frac{2}{(2n+1)} \frac{(n+m)!}{(n-m)!} \qquad (3.20)$$

We must also specify the summation over the variable v and the corresponding definition of X_v. In total there are approximately 400 to 500 different terms in the summation of v each consisting of a linear combination of six astronomical elements:

$$X_v = k_1 w_1 + k_2 w_2 + k_3 w_3 + k_4 w_4 - k_5 w_5 + k_6 w_6 \qquad (3.21)$$

where $k_1 \ldots k_6$ are integers and:

$$
\begin{array}{rcrcr}
w_2 & = & 218.3164 & + & 13.17639648\ T \\
w_3 & = & 280.4661 & + & 0.98564736\ T \\
w_4 & = & 83.3535 & + & 0.11140353\ T \\
w_5 & = & 125.0445 & - & 0.05295377\ T \\
w_6 & = & 282.9384 & + & 0.00004710\ T \\
\end{array}
$$

where T is provided in Julian days relative to January 1, 2000, 12:00 ephemeris time. (When working in UT this reference modified Julian date equals to 51544.4993.) Finally w_1 is computed as follows:

$$w_1 = 360 * U + w_3 - w_2 - 180.0$$

where U is given in fractions of days relative to midnight. In tidal literature one usually finds the classification of w_1 to w_6 as shown in table 3.1 where it must be remarked that w_5 is retrograde whereas all other elements are prograde. This explains the minus sign in equation (3.21).

The method used to compute tidal harmonics in CTE differs from the approach used here. In contrast to CTE, who used several convolution operators to separate tidal groups, we will show an algorithm that depends on a least-squares assumption and prior knowledge of all Doodson numbers in the summation over v. To obtain the coefficients $H^{(v)}$ for each Doodson number the following procedure is used:

Table 3.1. Classification of frequencies in tables of tidal harmonics

Here[1]	Frequency[2]	Cartwright. Doodson[3]	Explanation[4]
k_1, w_1	daily	τ, τ	mean time angle in lunar days
k_2, w_2	monthly	q, s	mean longitude of the moon
k_3, w_3	annual	q', h	mean longitude of the sun
k_4, w_4	8.85 yr	p, p	mean longitude of lunar perigee
k_5, w_5	18.61 yr	$N, -N'$	mean longitude of ascending lunar node
k_6, w_6	20926 yr	p', p_1	mean longitude of the sun at perihelion

[1] notation used in the Doodson number,
[2] frequency of the effect,
[3] notation used in tidal literature,
[4] explanation on the w_i variables.

- For each degree n and tidal species m (which equals k_1) the algorithm starts to collect all matching Doodson numbers.
- The following step is to generate values of:

$$U_{nm}^a(t) = \frac{\mu_b (R_e/R_{eb}(t))^n}{(2n+1)R_{eb}(t)} \times P_{nm}(\cos\theta_b(t)) \times \cos(m\lambda_b(t))$$

where t is running between 1990/1/1 00:00 and 2010/1/1 00:00 in a sufficiently dense number of steps to avoid undersampling. The DE200/LE200 positions of Sun and Moon are used to compute the distance $R_{eb}(t)$ between the astronomical body and the Earth's center and are transformed into Earth-fixed coordinates to obtain $\theta_b(t)$ and $\lambda_b(t)$.
- The following step is to apply a least-squares analysis of the $U_{nm}(t)$ quantity where the observations equations are as follows:

$$U_{nm}^a(t) = \sum_{v'} G^{(v')} \cos(X_{v'})$$

when $m+n$ is even and

$$U_{nm}^a(t) = \sum_{v'} G^{(v')} \sin(X_{v'})$$

whenever $m+n$ is odd. The v' symbol is used to indicate that we are only considering the appropriate subset of Doodson numbers to generate the $X_{v'}$ values.
- Finally the $G^{(v')}$ values need a scaling factor to convert them into numbers that have the same dimension as one finds in CTE. Partly this conversion is caused by a different normalization between surface harmonics used in CTE and eqns. (3.16), (3.17) and (3.18) here, although is it also required to take into account the factor g. As a result:

$$H^{(v')} = G^{(v')} g^{-1} f_{nm}^{-1} \Pi_{nm}^2$$

where Π_{nm} is the normalization factor in eq. (3.15) and f_{nm} the normalization factor used by CTE given in eqns. (3.19) and (3.20). In our algorithm g is computed as μ/R_e^2 where $\mu = 3.9860044 \times 10^{14}$ $[m^3/s^2]$ and $R_e = 6378137.0$ $[m]$.

Of all collected spectral lines we show in table 3.2 and 3.3 only those where $|H^{(v)}|$ exceeds the value of 0.0025. Tables 3.2 and 3.3 show in columns 2 to 7 the values of k_1 till k_6, in column 8 the degree n, in column 9 the coefficient H^v in equations (3.17) and (3.18), in column 10 the Darwin symbol provided that it exists, and in column 11 the Doodson number. Some remarks about the tables: a) The tables hold almost exactly in the time period indicated earlier in this section, but their first four decimal places are accurate enough for ordinary purposes for at least AD 1900-2000, b) There are small differences, mostly in the 5th digit behind the period, with respect to the values given in Cartwright (1993), c) In total we have used 484 spectral lines although a number of additional lines may be observed by means of a cryogenic gravimeter.

3.2 Love's theory of the deforming Earth

Imagine that the solid-Earth itself is somehow deforming under tidal forces, i.e. gradients of the tide generating potential. This is not unique to our planet, *all* bodies in the universe experience the same effect. Notorious are moons in the neighborhood of big planets where the tidal forces can exceed the maximal allowed stress causing the Moon to collapse.

Tidal deformation effects of the solid-Earth were extensively studied by Love (1927) who found that an applied astronomical tide potential:

$$U^a = \sum_n U_n^a = \sum_n U_n'(r) S_n \exp(j\sigma t) \qquad (3.22)$$

where S_n is a surface harmonic, will result in a deformation at the surface of the Earth:

$$\overline{u}_n(R) = g^{-1} \left[h_n(R) S_n \overline{e}_r + l_n(R) \nabla S_n \overline{e}_t \right] U_n'(R) \exp(j\sigma t) \qquad (3.23)$$

where $\overline{e}_r$ and $\overline{e}_t$ are radial and tangential unit vectors. The *indirect* potential caused by this solid-Earth tide effect will be:

$$\delta U(R) = k_n(R) U_n'(R) S_n \exp(j\sigma t) \qquad (3.24)$$

Equations (3.23) and (3.24) contain so-called Love numbers h_n, k_n and l_n describing the "geometric radial", "indirect potential" and "geometric tangential" effects. Finally we remark that Love numbers can be computed from geophysical Earth models and also from geodetic space technique such as VLBI, see table 3.4. Loading is described by separate Love numbers h_n', k_n' and l_n' that will be mentioned later on in these notes.

Table 3.2. Table of tidal harmonics for $m = k_1 = 0, 1$ and $n = 2$ only

	k_1	k_2	k_3	k_4	k_5	k_6	n	$H^{(v)}$	Darwin	Doodson
1	0	0	0	0	0	0	2	-0.31459	$M_0 + S_0$	055.555
2	0	0	0	0	1	0	2	0.02793		055.565
3	0	0	1	0	0	-1	2	-0.00492	Sa	056.554
4	0	0	2	0	0	0	2	-0.03099	Ssa	057.555
5	0	1	-2	1	0	0	2	-0.00673		063.655
6	0	1	0	-1	-1	0	2	0.00231		065.445
7	0	1	0	-1	0	0	2	-0.03518	Mm	065.455
8	0	1	0	-1	1	0	2	0.00228		065.465
9	0	2	-2	0	0	0	2	-0.00584		073.555
10	0	2	0	-2	0	0	2	-0.00288		075.355
11	0	2	0	0	0	0	2	-0.06660	Mf	075.555
12	0	2	0	0	1	0	2	-0.02761		075.565
13	0	2	0	0	2	0	2	-0.00258		075.575
14	0	3	-2	1	0	0	2	-0.00242		083.655
15	0	3	0	-1	0	0	2	-0.01275		085.455
16	0	3	0	-1	1	0	2	-0.00529		085.465
17	0	4	-2	0	0	0	2	-0.00204		093.555
18	1	-3	0	2	0	0	2	0.00664		125.755
19	1	-3	2	0	0	0	2	0.00801	σ_1	127.555
20	1	-2	0	1	-1	0	2	0.00947		135.645
21	1	-2	0	1	0	0	2	0.05019	Q_1	135.655
22	1	-2	2	-1	0	0	2	0.00953	ρ_1	137.455
23	1	-1	0	0	-1	0	2	0.04946		145.545
24	1	-1	0	0	0	0	2	0.26216	O_1	145.555
25	1	-1	2	0	0	0	2	-0.00343		147.555
26	1	0	0	-1	0	0	2	-0.00741		155.455
27	1	0	0	1	0	0	2	-0.02062	M_1	155.655
28	1	0	0	1	1	0	2	-0.00414		155.665
29	1	0	2	-1	0	0	2	-0.00394		157.455
30	1	1	-3	0	0	1	2	0.00713	π_1	162.556
31	1	1	-2	0	0	0	2	0.12199	P_1	163.555
32	1	1	-1	0	0	1	2	-0.00288	S_1	164.556
33	1	1	0	0	-1	0	2	0.00730		165.545
34	1	1	0	0	0	0	2	-0.36872	K_1	165.555
35	1	1	0	0	1	0	2	-0.05002		165.565
36	1	1	1	0	0	-1	2	-0.00292	ψ_1	166.554
37	1	1	2	0	0	0	2	-0.00525	ϕ_1	167.555
38	1	2	-2	1	0	0	2	-0.00394	τ_1	173.655
39	1	2	0	-1	0	0	2	-0.02062	J_1	175.455
40	1	2	0	-1	1	0	2	-0.00409		175.465
41	1	3	-2	0	0	0	2	-0.00342		183.555
42	1	3	0	0	0	0	2	-0.01128	OO_1	185.555
43	1	3	0	0	1	0	2	-0.00723		185.565
44	1	4	0	-1	0	0	2	-0.00216		195.455

Table 3.3. Table of tidal harmonics for $m = k_1 = 2$ and $n = 3$.

	k_1	k_2	k_3	k_4	k_5	k_6	n	$H^{(v)}$	Darwin	Doodson
45	2	-3	2	1	0	0	2	0.00467		227.655
46	2	-2	0	2	0	0	2	0.01601	$2N_2$	235.755
47	2	-2	2	0	0	0	2	0.01932	μ_2	237.555
48	2	-1	0	1	-1	0	2	-0.00451		245.645
49	2	-1	0	1	0	0	2	0.12099	N_2	245.655
50	2	-1	2	-1	0	0	2	0.02298	ν_2	247.455
51	2	0	-1	0	0	1	2	-0.00217		254.556
52	2	0	0	0	-1	0	2	-0.02358		255.545
53	2	0	0	0	0	0	2	0.63194	M_2	255.555
54	2	1	-2	1	0	0	2	-0.00466		263.655
55	2	1	0	-1	0	0	2	-0.01786	L_2	265.455
56	2	1	0	1	0	0	2	0.00447		265.655
57	2	2	-3	0	0	1	2	0.01719	T_2	272.556
58	2	2	-2	0	0	0	2	0.29401	S_2	273.555
59	2	2	-1	0	0	-1	2	-0.00246		274.554
60	2	2	0	0	0	0	2	0.07992	K_2	275.555
61	2	2	0	0	1	0	2	0.02382		275.565
62	2	2	0	0	2	0	2	0.00259		275.575
63	2	3	0	-1	0	0	2	0.00447		285.455
64	0	1	0	0	0	0	3	-0.00375		065.555
65	1	0	0	0	0	0	3	0.00399		155.555
66	2	-1	0	0	0	0	3	-0.00389		245.555
67	2	1	0	0	0	0	3	0.00359		265.555
68	3	-1	0	1	0	0	3	-0.00210		345.655
69	3	0	0	0	0	0	3	-0.00765		355.555

Table 3.4. Love numbers derived from two Earth models.

n	h_n	k_n	l_n	h_n	k_n	l_n
2	0.612[a]	0.303[a]	0.0855[a]	0.611[b]	0.304[b]	0.0832[b]
3	0.293[a]	0.0937[a]	0.0152[a]	0.289[b]	0.0942[b]	0.0145[b]
4	0.179[a]	0.0423[a]	0.0106[a]	0.175[b]	0.0429[b]	0.0103[b]

[a] Derived from the Dziewonski and Anderson (1981) model.
[b] Derived from the Gutenburg-Bullen model, see Farrell (1972).

3.2.1 Solid Earth tides. According to equations (3.23) and (3.24) the solid-Earth itself will deform under tidal forces. Well observable is the vertical effect resulting in height variations at geodetic stations. The so-called solid-Earth tide ζ_s follows directly from Love's theory:

$$\zeta_s = g^{-1} \sum_{n=2}^{\infty} h_n U_n^a \tag{3.25}$$

The correction itself has an amplitude of ≈ 30 cm and is probably accurate to within 1 percent so that one doesn't have to worry about errors in excess of a couple of millimeters for the altimetric experiment.

3.2.2 Long period equilibrium tides in the ocean. At periods substantially longer than 1 day the oceans are believed to be in equilibrium with respect to the astronomical tides. But also here the situation is more complicated than one immediately expects from equation (3.8) due to the existence of k_n in equation (3.24). For this reason long period equilibrium tides in the oceans are derived by:

$$\zeta_e = g^{-1} \sum_{n} (1 + k_n - h_n) U_n^a \tag{3.26}$$

essentially because the term $(1 + k_n)$ dictates the geometrical shape of the oceans due to the tide generating potential but also the indirect or induced potential $k_n U_n^a$. Still there is a need to include $-h_n U_n^a$ since ocean tides are always relative to the sea floor or land which is already experiencing the solid-Earth tide effect ζ_s described in equation (3.25). Again we emphasize that equation (3.26) is only representative for a long periodic response of the ocean tide which is believed to be in equilibrium. Hence equation (3.26) must only be applied to all $m = 0$ terms in the tide generating potential.

3.2.3 Permanent tidal deformations. In view of the direct (U_2^a) and the indirect tide potential $(k_2 U_2^a)$ at zero frequency we must be careful in defining their contribution at Doodson number 055.555 where it turns out that:

$$g^{-1} U_2^a = -0.19844 \times P_{20}(\sin\phi) \tag{3.27}$$

$$g^{-1} k_2 U_2^a = -0.06013 \times P_{20}(\sin\phi) \tag{3.28}$$

$$g^{-1}(1 + k_2) U_2^a = -0.25857 \times P_{20}(\sin\phi) \tag{3.29}$$

where we have assumed that $k_2 = 0.303$, and $H^{(v)} = -0.31459$ at Doodson number 055.555. The question "which equation goes where in geodesy" is not as trivial as one might think as was demonstrated by an animated discussion on this issue during the Summer school in Como. The situation is not clear and can only be understood if one carefully studies IAG resolutions summarizing the results of their study groups, or alternatively if one studies literature documenting gravity models or other results. In fact, three reference systems exist according to Rapp et al. (1991a):

- A *tide-free* system this means that U_2^a and $k_2 U_2^a$ are removed from a quantity and therefore that some value is assumed for k_2 at zero frequency.
- A *zero-tide* system: this means that U_2^a is removed but that $k_2 U_2^a$ is *not* removed from a quantity. Hence it means that no value is assumed for k_2 at zero-frequency.
- A *mean-tide* system: this means that $(1 + k_2)U_2^a$ is *not* removed from a quantity. Hence it also means that no value is assumed for k_2 at zero-frequency.

The problem with the tide-free definition is which value is assumed for k_2 at 055.555 since it can not be measured by any technique. The best possible estimates could come from the Earth's response at the 18.6 year line at Doodson number 055.565 or a geophysical theory. Finally Rapp et al. (1991a) also mention the term:

$$g^{-1} h_2 U_2^a = -0.12145 \times P_{20}(\sin \phi) \tag{3.30}$$

which assumes that $h_2 = 0.612$ at Doodson number 055.555 which is of relevance for the definition of geodetic station heights. Also this term could be used for defining a tide-free system or a zero-tide system, although Rapp et al. (1991a) don't explain its interpretation in the context of a mean-tide system.

Finally we mention that Rapp et al. (1991a) provide six recommendations with regard to permanent tidal deformations. These recommendations should be followed at any time to avoid misunderstandings on the issue of permanent tidal deformations of the Earth.

3.3 Ocean tides

Of course the oceans themselves will respond differently to the tide generating forces. Ocean tides are exactly the effect that one observes at the coast; i.e. the long periodic, diurnal and semi-diurnal motions between the sea surface and the land. In most regions on Earth the ocean tide amplitude is of the order of 0.5 meter whereas in some bays found along the coast of e.g. Normandy and Brittany (France), the Bristol Channel (UK) and Fundy Bay (Canada), the tidal amplitude is up to 5 meters. Ocean tides may have great consequences for daily life and also marine biology in coastal areas; some islands such as Mt. Saint Michel in Brittany can't be reached during high tide.

A map of the global M_2 ocean tide is given in figure 3.2 and 3.3, cf. Schrama and Ray (1994), from which one can see that there are regions without any water level variation which are called amphidromes where a tidal wave is continuously rotating about a fixed geographical location. If we ignore friction then the orientation of the rotation is determined by the balance between the pressure gradient and the Coriolis force, for which the reader is referred to appendix A. Cartwright (1993) mentions that it was Laplace who laid the foundations for modern tidal research. Laplace's main contributions were:

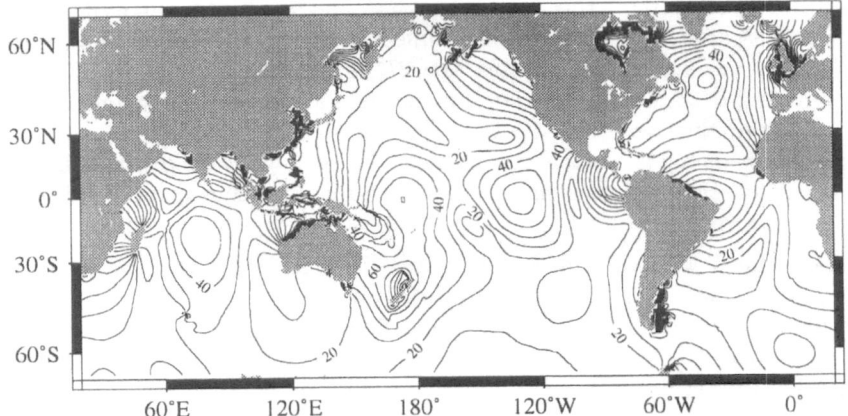

Fig. 3.2. M_2 ocean tide, amplitudes in centimeters, derived from the model of Schrama and Ray (1994). Figure provided by courtesy of R. Ray.

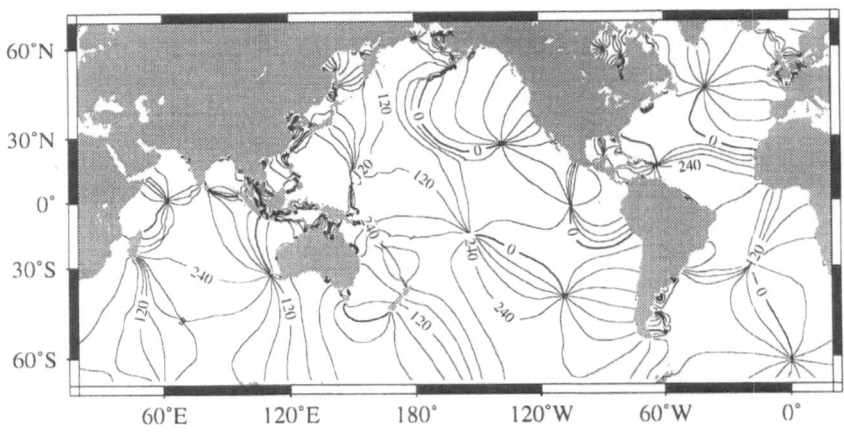

Fig. 3.3. M_2 ocean tide, phase in degrees, derived from the model of Schrama and Ray (1994). Figure provided by courtesy of R. Ray.

- The separation of tides into distinct *Species* of long period, daily and twice daily (and higher) frequencies.
- The (almost exact) dynamic equations linking the horizontal and vertical displacement of water particles with the horizontal components of the tide-raising force, see also appendix A.
- The hypothesis that, owing to the dominant linearity of these equations, the tide at any place will have the same spectral frequencies as those present in the generating force.

Laplace derived solutions for the dynamic equations only for the ocean and atmospheres covering a globe, but found them to be *strongly dependent on the assumed fluid depth*. Realistic bathymetry and continental boundaries rendered Laplace's solution mathematically intractable.

3.3.1 Harmonic Analysis methods. A direct consequence of the work of Laplace is that deep ocean tides respond at frequencies identical to the Doodson numbers encountered tables 3.2 and 3.3. This implies that at least 18.61 years of data would be required to separate two neighboring frequencies mostly because of the fact that main lines in the spectrum are modulated by smaller, but significant, side-lines. Fortunately extensive analysis conducted by Munk and Cartwright (1966) have shown that a smooth response of the sea level is likely. Therefore the more practical approach is to take at least two Doodson numbers and form an expression where only a year's worth of observations determine "amplitude and phase" of a constituent. However, this is only possible if one assumes a fixed *amplitude ratio* of a side-line with respect to a main-line where the ratio itself can be taken from the table of tidal harmonics. Consider for instance table 3.3 where M_2 is dominated by spectral lines at the Doodson numbers 255.555 and 255.545 and where the ratio of the amplitudes is approximately $-0.02358/0.63194 = -0.03731$. We will now seek an expression to model the M_2 constituent:

$$M_2(t) = C_{M_2} \left[\cos(2\omega_1 - \theta_{M_2}) + \alpha \cos(2\omega_1 + \omega_5 - \theta_{M_2}) \right]$$

where C_{M_2} and θ_{M_2} represent the amplitude and phase of the M_2 tide and where $\alpha = -0.03731$. This equation is written as:

$$M_2(t) = C_{M_2} f(t) \left\{ \cos(u(t)) \cos(2\omega_1 - \theta_{M_2}) - \sin(u(t)) \sin(2\omega_1 - \theta_{M_2}) \right\}$$

or

$$M_2(t) = C_{M_2} f(t) \cos(2\omega_1 + u(t) - \theta_{M_2}) \tag{3.31}$$

so that:

$$M_2(t) = A_{M_2} f(t) \cos(2\omega_1 + u(t)) + B_{M_2} f(t) \sin(2\omega_1 + u(t))$$

where

$$
\begin{aligned}
A_{M_2} &= C_{M_2} \cos(\theta_{M_2}) \\
B_{M_2} &= C_{M_2} \sin(\theta_{M_2})
\end{aligned}
$$

In literature the terms A_{M_2} and B_{M_2} are called "in-phase" and "quadrature" or "out-of-phase" coefficients of a tidal constituent, whereas the $f(t)$ and $u(t)$ coefficients are known as nodal modulation factors, stemming from the fact that ω_5 corresponds to the right ascension of the ascending node of the lunar orbit. In order to get convenient equations we work out the following system of equations: ($\Omega = \omega_5$):

$$f(t) = \{(1 + \alpha\cos(\Omega))^2 + (\alpha\sin(\Omega))^2\}^{1/2}$$

$$u(t) = \arctan\left(\frac{\alpha\sin(\Omega)}{1 + \alpha\cos(\Omega)}\right)$$

Finally a Taylor series around $\alpha = 0$ gives:

$$
\begin{aligned}
f(t) &= (1 + \frac{1}{4}\alpha^2 + \frac{1}{64}\alpha^4) + (\alpha - \frac{1}{8}\alpha^3 - \frac{1}{64}\alpha^5)\cos\Omega \\
&+ (-\frac{1}{4}\alpha^2 + \frac{1}{16}\alpha^4)\cos(2\Omega) + (\frac{1}{8}\alpha^3 - \frac{5}{128}\alpha^5)\cos(3\Omega) \\
&- \left(\frac{5}{64}\right)\cos(4\Omega) + \frac{7\alpha^5}{128}\cos(5\Omega) + O(\alpha^6) \\
u(t) &= \alpha\sin(\Omega) - \frac{1}{2}\alpha^2\sin(2\Omega) + \frac{1}{3}\alpha^3\sin(3\Omega) \\
&- \frac{1}{4}\alpha^4\sin(4\Omega) + \frac{1}{5}\alpha^5\sin(5\Omega) + O(\alpha^6)
\end{aligned}
$$

Since α is small it is possible to truncate these series at the quadratic term. The equations show that $f(t)$ and $u(t)$ are only slowly varying and that they only need to be computed once when e.g. working with a year worth of tide gauge data.

The Taylor series for the above mentioned nodal modulation factors were derived by means of the Maple software package and correspond reasonably with the expression provided to me by Richard Ray. However a more convenient way of finding the nodal modulation factors is to numerically compute at sufficiently dense steps values of the tide generating potential for a particular constituent at an arbitrary location on Earth over the full nodal cycle and to numerically estimate Fourier expressions like $f(\Omega) = \sum_n f_n \cos(n.\Omega)$ and $u(\Omega) = \sum_n u_n \sin(n.\Omega)$ with equations like eq. (3.31) as a point of reference.

3.3.2 Response method. The findings of Munk and Cartwright (1966) indicate that ocean tides $\zeta(t)$ can be predicted as a convolution of a smooth weight function and the tide generating potential U^a:

$$\hat{\zeta}(t) = \sum_s w(s)U^a(t - \tau_s) \tag{3.32}$$

with the weights $w(s)$ determined so that the prediction error $\zeta(t) - \hat{\zeta}(t)$ is a minimum in the least-squares sense with $\tau_s = 2s$ days as discussed in

Munk and Cartwright (1966). The weights $w(s)$ have a simple physical interpretation: they represent the sea level response at a point of observation to a unit impulse $U^a(t) = \delta(t)$, hence the name "response method". The actual input function $U^a(t)$ may be regarded as a sequence of such impulses. The scheme used in Munk and Cartwright (1966) is to expand $U^a(t)$ in spherical harmonics,

$$U^a(\theta, \lambda; t) = g \sum_{n=0}^{N} \sum_{m=0}^{n} [a_{nm}(t)U_{nm}(\theta, \lambda) + b_{nm}(t)V_{nm}(\theta, \lambda)] \qquad (3.33)$$

containing the complex spherical harmonics:

$$U_{nm} + jV_{nm} = (-1)^m \left[\frac{2n+1}{4\pi}\right]^{1/2} \left[\frac{(n-m)!}{(n+m)!}\right]^{1/2} P_{nm}(\cos\theta)\exp(jm\lambda)$$

and to compute the coefficients $a_{nm}(t)$ and $b_{nm}(t)$ for the desired time interval. The convergence of the spherical harmonics is rapid and just a few terms of n, m are sufficient. The m-values separate input functions according to species and the prediction formalism is:

$$\hat{\zeta}(t) = \sum_{n,m} \sum_{s} [u_{nm}(s)a_{nm}(t - \tau_s) + v_{nm}(s)b_{nm}(t - \tau_s)] \qquad (3.34)$$

where the prediction weights $w_{nm}(s) = u_{nm}(s) + jv_{nm}(s)$ are determined by least-squares methods, and tabulated for each port (these take the place of the tabulated C_k and θ_k in the harmonic method). For each year the global tide function $c_{nm}(t) = a_{nm}(t) + jb_{nm}(t)$ is computed. The tides are then predicted by forming weighted sums of c using the weights w appropriate to each port. The spectra of the numerically generated time series $c(t)$ have all the complexity of the Darwin-Doodson expansion; but there is no need for carrying out this expansion, as the series $c(t)$ serves as direct input into the convolution prediction. There is no need to set a lower bound on spectral lines; all lines are taken into account in an optimum sense. There is no need for the f, u factors, for the nodal variations (and even the 20926 y variation) is already built into $c(t)$. In this way the response method makes explicit and general what the harmonic method does anyway – in the process of applying the f, u factors. The response method leads to a more systematic procedure, better adapted to computer use. Its formalism is readily extended to include weakly nonlinear, and perhaps even meteorological effects. Once evaluated, the prediction weights may be easily converted to harmonic constants, as required by tidal maps.

3.4 Load tides

Any tide in the ocean will cause the bottom pressure to vary accordingly resulting in a load on the sea floor that in turn will deform the lithosphere.

Ocean load tides cause, in addition to the solid-Earth tide, a vertical displacement of geodetic stations as has been demonstrated by analysis of GPS and VLBI observations near the coast where vertical movements can be as large as several centimeters twice daily. Examples of the M_2 load tide are shown in figures 3.4 and 3.5 which are computed from an updated version of the ocean tide model developed by Schrama and Ray (1994). Since this model is constrained to the deep oceans and to latitudes between 66S and 66N, the hydrodynamical model of Le Provost et al. (1994) is used to fill in void areas. The procedure for computing the load tide from the altimetric tide (the altimeter will see the sum of the ocean and load tide), is described in Ray and Sanchez (1989).

The load tide may be expressed as a Green's function of angular distance from each incremental tidal load, effective up to 180°. But given global definition of the ocean tide it is more convenient to express it in terms of a sequence of load-Love numbers k'_n and h'_n times the spherical harmonics of degree n of the ocean tide. If $\zeta_n(\theta, \lambda; t)$ denote any n^{th} degree spherical harmonics of the tidal height ζ, the secondary potential and the bottom displacement due to elastic loading are $g(1 + k'_n)\alpha_n\zeta_n$ and $h'_n\alpha_n\zeta_n$ respectively where:

$$\alpha_n = \frac{3}{(2n+1)} \times \frac{\rho_w}{\rho_e} = \frac{0.563}{(2n+1)} \tag{3.35}$$

where ρ_w is the mean density of water and ρ_e the mean density of Earth, see also Cartwright (1993). The essential difference from the formulation of the body tide is that the spherical harmonic expansion of the ocean tide itself requires terms up to very high degree n, for adequate definition. Farrell (1972) values of the load Love numbers, based on the Gutenberg-Bullen Earth model, are frequently used. Table 3.5 is taken from Cartwright (1993) and lists a selection of both Farrell's values and those from a more advanced calculation by Pagiatakis (1990), based on the PREM model. In this table n in column 1 and α_n in column 2 correspond to equation (3.35), the values in column 3 and 4 correspond to the load-Love numbers computed by Farrell (1972) and the values in column 5 and 6 correspond to the load-Love numbers computed by Pagiatakis (1990).

3.4.1 On the formulation of the load tide.

The in-phase or quadrature ocean load tide maps symbolized by $H(\theta, \lambda)$ may be computed by a convolution on the sphere of an isotropic function $G(\psi)$ and an in-phase or quadrature ocean tide height function symbolized by $F(\theta, \lambda)$.

$$H(\theta, \lambda) = \int_{\sigma'} F(\theta', \lambda')G(\psi)\, d\sigma' \tag{3.36}$$

where $d\sigma' = \sin\psi\, d\psi\, d\alpha$ and ψ the spherical distance between θ, λ and θ', λ' and α the azimuth. In eq. (3.36) the F and G function are developed as:

$$F(\theta, \lambda) = \sum_{n=0}^{\infty} \sum_{m=0}^{n} \sum_{a=0}^{1} F_{nma} Y_{nma}(\theta, \lambda) \tag{3.37}$$

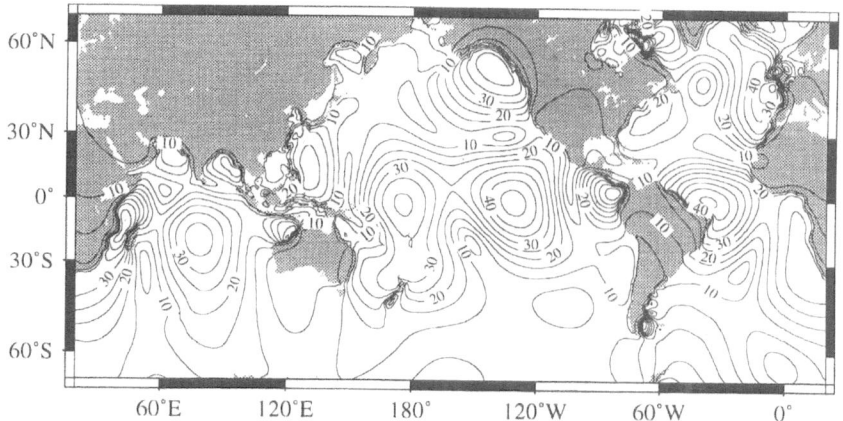

Fig. 3.4. M_2 load tide, amplitude in millimeters, derived from the model developed by Schrama and Ray (1994). Figure provided by courtesy of R. Ray.

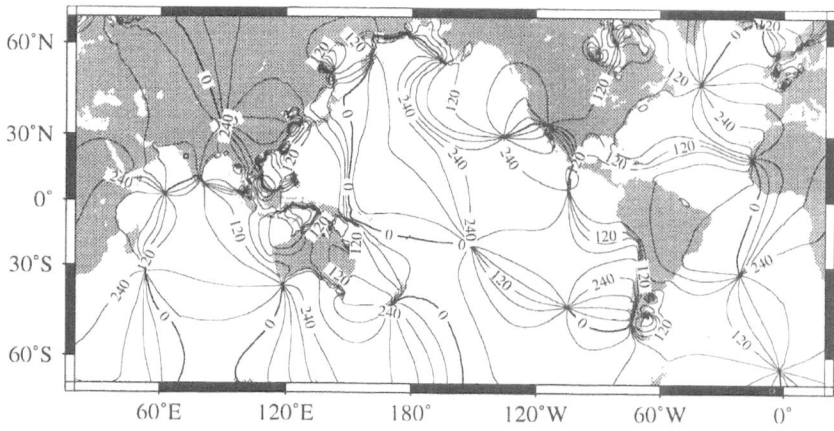

Fig. 3.5. M_2 load tide, phase in degrees, derived from the model developed by Schrama and Ray (1994). Figure provided by courtesy of R. Ray.

Table 3.5. Loading Love-numbers by two models

n	α_n	$-h'_n$	$-k'_n$	$-h'_n$	$-k'_n$
1	0.1876	0.290	0	0.295	0
2	0.1126	1.001	0.308	1.007	0.309
3	0.0804	1.052	0.195	1.065	0.199
4	0.0625	1.053	0.132	1.069	0.136
5	0.0512	1.088	0.103	1.103	0.103
6	0.0433	1.147	0.089	1.164	0.093
8	0.0331	1.291	0.076	1.313	0.079
10	0.0268	1.433	0.068	1.460	0.074
18	0.0152	1.893	0.053	1.952	0.057
30	0.0092	2.320[a]	0.040[a]	2.411	0.043
50	0.0056	2.700[a]	0.028[a]	2.777	0.030
100	0.0028	3.058	0.015	3.127	0.016

[a] data was interpolated at $n = 32, 56$

and

$$G(\psi) = \sum_{n=0}^{\infty} G_n P_n(\cos\psi) \tag{3.38}$$

Instead of numerically computing the expensive surface integral in eq. (3.36), it is more convenient to work out the following set of equations:

$$H(\theta, \lambda) = \sum_{n=0}^{\infty} \sum_{m=0}^{n} \sum_{a=0}^{1} H_{nma} Y_{nma}(\theta, \lambda) \tag{3.39}$$

where

$$H_{nma} = \frac{4\pi G_n}{2n+1} F_{nma} \tag{3.40}$$

3.5 Other tidal phenomena

Some examples which we won't pay too much attention to are for instance changes of the groundwater level which are associated with tides, atmospheric pressure variations of the order of a millibar or so at the twice per day frequency, tilt of the Earth's surface caused by (load) tides, and variations in Earth rotation due to tides. A good reference is the book of Lambeck (1988). Atmospheric tides do play a role in the processing of satellite altimeter data and also tide gauge data. The reason is that a decision must be made in which way one applies the inverted barometer correction, see section 2. The inverted barometer correction is normally proportional to air pressure variations, -1 cm sea level change per 1 millibar, which also implies that the so-called S_2 air tide is affecting the sea level.

3.6 Altimetry and tides

Collinear repeat track and crossover analysis of the altimeter data are known to undersample the diurnal and semi-diurnal frequencies thereby causing alias periods significantly longer than the natural periods of the tides. All altimeter satellites suffered from this problem, SEASAT's lifetime was too short for doing any serious tidal analysis, GEOSAT had several problems among which that the M_2 tide aliases to a period of about a year and finally ERS-1 is by definition not suited for tidal research because of the Sun-synchronous orbit causing all solar tides to be sampled in phase.

Finally we want to mention that TOPEX/Poseidon has stimulated the development of a series of new tide models more accurate than any previous global hydrodynamic model, see for instance Schrama and Ray (1994). The main reason for the success of T/P in modeling the *deep* ocean tides should be seen in the context of the design of the mission where the choice of the nominal orbit is such that all main tidal constituents alias to relatively short periods. A few of the results are tabulated in table 3.6 where the rms comparisons to 102 "ground-truth" stations in (cm) are shown. In this table the tide models developed by Schwiderski (1980) and Cartwright et al. (1991) were developed before TOPEX/Poseidon. Le Provost et al. (1994) ran a global finite element model that is free from TOPEX/Poseidon data. Egbert et al. (1994) also ran a finite element model while assimilating TOPEX/Poseidon data. Sanchez and Pavlis (pers.comm.) and Ray (pers.comm.) used so-called Proudman functions to model the tides, also incorporating TOPEX/Poseidon data. Schrama and Ray (1994) applied a straightforward harmonic analysis to the TOPEX/Poseidon data to determine improvements with respect to a number of tidal constituents.

Table 3.6. Ground truth comparison at 102 tide gauges, courtesy: R. Ray.

Authors	version	Q_1 [cm]	O_1 [cm]	P_1 [cm]	K_1 [cm]
Schwiderski	1980	0.34	1.23	0.61	1.44
Cartwright-Ray	1991		1.22	0.63	1.89
Le Provost et al.	meom94.1	0.28	1.04	0.46	1.23
Egbert et al.	tpxo.1		0.96		1.26
Egbert et al.	tpxo.2	0.29	0.98	0.45	1.32
Sanchez-Pavlis	gsfc94a	0.35	1.06	0.54	1.41
Ray et al.	1994	0.37	1.00	0.40	1.25
Schrama-Ray	1993.10		1.15		1.35
Schrama-Ray	1994.11		1.02		1.19

		N_2 [cm]	M_2 [cm]	S_2 [cm]	K_2 [cm]
Schwiderski	1980	1.19	3.84	1.66	0.59
Cartwright-Ray	1991	0.96	3.23	2.22	
Le Provost et al.	meom94.1	0.87	2.99	1.56	0.50
Egbert et al.	tpxo.1		2.30	1.55	
Egbert et al.	tpxo.2	0.76	2.27	1.26	0.56
Sanchez-Pavlis	gsfc94a	0.86	2.31	1.23	0.66
Ray et al.	1994	0.81	2.04	1.23	0.51
Schrama-Ray	1993.10		2.02	1.26	
Schrama-Ray	1994.11	0.85	1.85	1.20	

4. Ocean Dynamics

4.1 Equations of motion

The oceans can be seen as a thin rotating shell with an average thickness of approximately 5 km relative to a sphere with a radius of 6378 km. To understand the dynamics of fluid in this thin rotating shell we initially consider Newton's law $f = ma$ for a given water parcel. Pond and Pickard (1983) show that this parcel will experience a motion governed by the equations:

$$\frac{d\overline{v}}{dt} = -\frac{1}{\rho}\nabla p - 2\overline{\omega} \times \overline{v} + \overline{g} + \overline{f}. \tag{4.1}$$

where $\overline{v}$ is a velocity vector in a local coordinate system, ρ is the density function, p the pressure function, $\overline{\omega}$ is the Earth's spin vector, $\overline{g}$ is the sum of gravitational and centrifugal accelerations, ie. the gravity acceleration vector and $\overline{f}$ symbolizes additional accelerations which are for instance caused by friction.

This vector equation may be rewritten into three separate equations with the local coordinates x, y and z and the corresponding velocity components u, v and w. Here we follow the convention found in the literature and assign the x-axis direction corresponding with the u-velocity component to the local east, the y-axis direction and corresponding v-velocity component to the local north, and the z-axis including the w-velocity pointing out of the sea surface, causing depths with a minus sign. Accordingly, all vectors in equation (4.1) must be expressed in the local x, y, z coordinate frame. If ϕ corresponds to the latitude of the water parcel and Ω to the length of $\overline{\omega}$ then the following substitutions are allowed:

$$\begin{aligned}
\overline{\omega} &= (0, \Omega\cos\phi, \Omega\sin\phi)^T \\
\overline{g} &= (0, 0, -g)^T \\
\overline{f} &= (F_x, F_y, F_z)^T \\
\overline{v} &= (u, v, w)^T
\end{aligned}$$

so that the equations of motion become:

$$\begin{aligned}
\frac{du}{dt} &= -\frac{1}{\rho}\frac{\partial p}{\partial x} + F_x + 2\Omega\sin\phi\, v - 2\Omega\cos\phi\, w \\
\frac{dv}{dt} &= -\frac{1}{\rho}\frac{\partial p}{\partial y} + F_y - 2\Omega\sin\phi\, u \\
\frac{dw}{dt} &= -\frac{1}{\rho}\frac{\partial p}{\partial z} + F_z + 2\Omega\cos\phi\, u - g
\end{aligned} \tag{4.2}$$

see also Pond and Pickard (1983) whereas Veronis (1981) shows a slightly extended form of these equations including second-order terms. Providing that we forget about frictional terms eqns. (4.2) tell us nothing more than that the pressure gradient, the Coriolis force and the gravity vector are in balance with inertia, i.e. the left hand side of eqns. (4.2).

4.1.1 Non-linearity. The du/dt, dv/dt and dw/dt terms are a problem since they may lead to *non-linear behavior* of the equations of motion causing turbulence under certain conditions. We remark that these derivatives should be seen as absolute terms and that:

$$\frac{du}{dt} = \frac{\partial u}{\partial t} + u\frac{\partial u}{\partial x} + v\frac{\partial u}{\partial y} + w\frac{\partial u}{\partial z}$$
$$\frac{dv}{dt} = \frac{\partial v}{\partial t} + u\frac{\partial v}{\partial x} + v\frac{\partial v}{\partial y} + w\frac{\partial v}{\partial z} \qquad (4.3)$$
$$\frac{dw}{dt} = \frac{\partial w}{\partial t} + u\frac{\partial w}{\partial x} + v\frac{\partial w}{\partial y} + w\frac{\partial w}{\partial z}$$

could be substituted as a first-order approximation. In the literature terms like $\partial u/\partial t$ are normally considered as so-called "local accelerations" whereas advective terms like $u\partial u/\partial x + ...$ are considered as "field accelerations". The physical interpretation is that two types of accelerations may take place. In the first terms on the right hand side, accelerations occur locally at the coordinates (x, y, z) resulting in $\partial u/\partial t$, $\partial v/\partial t$, and $\partial w/\partial t$ whereas in the second case the velocity vector is changing with respect to the coordinates resulting in so-called advective terms. This effect is normally called non-linear because velocities appear as squares, (e.g. $u(\partial u/\partial x) = \frac{1}{2}[\partial(u^2)/\partial x]$) or as products between different velocity components. These non-linearities cause large-scale features to deform into smaller-scale features, implying an energy transfer from large-scale to small-scale phenomena (turbulence).

4.1.2 Friction. Another problem with equations (4.2) is that friction may appear in F_x, F_y and F_z. Based upon observational evidence, Stokes suggested that tangentional stresses are related to the velocity shear as:

$$\tau_{ij} = \mu \left(\frac{\partial u_i}{\partial x_j} + \frac{\partial u_j}{\partial x_i} \right) \qquad (4.4)$$

where μ is a molecular viscosity coefficient characteristic for a particular fluid. Frictional forces are obtained by:

$$F_x = \frac{\partial \tau_{ij}}{\partial x_j} = \mu \frac{\partial^2 u_i}{\partial x_j{}^2} + \mu \frac{\partial}{\partial x_i} \left(\frac{\partial u_j}{\partial x_j} \right) \qquad (4.5)$$

which is approximated by:

$$F_x = \mu \frac{\partial^2 u_i}{\partial x_j{}^2} \qquad (4.6)$$

by assuming an incompressible fluid. A separate issue is that viscosity $\nu(= \mu/\rho)$ may not be constant because of turbulence. In this case:

$$F_x = \frac{\partial \tau_{ij}}{\partial x_j} = \frac{\partial}{\partial x_j} \left(\mu \frac{\partial u_i}{\partial x_j} \right) \qquad (4.7)$$

although it should be remarked that this equation is also based on an assumption. As a general rule, no known oceanic motion is controlled by molecular viscosity, which is far too weak. In ocean dynamics, the "Reynolds stress" involving turbulence or eddy-viscosity always applies, as discussed by e.g. Pond and Pickard (1983). and Pedlosky (1987).

4.1.3 Turbulence and the Reynolds number. To judge whether turbulence occurs one should compute the so-called Reynolds number Re which is a measure for the ratio between non-linear terms and the frictional terms. Pond and Pickard (1983) mention that the Reynolds number may be approximated as $Re = \frac{UL}{\nu}$, where U and L are velocities and lengths at the characteristic scales at which the motions occurs. Moreover they remark that large Reynolds numbers, e.g. ones which are greater than 1000, usually indicate turbulence. An example of this phenomenon can be found in the Gulf stream area where L is of the order of 100 km, U of the order of 1 m/s and ν of the order of $10^{-6} m^2 s^{-1}$ so that $Re = \frac{UL}{\nu} \approx 10^{11}$. The effect displays itself as a meandering of the main stream which can be nicely demonstrated by infrared images of the area showing the turbulent flow of the Gulf stream occasionally releasing eddies that will live for considerable time in the open oceans. The same phenomenon can be observed in other western boundary regions of the oceans such as the Kuroshio current East of Japan and the Agulhas retroflection current south of Cape of Good Hope.

4.1.4 Ekman equations. Ekman dynamics tells us that the sea surface doesn't accelerate in the direction of the wind but rather 45° to the right on the Northern hemisphere, which was earlier observed by Ekman who studied the motion of icebergs. The Ekman equations model the balance between friction and the Coriolis force:

$$f\, v_e = -A_z \frac{\partial^2 u_e}{\partial z^2}$$
$$-f\, u_e = -A_z \frac{\partial^2 v_e}{\partial z^2} \tag{4.8}$$

where u_e and v_e are velocities in the Ekman layer. There are several assumptions underlying these equations: a) there are no boundaries, b) infinitely deep water to avoid bottom friction, c) a constant value for the eddy viscosity parameter A_z, d) a steady wind blowing for a long time, e) barotropic conditions (more on this later), f) the Coriolis parameter f (more on this later) is constant. The solution of the Ekman equations is:

$$u_e = \pm V_0 \cos\left(\frac{\pi}{4} + \frac{\pi}{D_e}z\right) \exp\left(\frac{\pi}{D_e}z\right)$$

$$v_e = V_0 \sin\left(\frac{\pi}{4} + \frac{\pi}{D_e}z\right) \exp\left(\frac{\pi}{D_e}z\right)$$

(+ for the Northern and − for the Southern hemisphere) where:

$$V_0 = (\sqrt{2}\pi\tau_{yn})/(D_e\rho|f|) \tag{4.9}$$

is the total Ekman surface current, τ_{yn} is the magnitude of the wind stress on the sea surface and:

$$D_e = \pi\sqrt{2A_z/|f|} \tag{4.10}$$

is the Ekman depth or *depth of the frictional influence* which is varying between 50 and 200 meter in the oceans depending on latitude and wind speed. The Ekman equations show that the surface water moves to 45° right (left) of the wind in the Northern (Southern) hemisphere. Moreover the equations show that the sub-surface currents will tend to spiral exponentially decreasing in velocity with increasing depth, in literature referred to as the Ekman spiral. The direction of the flow becomes opposite of the wind at $z = -D_e$ with a velocity of $\exp(-\pi) = 0.04$ relative to the surface.

4.1.5 General ocean-circulation. According to Wunsch (1993) the wind-driven general circulation of the ocean, away from the immediate vicinity of the equator, is not driven by the wind, nor the Ekman layer itself, but is rather caused by the *regional variations of the amount of fluid moving within the Ekman layer* which set the interior geostrophic fluid in motion. The Ekman layer itself is only of secondary interest for the general circulation and its surface effect is too small to be observed by altimetry. Yet its properties as a pump are a crucial element in understanding the general circulation. The underlying physics is probably best appreciated from the point of view taken by Stommel (1957) or Stommel (1984).

Alternatively a smaller part of the general circulation is driven by a thermohaline cycle. This phenomenon is caused by heating and cooling at the surface, precipitation and evaporation, runoff from land, and ice formation or melting.

4.1.6 Two branches in oceanography. The above mentioned examples for describing the dynamics of fluids is deliberately kept short and simple avoiding a detailed treatment of non-linearity and friction. We haven't really solved the problem by just translating $F = ma$ to a thin rotating shell on the sphere. To solve the equations of motion we should investigate frictional forces, wind effects and Ekman pumping, boundary conditions and non-linearity of the equations. "Then why at all do you bother to show these equations?" the reader may wonder. The reason is that we want to emphasize that two different approaches exist in dynamical oceanography.

The dynamical view would be to combine eqns. (4.2) together with all relevant continuity conditions given certain climatological models describing the wind stress field etc. and to design a well chosen finite element algorithm around this system. Such activities are described e.g. in the work of Semtner and Chervin (1992) and Semtner (1995) where extreme requirements in the order of a year of super computer time are mentioned necessary to integrate the resulting equations on a sufficiently dense three dimensional grid to a state of equilibrium. The nature of this work should be seen as an activity where the state of the oceans is *predicted ahead in time* which in the case of Semtner and Chervin (1992) and Semtner (1995) is resulting in an impressive movie revealing many interesting details about the motion of sea water.

Whether or not realistic results are obtained remains to be evaluated. A fact mentioned by Semtner (1995) is that their results will be influenced by the initial and forcing conditions put into their computations. Nevertheless it should be remarked that data assimilation activities, where certain model parameters are adjusted in an optimal sense to follow observational data, have become increasingly popular in this branch of oceanography.

The other view of dynamical oceanography is based on *descriptive* (or synoptic) methods for determining ocean circulation. Here the work consists of doing in-situ measurements of sea water density, flow, velocity etc. with respect to depth along a number of lines with the eventual goal of compiling a map of ocean currents. This process of measuring ocean circulation has resulted in several models of the mean dynamic topography and in some well observed areas the typical seasonal variations of ocean topography, see eg. Levitus (1982). An introduction to the assumptions underlying this process of measuring ocean circulation will be shown in the following section.

4.2 Observing ocean circulation

The first problem is the relevance of the terms encountered in the vertical part of eqns. (4.2). In a first-order approximation we notice that the term Ω is of the order of 10^{-4} rad/s, that u is of the order of 1 m/s and that g is 10 m/s^2 whereas the F_z term could for instance be caused by tidal effects and friction which are likely to be of the order of 10^{-5} to 10^{-6} m/s^2. The same will be true for vertical accelerations in the oceans so that dw/dt can be ignored in the third equation. The dominant terms remain and lead to the so-called hydrostatic equation:

$$\frac{dp}{dz} = -g\rho \tag{4.11}$$

which is known to be correct to about 1 part in 10^6, even when water is moving at the speeds of about 1 to 2 m/s as is typically encountered in western boundary currents.

Therefore equation (4.11) may be used in a straightforward vertical integration where we remark the values of gravity and density are nearly constant along the path of integration. The value of g will only change by approximately 0.03% between the surface and a depth of 1000 m. Since the compressibility modulus (K) of water is 20 kbar, (K appears in the equation $p = K\,dV/V$ where dV/V represents the relative volume change) the same simplification holds for variations in the density ρ. Therefore it is valid to assume an incompressible fluid since the resulting density variations are of the order of 0.5% for a depth of 1000 m. The hydrostatic equation is therefore confirming what we already know from a static fluid, namely that $p(z) = -g\rho z$. Hence the only purpose of the hydrostatic equation is to confirm that the pressure at a depth of 100 m is $-9.8 \times 1025 \times -100 = 1.045 \times 10^6$ Pa $= 1045$ kPa.

Finally if we look at the first equation in (4.2) it becomes clear that $-2\Omega\cos\psi\,w$ is likely to disappear because of the observational fact that vertical velocities (w) in the oceans are much smaller compared to the horizontal components u and v. The latter is not true for the Ekman layer, yet for the deeper oceans it is likely to be true. Some deep areas surrounded by higher sea bottom topography such as the Banda sea (Indonesia) are known to contain deep water which is more than 10000 years old.

4.2.1 Inertial motion. We maintain the horizontal accelerations, assume that $w = 0$, remove the pressure gradients by assuming no slope at the sea surface and ignore all friction terms and non-linear behavior of the equations. Under these assumptions the first two equations become:

$$\frac{du}{dt} = +fv$$
$$\frac{dv}{dt} = -fu \qquad (4.12)$$

where $f = 2\,\Omega\sin\phi$ is known as the Coriolis parameter, accompanied by the solution:

$$u(t) = +V_h\sin(ft)$$
$$v(t) = -V_h\cos(ft) \qquad (4.13)$$

see also Pond and Pickard (1983) pp. 64–65. They mention the example of a wind blowing steadily in one direction for a time, causing the water to acquire a speed V_h. When the wind stops the motion will continue without friction (to a first approximation) as a consequence of its inertia (properly its momentum) hence the term "inertial motion". As the solution indicates, the period of rotation is $2\pi/f$ indicating that inertial motions take place within a period of what is known as one-half Foucault pendulum day. (The length of one sidereal day divided by the sine of the latitude equals to the period required for the plane of a Foucault pendulum to rotate through a full circle.) The direction of rotation is clockwise viewed from above in the northern hemisphere and counterclockwise in the southern hemisphere. Sometimes the terms cyclonic and anticyclonic are used for motions in the ocean. A cyclonic motion will be in the direction of a cyclone, thus counterclockwise in the northern hemisphere and clockwise in the southern hemisphere. An anticyclonic system has high pressure in its center and winds circulating in the opposite direction.

4.2.2 Thermal wind equations. If we ignore the horizontal accelerations du/dt and dv/dt in (4.2), ignore all frictional terms and non-linear behavior of the equations, and assume that $-2\Omega\cos(\phi)w$ can be ignored in the first equation and $2\Omega\cos(\phi)u$ in the third equation then:

$$\frac{1}{\rho}\frac{dp}{dx} = +fv$$

$$\frac{1}{\rho}\frac{dp}{dy} = -fu \tag{4.14}$$

$$\frac{1}{\rho}\frac{dp}{dz} = -g$$

A well known manipulation of these equations is to differentiate the first two equations with respect to z:

$$+\frac{d\,(\rho f v)}{dz} = \frac{d}{dx}\left(\frac{dp}{dz}\right)$$

$$-\frac{d\,(\rho f u)}{dz} = \frac{d}{dy}\left(\frac{dp}{dz}\right) \tag{4.15}$$

which are called the thermal wind equations, because they were originally developed to show how temperature differences in the horizontal could lead to vertical variations in the geostrophic wind velocity. This can be seen by a substitution of the hydrostatic balance equation in eq. (4.15) resulting in the system:

$$+\frac{d\,(\rho f v)}{dz} = -\frac{d\,(g\rho)}{dx}$$

$$-\frac{d\,(\rho f u)}{dz} = -\frac{d\,(g\rho)}{dy} \tag{4.16}$$

The horizontal velocities u and v can be recovered by an integration over z:

$$u = \frac{+g}{f\rho_0}\int_{z_0}^{0}\frac{d\rho}{dy}\,dz + C_u$$

$$v = \frac{-g}{f\rho_0}\int_{z_0}^{0}\frac{d\rho}{dx}\,dz + C_v \tag{4.17}$$

where z_0 is a depth known as the level of no motion while C_u and C_v are integration constants which would cancel for a proper choice of z_0. The underlying idea behind the choice of z_0 is that the oceans are supposed to be at rest beneath a certain depth which is mostly based upon observational evidence and also taking into account the difficulties in realizing density measurements to extreme depths. In reality, values of z_0 around 1000 or 1500 meter are common in oceanography. The issue has led to animated discussions in the community because a poor choice for z_0 leads to "biased" estimates of the geostrophic velocities, but more importantly whether such a level of no motion exists at all. The z_0 parameter can not be chosen at the bottom of the oceans where friction will certainly play a role while we ignored that particular effect in the derivation of eqns. (4.17). Despite all the simplifications underlying the derivation of eqns. (4.17) they have enabled physical oceanographers to compute geostrophic velocities, simply by measuring horizontal density gradients of sea water.

The procedure itself is relatively straightforward to carry out. The traditional way is that one needs a ship expedition taking so-called CTD observations where conductivity and temperature are measured at certain depths by a probe released from the research vessel. Conductivity measurements indicate the salt content of sea water which normally varies between 0 grams per kg for fresh water and the extreme of 40 grams per kg for the Red sea in the winter although 90 procent of the oceans is closer to 35 grams per kg. Temperature of sea water is one of the important factors that are known to determine the climate in coastal regions. Whereas temperature in the deeper oceans will be closer to a few degrees centigrade surface values are more directly related to what we know from the atmosphere although the oceans appear to have a phase lag of about three months compared to astronomical definition of summer and winter. (More on this issue later in these notes). Eventually it is the combination of salinity and temperature that determines the in-situ density of the oceans. The technique of making CTD observations including the interpretation of density gradients is extensively discussed in the literature and needs no repetition in these notes. In the last 70 years several oceanographic organizations have been involved in making such measurements which has led to a climatological atlas of the oceans, cf. Levitus (1982).

By assuming a certain level of no motion these historical CTD data can be used to estimate geostrophic surface velocities relative to z_0 by integration of eqns (4.16) up to the surface of the oceans. Once these surface velocities u_s and v_s are known one can again apply the geostrophic relation and finally compute horizontal gradients of the surface pressure p_s:

$$f\, u_s = \frac{-1}{\rho_0}\frac{d\,p_s}{d\,y}, \quad f\, v_s = \frac{+1}{\rho_0}\frac{d\,p_s}{d\,x} \qquad (4.18)$$

which directly results in horizontal gradients of the dynamic topography η (because $\eta\, g = p_s/\rho$):

$$u_s = \frac{-g}{f}\frac{d\,\eta}{d\,y}, \quad v_s = \frac{+g}{f}\frac{d\,\eta}{d\,x} \qquad (4.19)$$

which are *relative to a chosen value of the level of no motion* in the oceans.

4.2.3 Isobaric and isopycnal surfaces and currents. In oceanography there is usually a distinction between what is known as *barotropic* or *baroclinic* conditions of a fluid. To explain both conditions and their effect on vertical integration of the thermal wind equations we have to introduce the concepts of *isobaric* and *isopycnal* surfaces. An *isobaric* surface is one on which the pressure is constant while an *isopycnal* or *isosteric* surface is one on which the density is constant. A *barotropic* condition occurs when the isobars and isopycnals are parallel in a fluid. This means that the density is only a function of the pressure, ie. $\rho = F(P)$. On the other hand, *baroclinic* conditions occur when the isobars are inclined relative to the *isosteric* surface. In fresh water a baroclinic condition normally occurs when density is a function of temperature and pressure, ie. $\rho = F(t, P)$. In the oceans baroclinic conditions occur when density is a function of the salt content, temperature and pressure, ie. $\rho = F(s, t, P)$.

In a barotropic fluid the slope of the isopycnals is small, typical values are of the order of 10^{-6} for a 0.1 m/s current at mid latitudes. This would correspond to a height change of 0.1 meter over a distance of 100 km. In order to clarify this barotropic situation we consider that the pressure distribution is in equilibrium:

$$0 = \frac{\partial p}{\partial x} = -g\frac{\partial}{\partial x}\int_{z}^{h} \rho \, dz = -g\rho_h\frac{\partial h}{\partial x} - g(h - z)\frac{\partial \rho}{\partial x} \qquad (4.20)$$

so that

$$\frac{-1}{\rho_h}\frac{\partial \rho}{\partial x} \approx \frac{1}{h - z}\frac{\partial h}{\partial x} \qquad (4.21)$$

To summarize the discussion, the vertical integration in equation (4.17) of the horizontal density gradients is in practice only applicable for detecting the baroclinic components of the flow. In barotropic conditions the isopycnal slopes are too small and other techniques (such as satellite altimetry) must be applied.

4.2.4 Hydrographic inversion methods. In reality density measurements obtained from modern hydrography underlying the geostrophic velocity computations are done on "long-lines" spanning the ocean. Such measurements will observe one component of the geostrophic velocity, namely that which is perpendicular to that line. The work of Wunsch (1977,1978) was motivated by an attempt to solve the problem of estimating the dynamic topography but now independent of the level no motion assumptions. Basically this method says that if we consider a "box" surrounded by hydrographic stations then if C is any quantity for which we are prepared to write a conservation statement, then transport of C into the box must be equal to what comes out unless one can identify known sources or drains. Wunsch (1977) was able to show that these equations of continuity will result in a system of linear equations:

$$E\,\overline{x} + \overline{n} = \overline{y} \qquad (4.22)$$

where $\overline{x}$ are unknowns to solve for, ie. nodes with velocity, flow of tracers, etc. on a discretized grid, $\overline{n}$ are noise parameters and $\overline{y}$ are observations and that this system can be solved under the assumption of a least-squares minimization. Based upon this procedure Martel and Wunsch (1993) computed for the North Atlantic region the absolute topography of the sea surface (up to an unknown constant) relative to the geoid.

4.2.5 Altimetric estimates of the dynamic topography. Interestingly enough the results of Martel and Wunsch (1993) may be compared to altimetric estimates of the mean dynamic topography in the same region. The technique applied in this approach is extensively described in e.g. Nerem et al. (1994a) who have used a straightforward isotropic local averaging operator on 1.5 year of high quality Topex/Poseidon altimetry data. For altimetry the estimated deviations of the mean sea surface are always relative to existing geoid models. In Nerem et al. (1994a) we have shown the altimetric version of the mean dynamic topography relative to the OSU91a geoid, cf. Rapp et al. (1991b) and in a similar procedure relative to a combination of the OSU91a and the JGM2 model, cf. Nerem et al. (1994b). A global picture of the dynamic topography estimated by altimetry is shown in figure 4.1 and a subsection using a more localized approximation technique corresponding to North Atlantic region of Martel and Wunsch (1993) is shown in figure 4.2. Although completely different methods and data were

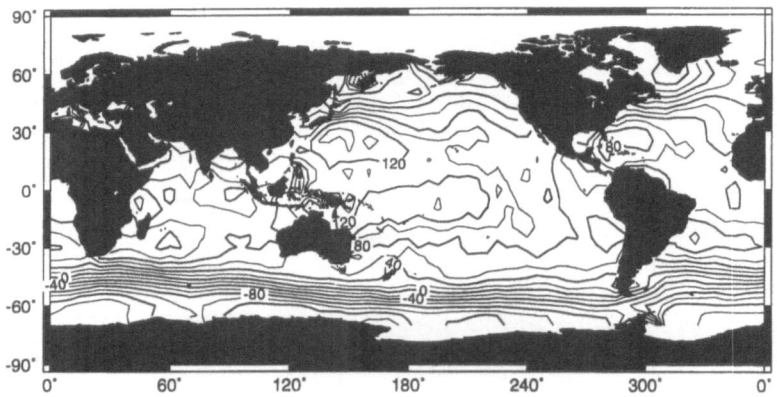

Fig. 4.1. Absolute topography of the sea surface relative to the OSU91a/JGM2 geoid, computed on $1 \times 1°$ bins from Topex/Poseidon altimetry, cycles 2 - 84. Spatial smoothing was performed in a $4°$ radius with a Gaussian half weight point at $2°$ radius, continental shelf data were ignored in the computations, CI=20 cm.

used to compute the estimates shown in Martel and Wunsch (1993) and 4.2 a number of similarities are encountered for which we would like to remark:

– Qualitatively the computed mean dynamic topography derived from the hydrographic method described by Martel and Wunsch (1993) is limited

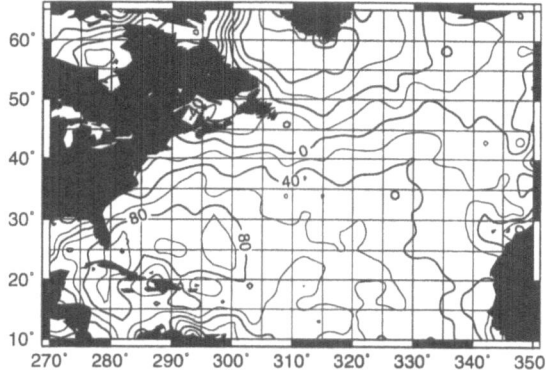

Fig. 4.2. Absolute topography of the sea surface in the gulf stream area relative to the OSU91a/JGM2 geoid computed as block mean averages of $3 \times 3°$ followed by a minimum curvature approximation on nodes spaced at 0.5° from Topex/Poseidon altimetry, cycles 2 - 84. CI=20 cm.

by the accuracy of the measurements and partly by an assumption of a reference level velocity estimate (ie. at the level of no motion.) Yet the formal error estimate of the dynamic topography of Martel and Wunsch (1993) is of the order of 5 cm, essentially because ocean transport is already well known from hydrography in the Gulf stream area.

– Altimetric estimates such as the one shown here in figure 4.2 depend on the choice of a reference geoid used in the computations. This explains several differences, also mentioned by Wunsch (1993), such as the increased circumpolar gyre and the increased dynamic topography at 25N and 62W in the altimetric estimate relative to the one showed by Martel and Wunsch (1993). In several publications such as Wunsch (1993) it was therefore suggested that the *marine geoid* could be improved by correcting the mean surface for a hydrographic estimate of the dynamic topography. Yet another possibility would be to reverse this thought and to compute better geoids in order to improve mean dynamic topography estimates, see for instance Rapp and Wang (1994).

– We notice that Topex/Poseidon altimetry yields a somewhat smoother estimate than the one shown in Martel and Wunsch (1993). This is due to the T/P intertrack spacing and the choice of our local averaging operator. One possibility to narrow the range of the averaging operator uses a blend ERS-1 and ERS-2 altimeter data with a proper correction for the increased orbit errors of the latter systems. Another possibility discussed by Wunsch (1993) would be to smooth the hydrographic estimate.

– Altimetry has shown that there is a strong time dependence in the dynamic topography. At the moment all altimetry was performed in the last 20 years meaning that our best estimates could tell something about "a mean" over this period. Hydrography has a much longer history and sea-water density estimates used by Levitus (1982) go back more than 70 years.

4.3 Temporal variations in the oceans

Apart from wind waves and tides, which were removed during the preprocessing of altimeter data, the sea level will exhibit excursions on various time and space scales. Hydrographic measurements of sea water density already display these features which can be seen in the monthly Levitus maps. Obviously such variations are hard to observe by means of in-situ measurements as was shown by (Nerem et al,1994a) where the annual variations in the southern oceans from the Levitus data are compared to altimetric estimates. Our choice for classifying sea level variations is based on the experiences with 20 years of altimeter data from which we have found that a) mesoscale variability, b) annual and semi-annual variations, c) large-scale waves in the ocean and possibly d) sea level rise and interdecadal sea level variations can be observed. In the following sections we want to explain briefly the physical phenomena underlying these sea level variations including the difficulties that were encountered in detecting them.

4.3.1 Mesoscale variability. As was mentioned before, the Reynolds number is of the order of 10^{11} in the Gulf stream area and also in other western boundary current regions. In this regime, the motion of fluid is turbulent thereby causing a meandering of the main stream occasionally forming closed loops, also known as eddies, which will continue to exist for several months in the open oceans. An example of mesoscale variability estimated from Topex/Poseidon altimetry data is shown in figure 4.3, where the effect is computed as a standard deviation of all Topex/Poseidon data between cycles 2 and 71 relative to a mean sea surface estimated for the same period. The figure confirms the earlier findings of Chelton et al. (1983) who have performed a collinear analysis of the SEASAT altimeter data. Variability maps as shown here are relatively easy to compute even when the altimeter orbits are heavily corrupted by unmodeled gravitational perturbations. In this case the requirement is that the orbit ground track pattern repeats itself thereby allowing a collinear analysis which is insensitive to gravitational orbit errors, see Rummel (1993). Therefore excellent variability maps have been produced with all altimeter missions confirming that western boundary currents regions are a source of variability. One of the nice examples is the Agulhas current south of Cape of Good Hope which demonstrates a retroflection mechanism causing the most energetic variability values. In this area we see a return of the south Indian ocean gyre whose western intensification point collides with the Antarctic circumpolar current (or ACC). Some of the eddies formed

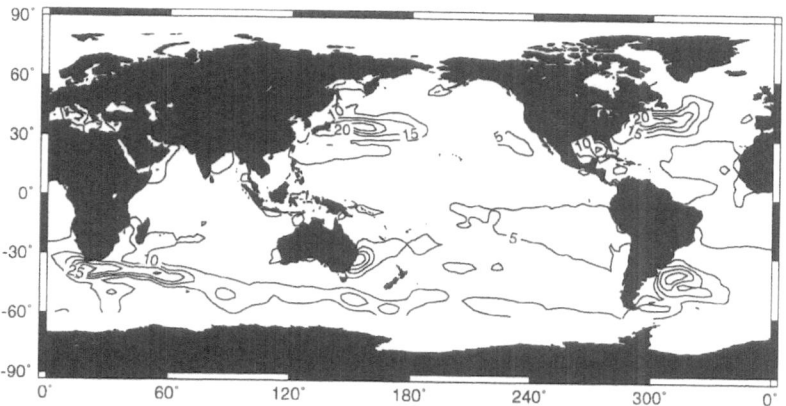

Fig. 4.3. Mesoscale variability estimated from Topex/Poseidon altimetry cycles 2 till 71. The values should be interpreted as standard deviations relative to a mean sea surface estimated from Topex/Poseidon altimetry. CI = 5 cm.

in this area are known to live for considerable time in the ACC causing a turbulence pattern that is clearly visible in the variability map. Other eddies generated in the Agulhas area don't follow the ACC but are picked up by the south Atlantic gyre explaining the extension of increased sea level variability west of Cape of Good Hope. Other examples of high variability are the Kuroshio current which shows a bifurcation with increased sea level variability east of Japan and a second part east of China. The Gulf stream area shows the same mechanism; most of the eddies are extending to the north-east and some are extending into the Gulf of Mexico. Weaker areas of variability are encountered east of Australia and east of the Patagonian shelf.

4.3.2 Annual variations. With the Topex/Poseidon system the annual signal is easily observed because of the excellent quality of the orbits and the relatively large magnitude of the signal compared to the inherent instrument noise, see also Nerem et al. (1994a). The same can not be said for other altimeters either because of their limited lifetime (ie. SEASAT lived only for three months) or because of large radial orbit errors corrupting the data (ie. GEOS-3, GEOSAT and ERS-1) or alternatively because the signals are hard to separate from aliased tidal signals, (eg. GEOSAT).

In the Topex/Poseidon data the annual cycle displays itself with a hemispherical distribution and is related to the annual heating cycle with the largest variations coinciding with the location of the western boundary currents. The phase of the annual cycle is an interesting quantity displaying that the entire northern hemisphere reaches its extreme roughly at the same time of the year, namely in the late summer or beginning of autumn. The opposite is seen in the southern hemisphere where the southern Indian ocean and the southern Pacific reach an extreme roughly in April whereas the southern Atlantic peaks around the middle of February. On average the annual cycle in the oceans is three months later than the astronomical heating and cooling cycle confirming that steric heating of sea water needs time to build up. Interestingly enough the phase of the annual cycle does not change in the western boundary regions on the northern hemisphere from which one may conclude that western boundary currents don't show an annual meandering pattern. A deviation from the global annual ocean response sketched above takes place in the equatorial regions. Here we clearly observe the seasonal oscillation of the equatorial ridge/trough system as described by Wyrtki (1974).

Another interesting way of observing the annual cycle is by means of radiometric data gathered from Earth observing sensors. Knudsen et al. (1996) show in their plate 1 the annual cycle estimated from the Along Track Scanning Radiometer (ATSR) which flies on the ERS-1 satellite. Their results should be interpreted as a sea surface temperature estimate which does show an interesting difference with sea surface height estimates obtained from altimetry. Whereas ATSR will observe the sea *surface* temperature cycle altimetry will see the steric heating effects also containing signals from greater depths in the ocean. Solar heating can of course only occur at the surface but will be mixed in the Ekman layer thereby also warming up the colder sub-surface waters. The evidence for this process is clearly seen in the altimetric annual sea level estimates revealing an oscillation for the entire Gulf stream whereas the ATSR doesn't provide evidence for this effect. The only explanation can be that the altimeter is observing the annual cycle of the warm sub-surface waters which are transported into the Gulf stream area from other regions in the oceans.

4.3.3 Semi-annual variations. Finally we mention that semi-annual signals also exist in the oceans, and notably in the tropical Indian ocean where the behavior of the southwest monsoon is clearly observed. The amplitude of the semi-annual variations approaches 9 cm in this region and is consistent with the results shown in Perigaud and Delecluse (1992) and the semi-annual estimates obtained from the Levitus maps. One of the problems in estimating the semi-annual cycle from the Topex/Poseidon altimeter data is that it is correlated with the K_1 tide especially at high latitudes, see also Schrama and Ray (1994).

4.3.4 Large scale waves in the ocean. One of the great successes of GEOSAT and Topex/Poseidon has been their capability in showing eastward propagating structures in the tropical Pacific which are associated with El Niño Southern Oscillation (ENSO) events, see also Philander (1990). ENSO events are hard to observe from the individual T/P cycle maps displaying the global sea level anomalies relative to a mean but do become visible in so-called Hövmöller diagrams where longitude is displayed along the horizontal axis and time along the vertical axis at a certain chosen latitude. In

Nerem et al. (1994a) we have computed several of these time-longitude diagrams and we could find evidence for lineations which are a clear sign for propagating structures.

In the Topex/Poseidon Hövmöller diagram at the equator one can clearly identify a Kelvin wave starting at cycles 2-5 in the western Pacific moving linearly in time at a speed of about 238 cm/s until it reaches the eastern boundary of the Pacific basin. The phase speed is consistent with previous independent observations and models of Kubota and O'Brien (1988) as well as with contemporary observations from the TOGA Tropical Atmosphere Ocean observing system as discussed by Busalacchi et al. (1994).

Consistent patterns of westward propagating disturbances can also be identified in the Topex/Poseidon time-longitude plots, especially at low-latitudes, but not at the equator. These features must be identified with Rossby waves which originate from a reflection of the equatorial Kelvin wave on the eastern basin of the Pacific. In Nerem et al. (1994a) we have compared the observed phase speed with the theoretical phase speed for long-wavelength baroclinic Rossby waves as described by Gill (1982) and did notice a strong correlation at latitudes till approximately 25°. At higher latitude, the Rossby wave phase speeds are currently found to be 2–3 times faster than theoretical predictions based on standard theory, cf. Chelton and Schlax (1996), meaning that the response of high-latitude ocean to tropical events such as El Ninõ is much faster than previously thought, see also Fu et al. (1996).

4.3.5 Altimetric estimates of global sea level rise. Several sources have reported on the possibility of detecting global sea level rise by means of satellite altimetry. According to Nerem (1995) the estimated rate of global mean sea level rise using 2 years of Topex/Poseidon altimeter data is +5.8 mm/year with a formal error of 2.5 mm/year. Unfortunately there are several problems with such estimates: a) Callahan (pres.comm.) recently explained the erroneous implementation of an oscillator drift correction algorithm used for the generation of the T/P altimeter data records which erroneously increased the estimated rate of sea level rise. A proper application of the oscillator drift correction currently suggests a decreased or even no global trend. b) Any estimate of sea level rise may be associated with a short-term trend which may take place within several years as suggested by sea surface temperature estimates, c) Altimetric trend estimates do show spatial patterns which are dominated by regional interannual oscillations in the ocean, see also Fu et al. (1996).

5. Some closing remarks

Where will altimetry go and what will it mean for geodesy as a science in the near future? Clearly a significant part of realizing a satellite altimetry experiment involves space geodesy where a lot of progress has been made in the last few years. One of the main priorities in the past, namely the computation of accurate satellite orbits is an accomplished goal for TOPEX/Poseidon. The success of precision orbit determination of this mission is a result of the choice of the orbit, satellite gravity model improvements, and new tracking systems such as the French DORIS Doppler tracking system and GPS. Before the launch of TOPEX/Poseidon the goal within the US space agency, NASA, was to strive for a radial orbit accuracy of 13 cm. With the JGM-2 gravity model and state of the art tracking systems this goal is easily accomplished and 2.0 cm accuracy is currently realized, see also Tapley et al. (1994). Still the same can not be said for other altimeter missions such as ERS-1 where the radial orbit accuracy is more like 15 cm because of the failure of the PRARE tracking system, although the situation is likely to improve with ERS-2. Fortunately ERS-1 orbits can be improved radially via the altimeter itself by referencing to the TOPEX/Poseidon altimeter profiles. We conclude that the precision orbit determination activities are not yet finished and that there are still open challenges such as to further improve the orbit accuracy and also extend such computations for older altimeter satellites.

Tides are an important subject in geodesy and play a role within nearly every observation equation and even affect the definition of coordinate systems. In the context of altimetry we think that the problem of a reference ocean tide model still remains as an unfinished activity despite all the improvements we have seen from TOPEX/Poseidon. The problem of correlated constituents such as K_1 and Ssa won't be resolved until the timespan of the TOPEX/Poseidon altimeter dataset is increased. Moreover most altimetrically derived ocean tide models are only valid in the deep oceans and tides still remain a limiting factor for accurate altimeter applications in shallow waters.

We have seen that the sea surface observed by altimetry significantly differs from that of the marine geoid. Actually we don't yet fully understand this difference which obviously consists of an average field which we denoted by the mean dynamic topography including many non-tidal variations which occur on time scales much longer than previously thought. By means of satellite altimetry our mean sea level estimates are only an "average" estimate in the sense that they describe the sea level over the last 20 years. Long-term fluctuations of the sea level are not well understood. By means of altimetry our only hope is to extend the data record either by improving the older data or by collecting new data. This is one of the reasons that a phenomenon such as global sea level rise is currently difficult to detect by means of satellite altimeter data. Only under the above given time constraints it is possible to relate the mean sea level to a marine geoid.

We conclude these lectures by suggesting three closely related research strategies that all focus on the interaction between a marine geoid, the mean dynamic topography and the mean sea surface:

- A first strategy would be to demand adequate hydrographic knowledge of the mean dynamic ocean topography which is related to ocean circulation. In this case the marine geoid could be determined by subtracting such an estimate from an altimetrically determined mean sea surface. Although different hydrographic estimates of the mean dynamic topography exist, promising formal error estimates of the order of 5 cm are mentioned by Martel and Wunsch (1993). Subtracting their estimates from the altimetric mean sea level to obtain a marine geoid could therefore be a promising technique. An obvious drawback of this method is the geographical limitation of hydrographic estimates, ie. the Northern Atlantic is probably better observed than most Southern oceans. However this may be a technological problem and hydrographic sampling or the ability to run large global ocean circulation models may improve in the future.

- A second strategy would be to estimate the mean dynamic topography by subtracting a high precision gravimetric geoid from an altimetrically determined mean sea surface, cf. Rapp and Wang (1994). However in many areas there are no marine gravimetric geoids of sufficient accuracy so that the resolution of the mean dynamic topography is restricted to the longer wavelengths where gravity information is mostly constrained by satellite tracking information, see also Schrama (1989) and Nerem et al. (1994a). Possibly more realistic estimates of the mean dynamic topography may be obtained from the recent cooperation between the US Defense Mapping Agency, the Ohio State University and the Goddard Space Flight Center to use declassified gravity data over previously unchartered areas such as the former Soviet Union in their new model. An alternative option would be to realize a gravity explorer mission such as GOCE, an acronym that stands for a candidate Earth explorer mission with the title "Gravity Field and Steady-State Ocean Circulation", within the European Space Agency, ESA.

- A third strategy would be to extend the hydrographic estimation technique and to assimilate altimeter data into the model. Various studies exist on this issue; noteworthy is the paper by Wunsch and Gaposchkin (1980) and more recently by Ganachaud et al. (1996), Minster and Le Grand (1996) and Hasselmann and Giering (1996) which are closely related to ESA's decision to initiate the GOCE project. Minster and Le Grand (1996) and Hasselmann and Giering (1996) remark that the global picture of the ocean currents is already known to oceanographers and that the interest in getting a better estimate from altimetry and a dedicated gravity mission should come from the assimilation of the observed dynamic topography into ocean circulation models. After this assimilation procedure the adjusted dynamic parameters will contain vital information increasing our understanding of the coupled ocean atmosphere system.

Acknowledgments -- I'm indebted to Dick Rapp who provided essential remarks on the issue of the Earth's permanent tidal deformations, Kees Vreugdenhil, David Cartwright, Steve Nerem and Srinivas Bettadpur who helped to review these notes, Axel Smits who provided figures 2.1 and 3.1, Richard Ray who provided figures 3.2, 3.3, 3.4 and 3.5, and table 3.6, Paul Wessel and Walter H.F. Smith for making the generic mapping tools (GMT) available to the "public domain" and the European Space Agency for their financial support. These notes were typeset in LaTeX using the clmult01 package provided by Frank Holzwarth (Springer Verlag).

A. Laplace Tidal Equations

To demonstrate these equations we consider a box of water with the ground plane dimensions dx times dy and height h representing the mean depth of the ocean. Moreover let u_1 be the mean velocity of water entering the box via the $dy \times h$ plane from the west and u_2 the mean velocity of water leaving the box via the $dy \times h$ plane to the east. Also let v_1 be the mean velocity of water entering the box via the $dx \times h$ plane from the south and v_2 the mean velocity of water leaving the $dx \times h$ plane to the north. In case there are no additional sources or drains we find that:

$$h \, \partial y \, (u_2 - u_1) + h \, \partial x \, (v_2 - v_1) + \frac{\partial Volume}{\partial t} = 0 \tag{A.1}$$

where the volume is computed as $dx \, dy \, h$. Take ζ as the surface elevation due to the in-flux of water and: $\partial Volume / \partial t = dx \, dy \, (\partial \zeta / \partial t)$. If the latter equation is substituted in eq.(A.1) and all terms are divided by $dx \, dy$ we find:

$$h \left(\frac{\partial u}{\partial x} + \frac{\partial v}{\partial y} \right) + \frac{\partial \zeta}{\partial t} = 0 \tag{A.2}$$

The latter equation should now be combined with eq. (4.2) where the third equation can be simplified resulting in the hydrostatic approximation $p = g \, \rho \, \zeta$ resulting from a requirement that the pressure p is computed at a horizontal reference surface. After differentiation we get the horizontal pressure gradients:

$$\frac{-1}{\rho} \frac{\partial p}{\partial x} = \frac{\partial(-g\zeta)}{\partial x} \quad \text{and} \quad \frac{-1}{\rho} \frac{\partial p}{\partial y} = \frac{\partial(-g\zeta)}{\partial y} \tag{A.3}$$

Moreover for the forcing terms F_x and F_y in eq. (4.2) we substitute the horizontal gradients:

$$F_x = \frac{\partial U^a}{\partial x} + G_x \quad \text{and} \quad F_y = \frac{\partial U^a}{\partial y} + G_y \tag{A.4}$$

where U^a is the total tide generating potential and G_x and G_y horizontal accelerations due to friction. Substitution of eqns. (A.3) and (A.4) in eqn. (4.2) and elimination of the term $2\Omega \cos(\phi)w$ in the first equation lead to the Laplace tidal equations:

$$
\begin{aligned}
\frac{du}{dt} &= \frac{\partial}{\partial x}(-g\zeta + U^a) + f\,v + G_x \\
\frac{dv}{dt} &= \frac{\partial}{\partial y}(-g\zeta + U^a) - f\,u + G_y \\
\frac{d\zeta}{dt} &= -h\left(\frac{du}{dx} + \frac{dv}{dy} \right)
\end{aligned}
\tag{A.5}
$$

where it must be remarked that we ignored the effect of tidal loading for which the reader is referred to Cartwright (1993).

References

Bertiger W.I. et al. (1994): GPS precise tracking of TOPEX/Poseidon: results and implications, JGR Oceans, Vol 99, No C12, pp. 24449–24464.

Brown G.S. (1977): The average impulse response of a rough surface and its applications, IEEE transactions Antennas Propag. AP-25, 67–74.

Busalacchi A.J., M.J. McPhaden and J. Picaut (1994): Variability in equatorial Pacific sea surface topography during the verification phase of the TOPEX/Poseidon mission, JGR, Vol 99, No C12, pp 24725–24738.

Cartwright D.E. and R..J. Tayler (1971): New computations of the tide-generating potential. Geoph.J.R.Astr.Soc., 23, 45–74.

Cartwright D.E. and A.C. Edden (1973): Corrected tables of tidal harmonics, Geophys.J.R.astr.Soc. 33, 253–264.

Cartwright D.E., R.D. Ray and B.V. Sanchez (1991): Ocean tide maps and spherical harmonic coefficients from Geosat altimetry. NASA TM 104544, Greenbelt Md. 20771.

Cartwright D.E. (1993): Theory of Ocean Tides with Application to Satellite Altimetry, Lectures Notes in Earth Sciences, 50, "Satellite Altimetry in Geodesy and Oceanography", Springer.

Cheney R.E., J.G. Marsh and B.D. Beckly (1983): Global mesoscale variability from collinear tracks of SEASAT altimeter data, JGR, Vol 88 No C7, pp. 4343–4354.

Chelton D.B. and M.G. Schlax (1996): Global Observation of Oceanic Rossby waves, Science, in press.

Darwin G.H. (1883): The harmonic analysis of tidal observations, pp. 49-118 of British Association for the Advancement of Science – Report for 1883.

Doodson A.T. (1921): The harmonic development of the tide-generating potential, Proc.R.Soc. London, A,100, 305–329.

Dziewonski A.M. and D.L. Anderson (1981): Preliminary reference Earth model (PREM), Phys. Earth Planet. Int., 25, 297–356.

Egbert G.D., A.F. Bennett and M.G.G. Foreman (1994): TOPEX/Poseidon tides estimated using a global inverse model, JGR Oceans Vol 99 No C12 pp 24821–24852.

Farrell W.E. (1972): Deformation of the Earth by surface loads, Rev.Geophys. & Space Phys. 10,(3),761–737, 1972.

Fu L.L. and G. Pihos (1994): Sea level response to atmospheric pressure forcing using TOPEX/POSEIDON data, JGR oceans, Vol 99, No. C12, pp. 24633–24642.

Fu L.L., C.J. Koblinsky, J.F. Minster and J. Picaut (1996): Reflecting on the first three years of TOPEX/Poseidon, EOS, Vol 77, No. 12.

Hayne G.S., D.W. Hancock and C.L. Purdy (1994): The correction for significant wave height and attitude effects in the TOPEX radar altimeter, JGR oceans, Vol 99, No C12, pp. 24941–24955.

Hasselmann K. and R. Giering (1996): Impact of Geoid on Ocean Circulation Retrieval, Part 1: The global ocean circulation, ESTEC contract No. 11528/95/NL/CN.

Gaspar P. et al. (1994): Estimating the sea state bias of the TOPEX and Poseidon altimeters from cross-over differences, JGR oceans, Vol 99 No C12 pp 24981–24994.

Ganachaud A., C. Wunsch, M.C. Kim and B. Tapley (1996): Combination of TOPEX/Poseidon data with a hydrographic inversion for the determination of oceanic general circulation, to be submitted.

Gill A.E. (1982): Atmosphere-Ocean Dynamics, Academic Press, San Diego, Ca., 662 pp.

Knudsen P., O.B. Anderson and T. Knudsen (1996): ATSR sea surface temperature data in a global analysis with TOPEX Poseidon Altimetry, Geophysical Research Letters, Vol 23, No. 8, pages 821–824, April 15.

Kubota M., and J.J. O'Brien (1988): Variability of the upper tropical Pacific Ocean model, JGR, Vol 93 pp. 13930–13940.

Lambeck K. (1988): Geophysical Geodesy, The slow deformations of the Earth, Oxford Science Publications.

Lerch F.J., C.A. Wagner and S.M. Klosko (1981): Goddard Earth Model for Oceanographic Applications, (GEM10B and GEM10C), Marine Geodesy, Vol 5, pp 145–187.

Le Provost C., M.L. Genco, F. Lyard, P. Vincent, and P. Canceil (1994): Spectroscopy of the world ocean tides from a finite element hydrodynamical model. JGR Oceans, Vol 99 No C12 pp 24777–24797.

Levitus S. (1982): Climatological Atlas of the World Oceans, NOAA professional paper 13, 173 pp.

Love A.E.H. (1927): A treatise on the mathematical theory of elasticity (4th edn). Cambridge University Press (reprinted Dover 1944, New York).

Marsh J.G. and R.G. Williamson (1982): Seasat altimeter timing bias estimation, JGR Vol 87, No. C5 pp. 3232–3238.

Martel F. and C. Wunsch (1993): Combined inversion of hydrography, current meter data and altimetric elevations for the North Atlantic circulation, Manuscripta Geodaetica 18:219–226.

Minster J.F. and P. Le Grand (1996): Impact of Geoid on Ocean Circulation Retrieval, ESTEC Contract No. 11528/95/NL/CN.

Munk W.H. and D.E. Cartwright (1966): Tidal Spectroscopy and Prediction, Phil.Trans.R.Soc London, A,259,533–581.

Nerem R.S., E.J. Schrama, C.J. Koblinskv and B.D. Beckly, (1994a): A preliminary evaluation of ocean topography from the TOPEX Poseidon mission, JGR oceans, Vol 99, No C12, pp. 24565–24583.

Nerem R.S. et al. (1994b): Gravity Model Improvement for TOPEX Poseidon: Joint Gravity Models 1 and 2, 24421–24447, Vol 99 No. C12.

Nerem R.S. (1995), Measuring global mean sea level variations using TOPEX Poseidon altimeter data, JGR Vol 100, No C12, pp. 25135–25151.

Pagiatakis S.D. (1990): The response of a realistic earth to ocean tide loading, Geophys.J.Internat. 103,541–560.

Pedlosky J. (1987): Geophysical fluid dynamics, Springer.

Perigaud C. and P. Delecluse (1992): Annual sea-level variations in the tropical Indian Ocean from Geosat and shallow water simulations, JGR, Vol 97, 20169–20179.

Philander S.G. (1990): El Ninō, La Ninā, and the Southern Oscillation, Academic Press.

Pond and Pickard (1983): Introductory Dynamical Oceanography, 2nd edition, Pergamon Press.

Rapp R.H, R.S. Nerem, C.K. Shum, S.M. Klosko and R.G. Williamson, (1991a): Consideration of Permanent Tidal Deformation in the Orbit Determination and Data Analysis for the TOPEX/Poseidon Mission, NASA TM 100775, Greenbelt Md.

Rapp R.H., Y.M. Wang and N.K. Pavlis (1991b): The Ohio State 1991 geopotential and sea surface topography harmonic coefficient models, report 410, Dept. of Geodetic Science and Surveying, The Ohio State University, Columbus Ohio.

454

Rapp R.H. and Y.M. Wang (1994): Dynamic topography estimates using Geosat data and a gravimetric geoid in the Gulf Stream region, Geophys. J. Int. **117** (**2**), 511-528.

Ray R.D. and B.V. Sanchez (1989): Radial deformation of the Earth by oceanic tidal loading, NASA TM 100743, Greenbelt Md. 20771.

Rummel R. (1993): Principle of Satellite Altimetry and Elimination of Radial Orbit Errors, Lecture notes in Earth Sciences, 50, Springer.

Schrama E.J.O. (1989): The role of orbit errors in processing of satellite altimeter data, thesis Delft University of Technology, Netherlands Geodetic Commission, report 33.

Schrama E.J.O. (1992): Some remarks on the definition of geographically correlated orbit errors, consequences for satellite altimetry, Manuscripta Geodetica, Vol 17, pp. 282-294.

Schrama E.J.O. and R.D. Ray (1994): A preliminary tidal analysis of TOPEX-Poseidon altimetry, JGR Vol 99, No C12, pp 24799-24808.

Schwiderski E.W. (1980): On Charting Global Ocean Tides, Reviews of Geophysics and Space Physics, Vol 18, No 1, pp 243-268.

Semtner A.J. and R.M. Chervin (1992): Ocean general circulation from a global eddy-resolving model, JGR Vol 97, 5493-5550.

Semtner A.J. (1995): Modeling Ocean Circulation, Science v. 269,#5231.

Standish E.M. (1990): The Observational Basis for JPL's DE200, the Planetary Ephemeris of the Astronomical Almanac," Astron. Astrophys., 233, pp. 252-271.

Stommel H. (1957): A survey of ocean current theory, Deep-Sea Res., 4, 149-184.

Stommel H. (1984): The delicate interplay between wind-stress and buoyancy input in ocean circulation: the Goldsbrough variations, Tellus, 36A, 111-119.

Tapley B.D. and 14 others, Precision orbit determination for TOPEX/Poseidon, JGR Vol 99, No C12, pp. 24,383-24,404.

Veronis G. (1981): Dynamics of large-scale ocean circulation, Evolution of Physical Oceanography, Scientific Surveys in Honor of Henry Stommel, B.A. Warren and C. Wunsch, eds., The MIT Press, Cambridge, Ma, 140-183.

Wunsch C. (1977): Determining the general circulation of the oceans: A preliminary discussion. Science 196, 871-875.

Wunsch C. (1978): The North Atlantic general circulation west of 50° determined by inverse methods. Revs. Geophys. and Space Phys., 16, 583-620.

Wunsch C. and E.M. Gaposchkin (1980): On using satellite altimetry to determine the general circulation of the ocean with application to geoid improvements, Reviews of Geophysiscs and Space Physics, **18**, 725-745.

Wunsch C. (1993): Physics of Ocean Circulation, Lecture Notes in Earth Sciences, 50, "Satellite Altimetry in Geodesy and Oceanography", Springer Verlag.

Wyrtki K. (1974): Sea level and the seasonal fluctuations of the equatorial currents in the western Pacific Ocean, J.Phys.Oceanogr., 4(1), 91-103.

PART IV: SEMINARS

Stochastic Boundary Value Problem Theory : An Elementary Example

F. Sacerdote

1 Introduction

The general theory of boundary-value problems for the Poisson equation with stochastic right-hand side and boundary condition is comprehensively illustrated in the present volume (Rozanov and Sansò 1996). A former discussion connected to the formulation of geodetic overdetermined boundary-value problems was given in (Sansò 1988). Here a very elementary one-dimensional example is presented, in order to focus the main mathematical features of such problems, in a context which globally maintains their essential character, and enables to simplify as much as possible the technical apparatus of the proofs.

More precisely, the equation $u'' = \nu$ in the open interval $I =]0, 1[$ is investigated, where ν is a suitable random process on I. As the boundary contains only two isolated points, the mathematical structure of the boundary condition is very simple; for example, one may formulate a Dirichlet problem by setting $u(0) = \omega_0$, $u(1) = \omega_1$, where ω_i, $i = 0, 1$ are two random variables. A similar structure can be imposed for any kind of admissible boundary conditions, as will be illustrated in detail in the sequel.

It is useful to recall some properties of the corresponding differential equation in the usual sense, i.e. $u'' = f$, where now f is a "suitable" function. A Schwarz distribution u is said to be a solution of the differential equation if, for each test function $\phi \in C_0^\infty(I)$ the equality

$$< u, \phi'' > = \int_0^1 f\phi\, dt \tag{1.1}$$

holds, where the coupling $< \cdot, \cdot >$ describes the application of the distribution to a test function.

It is well-known that a sufficiently regular function (for example, square integrable) w can be viewed as a distribution with the coupling rule

$$< w, \psi > = \int_0^1 w\psi\, dt \quad , \quad \psi \in C_0^\infty(I)$$

If w is twice continuously differentiable, integrating by parts twice one obtains $< w'', \phi > = < w, \phi'' >$ $\forall \phi \in C_0^\infty(I)$. Consequently, any solution in the ordinary sense is also solution in the distribution sense.

Remark 1.1: The map $\phi \mapsto \psi = \phi''$ $(\phi \in C_0^\infty(I))$ is injective, as ϕ is the unique solution of the equation $\phi'' = \psi$ vanishing on the boundary. Therefore the inverse map D^{-2} is defined: $\phi = D^{-2}\psi$. The set of the functions ψ above defined does not cover the whole $C_0^\infty(I)$, but is a proper subspace; indeed it is subjected to linear constraints:

$$\int_0^1 \psi(t)dt = \phi'(1) - \phi'(0) = 0 \quad ;$$

$$0 = \phi(1) - \phi(0) = \int_0^1 \phi'(t)dt = \int_0^1 dt \int_0^t \psi(\tau)d\tau = \quad (1.2a)$$

$$= \int_0^1 d\tau \psi(\tau) \int_\tau^1 dt = \int_0^1 d\tau \psi(\tau)(1 - \tau)$$

From the two equalities in (1.2a) it follows

$$\int_0^1 t\psi(t)dt = 0 \quad (1.2b).$$

Equation (1.1) can be rewritten in the form

$$< u, \psi > = \int_0^1 f D^{-2}\psi dt \quad \forall \psi : \psi = \phi'' , \phi \in C_0^\infty(I) \quad (1.3)$$

Assume now f to be an arbitrary function in $L^2(I)$; the right-hand side of equation (1.1) can be immediately extended by continuity to any $\phi \in L^2(I)$, as $C_0^\infty(I)$ is dense in $L^2(I)$.

The problem is to introduce a suitable norm H for which the functional defined in (1.3) is bounded:

$$| < u, \psi > | \le c\|\psi\|_H \quad (1.4) .$$

This functional is now applied to the closure in H of the set of functions ψ , that may be a proper subspace of H ; if it is extended to the whole H , a solution is defined as an element of the dual space H^* . It turns out that Sobolev spaces are a natural frame to develop this theory, as illustrated in detail in the following sections.

Within this frame the definition of solution of the differential equation $u'' = \nu$ with stochastic right-hand side is quite straightforward. Indeed, one may assume ν to be a stochastic mean-square bounded functional on $L^2(I)$:

$$E| < \nu, \phi > |^2 \le c\|\phi\|_{L^2}^2 \quad (1.5) ;$$

similarly to (1.3) the differential equation can be written in the form

$$< u, \psi > = < \nu, D^{-2}\psi > \qquad (1.6)$$

Now u is a stochastic functional, but the space of test functions is exactly the same as above; the condition expressed by (1.4) here has simply to be replaced by the requirement for u to be a mean-square bounded stochastic functional:

$$E| < u, \psi > |^2 \leq c||\psi||_H^2 . \qquad (1.7)$$

Hence the problem is to study the extensions of u as a bounded stochastic functional on the whole H in the sense expressed by inequality (1.7). It will be seen that some of these extensions are equivalent to boundary conditions corresponding to well-posed problems; obviously on such extensions the discussion will be especially focused.

2 Some remarks on Sobolev spaces

Definitions and properties of Sobolev spaces $W^{k,p}$, where k is a positive integer and $p \geq 1$ are reported in all books of functional analysis (see for example Dautray and Lions 1988, vol.2, chap.IV). Here only the spaces $W^{k,2} \equiv H^k$ in an open interval I (possibly the whole real line) are introduced. A frequently used definition is

$$H^k(I) = \{f \in L^2(I) \mid f^{(j)} \in L^2(I) , \ j = 0, \cdots, k\} \qquad (2.1)$$

where $f^{(j)}$ is the j-th order distribution derivative. [1] The norm is defined as

$$||f||_k^2 = \sum_{j=0}^{k} ||f^{(j)}||_{L^2(I)}^2 \qquad (2.2)$$

It can be shown that the same spaces $H^k(I)$ are obtained if distribution derivatives are replaced by strong derivatives[2]

This definition is obviously meaningful only for positive integer k and its generalization to an arbitrary real index α is not straightforward.

If the domain is the whole real line, the space $\hat{H}^\alpha(\mathbf{R})$ is defined as the closure of $C_0^\infty(\mathbf{R})$ with respect to the norm

---

[1] Recall that the first-order derivative of a distribution v is defined by v' : $< v', \phi > = - < v, \phi' > \ \forall \phi \in C_0^\infty$.

[2] f^j is the j-th order strong derivative of f if there exists a sequence $\{f_n\}$, $f_n \in C^j(I)$ convergent in L^2 to f , such that the sequence of the j-order ordinary derivatives $f_n^{(j)}$ converges in L^2 to $f^{(j)}$.

$$\|\phi\|_\alpha^2 = \int_{-\infty}^{+\infty} (1+\lambda^2)^\alpha |\hat{\phi}(\lambda)|^2 d\lambda \qquad (2.3)$$

where $\hat{\phi}(\lambda)$ is the Fourier transform of $\phi(t)$.

Remark 2.1: For positive integer k the space $\hat{H}^k(\mathbf{R})$ coincides with $H^k(\mathbf{R})$, and the norms defined in (2.2) and (2.3) are equivalent. Indeed, if $f \in L^2(\mathbf{R})$, its Fourier transform $\hat{f}$ belongs to $L^2(\mathbf{R})$ too; furthermore $\widehat{f^{(j)}} = (i\lambda)^j \hat{f}$. Therefore $f^{(k)} \in L^2(\mathbf{R})$ if and only if $\lambda^{2k}|\hat{f}|^2$ has finite integral over the real line, and consequently the integral in (2.3) converges. From now on, without ambiguity, the notation $H^\alpha(\mathbf{R})$ instead of $\hat{H}^\alpha(\mathbf{R})$ will be used.

Remark 2.2: For arbitrary $\alpha \in \mathbf{R}$, $H^\alpha(\mathbf{R})$ and $H^{-\alpha}(\mathbf{R})$ are mutually dual; their coupling is defined by

$$< f, g >= \int_{-\infty}^{+\infty} \overline{\hat{f}(\lambda)} \hat{g}(\lambda) d\lambda \qquad f \in H^\alpha(\mathbf{R}) , \; g \in H^{-\alpha}(\mathbf{R}) \qquad (2.4)$$

Indeed, it is straightforward to prove the inequality

$$| < f, g > |^2 \le \int_{-\infty}^{+\infty} |\hat{f}(\lambda)|^2 (1+\lambda^2)^\alpha d\lambda \int_{-\infty}^{+\infty} |\hat{g}(\lambda)|^2 (1+\lambda^2)^{-\alpha} d\lambda \qquad (2.5)$$

Furthermore

$$\|g\|_{-\alpha} = \sup_{f \in H^\alpha, \|f\|_\alpha = 1} | < f, g > | \qquad (2.6)$$

Remark 2.3: It is clear from (2.3) that the elements in H^α for $\alpha > 0$ are functions in $L^2(\mathbf{R})$; for $\alpha < 0$, on the contrary, they may be distributions. For example, it follows from (2.3) that $f(t) = \delta_0(t)$, the Dirac delta localized at $t = 0$, belongs to $H^\alpha(\mathbf{R})$ for $\alpha < -1/2$. Indeed, $\delta_0(t)$ is the H^α-limit of sequences of strictly positive functions $\phi_n \in C_0^\infty(\mathbf{R})$, with supports in intervals $[a_n, b_n]$ with $\lim a_n = \lim b_n = 0$, and such that $\int_{-\infty}^{+\infty} \phi_n(t) dt = 1$. It is easy to see that $\lim_{n\to\infty} < \phi_n, \psi >= \psi(0) \; \forall \psi \in C_0^\infty(\mathbf{R})$.

Remark 2.4: By Sobolev embedding theorem, for any open set $\Omega \subseteq \mathbf{R}^n$, $H^k(\Omega) \subset C^r(\Omega)$ for indexes k , r such that $0 < n/(k-r) < 2$. This result is related to the one illustrated in the previous remark, i.e. $\delta_0 \in H^{-\alpha}(\mathbf{R})$, $\alpha > n/2$ and, more generally, the r-th order derivative $\delta_0^{(r)} \in H^k(\mathbf{R})$, $k = \alpha + r \Rightarrow k - r > n/2$. Indeed, δ_0 is applied to continuous functions ($r = 0$); $\delta_0^{(r)}$ is applied to functions r times continuously differentiable.

3 Some properties of weak solutions of ordinary differential equations

In view of the distribution form of the differential equation $u'' = f$, given by (1.1), (1.3) and of the inequality (1.4), that is required to hold for a suitable norm H, the natural choice of a space for ψ in (1.4) in the frame of Sobolev spaces is H^{-2}, as ψ is the second derivative of ϕ belonging to L^2, and the second derivative operator is bounded from L^2 to H^{-2}, as it is clear from the norm definition (2.3) and from Remark 2.1:

$$\|\phi''\|_{-2} \leq c\|\phi\|_{L^2} \tag{3.1}$$

As Sobolev spaces in their most general form have been introduced for functions defined on the whole real line, now it is necessary to extend the functions in $L^2(I)$ to $\mathbf{R}$ by giving zero values outside I; with this procedure a proper subspace of $L^2(\mathbf{R})$ is obtained.

It is remarkable that this extension is not admissible for $H^k(I)$ with positive integer k: a function vanishing outside I, whose restriction to I belongs to $H^k(I)$, generally does not belong to $H^k(\mathbf{R})$. This is immediately clear from the fact that, by Sobolev embedding theorem, functions in $H^k(\mathbf{R})$ for all integers $k \geq 1$ must be continuous. Therefore it is necessary to be very careful in making use of the spaces H^α with arbitrary α, above defined on $\mathbf{R}$, to study problems on bounded intervals.

Coming back to inequality (1.4), the aim is to prove that it holds when ψ is equipped with the H^{-2}-norm. From (1.3) it is clear that

$$|<u, \psi>| \leq c\|D^{-2}\psi\|_{L^2} \tag{3.2}$$

Therefore (1.4) holds if $\|D^{-2}\psi\|_{L^2} \leq c\|\psi\|_{-2}$, i.e. $\|\phi\|_{L^2} \leq c\|\phi''\|_{-2}$ or, written explicitly,

$$\int_{-\infty}^{+\infty} |\hat{\phi}(\lambda)|^2 d\lambda \leq C \int_{-\infty}^{+\infty} \frac{\lambda^4}{(1+\lambda^2)^2} |\hat{\phi}(\lambda)|^2 d\lambda \tag{3.3}$$

This inequality does not hold in the general case. Indeed it is possible to choose a sequence ϕ_n in L^2 such that

$$\int_{-\infty}^{+\infty} |\hat{\phi}_n(\lambda)|^2 d\lambda = 1 \quad ; \quad \int_{-\infty}^{+\infty} \frac{\lambda^4}{(1+\lambda^2)^2} |\hat{\phi}_n(\lambda)|^2 d\lambda \to 0 \tag{3.4}$$

Yet, this situation can be achieved only if the contribution to the first integral is, for increasing n, more and more concentrated around the origin, i.e. if, for arbitrary $\bar{\lambda} > 0$, $\epsilon > 0$, the inequality $\int_{-\bar{\lambda}}^{\bar{\lambda}} |\hat{\phi}_n(\lambda)|^2 d\lambda > 1 - \epsilon$ holds for sufficiently large n; this case cannot occur if the supports of ϕ_n are all contained in a fixed interval, as a consequence of the so-called uncertainty principle.

This result may seem at first sight surprising, as, for example, a non-vanishing constant function in $L^2(I)$ has a second derivative that is identically zero in I. Yet, it has to be remarked again that the space here considered is not $L^2(I)$, but the subspace of $L^2(\mathbf{R})$ whose elements vanish identically outside I; discontinuities at the boundary points of I imply that the second-order distribution derivatives at these points are first-order derivatives of the δ distribution, whose H^{-2}-norm obviously does not vanish.

The above discussion shows that the inequality (1.4) holds for $H \equiv H^{-2}(\mathbf{R})$; consequently the functional u is defined and bounded on the subspace $\mathcal{D}^2 = [\overline{D^2 C_0^\infty(I)}]_{-2}$ i.e. the closure with respect to the H^{-2}-norm of the space of the second-order derivatives of C_0^∞ functions with support in I.

The problem is now to study the extensions of u; yet, as the aim is to obtain the solutions of a differential equation on the interval I, it is not necessary to extend the functional to the whole $H^{-2}(\mathbf{R})$, but only to those distributions in $H^{-2}(\mathbf{R})$ that vanish when applied to functions different from zero only outside I.

To make this statement more precise, it is useful to introduce the concept of support of a distribution $x \in H^{-2}(\mathbf{R})$. First let the subset of the real line $\mathcal{O}_x$ be defined as follows: a point t belongs to $\mathcal{O}_x$ if there exists a neighbourhood Ω of t such that $< x, \phi > = 0$ for any $\phi \in C_0^\infty(\Omega)$. Then the support of x is the complement of $\mathcal{O}_x$, i.e. $\mathbf{R} \setminus \mathcal{O}_x$.

Then the subspace of $H^{-2}(\mathbf{R})$ to which the functional u has to be extended is precisely the space of the distributions whose support is contained in the closure $\bar{I}$ of the interval I. Let it be called $\widehat{H}^{-2}(\bar{I})$; it has to be distinguished from $H^{-2}(I)$ as defined in most books of functional analysis.[3]

A remarkable property of $\widehat{H}^{-2}(\bar{I})$ is that, if ϕ_1 and ϕ_2 are two functions in $C_0^\infty(\mathbf{R})$ such that $\phi_1 = \phi_2$ in I, then $< x, \phi_1 > = < x, \phi_2 >$ for every $x \in \widehat{H}^{-2}(\bar{I})$.

In view of this property, it is useful to collect all functions in $H^2(\mathbf{R})$ that are identically equal on I into equivalence classes. As $C_0^\infty(\mathbf{R})$ is dense in $H^2(\mathbf{R})$, it follows immediately that any functional in $\widehat{H}^{-2}(\bar{I})$ has the same value on all elements of one of these equivalence classes; therefore it can be viewed as a bounded functional on the space of the equivalence classes, here denoted as $\widehat{H}^2(I)$ and equipped with the norm

$$\|\mathbf{f}\| = \inf_{f \in \mathbf{f}} \|f\|_2 \tag{3.5}$$

It turns out that $\widehat{H}^{-2}(\bar{I})$ and $\widehat{H}^2(I)$ are mutually dual.

---

[3] Usually $H^{-2}(I)$ is defined as the dual of $H_0^2(I)$, that is the closure of $C_0^\infty(I)$ in $H^2(I)$; therefore it does not contain distributions concentrated on the boundary points of I, as $\widehat{H}^{-2}(\bar{I})$ does.

It is worth investigating the relation between $\widehat{H}^2(I)$ and the space $H^2(I)$ defined at the beginning of section 2. As a matter of fact, there is a one-to-one mapping between the elements of the two spaces. Indeed, every equivalence class can be identified with its common value on I, that is actually an element of $H^2(I)$. The converse can be proved by making use of the embedding theorem, according to which a function in $H^2(I)$ is continuously differentiable in $\bar{I}$ and consequently can be extended to a function in $H^2(\mathbf{R})$ with bounded support, for example by using piecewise low-degree polynomials. It is remarkable that this property does not hold for an arbitrary subset of $\mathbf{R}$: for example a function in $H^2(I \setminus a)$, where a is an arbitrary point in I, having different left and right limits for $t \to a$ cannot be extended to a function in $H^2(I)$.

As for the norms in the two spaces, they are certainly not equal. It can be proved that the inf in (3.5) is actually a minimum, that corresponds to a well defined element of the equivalence class, and cannot be equal to the H^2-norm of the restriction to I unless the extension vanishing outside I is admissible. Yet, it is not difficult to see that the two norms are equivalent. Without going into technical details, starting from any function in $H^2(I)$, it is possible to obtain a suitable extension with piecewise low-degree polynomials, as mentioned above, whose norm in $H^2(\mathbf{R})$ is linearly bounded by the values of the function and its first-order derivative at the boundary points. As these values are bounded functionals in $H^2(I)$, the norm equivalence is proved.

Now the main tools to investigate the differential equation and its boundary-value problems have been illustrated. As already mentioned above, the functional u defined by equation (1.1) as generalized solution of the differential equation and satisfying inequality (1.4) has to be extended from the subspace $\mathcal{D}^2$ to the whole $\widehat{H}^{-2}(\bar{I})$. In the present example the study of the admissible extensions is very simple, as the codimension of $\mathcal{D}^2$ with respect to $\widehat{H}^{-2}(\bar{I})$ is finite. Indeed, the subspace of $\widehat{H}^2(I)$, the dual of $\widehat{H}^{-2}(\bar{I})$, which vanishes when applied to all $\psi \in \mathcal{D}^2$ is two-dimensional, being made of constant and linear functions: $\phi(t) = a + bt$, as observed in (1.2a), (1.2b). Hence a 2-dimensional direct complement V of $\mathcal{D}^2$ can be defined (not uniquely):

$$\widehat{H}^{-2}(\bar{I}) = \mathcal{D}^2 \oplus V \qquad , \qquad \dim V = 2 \qquad (3.6)$$

such that any element x of $\widehat{H}^{-2}(\bar{I})$ can be written in a unique way as $x = x_1 + x_2$, where $x_1 \in \mathcal{D}^2$, $x_2 \in V$; furthermore, as V is a finite-dimensional closed subspace, the projection operator onto V defined by $Px = x_2$ is certainly bounded, as well as the projection $I - P$ onto $\mathcal{D}^2$.

The procedure to define V consists in choosing as basis for V two linearly independent elements of $\widehat{H}^{-2}(\bar{I})$ non-vanishing and acting independently as functionals on the subspace $\{a + bt\} \in \widehat{H}^2(I)$. Any extension of

u defined in equation (1.1) as a functional on $\hat{H}^{-2}(\bar{I})$, obtained by defining its action on the basis of V, identifies a particular generalized solution of the differential equation. Therefore, in order to investigate classical boundary-value and initial-value problems, it is important to show that distributions with support on the boundary of I can be chosen as basis elements of V. To any admissible choice corresponds a well defined boundary-value or initial-value problem whose solution is unique.

In dimension 1 the situation is indeed very simple. The boundary of I contains only 2 points, i.e. 0 and 1 ; furthermore a well-known theorem on distributions (Bremermann 1965, cap.4) states that any distribution with point support is a linear combination of δ and its derivatives. On the other hand, by Sobolev embedding theorem, $H^2(I) \subset C^r(I)$ for $r < 3/2$; therefore only δ and its first derivative, that are essentially the evaluation functionals for the functions $u \in H^2(I)$ and their first derivatives (both continuous), belong to $\hat{H}^{-2}(\bar{I})$. Hence the 4 functionals δ_0, δ_0', δ_1, δ_1' form a complete basis for the subspace $\hat{H}^{-2}(\partial I)$ of distributions in $\hat{H}^{-2}(\bar{I})$ with support on the boundary of I.

Now, a basis for a direct complement V of $\mathcal{D}^2$ can be chosen by selecting a pair out of the 4 elements mentioned above, satisfying the requirements previously illustrated. It turns out that only the pair $\{\delta_0', \delta_1'\}$ does not fit such requirements; indeed

$$< \delta_0', a + bt >=< \delta_1', a + bt >= b \qquad (3.7)$$

so that δ_0' and δ_1' do not act independently on the 2-dimensional subspace of linear functions. This corresponds to the well-known fact that the Neumann problem

$$\begin{cases} u'' = f & \text{in } I \\ u'(0) = a_0 , \ u'(1) = a_1 \end{cases} \qquad (3.8)$$

is not well-posed. Any other pair does work. For example, for the pair $\{\delta_0, \delta_1\}$ one obtains

$$< \delta_0, a + bt >= a \quad , \quad < \delta_1, a + bt >= a + b$$

and the corresponding (Dirichlet) boundary-value problem

$$\begin{cases} u'' = f & \text{in } I \\ u(0) = a_0 , \ u(1) = a_1 \end{cases} \qquad (3.9)$$

is well-posed. Similar conclusions can be drawn for the pairs $\{\delta_0, \delta_0'\}$ and $\{\delta_1, \delta_1'\}$, corresponding to Cauchy problems, and $\{\delta_0, \delta_1'\}$, $\{\delta_0', \delta_1\}$, corresponding to mixed problems.

4 Application to stochastic boundary-value problems

Now, it is possible to apply the results of the previous section to stochastic differential equations, following the scheme outlined in the introduction. If the right-hand side ν of the differential equation $u'' = \nu$ fulfils inequality (1.5), then the stochastic functional defined on $\mathcal{D}^2$ by (1.6) is such that

$$E| < u, \psi > |^2 \leq c \|\psi\|^2_{\widehat{H}^{-2}(\bar{I})} \tag{4.1}$$

As illustrated in the previous section, u can be extended as a continuous functional on the whole $\widehat{H}^{-2}(\bar{I})$ simply by setting, for example (Dirichlet problem),

$$\begin{aligned} < u, \delta_0 > \equiv u(0) = \omega_0 \\ < u, \delta_1 > \equiv u(1) = \omega_1 \end{aligned} \tag{4.2}$$

where ω_0 and ω_1 are random variables. A similar procedure is adopted for other kinds of admissible boundary or initial value problems.

Then, if x is an arbitrary element of $\widehat{H}^{-2}(\bar{I})$, it can be expressed as $x = \psi + c_0 \delta_0 + c_1 \delta_1$, where $\psi \in \mathcal{D}^2$. Furthermore, by virtue of the boundedness of the projection operators,

$$\begin{aligned} \|\psi\|_{\widehat{H}^{-2}(\bar{I})} &\leq c_\psi \|x\|_{\widehat{H}^{-2}(\bar{I})} \\ |c_0| \|\delta_0\|_{\widehat{H}^{-2}(\bar{I})} &\leq \bar{c}_0 \|x\|_{\widehat{H}^{-2}(\bar{I})} \\ |c_1| \|\delta_1\|_{\widehat{H}^{-2}(\bar{I})} &\leq \bar{c}_1 \|x\|_{\widehat{H}^{-2}(\bar{I})} \end{aligned} \tag{4.3}$$

From these inequalities it is easy to obtain

$$E| < u, x > |^2 \leq \bar{C} \|x\|^2_{\widehat{H}^{-2}(\bar{I})} \quad \forall x \in \widehat{H}^{-2}(\bar{I}) \tag{4.4}$$

Hence u is defined as a bounded stochastic functional on $\widehat{H}^{-2}(\bar{I})$; equation (1.6) identifies it as a solution of the differential equation, while (4.2) defines the boundary conditions. From the construction itself it is clear that u is uniquely defined.

Yet, this definition of solution is quite abstract: u is a random process whose samples are functionals, whose continuity properties are defined only in an average sense. Therefore it is convenient to investigate more deeply its properties, in order to obtain information on the features of individual samples. A good starting point is a mean-square continuity result:

$$E|u(t_2) - u(t_1)|^2 = E|<u, \delta_{t_2} - \delta_{t_1}>|^2 = C\|\delta_{t_2} - \delta_{t_1}\|^2_{\tilde{H}^{-2}(\tilde{I})} =$$

$$=C\int_{-\infty}^{+\infty} |e^{i\lambda t_2} - e^{i\lambda t_1}|^2(1+\lambda^2)^{-2}d\lambda =$$

$$=C\int_{-\infty}^{+\infty} |e^{i\lambda(t_2-t_1)} - 1|^2(1+\lambda^2)^{-2}d\lambda \le$$

$$\le \bar{C}(t_2-t_1)^2\int_{-\infty}^{+\infty} \lambda^2(1+\lambda^2)^{-2}d\lambda$$

(4.5)

where the last integral is convergent. Hence it is possible to apply the continuity theorem for random processes (see for example Baldi 1984), which, for the 1-dimensional case, states that, if the inequality

$$E|u(t_2) - u(t_1)|^\beta \le k|t_2 - t_1|^\alpha$$

holds for some $\alpha > 1$, then there exists a version of the process $u(t)$ Hölder continuous of degree γ for any $\gamma < (\alpha - 1)/\beta$. In the present case, with $\alpha = 2$, $\beta = 2$, the samples of the process belong to $\mathcal{C}^{1/2}$; therefore they are something more regular than continuous functions, although not differentiable. It is remarkable that this proof relies on the convergency of the integral in (4.5); hence it is not applicable in dimension higher than 1.

5   Concluding remarks

The example discussed in the present paper, being extremely elementary, allows to illustrate the procedure adopted for the transition from deterministic to stochastic boundary-value problems without technical difficulties, that in the general case are mainly related to a correct description of the functional spaces in which the well-posedness of the deterministic problem is established. The main point is the definition of the complementary subspace V (see (3.6)) to which the functional u (either deterministic or stochastic) has to be extended in order to be considered as a solution belonging to a well defined functional space. In the present example the discussion was very simple, as V was finite-dimensional; in $\mathbf{R}^n$ the dimension of V is generally infinite and its description is not so elementary. Yet, it has to be expected that the use of well-posedness results for deterministic boundary-value problems, widely available in the mathematical literature, may help to apply the present procedure to a large class of stochastic BVP's, including the ones relevant in the geodetic theory. Obviously, suitable results on random processes have to be used in order to obtain information on the properties of the samples of the stochastic solutions in the various cases.

References

Baldi, P. (1984) - Equazioni differenziali stocastiche e applicazioni, Quaderni dell'Unione Matematica Italiana, **28**, Pitagora.

Bremermann, H. (1965) - Distributions, Complex Variables and Fourier Transforms, Addison-Wesley.

Dautray, R. and J.-L. Lions (1988) - Mathematical Analysis and Numerical Methods for Science and Technology, Springer-Verlag.

Rozanov, Yu.A. and F. Sansò (1996) - Boundary value problems for harmonic random fields, in Proceedings of the Int. Summer School of Theor. Geodesy "Boundary value problems and the modeling of the earth's gravity field in view of the one centimeter geoid", Como.

Sansò, F. (1988) - The Wiener integral and the overdetermined boundary value problems of physical geodesy, Man. Geod., **13**, 75-98.

Variational Methods for Geodetic Boundary Value Problems

P. Holota

1. Introduction

Considering applications in geodesy, we will construct a Sobolev weight space for an unbounded solution domain first and deduce quantitative estimates for equivalent norms and for traces of functions on a lipschitzian boundary. Then, following the concept of the so-called weak solution and its natural tie to fundamental functional-analytic aspects, we give a generalized formulation for the linear Molodensky problem and for the linear gravimetric boundary value problem. A special decomposition of the Laplace operator will be shown to express the oblique derivative in the respective boundary conditions.

The ellipticity of bilinear forms associated with the boundary value problems in question and the role of the Lax-Milgram theorem in the solvability studies of these problems will be also mentioned. For details we refer to (Holota 1996a,b), where in addition a geometrical interpretation of the ellipticity conditions together with some basic aspect of a numerical solution and the convergence of Galerkin approximations are discussed. Modifications of these concepts for the invariance of the Molodensky problem under translation have been approached in (Holota 1996a).

The applications considered are associated with the need to know the gravity field and figure of the Earth. Note that by gravity we mean the resultant of the attractive force of the masses of the Earth, also called gravitation, and the centrifugal force of the Earth's rotation.

In general we have the following situation (idealized up to a certain degree). We consider 3-dimensional Euclidean space $\boldsymbol{R}^3$, where x_i, $i = 1, 2, 3$, are rectangular Cartesian coordinates with the origin at the center of gravity of the Earth. (For the general point $\boldsymbol{x} = (x_1, x_2, x_3)$ we put $|\boldsymbol{x}| = (\sum_{i=1}^{3} x_i^2)^{1/2}$.) We suppose that the Earth is a rigid body rotating together with the above system of coordinates with a known constant angular velocity ω around the x_3-axis. Finally, denoting by W and Γ the Earth's gravity potential and the surface of the Earth, respectively, we know that

$$\Delta W = 2\omega^2 \quad \text{outside} \quad \Gamma \,, \tag{1}$$

where Δ means the Laplace operator. At the surface of the Earth the potential

W has to satisfy boundary conditions which correspond to surface measurements done. (We will assume that hypothetically the boundary data cover Γ continuously, which obviously is an idealized case.)

The formulations above are rather general, but still more, in the Molodensky problem, e.g., the shape of Γ is taken for an unknown. Thus in this case the problem is to determine not only W outside Γ, but also the geometry of the embedding of Γ in $\mathbf{R}^3$. The problem in this case is a free boundary value problem, which, as such, is non-linear. Moreover, it is obvious that also the possibility to define location on Γ (i.e., to parametrize Γ) is a problem of crucial importance.

We will reduce explanations related to the linearization of the problem to a minimum. Instead we give some attention to the formulation of the linear Molodensky problem in the investigation of the external gravitational field and figure of the Earth.

Mathematically, solutions of free boundary value problems represent an attractive field of research. In the last two decades, however, new aspects appeared which initiated a discussion on alternative mathematical concepts underlying the gravity field determination from surface gravity data.

In (Backus 1968), (Koch 1971), (Koch and Pope 1972), (Bjerhammar and Svensson 1983), (Sacerdote and Sansò 1989) and (Grafarend 1989) the attention was paid to a non-linear gravimetric boundary value problem. The physical surface of the Earth is assumed to be known, while gravity measurements of the length of the gravity vector yield a boundary condition. The problem then is to determine gravity potential in the Earth's exterior.

We can agree with (Backus 1968) that in modern gravity surveys field intensities are measured quite accurately, while field directions are often not measured at all. On the other hand, however, the crucial point is the assumption concerning the figure of the Earth. Usually a reference is made to considerable advances in space geodetic measurements yielding possibilities to determine the figure of the Earth by methods which mostly are geometric in nature (e.g. by means of the global positioning system based on artificial satellites). In general, however, these aspects and also the density and accuracy of geodetic and gravimetric surveys need some discussion.

As regards our main intention the concept of the so-called classical solution seems to have a special position in geodesy. Indeed, by definition the classical solution is a sufficiently smooth function satisfying the partial differential equation in question and the respective boundary conditions pointwise. However, the classical definition is far too narrow and does not have the desired degree of universality. (Take e.g. the irregularity of the boundary, an oblique derivative boundary condition, mixed boundary value problems with applications in satellite altimetry etc.) Therefore, we try to look for a measurable function satisfying a certain integral identity associated with the boundary value problem in question. Note that this approach yields a system of linear equations for an approximation solution of the problem in a straight and transparent way.

2. Elementary Notions and Tools

2.1 Sobolev Weight Space

In the theory of elliptic partial differential equations variational (direct) methods and the concept of the weak (or generalized) solution are mostly considered for bounded solution domains. However, our aim is to develope a variational method for the solution of geodetic boundary value problems which are given for an unbounded domain Ω. Therefore, for Ω we will first formulate a basic apparatus as an analogue of the theory associated with bounded domains discussed e.g. in (Nečas 1967) and (Rektorys 1974).

Let $\varepsilon(\bar{\Omega})$ be a space of functions having derivatives of all orders continuous in Ω and continuously extendable to the closure $\bar{\Omega}$ of Ω, i.e. to $\bar{\Omega} = \Omega \cup \partial\Omega$, where $\partial\Omega$ is the boundary of Ω. Moreover, the functions from $\varepsilon(\bar{\Omega})$ are supposed to be equal zero in a neighbourhood of infinity (or to have a bounded support); naturally, these neighbourhoods vary from one function to another.

In $\varepsilon(\bar{\Omega})$, supposing that $|x| > 0$ for all $x \in \Omega$, we define an inner product:

$$(u,v)_1 \equiv \int_\Omega \frac{uv}{|x|^2}\, dx + \sum_{i=1}^{3} \int_\Omega \frac{\partial u}{\partial x_i} \frac{\partial v}{\partial x_i}\, dx \ . \tag{2}$$

It induces the norm:

$$(u,u)_1^{1/2} \equiv \|u\|_1 \ . \tag{3}$$

The completion of $\varepsilon(\bar{\Omega})$ in this norm represents a Sobolev weight space and will be denoted by $W_2^{(1)}$ in this paper. Roughly speaking, $W_2^{(1)}$ is a space produced by functions which are square integrable on Ω under the weight $|x|^{-2}$ and have derivatives of the first order in a certain generalized sense. Moreover, recalling the notion of the absolute continuity, we could show in a relatively easy way that $W_2^{(1)}$ cannot contain functions with discontinuities as e.g. a jump. On the other hand it is obvious that harmonic functions with their characteristic regularity at infinity belong to $W_2^{(1)}$.

2.2 Lipschitz Boundary

Intuitively, the existence and properties of boundary values of functions defined on Ω are influenced not only by the functions themselves, but also by the smoothness of $\partial\Omega$. In this paper, putting $\Omega' = \mathbf{R}^3 - \bar{\Omega}$, we will suppose that Ω' is the so-called domain with a Lipschitz boundary. For definition we refer to (Nečas 1967, Chap. 1), (Rektorys 1974, Chap. 28) or (Kufner et al. 1977, Chap. 6) etc. It can be stated that the Lipschitz boundary is already general enough to represent, with a small degree of idealization, boundaries of regularity comparable with the smoothness of the Earth's topography. Domains with Lipschitz boundaries are, e.g., the sphere, ellipsoid, cube, polyhedron as well as substantially more general domains with smooth or piecewise smooth boundaries etc. Let us note, however, that among domains with the Lipschitz boundary one cannot range those having singularities like highly sharp edges, vertices or turning points analogous to the two dimensional case. Moreover, it can be shown that the Lipschitz boundary has an outer (inner) normal almost everywhere, see (Nečas 1967, Chap. 2, Lemma 4.2) or (Kufner et al. 1977, Theorem 6.2.14). In the sequel we will continue using n for the (unit) outer normal of $\partial \Omega' \equiv \partial \Omega$.

2.3 Trace Theorem

It is obvious that every function from $\varepsilon(\bar{\Omega})$ has uniquely defined values on $\partial \Omega$. Let u be a function from $\varepsilon(\bar{\Omega})$. A restriction of $u(x)$ for $x \in \partial \Omega$ is usually referred to as the trace of u on $\partial \Omega$. Evidently, the trace of u is continuous and square integrable on $\partial \Omega$. In case of bounded domains the possibility to extend the notion of the trace to all functions from a Sobolev space is demonstrated in (Nečas 1967) Chap. 1, Theorem 1.2, Chap. 2, Theorem 4.2, Chap. 6, Theorem 2.2; (Kufner et al. 1977, Theorem 6.4.1 – 6.4.4) or (Nečas and Hlaváček 1981, Sect. 6.2). For our unbounded domain Ω the extension of the notion of the trace to all functions from $W_2^{(1)}$ can be constructed in a similar way. Denoting by $L_2(\partial \Omega)$ the space of square integrable functions on $\partial \Omega$, we give below the respective proof. However, here and in the sequel we will confine ourselves to domains Ω such that Ω' is *starshaped at the origin*.

Remark 2.1. Recall that the bounded domain Ω' is said to be starshaped at the origin if there exists a continuous and positive function h given on the unit sphere such that $\Omega' = \{x \in R^3; |x| < h(x/|x|) \text{ if } x \neq 0\} \cup \{0\}$, see (Kufner et al. 1977, Sect. 5.5).

This restriction will enable us to apply a relatively simple technique and to derive an effective quantitative estimate for the norm of the basic mapping involved. Below we will denote by $\langle \, , \, \rangle$ the inner (scalar) product of two vectors in R^3.

Theorem 1 *(Trace theorem). Let Ω be an unbounded domain and $\Omega' = R^3 - \bar{\Omega}$ be a starshaped domain at the origin with the Lipschitz boundary such that $\langle x, n \rangle > 0$ for almost all $x \in \partial \Omega$. Then there exists a continuous linear mapping Z from $W_2^{(1)}$ into $L_2(\partial \Omega)$ such that $Zu = u$ for all $u \in \varepsilon(\bar{\Omega})$ and*

$$\|Zu\|_{L_2(\partial \Omega)} = \|u\|_{L_2(\partial \Omega)} \leq \beta \|u\|_1 \,, \tag{4}$$

where β is a positive constant given by $\beta = 1/\sqrt{b}$ and

$$b = \inf_{\partial \Omega} \langle \frac{x}{|x|^2}, n \rangle = \inf_{\partial \Omega} \left[\frac{\cos(x, n)}{|x|} \right] \,. \tag{5}$$

Zu is called the trace of the function u on $\partial \Omega$. (Usually it is denoted just by u.)

Proof. It is enough to show that (4) is valid for all $u \in \varepsilon(\bar{\Omega})$. The proof can then be easily completed by means of a continuous extension of the mapping Z from the dense subspace $\varepsilon(\bar{\Omega})$ to the whole space $W_2^{(1)}$. Let

$$g = |\boldsymbol{x}|^{-1} \tag{6}$$

be an auxiliary function. Moreover, for $u \in \varepsilon(\bar{\Omega})$ let

$$v = ug^{-1} \quad \text{(i.e., } u = gv\text{)}. \tag{7}$$

Now an elementary calculation yields

$$\left(\frac{\partial u}{\partial x_i}\right)^2 = \left(\frac{\partial gv}{\partial x_i}\right)^2 = g\left(\frac{\partial v}{\partial x_i}\right)^2 + 2gv\frac{\partial g}{\partial x_i}\frac{\partial v}{\partial x_i} + v^2\left(\frac{\partial g}{\partial x_i}\right)^2 =$$

$$= g^2\left(\frac{\partial v}{\partial x_i}\right)^2 + \frac{\partial}{\partial x_i}\left(v^2 g\frac{\partial g}{\partial x_i}\right) - v^2 g\frac{\partial^2 g}{\partial x_i^2} . \tag{8}$$

and it is immediately obvious that

$$\left(\frac{\partial u}{\partial x_i}\right)^2 \geq \frac{\partial}{\partial x_i}\left(v^2 g\frac{\partial g}{\partial x_i}\right) - v^2 g\frac{\partial^2 g}{\partial x_i^2} . \tag{9}$$

In consequence

$$|\boldsymbol{grad}\, u|^2 \geq \sum_{i=1}^{3}\frac{\partial}{\partial x_i}\left(v^2 g\frac{\partial g}{\partial x_i}\right) - v^2 g \Delta g \tag{10}$$

and recalling that $\Delta g = 0$ in Ω, we have

$$|\boldsymbol{grad}\, u|^2 \geq \sum_{i=1}^{3}\frac{\partial}{\partial x_i}\left(v^2 g\frac{\partial g}{\partial x_i}\right) \quad \text{in} \quad \Omega . \tag{11}$$

Before the next step let us denote by R a positive constant such that $\Omega' \subset S_R$ for $S_R = \{\boldsymbol{x} \in \boldsymbol{R}^3, \ |\boldsymbol{x}| < R\}$. Integrating now the last inequality over $S_R \cap \Omega$ and using Green's theorem, we obtain

$$\int_{S_R \cap \Omega} |\boldsymbol{grad}\, u|^2 \, d\boldsymbol{x} \geq \int_{\partial S_R} v^2 g\frac{\partial g}{\partial n} \, dS - \int_{\partial \Omega} v^2 g\frac{\partial g}{\partial n} \, dS =$$

$$= \int_{\partial S_R} u^2 g^{-1}\frac{\partial g}{\partial n} \, dS - \int_{\partial \Omega} u^2 g^{-1}\frac{\partial g}{\partial n} \, dS , \tag{12}$$

where $\partial/\partial n$ denotes the derivative in the direction of the (unit) outer normal $\boldsymbol{n}$ of ∂S_R or $\partial \Omega'$, respectively. For $R \to \infty$ inequality (12) reduces to

$$-\int_{\partial \Omega} u^2 g^{-1}\frac{\partial g}{\partial n} \, dS \leq \int_{\Omega} |\boldsymbol{grad}\, u|^2 \, d\boldsymbol{x} \tag{13}$$

since, according to the definition, for every given function $u \in \varepsilon(\bar{\Omega})$ there exist a constant R_u such that

$$u = 0 \quad \text{and} \quad \frac{\partial^k u}{\partial x_i^k} = 0 , \tag{14}$$

in $\boldsymbol{R}^3 - S_{R_u}$, where $S_{R_u} = \{\boldsymbol{x} \in \boldsymbol{R}^3 , \ |\boldsymbol{x}| < R_u\}$ and $i = 1, 2, 3$, $k = 1, 2, 3, \dots$. Moreover, we easily deduce that

$$\frac{\partial g}{\partial n} = -\langle \frac{\boldsymbol{x}}{|\boldsymbol{x}|^3} , \boldsymbol{n} \rangle \tag{15}$$

which enables us to give (13) the final form

$$\int_{\partial\Omega} \langle \frac{\boldsymbol{x}}{|\boldsymbol{x}|^2} , \boldsymbol{n} \rangle u^2 \, dS \leq \int_\Omega |grad \, u|^2 \, d\boldsymbol{x} \ . \tag{16}$$

Using now our notation (5), we immediately arrive at

$$\int_{\partial\Omega} u^2 \, dS \leq \beta^2 \int_\Omega |grad \, u|^2 \, d\boldsymbol{x} \tag{17}$$

as we had to show, cf. (4). $\qquad\square$

Remark 2.2. Apart from the particular choice of g the fundamental trick (7)–(9) is the same as in (Rektorys 1974, Chap. 18). We have actually proved a stronger inequality than (4).

2.4 Equivalent Norms

Theorem 2. *Let Ω be an unbounded domain such that $\Omega' = \boldsymbol{R}^3 - \bar{\Omega}$ is a starshaped domain at the origin with the Lipschitz boundary. Then*

$$\int_\Omega \frac{u^2}{|\boldsymbol{x}|^2} \, d\boldsymbol{x} \leq 4 \int_\Omega |grad \, u|^2 \, d\boldsymbol{x} \tag{18}$$

for all $u \in W_2^{(1)}$.

Proof. A similar reasoning can be followed as in the proof of Theorem 1. However, for the auxiliary function we take

$$g = |\boldsymbol{x}|^{-(1+\varepsilon)} \ . \tag{19}$$

Let u be again a function from $\varepsilon(\bar{\Omega})$. Repeating literally the steps as in (7)–(10) and taking into consideration that

$$\Delta g = \frac{\varepsilon(1+\varepsilon)}{|\boldsymbol{x}|^2} g \quad \text{in} \quad \Omega , \tag{20}$$

we arrive at

$$|grad \, u|^2 \geq \sum_{i=1}^3 \frac{\partial}{\partial x_i}\left(v^2 g \frac{\partial g}{\partial x_i}\right) - \frac{\varepsilon(1+\varepsilon)}{|\boldsymbol{x}|^2} v^2 g^2 \ . \tag{21}$$

Integrating the last inequality over $S_R \cap \Omega$ and using Green's theorem, we obtain

$$\int_{S_R\cap\Omega} |grad \, u|^2 \, d\boldsymbol{x} \geq \int_{\partial S_R} u^2 \, g^{-1} \frac{\partial g}{\partial n} \, dS -$$

$$- \int_{\partial\Omega} u^2 \, g^{-1} \frac{\partial g}{\partial n} \, dS - \varepsilon(1+\varepsilon) \int_{S_R\cap\Omega} \frac{1}{|\boldsymbol{x}|^2} u^2 \, d\boldsymbol{x} \ , \tag{22}$$

where we have already applied (7). For $R \rightarrow \infty$ (20) reduces to

$$-\varepsilon(1+\varepsilon) \int_\Omega \frac{u^2}{|\boldsymbol{x}|^2}\, d\boldsymbol{x} - \int_{\partial\Omega} u^2\, g^{-1} \frac{\partial g}{\partial n}\, dS \leq \int_\Omega |\boldsymbol{grad}\, u|^2 \qquad (23)$$

since for every $u \in \varepsilon(\bar{\Omega})$ there exists a neighbourhood of infinity, where u equals zero. (Obviously, the same is true for all its derivatives.) In addition

$$g^{-1} \frac{\partial g}{\partial n} = -(1+\varepsilon) \langle \frac{\boldsymbol{x}}{|\boldsymbol{x}|^2}\, ,\, \boldsymbol{n} \rangle \,. \qquad (24)$$

Putting now $\varepsilon = -1/2$, we obtain

$$\frac{1}{4} \int_\Omega \frac{u^2}{|\boldsymbol{x}|^2}\, d\boldsymbol{x} + \frac{1}{2} \int_{\partial\Omega} \langle \frac{\boldsymbol{x}}{|\boldsymbol{x}|^2}\, ,\, \boldsymbol{n} \rangle\, u^2\, dS \leq \int_\Omega |\boldsymbol{grad}\, u|^2\, d\boldsymbol{x} \qquad (25)$$

which immediately yields (18) since according to our assumptions $\langle \boldsymbol{x}, \boldsymbol{n} \rangle$ is not negative for our domain Ω. Moreover, the extension of the proof to all functions from $W_2^{(1)}$ easily follows from the density of the subspace $\varepsilon(\bar{\Omega})$ in $W_2^{(1)}$ and completes the proof. $\qquad \square$

Theorem 3. *Let Ω be an unbounded domain such that $\Omega' = \boldsymbol{R}^3 - \bar{\Omega}$ is a starshaped domain at the origin with the Lipschitz boundary. Then for $u \in W_2^{(1)}$ the norm $\|u\|_1$ and*

$$\|u\| = \left(\int_\Omega |\boldsymbol{grad}\, u|^2\, d\boldsymbol{x} \right)^{1/2} \qquad (26)$$

are equivalent.

Proof. It is obvious that for $u \in W_2^{(1)}$

$$\|u\| \leq \|u\|_1 \leq \sqrt{5}\, \|u\| \,. \qquad (27)$$

All assertions easily follow from Theorem 2. $\qquad \square$

Remark 2.3. In (25) we actually have another proof of the trace theorem, but (25) yields somewhat weaker estimate then (17), only $\|u\|_{L_2(\partial\Omega)} \leq 2\beta\, \|u\|$.

3. Geodetic Boundary Value Problems

3.1 Molodensky Problem – an Elementary Exposition

By $\bar{W}$ and $\bar{g}$ we will denote restrictions of W and $\boldsymbol{g} = \boldsymbol{grad}\, W$ to Γ, respectively. The potential W is assumed to be known on Γ (possibly apart from an additive constant for which, theoretically, one can compensate by measuring at least one distance or by using satellite derived positions of surface points etc.). In addition also the gradient $\boldsymbol{g}$ is assumed to be known at every point of Γ.

On the other hand, Γ, i.e. the shape of the Earth's surface, is not known, but we will suppose that there exist a smooth and one-to-one correspondence between Γ and the unit sphere $\sigma = \{\boldsymbol{x} \in \boldsymbol{R}^3; x_1^2 + x_2^2 + x_3^2 = 1\}$. (In practice this may be materialized e.g. by means of observations of astronomical latitude and longitude on $\varGamma$.) In consequence the data $\bar{W}$ and $\bar{g}$ measured on $\varGamma$ may be considered functions defined on σ, i.e. $\bar{W} : \sigma \to R$ and $\bar{g} : \sigma \to R^3$. We will assume that they are corrected for gravitational interactions with extraterrestrial masses (such as the Moon, the Sun and the planets), for effects of precession, nutation and so on.

The problem now is to find a differentiable embedding of σ into R^3, i.e. a mapping $S : \sigma \to R^3$, and a function W such that: $\varDelta W = 2\omega^2$ in the exterior of $S(\sigma)$, where $S(\sigma) = \{x \in R^3; x = S(u), u \in \sigma\}$, and that for $u \in \sigma$

$$W[S(u)] = \bar{W}(u) \quad \text{and} \quad g[S(u)] = \bar{g}(u) , \tag{28}$$

cf. (Hörmander 1975). Moreover, for physical reasons we have to suppose that $W(x) = V(x) + (\omega^2/2)(x_1^2 + x_2^2)$, where V is a harmonic function (gravitational potential) outside $S(\sigma)$ such that $V(x) = c/|x| + O(|x|^{-3})$ for $x \to \infty$. This asymptotic condition actually means that V does not contain first degree harmonics. Note that $c = GM$, where G is the Newton gravitational constant and M is the total mass of the Earth. Thus, c should be also considered as an unknown.

The problem above has a solution, but under certain conditions. Even more, its solution is unique, at least locally. This, however, goes beyond the scope of this publication of a limited extent. Therefore, we rather refer to (Hörmander 1975). Here we will confine ourselves to the linearized problem.

We denote by U the so called normal potential of gravity of the Earth and by $\varDelta W$ and $\varDelta g$ the potential and the (vectorial) gravity anomaly, respectively. These anomalies are related to an adopted model of the solution of the non-linear Molodensky problem, i.e. to the potential U and to an adopted model of the Earth's surface. The latter is actually given by an approximation S' of the embedding S. In this paper we will suppose that the surface $S'(\sigma)$ is a boundary of a simple connected domain. Now recall that the linear Molodensky problem means to find T (a harmonic perturbation of U) which meets certain boundary condition at $S'(\sigma)$. We will denote the exterior of $S'(\sigma)$ by Ω and as usual we will call the boundary $\partial\Omega$ of Ω the telluroid. Obviously, $S'(\sigma) \equiv \partial\Omega$. Following (Krarup 1973), (Hörmander 1975) or (Moritz 1980), we can give the following formulation. Our problem is to find T such that

$$\varDelta T = 0 \quad \text{in} \quad \Omega , \tag{29}$$

$$\langle h, grad\, T \rangle + T = \varDelta W + \langle h, \varDelta g \rangle \quad \text{on} \quad \partial\Omega \tag{30}$$

and

$$T(x) = c^*/|x| + O(|x|^{-3}) \quad \text{for} \quad x \to \infty , \tag{31}$$

where c^* is a constant. The vector

$$h = -[M_{ij}]^{-1} grad\, U , \tag{32}$$

provided that for the Hessian $[M_{ij}] = [\partial^2 U/\partial x_i \partial x_j]$ (sometimes called the Marussi tensor) the following condition holds: $[M_{ij}(\boldsymbol{x})] \neq 0$, $\boldsymbol{x} \in \partial\Omega$. For standard choice of U the vector $\boldsymbol{h}$ is close to $\boldsymbol{x}/2$.

Remark 3.1. Recall that an essential part of the linear Molodensky problem is also to improve the shape of the adopted model of the Earth's surface, i.e., to find a (vectorial) perturbation $\boldsymbol{\zeta}$ of the figure of the telluroid. Note that this quantity is usually interpreted in terms of quasigeoid undulations and deflections of the vertical. Here, however, we give only the respective formula: $\boldsymbol{\zeta}(\boldsymbol{x}) = [M_{ij}(\boldsymbol{x})]^{-1}[\Delta g(\boldsymbol{x}) - \boldsymbol{grad}\,T(\boldsymbol{x})]$, $\boldsymbol{x} \in \partial\Omega$, and for details refer to (Krarup 1973), (Hörmander 1975) or (Moritz 1980) again.

3.2 Linear Gravimetric Boundary Value Problem

Following our notation, we put $g = |\boldsymbol{grad}\,W|$ for the measured gravity (corrected for gravitational interaction with the Moon, the Sun and the planets, for the precession, nutation and so on, according to our assumption). By analogy $\gamma = |\boldsymbol{grad}\,U|$ denotes the normal gravity. Moreover,

$$T(\boldsymbol{x}) = W(\boldsymbol{x}) - U(\boldsymbol{x}) \tag{33}$$

is the disturbing potential and

$$\delta g(\boldsymbol{x}) = g(\boldsymbol{x}) - \gamma(\boldsymbol{x}) \tag{34}$$

is usually known as the gravity disturbance.

The linear gravimetric boundary value problem is an oblique derivative problem, cf. (Koch and Pope 1972), (Bjerhammar and Svensson 1983) or (Grafarend 1989). Its solution domain is represented by the exterior of the Earth. We will denote it by Ω. Now the problem may be formulated as follows:

$$\Delta T = 0 \quad \text{in} \quad \Omega\,, \tag{35}$$

$$\langle s, \boldsymbol{grad}\,T \rangle = -\delta g \quad \text{on} \quad \partial\Omega\,, \tag{36}$$

where

$$s = -\frac{1}{\gamma}\,\boldsymbol{grad}\,U\,. \tag{37}$$

In addition T is assumed regular at infinity, i.e.,

$$T = O(|\boldsymbol{x}|^{-1}) \quad \text{as} \quad \boldsymbol{x} \to \infty\,. \tag{38}$$

The oblique derivative problem is sometimes known as the generalized Neumann problem, see (Bers et al. 1964), Part II, Sect. 2.4. Moreover, an analogue of the famous Schauder theory (for second order linear elliptic equations) has been developed for the (internal) regular oblique derivative problem, see (Gilbarg and Trudinger 1983), Sect. 6.7. In our case the problem is formulated for a highly irregular boundary given by the Earth's topography.

4. Generalized Oblique Derivative Boundary Value Problem

Let Ω be our unbounded domain such that Ω' is a domain with the Lipschitz boundary. Let $a = (a_1, a_2, a_3)$ be a vector field. We shall suppose that the components a_i and also $|x|\,(curl\,a)_i$, $i = 1, 2, 3$, are Lebesgue measurable functions which are defined and bounded almost everywhere on Ω. (Within a standard notation it means that they belong to the space $L_\infty(\Omega)$.) Denoting with $\times$ the vector product, we define by

$$A(v, u) = \int_\Omega \langle grad\,v, grad\,u \rangle\, dx - \int_\Omega \langle grad\,v, a \times grad\,u \rangle\, dx -$$

$$- \int_\Omega v\, \langle curl\,a, grad\,u \rangle\, dx \tag{39}$$

a bilinear form on $W_2^{(1)} \times W_2^{(1)}$. (We believe that the conventional notation $\times$ used now for the Cartesian product will not be a cause of any confusion.)

Moreover, on $W_2^{(1)} \times W_2^{(1)}$ let us define by

$$a(v, u) = \int_{\partial\Omega} \chi Z v Z u\, dS \tag{40}$$

a boundary bilinear form, where Z is the operator of traces and $\chi(x)$ is a Lebesgue measurable function defined and bounded almost everywhere on $\partial\Omega$. In a usual notation this means that $\chi \in L_\infty(\partial\Omega)$.

Putting now

$$((v, u)) = A(v, u) + a(v, u)\,, \tag{41}$$

we can show easily that this bilinear form is continuous on $W_2^{(1)} \times W_2^{(1)}$, i.e., we can find a constant m such that

$$|((v, u))| \le m\,\|v\|_1 \|u\|_1 \quad \text{for all} \quad v, u \in W_2^{(1)}\,. \tag{42}$$

Finally, let f be a function from $L_2(\partial\Omega)$.

We are now in the position to give a fundamental definition. A function $u \in W_2^{(1)}$ is the weak (or generalized) solution of an oblique derivative boundary value problem if

$$((v, u)) = \int_{\partial\Omega} v f\, dS \tag{43}$$

holds for all $v \in W_2^{(1)}$.

To make this clear we will suppose that the functions a_i, the boundary $\partial\Omega$ as well as the solution u are sufficiently smooth. Then, using Green's theorem and the identity: $curl\,ua = u\,curl\,a - a \times grad\,u$ (which is well-known in the field theory), we arrive at

$$((v, u)) = -\int_\Omega v\, \Delta u\, dx - \int_{\partial\Omega} v\,(\langle n + a \times n, grad\,u \rangle - \chi u)\, dS\,. \tag{44}$$

Finally, supposing that also f is sufficiently smooth and applying the usual reasoning to integrals from continuous functions, we obtain from (44)

$$\Delta u = 0 \quad \text{in} \quad \Omega \tag{45}$$

and

$$\langle \boldsymbol{\sigma}, \boldsymbol{grad}\, u \rangle - \chi u = -f \quad \text{on} \quad \partial\Omega , \tag{46}$$

where the vector

$$\boldsymbol{\sigma} = \boldsymbol{n} + \boldsymbol{a} \times \boldsymbol{n} \tag{47}$$

is oriented towards the exterior of the domain Ω' and is never tangential to its boundary. Indeed, we have

$$\langle \boldsymbol{\sigma}, \boldsymbol{n} \rangle = 1 . \tag{48}$$

It is now obvious that the above definition of the weak solution actually represents a meaningful generalization of the classical notion of the solution of an oblique derivative boundary value problem.

Remark 4.1. Inspecting the structure of (44), we can see that it actually represents a special decomposition of the Laplace operator, such that in an implicit way it expresses the oblique derivative in the respective boundary condition.

In particular, recalling the linear Molodensky problem (29)–(31) and assuming explicitly that

$$\langle \boldsymbol{h}, \boldsymbol{n} \rangle \neq 0 \quad \text{on} \quad \partial\Omega , \tag{49}$$

we can put

$$\boldsymbol{\sigma} = \boldsymbol{h}\langle \boldsymbol{h}, \boldsymbol{n} \rangle^{-1} , \ \chi = -\langle \boldsymbol{h}, \boldsymbol{n} \rangle^{-1} , \ f = -\langle \boldsymbol{h}, \boldsymbol{n} \rangle^{-1}[\Delta W + \langle \boldsymbol{h}, \Delta \boldsymbol{g} \rangle] \tag{50}$$

and convert boundary condition (29) into (46).

In the case of the linear gravimetric boundary value problem, putting

$$\boldsymbol{\sigma} = \boldsymbol{s}\langle \boldsymbol{s}, \boldsymbol{n} \rangle^{-1} , \ \chi = 0 \quad \text{and} \quad f = \langle \boldsymbol{s}, \boldsymbol{n} \rangle^{-1}\delta g , \tag{51}$$

we have a similar conversion between (36) and (46).

Remark 4.2. Our initial supposition concerning the regularity of u may be interpreted as a consequence of an adequate smoothness of all those data that define the problem in question. Problems like this constitute the topic of the regularity theory of weak solutions. A summary on general regularity results may be found in (Rektorys 1974, Chap. 46) or (Kufner et al. 1977, Sect. 4.11.1). A detailed theory is in (Nečas 1967, Chap. 4).

Finally, we deduce some additional relations between the vectors $\boldsymbol{a}, \boldsymbol{\sigma}, \boldsymbol{n}$, which hold on $\partial\Omega$ and result from (47). A direct calculation yields:

$$\langle \boldsymbol{\sigma} - \boldsymbol{n}, \boldsymbol{a} \rangle = 0 , \quad \boldsymbol{n} \times \boldsymbol{\sigma} = \boldsymbol{a} - \langle \boldsymbol{n}, \boldsymbol{a} \rangle \boldsymbol{n} , \tag{52}$$

$$|\sigma - n|^2 = |\sigma|^2 - 1 = \tan^2(\sigma, n) = |a|^2 - \langle n, a \rangle^2 . \tag{53}$$

Hence, it is obvious from (52) that for σ given, the trace on $\partial\Omega$ of the vector a (i.e. its restriction to $\partial\Omega$) lies in the plane perpendicular to the vector $\sigma - n$. This plane contains the vector n since $\langle \sigma - n, n \rangle = 0$ in view of (48). In addition from (53) we see that the modulus and the position of the trace of a in this plane are related in conformity with $|\sigma|^2 - 1 = |a|^2 \sin^2(n, a)$. This indeterminacy results from the singularity of the system (47). Its determinant equals zero.

5. Lax-Milgram Theorem and the Ellipticity

In this section our considerations will be based on a simple generalization of the well-known F. Riesz representation theorem.

Theorem 4 (*Lax-Milgram*). *Let H be a Hilbert space equipped by the inner product (v, u) and $B(v, u)$ be a continuous bilinear form defined on $H \times H$. Suppose that for all $v \in H$*

$$B(v, v) \geq \alpha \|v\|^2 , \quad \alpha = const. > 0 , \tag{54}$$

where $\|v\|^2 = (v, v)$. Then every linear continuous functional F on H can be represented by

$$Fv = B(v, u) , \quad v \in H , \tag{55}$$

where the element $u \in H$ is uniquely given by the functional F. Moreover,

$$\|u\| \leq \alpha^{-1} \|F\| , \tag{56}$$

where $\|F\|$ is the norm of the functional F.

The *proof* of this theorem can be found e.g. in (Nečas 1967), Chap. 1, Lemma 3.1 or (Rektorys 1974, Theorem 33.1). □

The Lax-Milgram theorem enables us to study the existence, uniqueness and stability of the weak solution of the generalized boundary value problem, i.e., it gives us also an aid to examine the linear Molodensky problem and the linear gravimetric boundary value problem. For this purpose we give the following definition.

The bilinear form $((v, u))$ is said to be $W_2^{(1)}$-elliptic if

$$((v, v)) \geq \alpha \|v\|_1^2 , \quad \alpha = const. > 0 , \tag{57}$$

holds for all $v \in W_2^{(1)}$.

We immediately see that the assumptions of the Lax-Milgram theorem will be met if we show that $((v, u))$ is $W_2^{(1)}$- elliptic and that the integral

$$Fv = \int_{\partial\Omega} vf \, dS \tag{58}$$

is a continuous functional on $W_2^{(1)}$.

From Hölder's inequality it follows immediately that

$$|Fv| \leq \|f\|_{L_2(\partial\Omega)} \|v\|_{L_2(\partial\Omega)} \tag{59}$$

for all $v \in W_2^{(1)}$. Moreover, recalling Theorem 1, we can write for all $v \in W_2^{(1)}$

$$|Fv| \leq \beta \|g\|_{L_2(\partial\Omega)} \|v\|_1 \tag{60}$$

with $\beta = 1/\sqrt{b}$ and b given by (5). From here it is clear that Fv is a continuous functional on $W_2^{(1)}$ as required in the Lax-Milgram theorem and it is obvious that

$$\|F\| \leq \beta \|f\|_{L_2(\partial\Omega)} . \tag{61}$$

Thus we still have to examine the $W_2^{(1)}$–ellipticity of the bilinear form $((v, u))$, i.e., to examine inequality (57). This, however, needs some computation and reasoning which cannot be reproduced here, mainly for page limitation. Therefore, we mention this problems in brief only in the last section.

6. Conclusion

In order to complete the discussion above we give two references. In the case of the Molodensky problem the details can be found in (Holota 1996a), where the results of the reasoning concern especially:

- problems associated with the ellipticity of the linear Molodensky problem violated by the invariance of the Molodensky problem under translation;

- decomposition of the Sobolev weight space $W_2^{(1)}$ and the Q–ellipticity of the linear Molodennsky problem (as an alternative to the $W_2^{(1)}$–ellipticity broken in a supplementary space spanned by the elementary potential and by the components of the normal gravity vector);

- some geometrical interpretations associated with the ellipticity of the problem;

- solvability conditions and the interpretation of the generalized Molodensky problem in terms of function bases.

In the case of the linear gravimetric boundary problem similar discussion can be found in (Holota 1996b). The situation here is somewhat simpler since under certain quantitative bounds concerning mainly the topography of the Earth's surface the linear gravimetric problem is $W_2^{(1)}$-elliptic. In the mentioned paper also a geometrical interpretation for the constant α appearing in the ellipticity inequality has been found and the convergence of approximation solutions constructed by means of the Galerkin method has been proved.

Acknowledgements. The work on this paper has been supported by the Grant Agency of the Czech Republic through Grant No. 205/96/0956. This support is gratefully acknowledged.

References

Backus G.E. (1968): Application of a non-linear boundary-value problem for Laplace's equation to gravity and geomagnetic intensity surveys. Quart. Journ. Mech. and Applied Math., Vol. XXI, Pt. 2, 195–221.

Bers L., John F., Schechter M. (1964): Partial differential equations. John Wiley and Sons, Inc., New York London Sydney.

Bjerhammar A., Svensson L. (1983): On the geodetic boundary-value problem for a fixed boundary surface – A satellite approach. Bull. Géod. 57, 382–393.

Gilbarg D., Trudinger N.S. (1983): Elliptic partial differential equations of second order. Springer-Vlg., Berlin etc.

Grafarend E.W. (1989): The geoid and the gravimetric boundary-value problem. Report No. 18 from the Dept. of Geod., The Royal Inst. of Technology, Stockholm.

Holota P. (1996a): Variational methods for quasigeoid determination. Proc. Session G7, XXI Gen. Assembly of EGS. The Hague, 1996. Publ. of the Finnish Geod. Inst., Helsinki (in print).

Holota P. (1996b): Coerciveness of the linear gravimetric boundary-value problem and a geometrical interpretation. Journal of Geodesy (submitted).

Hörmander L. (1975): The boundary problems of physical geodesy. The Royal Inst. of Technology, Division of Geodesy, Stockholm, 1975; also in: Archive for Rational Mechanics and Analysis 62(1976), 1–52.

Koch K.R. (1971): Die geodätische Randwertaufgabe bei bekannter Erdoberfläche, ZfV 96, 218–224.

Koch K.R., Pope A.J. (1972): Uniqueness and existence for the geodetic boundary-value problem using the known surface of the Earth. Bull. Géod. 106, 467–476.

Krarup T. (1973): Letters on Molodensky's problem III: Mathematical foundation of Molodensky's problem. Unpublished manuscript communicated to the members of IAG special study group 4.31.

Kufner A., John O., Fučík S. (1977): Function spaces. Academia, Prague.

Moritz H. (1980): Sovremennaya fizicheskaya geodeziya. Nedra Publishers, Moscow, 1983. English original "Advanced physical geodesy" published by H. Wichmann Vlg., Karlsruhe and Abacus Press, Tunbridge, Wells Kent.

Nečas J. (1967): Les méthodes directes en théorie des équations elliptiques. Academia, Prague.

Nečas J., Hlaváček I. (1981): Mathematical theory of elastic and elasto-plastic Bodies: An introduction. Elsevier Sci. Publishing Company, Amsterdam Oxford New York and SNTL Publishers of Technical Literature, Prague.

Rektorys K. (1974): Variační metody v inženýrských problémech a v problémech matematické fyziky. SNTL Publishers of Technical Literature, Prague 1974; also in English: Variational methods. Reidel Co., Dordrecht Boston, 1977.

Sacerdote F. and Sansò F. (1989): On the analysis of the fixed-boundary gravimetric boundary-value problem. Proc. of the 2nd Hotine-Marussi Symp. on Math. Geod., Pisa, Politecnico di Milano, 507–516.

Topics on Boundary Element Methods

R. Klees

1. Introduction

Geodetic boundary value problems form the mathematical foundation of global gravity field determination. Today, they are solved numerically by first reducing them formally to the Neumann or to the Robin boundary value problem of potential theory for spherical boundary surfaces; depending on whether the boundary is a priori known or not. This allows to express the solution analytically as a convolution integral over a spherical surface, e.g. Stokes' integral, Hotine's integral, or the formulas of Vening-Meinesz. Then, the integrals can be evaluated pointwise using numerical integration techniques or Fast Fourier Techniques. Alternatively, we may look for a parametrization of the solution in terms of surface spherical harmonics. Since they define an orthogonal system on the sphere with respect to the L^2-scalar product, the parameters can be calculated without solving a linear system of equations. This reduces the numerical effort considerably and enables the determination of the earth's gravity field with a resolution of 0.5° and higher. An alternative to integral formulas are least-squares techniques. After expressing the solution in terms of spherical harmonics, the boundary conditions define the mathematical model, i.e. the connection between observations on the boundary (boundary data) and the unknown spherical harmonic coefficients. Any type of boundary data can easily be taken into account by a weighted least-squares adjustment. The huge number of observations and parameters, however, makes the rigorous least squares procedure impractical unless some assumptions and/or approximations are made on the shape of the normal equations (e.g. Wenzel 1985, Bosch 1987, Rapp 1993). Least-squares collocation (e.g. Moritz 1989) may give optimal results but requires the solution of a linear system of equations the size of which grows proportional to the number of observations. They are therefore not applied to global gravity field determination but exclusively used for local and regional applications.

Most high resolution global gravity models available today have been calculated using integration techniques, sometimes combined with a least squares approach for a subset of parameters (e.g. Rapp & Pavlis 1990, Rapp 1993). One main drawback of that approach from a mathematical point of view is the reduction of the original boundary condition defined on a boundary with a complex geometry into Neumann or Robin boundary conditions on spherical surfaces. Unfortunately, most of the mathematical analysis of the different approximation steps has still to be done.

BVPs arise, however, not only in geodesy but also in other fields of engineering science. Here, deterministic methods have been developed for decades and applied successfully. The most popular methods in engineering applications are Finite Element Methods and Finite Difference Methods. Applications of Finite Difference Methods to three-dimensional BVPs for unbounded domains with complex boundary geometries have not been considered yet for those boundary conditions that are of interest for geodetic applications. This might become a topic for future research. Finite Element Methods, which appear to be the most widely used techniques among scientists and engineers, have already been applied to GBVPs, see Meissl (1981) and Baker (1988). However, the modelling techniques they use do not represent the state-of-the-art of Finite Element Methods anymore and do by the way not exploit the capabilities of modern computer hardware such as vector pipeline machines and parallel computers. Therefore, the capabilities of Finite Element Methods have not been exhausted for geodetic applications, which should give reasons enough to intensify the research in this field. Those who are interested in details should read e.g. Babuška & Aziz (1972), Ciarlet (1978), Ciarlet & Lions (1991), and Brenner & Scott (1994).

The oldest method, however, to treat BVPs are integral equation techniques. Their long history goes back to 1903 when Fredholm published his work on integral equations in potential theory. If the fundamental solution of the partial differential equation is explicitly known, as e.g. for Laplace's equation, it is easy to reduce the BVP to an equivalent integral equation on the boundary of the domain using (generalized) Green identities or source distributions on the boundary. Thereafter, the boundary integral equation can be discretized using finite elements on the boundary. This yields a linear system of equations which can be solved. Its solution approximates the theoretical solution of the integral equation. Finally, the solution of the BVP in the domain is obtained using integral representation formulas.

Before the seventies, equivalent integral equation formulation of BVPs were only used to derive existence and uniqueness results, but not to solve them numerically because of the computational complexity. But since the seventies these methods have become more and more popular and have been used to approximately solve boundary integral equations numerically. Today, numerical techniques to solve boundary integral equations using finite elements are mostly called 'Boundary Element Methods' (BEM). They have been introduced in geodesy by Koch (1967, 1970, 1971, 1972). In the meantime they have reached the same popularity as Finite Element Methods. Compared with the latter and with Finite Difference Methods they have the special property that the approximate solution is always an exact solution of the partial differential equation in the domain, parameterized by a finite set of parameters living on the boundary of the domain (Costabel 1987). They are motivated by the major drawback of the classical Finite Element and Finite Difference Methods, namely to require that the complete domain is discretized which is a serious problem for three-dimensional BVPs in unbounded domains. In addition, practical experience shows that boundary element techniques yield good accuracies even for fairly coarse discretizations. In view of the solution to GBVPs they also have the advantage that they are flexible w.r.t. the geometry of the boundary surface and the data distribution and allow local refinements. As main drawbacks we have to mention the complex mathematical analysis for making right choices (representation formula, numerical integration, approximation of the boundary surface), the error analysis which is not yet complete, the problem of singularities for non-smooth boundaries, and the computational load (numerical integration, solution of the linear system of equations, solution in the domain).

The aim of the lecture is to describe the basic principles of Boundary Element Methods in view of their applications to GBVPs. We show how to apply them to BVPs of potential theory and to (linearized) GBVPs such as the problem of Stokes and Molodensky, and the fixed gravimetric BVP. We do not focus on mixed GBVPs or BVPs using second derivatives of the gravitational potential, although boundary element techniques are suitable tools to solve them. We assume that the reader is familiar with the concept of the Lebesgue integral, Sobolev spaces, weak derivative etc. and (some other) basics of functional analysis; most of them will be provided by other lecturers.

The structure of the paper is the following: Section 2 deals with the mathematical model of the BVP which defines the starting point of every BEM application. We will show that the Poincaré BVP for the Laplace operator in unbounded domains (see eq. (2.1)) is important, because it includes the Dirichlet, the Neumann, and the Robin problem as special cases. The linearized Stokes and Molodensky BVPs as well as the linearized fixed gravimetric BVP can be considered as special cases, too. In Section 3 we discuss how to represent the solution of the BVP in the domain by means of boundary potentials. Thereafter, the transformation of the BVP into an integral equation over the boundary of the domain will be discussed (Section 4). We show that for every BVP there are different equivalent integral equation formulations defining integral operators with different mapping properties. The definition and construction of suitable finite dimensional function spaces in which we look for the approximate solution of the integral equation is the topic of Section 5. Thereafter, we discuss different methods for discretizing the boundary integral equation (Section 6). In Section 7 we briefly discuss the solution of the linear system of equations. Like any other numerical technique for solving BVPs, BEM also have their bottlenecks. In Section 8 we discuss the panel clustering technique and the multiscale Galerkin approach which aim at reducing the numerical effort and making BEM applicable to high resolution global gravity field determination. Besides, we briefly point out how to compute efficiently the entries of the Galerkin matrix that are defined by singular surface integrals. In Section 9 we discuss the problem of how to efficiently evaluate the representation formula when the solution of the BVP in the domain has to be computed. A state-of-the-art overview about error estimates is given in Section 10. We focus on the Galerkin discretization error and the error of numerical integration. In Section 11 we give a list of the most important references.

2. Mathematical model

The starting point of every boundary element application is the mathematical model of the BVP, consisting of a field equation and a boundary condition which connects the unknown function and the given data. The field equation enables the determination of the unknown function defined in a domain given only functionals on the boundary of the domain. For GBVPs which aim at the determination of the gravitational potential of the earth outside the earth's surface, Laplace's equation is mostly chosen as field equation. Deviations from the physical reality are taken into account by small corrections to the boundary data. Sometimes Poisson's equation is chosen as field equation, see Grafarend (1989) and Engels (1991). When applying BEM to these problems, special attention must be given to the treatment of the domain integral which arises in the integral equation. The problem will not be discussed further, some references will be given later.

The given boundary data are linear and/or non-linear functionals of the unknown gravity potential. An example for a non-linear functional of the earth's gravity potential is the gravity, defined as the magnitude of the gradient of the gravity potential. Other examples are gravity anomalies and gravity disturbances. An example of a linear functional is the gravity potential itself or the gravity vector, i.e. the gradient of the gravity potential. For the astronomical variants of Stokes' and Molodensky's BVP, the geometry of the boundary surface is not known. For their geodetic variants it is only partially known. The missing information about the geometry must then be determined from the boundary data. Therefore, more then one functional must be given on the boundary to uniquely determine geometry and potential. If, however, the boundary surface is assumed to be given, as e.g. for the fixed gravimetric BVP, one functional is sufficient. Though the mentioned GBVPs are non-linear, no methods have been investigated or even used that directly operate on the non-linear problems. Instead, using an approximation of the gravity potential and, if the geometry is unknown also for the earth's surface, all mentioned BVPs are linearized resulting in BVPs for harmonic functions with linear boundary conditions and fixed boundary surfaces. The surface can be a sphere, an ellipsoid of revolution, a telluroid, or even the earth's surface, depending on the BVP and the order of the approximation.

The classical BVPs of potential theory (Dirichlet, Neumann, Robin) but also the linearized GBVPs (Stokes, Molodensky, fixed gravimetric) are special cases of the exterior Poincaré problem for Laplace's equation:

$$\Delta u = 0 \text{ on } D^c \subset \mathbb{R}^3$$

$$Ru := a \cdot u + b \cdot D_\tau u = g \text{ on } S := \partial D$$

(2.1)

Here, $D \subset \mathbb{R}^3$ denotes a smooth and bounded domain, D^c its complement; a, b, and g are functions on the boundary surface S, and τ is a given unit vector field on S. Since the domain D^c is unbounded, u must also fulfil certain conditions at infinity which are necessary to get a unique solution of the BVP. They depend on the behaviour of the coefficients of the partial differential operator in a neighborhood of infinity. For harmonic functions the conditions are equivalent to

$$u(x) \to 0 \text{ for } |x| \to \infty.$$

(2.2)

Sometimes, also the function g has to fulfil some conditions to guarantee unique solvability of the BVP (e.g. Günter 1957; Miranda 1970; Hörmander 1976, Witsch 1985, Sacerdote & Sansò 1986). Depending on the choice of a, b, τ, and S, we can derive the BVPs of importance to us (cf. Table 2.1). In case of the linearized GBVPs u denotes the disturbing potential. The definition of S, τ, b and the inhomogeneity g are specified in Table 2.2. It should be noted that the vector and scalar Stokes BVP and the vector and scalar Molodensky BVP lead formally to the same linear BVP (cf. Table 2.2). However the definition of the boundary surface S and the unit vector field τ are different. In case of the linearized Stokes BVPs, S is the ellipsoid and in case of the linearized Molodensky BVPs, the telluroid. For the linearized scalar Stokes problem, τ is the normal unit vector field on the boundary S, whereas for the linearized vector Stokes problem and the two Molodensky problems, τ deviates from the normal unit vector field on S, i.e. τ is oblique. The linearized fixed gravimetric BVP is a classical oblique BVP; τ and g are specified in Table 2.2.

Table 2.1. BVPs as special case of Poincaré's BVP for Laplace's equation (n=normal unit vector field on S, τ oblique unit vector field on S)

a	b	τ	S	name of BVP
1	0		arbitrary	Dirichlet
0	1	τ=n	arbitrary	Neumann
a(x)	1	τ=n	arbitrary	Robin
0	1	$\tau \neq$ n	earth's surface	linearized fixed gravimetric
1	b(x)	$\tau \neq$ n/ τ=n	ellipsoid	linearized Stokes (vector/scalar)
1	b(x)	$\tau \neq$ n	telluroid	linearized Molodensky (vector/scalar)

For boundary element applications it is convenient to have a homogeneous differential equation (as Laplace's equation) and an inhomogeneous boundary condition (i.e. g≠0, cf. (2.1)). Homogeneous differential equations are not a necessary condition for the application of boundary element methods. Also inhomogeneous differential equations, e.g. Poissons's equation, can be treated. The inhomogeneity, however, gives rise to a Newton integral with the inhomogeneity as source density. To evaluate this domain integral, the domain could be discretized into cells. This, however, would increase considerably the numerical expenditure and the method would loose some of its advantages in relation to Finite Element Methods or other techniques. The so-called 'Dual Reciprocity Method' is an alternative which is generally applicable. It enables a "boundary only" solution without discretizing the domain into cells. The method is beyond the scope of this introduction and for those who are interested in I want to refer to Partridge et al. (1991) and the numerous references given there.

A necessary condition, however, to apply Boundary Element Methods is that the fundamental solution of the differential equation is given (see Section 3). For instance, the translation-invariant fundamental solution of Laplace's equation in $\mathbb{R}^3$ is

$$s(x-y) = \frac{1}{4\pi \cdot |x-y|}, \ x,y \in \mathbb{R}^3. \tag{2.3}$$

For the definition of fundamental solutions, we refer to Mizohata (1973).

Table 2.2. Boundary conditions of various GBVPs after linearization with respect to a reference gravity field

name of BVP	τ	b	g
linearized scalar Molodensky	$\dfrac{\gamma}{\|\gamma\|}$	$-\|\gamma\|\cdot\dfrac{<\tau,n_E>}{<\tau,\nabla\gamma\cdot n_E>}$	$b\cdot\Delta\gamma+\Delta C$
linearized vector Molodensky	$\dfrac{M^{-1}\cdot\gamma}{\|M^{-1}\cdot\gamma\|}$	$-\|M^{-1}\cdot\gamma\|$	$\Delta C+b\cdot<\tau,\dfrac{\delta\gamma}{\gamma}>\Delta\gamma$
simple Molodensky	$\dfrac{x}{\|x\|}$	$-\|\gamma\|\cdot\dfrac{\dfrac{\|x\|}{2}}{<n_E,\mathrm{grad}\,\gamma\cdot n_E>}$	$-\dfrac{\|x\|}{2}\Delta\gamma+\dfrac{\|x\|}{R}\Delta C$
linearized scalar Stokes	n_E	1	$b\Delta\gamma+\Delta C$
linearized fixed gravimetric	$\dfrac{\gamma}{\|\gamma\|}$	1	$\dfrac{\|\Gamma\|^2-\|\gamma\|^2}{2\|\gamma\|}$

Γ gravity vector, γ reference gravity vector, M Marussi tensor, n_E ellipsoidal normal unit vector, $\Delta\gamma$ scalar gravity anomaly; $\delta\gamma$ gravity anomaly vector; Marussi telluroid mapping; ΔC potential anomaly (cf. Heck 1991).

3. Representation formulae

In the next step we have to represent the solution of the BVP in the domain in terms of potentials generated by boundary charges supported by S. That is the reason why the fundamental solution must be known. Examples from potential theory are:

a) Green's third identity

$$u(x)=\int_S D_{n_y}s(x-y)u(y)dS_y - \int_S s(x-y)D_{n_y}u(y)dS_y,\ x\in D^c. \tag{3.1}$$

Table 3a: Integral equations for the linearized fixed gravimetric boundary value problem

boundary condition: $(D_\tau u)|_S = g$, integral equation: $(\lambda I - K)v = h$, $(Kv)(x) = \int_S k(x,y)v(y)\,dS_y$, $x \in S$

	generalized Green	single layer	double layer
representation formula $\lambda(x)$	$1/2$	$-\tfrac{1}{2}\langle n_x, \tau_x\rangle$	$\tfrac{1}{2}\langle \tau_x, \bar{\nabla}\rangle$
unknown $v(x)$	$u_{|S}(x)$	$\epsilon(x)$	$\mu(x)$
kernel function $k(x,y)$	$P_{v_y}\, s(y-x)$	$-D_{\tau_x} s(y-x)$	$-D_{\tau_x} D_{n_y}\, s(y-x)$
type kernel function $k(x,y)$	strongly singular	strongly singular	hyper-singular
right-hand side $h(x)$	$-\int_S \dfrac{s(y-x)}{\langle n_y,\tau_y\rangle} g(y)\,dS_y$	$g(x)$	$g(x)$

Table 3b: Integral equations for the linearized scalar Stokes boundary value problem

boundary condition: $(u + b \cdot D_a u)|_S = g$; integral equation: $(\lambda I - K)v = h$, $(Kv)(x) = \int_S k(x,y)v(y)\,dS_y$, $x \in S$

representation formula	Green		single layer	double layer
$\lambda(x)$	½	$-\tfrac{1}{2}\,b(x)$	$-\tfrac{1}{2}\,b(x)$	½
unknown $v(x)$	$u\big|_S(x)$	$D_a u\big|_S(x)$	$\epsilon(x)$	$\mu(x)$
kernel function $k(x,y)$	$D_{n_y}\,s(y-x) + \dfrac{s(y-x)}{b(y)}$	$-\big(s(y-x) + b(y)\cdot D_{n_y}\,s(y-x)\big)$	$-\big(s(y-x) + b(x)\cdot D_{a_x}\,s(y-x)\big)$	$\big(b(x)\cdot D_{a_x} D_{a_y}\,s(y-x) - D_{a_y}\,s(y-x)\big)$
type kernel function $k(x,y)$	weakly singular	weakly singular	weakly singular	hyper-singular
right-hand side $h(x)$	$-\displaystyle\int_S s(y-x)\cdot \dfrac{g(y)}{b(y)}\,dS_y$	$-\tfrac{1}{2}g(x) + \displaystyle\int_S g(y) D_{a_y}\,s(y-x)\,dS_y$	$g(x)$	$g(x)$

Table 3c: Integral equations for the linearized vector Molodensky and vector Stokes boundary value problem

boundary condition: $(u + b \cdot D_\tau u)|_S = g$, integral equation: $(\lambda I + \bar{D}_\tau - K) v = h$, $(Kv)(x) = \int_S k(x,y) v(y) \, dS_y$, $x \in S$

representation formula	generalized Green-identity	single layer	double layer
$\lambda(x)$	$3/2$	$-\tfrac{1}{2} b(x) <n_x, \tau_x>$	$-\tfrac{1}{2}$
$\bar{D}_\tau$	-	-	$-\tfrac{1}{2} <\tau_x, \bar{\nabla}>$
unknown $v(x)$	$u\vert_S(x)$	$\epsilon(x)$	$\mu(x)$
kernel function $k(x,y)$	$-\chi(y) \cdot s(y-x) + \dfrac{1}{\alpha(y)} D_{v_y} s(y-x) + \dfrac{s(y-x)}{\alpha(y) b(y)}$	$-s(y-x) - b(x) \cdot D_{\tau_x} s(y-x)$	$-D_{n_y} s(y-x) - b(x) D_{\tau_x} D_{n_y} s(y-x)$
type kernel function $k(x,y)$	strongly singular	strongly singular	hyper-singular
right-hand side $h(x)$	$-\displaystyle\int_S s(y-x) \cdot \dfrac{g(y)}{\alpha(y) b(y)} \, dS_y$	$g(x)$	$g(x)$

b) the generalized Green-identity

$$u(x) = \int_S P_{v_y} s(x-y) u(y) dS_y \; - \; \int_S s(x-y) P_{\tau_y} u(y) dS_y, \; x \in D^c \; ,$$

(3.2)

$$P_v := \frac{1}{<n,\tau>} D_v - \chi I, \; P_\tau := \frac{1}{<n,\tau>} D_\tau, \; v := 2<n,\tau> n - \tau \; .$$

Here, τ is an arbitrary unit vector field on the boundary. Note that v is oblique if τ does. The function χ depends on the curvature of the boundary S (see Miranda 1970; Klees 1992).

c) the single layer representation

$$u(x) = \int_S s(x-y) \epsilon(y) dS_y, \; x \in D^c.$$

(3.3)

d) the double layer representation

$$u(x) = \int_S D_{n_y} s(x-y) \mu(y) dS_y, \; x \in D^c.$$

(3.4)

When $y \in S$ we define

$$u(y) := \lim_{x \to y, x \in D^c} u(x), \; D_{e(y)} u(y) := \lim_{t \to +0} \frac{u(y+t \cdot e) - u(y)}{t},$$

(3.5)

where e represents any of the unit vectors n, τ, or v. The limits are taken in the pointwise sense. The normal unit vector n is always directed into D^c, and τ points always into D. Note that (3.1) can be expressed as a linear combination of (3.3) and (3.4) with $\epsilon = D_n u$ and $\mu = u$. In the next section we will see that Green-type representation formulae lead to integral equations that only involve physical boundary quantities as unknowns, i.e. values of the harmonic function and its directional derivative. Therefore, the method is called "direct integral equation method". It differs from the classical "source density approach", i.e. the use of single and double layer representation, which leads to integral equations for source densities. After they have been solved for, the unknown potential can be calculated everywhere in D^c. Thus, these methods are called "indirect integral equation methods". Brovar (1963, 1964) derived alternative representation formulae with properties in D^c similar to the potential of a single layer, a double layer and the potential of a volume mass distribution. They may be used as alternative to derive integral equations to GBVPs.

4. Jump relations and boundary integral equations

The representation formula yields a description of the solution of the BVP in the domain D^c. Not yet determined are the boundary potentials; they have to be

chosen such that the boundary condition is fulfilled. We simply insert the representation formula into the boundary condition and take into account the jump relations of boundary potentials (e.g. Courant & Hilbert 1962; Miranda 1970). This gives the desired linear boundary integral equation.

Let $H^t(S)$, $t \in \mathbb{R}^+$, denote the usual Sobolev-Slobodeckij space and $H^{-t}(S)$ the dual of $H^t(S)$ with respect to the $L^2(S)$ inner product (cf. Adams 1975). The important jump relations and the mapping properties of the corresponding operators are:

a) single layer potential ($\epsilon \in H^t(S) \Rightarrow u \in H^{t+1}(S)$, $\nabla u \in H^t(S)$, $t \in \mathbb{R}$)

$$\lim_{z \to x, z \in D} u(z) = \int_S s(x-y)\epsilon(y)\,dS_y, \quad x \in S,$$

$$\lim_{z \to x, z \in D} \nabla u(z) = -\frac{1}{2}\epsilon(x)n(x) + \int_S \nabla_x s(x-y)\epsilon(y)\,dS_y, \quad x \in S. \tag{4.1}$$

b) double layer potential ($\mu \in H^t(S) \Rightarrow u \in H^t(S)$, $\nabla u \in H^{t-1}(S)$, $t \in \mathbb{R}$)

$$\lim_{z \to x, z \in D^c} u(z) = \frac{1}{2}\mu(x) + \int_S D_{n_y} s(x-y)\mu(y)\,dS_y, \quad x \in S,$$

$$\lim_{z \to x, z \in D^c} \nabla u(z) = \frac{1}{2}\bar{\nabla}\mu(x) - \int_S \left[\nabla_x s(x-y) \wedge \left(\bar{\nabla}\mu(y) \wedge n(y)\right)\right]dS_y, \quad x \in S \tag{4.2}$$

$$= \frac{1}{2}\bar{\nabla}\mu(x) - \int_S \nabla_x\big(D_{n_y} s(x-y)\big)\mu(y)\,dS_y, \quad x \in S.$$

(c) Green's third identity ($u \in H^t(S) \Rightarrow u \in H^t(S)$, $\nabla u \in H^{t-1}(S)$, $t \in \mathbb{R}$)

$$\lim_{z \to x, z \in D^c} u(z) = \frac{1}{2}u(x) - \int_S s(x-y)D_{n_y}u(y)\,dS_y + \int_S D_{n_y} s(x-y)u(y)\,dS_y, \quad x \in S,$$

$$\lim_{z \to x, z \in D^c} \nabla u(z) = \frac{1}{2}\bar{\nabla}u(x) - \int_S \left[\nabla_x s(x-y) \wedge \left(\bar{\nabla}u(y) \wedge n(y)\right)\right]dS_y \tag{4.3}$$

$$+ \frac{1}{2}D_{n_x}u(x)n(x) - \int_S \nabla_x s(x-y)D_{n_y}u(y)\,dS_y, \quad x \in S.$$

(d) Generalized Green-identity ($u \in H^t(S) \Rightarrow u \in H^t(S)$, $t \in \mathbb{R}$)

$$\lim_{z \to x, z \in D^c} u(z) = \frac{1}{2}u(x) + \int_S P_{v_y} s(x-y)u(y)\,dS_y - \int_S s(x-y)P_{\tau_y}u(y)\,dS_y, \quad x \in S. \tag{4.4}$$

Remarks: (1) We want to note that the first integrals on the right hand side of (4.1) and (4.2) exist as improper integrals, if the boundary surface is at least piecewise C^1 and the source density is at least bounded and at x continuous. The second integral in (4.1) exists as Cauchy principal value, and in (4.2) as Hadamard integral or finite part integral, if in addition the boundary surface is from C^{1+s}, $0<s\leq 1$ at x, and the source densities at x are from C^s and C^{1+s}, respectively (cf. Hackbusch 1989).

(2) Since μ is a function on the boundary S, only tangential derivatives exist. Thus, $\bar{\nabla}\mu$ denotes the surface gradient of μ. If we formally set the derivative into the direction of the normal unit vector n equal to zero, the surface gradient can be expressed by the usual gradient of μ, since the latter is fully defined. Thus, $<\nabla\mu,\tau>$ is nothing else but the directional derivative of μ into the direction of the tangential component of τ. If $\tau=n$ at a point x on S then $<\nabla\mu,\tau>=0$.

Tables 3a-3c contain some integral equations for the GBVPs listed in Table 2.2. The examples show that we obtain, depending on the representation formula, different boundary integral equations with different mapping properties to one and the same BVP. Hartley they can be written as

$$(Av)(x):=\lambda(x)v(x)-(Kv)(x) = f(x), \ x\in S. \tag{4.5}$$

If $\lambda(x) \equiv 0$ then (4.5) is called integral equation of the first kind; if $\lambda(x) \neq 0$ for all $x\in S$, then (4.5) is called integral equation of the second kind. The unknown function v denotes the (disturbing) potential or its directional derivative restricted to the boundary (direct methods) or source densities on the boundary (indirect methods). Usually the operator K has the form

$$(Kv)(x) = \int_S k(x,y-x)v(y)dS_y \tag{4.6}$$

with a certain kernel function $k(x,z)$. $k(x,z)$ is piecewise smooth in x, dependent of the smoothness of the boundary S, and singular for $z=0$. In a neighbourhood U of $z=0$, it is bounded by

$$C_1(x)\cdot |z|^{-s}\leq k(x,z)\leq C_2(x)\cdot |z|^{-s} \text{ for } z\in U; \tag{4.7}$$

s denotes the order of the singularity. For $0<s<2$, $k(x,z)$ is called weakly singular, and the integral $(Kv)(x)$ exists as improper integral if the surface does not contain edges or corners. For $s=2$, $k(x,z)$ is called strongly (or Cauchy) singular and $(Kv)(x)$ may exists as Cauchy principal integral. For $s>2$, $k(x,z)$ is called hyper-singular, and $(Kv)(x)$ is defined as finite part integral. For the definition of the Cauchy principal value and the finite part integral see Section 9.

Obviously, each different representation formula yields a different integral equation, and we may obtain a variety of integral equations for one and the same BVP. Therefore, a convenient choice has to be made, depending on the application. Two remarks should be allowed:

(1) The equivalence between the BVP and each integral equation is not automatically guaranteed but must be proved. Besides, existence and uniqueness of the integral equation have to be proved, as well. When Green-type identities are used the equivalence follows automatically. If the integral operator is a strongly elliptic pseudo-differential operator, existence follows from uniqueness due to Fredholm's alternative (cf. Section 10). For most of the integral equations in Table 3a-3c equivalence, uniqueness, and existence have not been proved yet.

(2) Assuming that equivalence, existence, and uniqueness have been proved, a right choice of the integral equation mainly depends on numerical aspects. For instance, integral equations of the second kind ($\lambda(x) \neq 0$) are normally preferred to first kind equations ($\lambda(x) \equiv 0$) because of stability reasons. Weakly singular integral operators are preferred to strongly singular or hyper-singular integral operators because of the problems of numerical integration. However, we will show in Section 9 that fast numerical algorithms benefit from a high order of singularity. Integral equations based on Green-type identities immediately yield the missing Cauchy data (potential values or directional derivatives) on the boundary without any further calculations, whereas indirect formulations only yield a function with no phyiscal meaning from which the Cauchy data must be calculated. On the other hand, the inhomogeneous term of the integral equation is much more difficult to calculate when Green-type identities are used because it always involves integration of the boundary data over the boundary. The corresponding integral operator may be a smoothing operator, which is an advantage for geodetic applications; if it acts like a differentiation, the corresponding integral equation should not be used. Moreover, sometimes the solution must have a higher regularity, as e.g. for the Dirichlet problem where the normal derivative of the solution must exist, an assumption which is not necessary for indirect formulations.

One major drawback of the integral equation approach to solving BVPs should be mentioned: it is the strong smoothness assumption on the boundary surface. For the integral equations mentioned here at least boundary surfaces of class C^{1+s} are needed, with $0 < s \leq 1$. If the boundary is allowed to have edges and corners the integral equations and the mapping properties of the integral operators will change, see e.g. Fabes (1988) and Král & Wendland (1988).

5. Construction of the approximating subspace

There are various methods to approximately solve an integral equation $Av=f$. If $v \in H$ and H is some Hilbert space, one mostly looks for an approximation $v_h \in V_h \subset H$, where V_h denotes some finite dimensional subspace of H. We can use e.g. a truncated series of spherical harmonics or global polynomials. However, proper Boundary Element Methods use always functions with finite support on the boundary, so-called finite element functions. They are constructed in three steps:

Step 1: The boundary surface S is partitioned into finitely many patches $\{S_i: i=1,...,N\}$, called panels, finite elements, or boundary elements. Mostly triangular elements are used because they are easy to handle and more flexible than e.g, rectangular elements. Therefore, we will assume in the following that the panels S_i are triangles. Each S_i can be considered to be the image of a C^r-application $x=\chi_i(u)$, where u is defined on the (standard) triangle $T:= \{u=(u,v)^T: 0 \leq u \leq 1, 0 \leq v \leq u\}$ (cf. Figure 5.1). Thus, $S_i=\chi_i(T)$. Mostly, the mappings $\chi_i(u)$ are polynomials of degree r in each component:

$$\chi_{i,j} = \sum_{|\alpha| \leq r} d_{\alpha,j} u^\alpha, \quad j \in \{1,2,3\}. \tag{5.1}$$

Step 2: Let P_k denote the space of polynomials of total degree k. On T we define algebraic polynomials $p \in P_k$ with $n_k=1/2(k+1)(k+2)$ degrees of freedom:

$$p(u) = \sum_{1 \leq j \leq n_k} p_j b_j(u), \tag{5.2}$$

where $\{b_j: j=1,...,n_k\}$ denote the basis functions, which are sometimes called trial functions. The basis is mostly taken from the nodal point representation of classical Finite Element Methods, e.g. Schwarz (1980), Zienkiewicz (1983), Zienkiewicz & Morgan (1983), Brenner & Scott (1994). Then, $b_j(u_i)=\delta_{ji}$, i.e. the basis functions b_j define a Lagrangian basis.

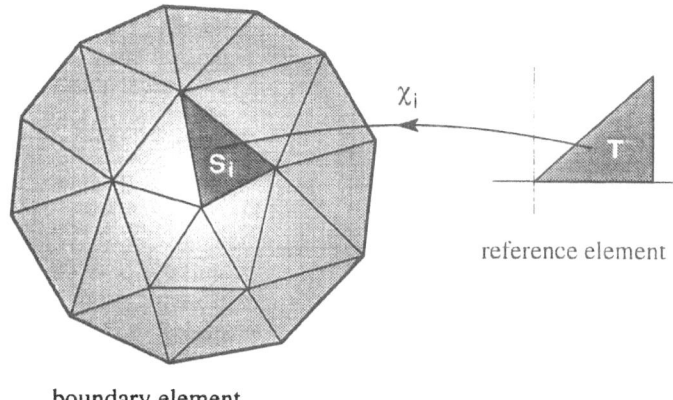

boundary element

Fig. 5.1. Triangulation of the boundary and maping χ_i of the reference triangle T onto the panel S_i

Step 3: The approximating subspace V_h is now defined by lifting p from the parameter domain T onto S via χ_i^{-1}:

$$v_{h_{|S_i}} = p \circ \chi_i^{-1} \tag{5.3}$$

Thus, v_h is piecewise polynomial of degree k on each panel S_i; globally it may be of class $C^{t-1}(S)$ for some $t \geq 0$. When $(k,t)=(0,0)$, $V_h \in H^0(S)$ consits of piecewise constant functions on each S_i. When $t=1$, $k=\{1,2,3\}$, $V_h \in H^{1/2}(S)$, and on each panel S_i, v_h is a {linear, quadratic, cubic} polynomial. The nodes of some very popular triangular elements are shown in Figure 5.2. For the definition of the corresponding basis functions, see e.g. Schwarz (1980), Zienkiewicz (1983).

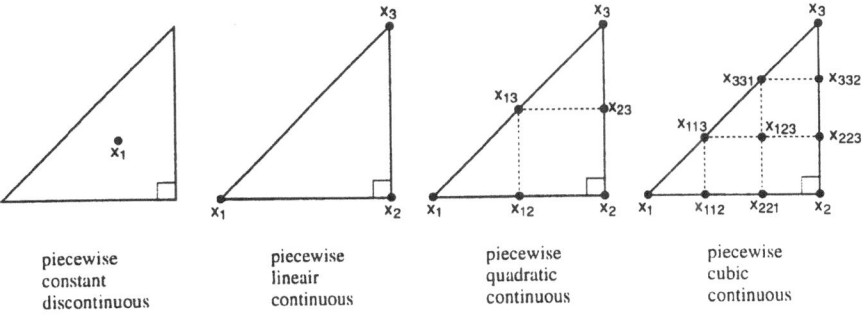

| piecewise constant discontinuous | piecewise lineair continuous | piecewise quadratic continuous | piecewise cubic continuous |

Fig. 5.2. Nodes of various Lagrangian triangular elements

The breadth of the range of possible basis functions of the finite dimensional subspace gives a great flexibility to Boundary Element Methods. A proper choice however must also take into consideration the numerical effort related to it and the approximation properties which determine the order of convergence of Galerkin methods (cf. Section 10).

If the boundary surface is not explicitly known, i.e. if no parameter representation is available, we also have to approximate the boundary surface itself. Then we use the coordinate representation of the *approximate* boundary. Approximations of the boundary surface may use global basis functions, e.g. spherical harmonics or polynomials. Mostly local basis functions are preferred, e.g. planar elements (then, the approximate surface is a polyhedron, and only the corner points of the polyhedron are on the actual boundary), or piecewise quadratic or cubic polynomials, splines etc. Often, the same type but different orders of basis functions are used to approximate the boundary *and* the unknown function. A very popular choice for integral equations with strongly singular kernel are piecewise linear polynomials for the approximation of the boundary surface (polyhedron) and piecewise constant functions to approximate the solution. For hyper-singular kernels, polyeder surfaces combined with continuous piecewise linear trial functions are used.

Higher order approximations of the solution of the integral equation requires at the same time also the improvement of the approximation of the boundary surface. For instance, a cubic spline interpolation of the finite element corner points results in curved rectangles which better approximate the boundary surface then a polyhedron. However, at the same time we should also improve the approximation of the solution of the integral equation by using e.g. on each panel quadratic or cubic trial functions. Thus, the approximation of the boundary and the unknown function are related to each other. It is of little use to combine a high degree approximation of the boundary with a low degree approximation of the unknown function and conversely. A careful error analysis is necessary to show what the relation is. However, it has not been completely developed yet, and results are only available for some integral operators (cf. Section 10).

6. Discretization of the integral equation

To determine the finite number of parameters defining the approximate solution of the integral equation we may choose between three different principles. The background is that when inserting the approximate solution into our integral equation, the equation is no longer fulfilled, since the theoretical solution belongs to an infinite dimensional function space and the approximate solution does not, but to a finite dimensional subspace. Let us assume that an integral equation

$$Av = f \quad \text{on } S \tag{6.1}$$

with $v \in H$ is given and that we look for an approximate solution

$$v_h(x) = \sum_{i=1}^{N} v_{h,i} \cdot b_i(x) \qquad (6.2)$$

with $v_h \in V_h$ and basis functions $\{b_i(x): i=1,...,N\}$ spanning V_h. h may denote the maximum diameter of all panels of a given discretization. Inserting (6.2) into (6.1) gives a residuum $Av_h - f$ which is generally unequal to zero. The unknown coefficients $\{v_{h,i}: i=1,...,N\}$ can now be determined in such a way that the residuum fulfils certain conditions:

a) We choose a suitable set of points $\{x_k: k=1,...,N\} \subset S$ and require that the residuum becomes zero in these points. Thus, the approximate solution v_h fulfils

$$Av_h(x_k) = f(x_k) \qquad (k=1,...,N). \qquad (6.3)$$

The method is called (exact) collocation and is frequently used in numerical mathematics. The points $\{x_k: k=1,...,N\}$ are called collocation points.

b) We multiply the residuum by a test function $\chi(x)$ from some N-dimensional function space, integrate over the boundary, and require that the integral becomes zero. This method is called Galerkin method. Most applications use the same function space as for the approximation of the solution ("Galerkin-Bubnov method"). Then, we obtain

$$<Av_h, b_k>_{L^2(S)} = <f, b_k>_{L^2(S)} \qquad (k=1,...,N). \qquad (6.4)$$

c) The (continuous) least squares method minimizes the residuum $Av_h - f$ w.r.t. the $L^2(S)$ norm:

$$<Av_h - f, Av_h - f>_{L^2(S)} = \underline{minimum}. \qquad (6.5)$$

Each equation defines a linear system of equations which approximates the boundary integral equation. Table 6.1 contains the definition of the matrix entries and of the right-hand side. The computation of the matrix entries forms one of the bottlenecks of Boundary Element Methods because it requires numerical integration. What cubature formulae must be used depends on the properties of the kernel $k(x, y-x)$ of the operator A which is closely related to the choice of the boundary integral equation. Therefore, one main topic of boundary element research is the construction of efficient cubature formulas to compute surface integrals with point singularities. We can only touch this problem; more details can be found in the references.

Table 6.1. Definition of the matrix entries and the right-hand side for different discretization methods ($<.,.>$ denotes the $L^2(S)$ inner product)

	collocation	Galerkin	least-squares
matrix entry	$Ab_i(x_k)$	$< Ab_i, b_k >$	$< Ab_i, Ab_k >$
right hand side	$f(x_k)$	$< f, b_k >$	$< f, Ab_k >$

From the numerical integration point of view, collocation is the easiest method because the matrix entries are defined as surface integrals and the right-hand side involves only the evaluation of a function at the collocation points. For Galerkin and least-squares we have to evaluate double and triple surface integrals. In addition, the right-hand side requires numerical integration, as well. Therefore, collocation is preferred in practical applications. However, it has been observed that efficient cubature formulas are much more difficult to develop for collocation than for Galerkin. Besides, collocation has also some other drawbacks from a numerical point of view: it always results in a non-symmetric coefficient matrix, independently whether the integral equation operator A is self-adjoint or not. Compared with that, the least-squares method always gives a symmetric coefficient matrix, which is even positive definite if the normal equations are uniquely solvable. The same would also be true for least-squares collocation. For Galerkin the coefficient matrix has mostly no special structure. We only may try to select the type of integral equation very carefully which sometimes may result in a positive self-adjoint operator A. Then, the coefficient matrix will be positive definite and symmetric, as well.

Main difficulties arise when $x_i \in$ supp b_k (collocation) or supp $b_i \cap$ supp $b_k \neq \emptyset$ (Galerkin, least squares), since when $x \rightarrow y$ the kernel functions are of the order $O(|x-y|^s)$, $s=\{1,2,3\}$. $O(N)$ integrals of this type have to be calculated. Cubature formulas for the collocation matrix entries have been developed, e.g. by Guiggiani & Gigante (1990), Klees (1992), and Schwab & Wendland (1992). The computation of the Galerkin matrix entries is easier because of the second integration. Here, very efficient numerical cubatures have been developed by Hackbusch & Sauter (1992), Sauter (1992), and Schwab & Wendland (1992).

For collocation, special attention needs also the case $0 <$ dist $(x_i, $supp $b_k) \leq C \cdot h$ with some small positive C. Integrals of this type massively arise in collocation methods; their number increases proportional to N^2. Although, the corresponding matrix entries are defined by regular integrals, the numerical behaviour of the integrand is complicated and standard cubature formulas fail. This case has been treated in Klees (1992) and Hackbusch & Sauter (1993).

7. Linear system of equations

Not only the calculation of the coefficients of the linear system of equations, but also its solution is a real challenge from the numerical point of view. First of all, the number of unknowns may be extremely high, exceeding 10^5. Secondly, the coefficient matrix is always full so that fast sparse equation solvers cannot be used. Table 7.1 gives an impression of the numerical effort typical for geodetic applications.

Several fast *direct* (out-of-core) solvers for non-symmetric dense linear systems of equations have been developed exploiting the capabilities of modern computer architectures (e.g. Geers & Klees 1993). Nonetheless, solving the N×N matrix equation requires $O(N^3)$ arithmetic operations, and the storage requirements are of the order of $O(N^2)$. Sometimes, *conjugate gradient methods* such as GMRES seem to converge after a few iterations for integral equations resulting from GBVPs (Ballani, Barthelmes & Klees 1995; Geers & Klees 1995). Alternatively, Prasad, Keyes & Kane (1993) applied a *fast wavelet transform* to the rows and columns of the equation matrix to change into the wavelet basis. Then small, nonzero entries are dropped and the resulting sparse system is solved using fast sparse equation solvers. Nonetheless, for practical geodetic applications methods are needed which solve the linear system in $O(N \cdot log^s N)$ arithmetic operations for some small s>0. In addition, the storage requirements must be reduced to the same order. This will be discussed in Section 9.

Table 7.1. Dimension of the linear system of equations and storage requirements for different resolutions. Triangular elements and piecewise constant trial functions (i.e. k=0) are assumed.

resolution (half wavelength)	5°	2.5°	1°	0.5°
side length of triangle	850 km	420 km	170 km	85 km
number of triangles	1650	6600	41250	165000
number of unknowns (k=0)	1650	6600	41250	165000
storage requirements	21 MB	340 MB	13 GB	213 GB

8. Fast numerical algorithms

The bottleneck of every integral equation approach is twofold. Firstly, since the equation matrix is dense, we have to compute N^2 matrix entries, when N denotes the dimension of V_h, i.e. the degree of freedom on the boundary surface. Their computation is rather involved because of the singular and nearly singular behavior of the kernel function in the neighborhood of y=x. Secondly, the generation of the matrix requires $O(N^2)$ arithmetic operations, and the solution of the linear system of equations using a direct equation solver requires $O(N^3)$ arithmetic operations. Even when the condition number is bounded by a constant independent on N and, therefore, iterative solvers may only need a few iteration steps, each matrix-vector multiplication still requires $O(N^2)$ arithmetic operations. Moreover, storing all coefficients of the linear system of equations requires $N \cdot (N+1)$ words. Even when using supercomputers, we may exceed the capacity of every available hardware very soon when applying the technique to GBVPs without additional optimization (cf. Table 7.1).

To make the boundary element approach applicable to GBVPs one should be able to solve the linear system of equations with no more than $O(N \cdot \log^s N)$ storage and arithmetic operations for some small $s \geq 0$. Several methods have been proposed to reach this goal. Rokhlin (1983) introduces a multipole expansion (see also Greengard & Rokhlin 1987), and Hackbusch & Nowak (1989) the panel clustering (see also Sauter 1991, 1992). Both methods exploit the smoothness of the kernel function of the boundary integral operator by using certain Taylor and Laurent series expansions for the entries of the matrix which are far away from the main diagonal. This results in an algorithm for the multiplication of the matrix by a vector with $O(N \cdot \log^s N)$ operations. Other methods have been proposed by Brandt & Lubrecht (1990) and Harten & Yad-Shalom (1992). Just recently, Petersdorff, Schwab & Schneider (1994), Dahmen, Prößdorf & Schneider (1995), and Petersdorff & Schwab (1995) developed wavelet techniques to approximate the boundary integral operator by orthogonal and compactly supported wavelets. This approach goes back to Beylkin, Coifman & Rokhlin (1991).

In the following we briefly want to discuss the basic principle of panel clustering and multiscale Galerkin Boundary Element Methods because they seem to be the most promising methods for GBVPs. For more details we refer to the literature mentioned before.

8.1 Panel Clustering

The starting point is the integral equation (4.5) and the discretized integral equation after applying the Galerkin (6.4) principle which can be written in the form

$$(M - K)\underline{v} = \underline{f},$$

(8.1)

$$M_{ij} := <\lambda b_j, b_i>, \quad K_{ij} := <K b_j, b_i>, \quad v_i := v_{h,i}, \quad f_i := <f, b_i>, \quad i,j = 1...N.$$

$\underline{M}$ is sparse since the basis functions b_k have finite support but $\underline{K}$ is not. Therefore, most of the time we have to spend for the matrix-vector multiplication $\underline{K}\underline{v}$. The i-th element of the vector $\underline{K}\underline{v}$ is:

$$(\underline{K}\underline{v}_h)_i = \sum_j \left(\int_{B_i} b_i(x) \int_{B_j} k(x,y-x)b_j(y)\,dS_y\,dS_x \right) v_{h,j}.$$ (8.2)

B_k is the support of the basis function b_k. Since

$$k(x,y-x) = O(|y-x|^{-s}), \quad s \in \{1,2,3\},$$ (8.3)

the coefficients K_{ij} will be very small for large distances $\mathrm{dist}(B_i, B_j)$. Let us assume that the "small" matrix entries are contained in the matrix $\underline{K}^{far}$, and all other elements in $\underline{K}^{near}$. Then, we can formally write $\underline{K}=\underline{K}^{near}+\underline{K}^{far}$, and obtain

$$(\underline{M} - \underline{K}^{near})\underline{v} = \underline{f} + \underline{K}^{far}\underline{v}.$$ (8.4)

Since the contribution $\underline{K}^{far}\underline{v}$ to $\underline{K}\underline{v}$ is "much smaller" than that of $\underline{K}^{near}\underline{v}$, we may approximate the former one by (a) approximating the entries of $\underline{K}^{far}$, and (b) approximating $\underline{v}$:

$$(\underline{M} - \underline{K}^{near})\underline{v} \approx \underline{f} + \tilde{\underline{K}}^{far}\tilde{\underline{v}}.$$ (8.5)

(b) can be realized within an iterative scheme:

$$(\underline{M} - \underline{K}^{near})\underline{v}^{(n)} \approx \underline{f} + \tilde{\underline{K}}^{far}\underline{v}^{(n-1)}, \quad n=1,2,\dots, \quad \underline{v}^{(0)}=\underline{0},$$ (8.6)

where conjugate gradient methods may be used to solve the sparse system of equations. The crucial point now is how to speed up the matrix-vector multiplication $\tilde{\underline{K}}^{far}\underline{v}^{(n-1)}$?

Before giving more details let us first illustrate the basic principle using a very simple example. Let us assume that our surface is discretized into 4 panels (see Figure 8.1) and that piecewise constant trial functions are used.

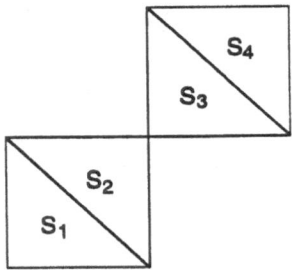

Fig. 8.1. Simple example of panel clustering

Then, the Galerkin system reads

$$
\begin{pmatrix}
M_{11}-K_{11} & K_{12} & K_{13} & K_{14} \\
K_{21} & M_{22}-K_{22} & K_{23} & K_{24} \\
K_{31} & K_{32} & M_{33}-K_{33} & K_{34} \\
K_{41} & K_{42} & K_{43} & M_{44}-K_{44}
\end{pmatrix}
\cdot
\begin{pmatrix}
v_{h,1} \\ v_{h,2} \\ v_{h,3} \\ v_{h,4}
\end{pmatrix}
=
\begin{pmatrix}
f_1 \\ f_2 \\ f_3 \\ f_4
\end{pmatrix},
\tag{8.7}
$$

$$
M_{ii} = \int_{S_i} \lambda(x)\,dS_x, \quad K_{ij} = \int\int_{S_i S_j} k(x,y-x)\,dS_y\,dS_x, \quad f_i = \int_{S_i} f(x)\,dS_x, \quad i,j=1,\dots,4.
\tag{8.8}
$$

Let us consider the first row and let us assume that the panels S_3 and S_4 are neighbours and "far away" from S_1. Then, the matrix elements K_{13} and K_{14} are very small and are elements of the "far-field" with respect to S_1, i.e. they belong to $\underline{K}^{far}$ and we may write for the far-field contribution to the first row, $K_1^{far}v_h$:

$$
K_1^{far}v_h := K_{13}v_{h,3} + K_{14}v_{h,4} = v_{h,3}\int\int_{S_1 S_3} k(x,y-x)\,dS_y\,dS_x + v_{h,4}\int\int_{S_1 S_4} k(x,y-x)\,dS_y\,dS_x.
\tag{8.9}
$$

K_{13} and K_{14} are now approximated, e.g. by replacing the kernel function $k(x,z)$ by its Taylor series expansion up to order M around the centre z_m of $S_3 \cup S_4$:

$$
k(x,y-x) \approx \sum_{|\delta|<M} k_\delta(x,z_m)\cdot y^\delta,
$$

$$
K_1^{far}v_h \approx \tilde{K}_1^{far}v_h = \sum_{|\delta|<M} \int_{S_1} k_\delta(x,z_m)\cdot\left[v_{h,3}\int_{S_3} y^\delta dS_y + v_{h,4}\int_{S_4} y^\delta dS_y\right].
\tag{8.10}
$$

Note that the integrals over S_3, S_4 are now independent on $x \in S_1$. Then, we obtain the approximate linear system

$$
\begin{pmatrix}
M_{11} - K_{11} & K_{12} & 0 & 0 \\
K_{21} & M_{22} - K_{22} & K_{23} & K_{24} \\
K_{31} & K_{32} & M_{33} - K_{33} & K_{34} \\
K_{41} & K_{42} & K_{43} & M_{44} - K_{44}
\end{pmatrix}
\cdot
\begin{pmatrix}
v_{h,1} \\
v_{h,2} \\
v_{h,3} \\
v_{h,4}
\end{pmatrix}
=
\begin{pmatrix}
f_1 - \tilde{K}_1^{far} v_h \\
f_2 \\
f_3 \\
f_4
\end{pmatrix}
.
\tag{8.11}
$$

Note that two elements of the first row are now equal to zero. Similar, we proceed for the other rows, and obtain finally

$$
\begin{pmatrix}
M_{11} - K_{11} & K_{12} & 0 & 0 \\
K_{21} & M_{22} - K_{22} & K_{23} & 0 \\
0 & K_{32} & M_{33} - K_{33} & K_{34} \\
0 & 0 & K_{43} & M_{44} - K_{44}
\end{pmatrix}
\cdot
\begin{pmatrix}
v_{h,1} \\
v_{h,2} \\
v_{h,3} \\
v_{h,4}
\end{pmatrix}
=
\begin{pmatrix}
f_1 - \tilde{K}_1^{far} v_h \\
f_2 - \tilde{K}_2^{far} v_h \\
f_3 - \tilde{K}_3^{far} v_h \\
f_4 - \tilde{K}_4^{far} v_h
\end{pmatrix}
.
\tag{8.12}
$$

Since the elements $\{\tilde{K}_i^{far} v_h : i=1,...,4\}$ are "small", it is allowed to use instead of v_h any reasonable approximation $\tilde{v}_h$, e.g. by using the solution of the previous iteration step within the following iterative scheme (8.6) ($\underline{v}_h^{(0)} = \underline{0}$):

$$
\begin{pmatrix}
M_{11} - K_{11} & K_{12} & 0 & 0 \\
K_{21} & M_{22} - K_{22} & K_{23} & 0 \\
0 & K_{32} & M_{33} - K_{33} & K_{34} \\
0 & 0 & K_{43} & M_{44} - K_{44}
\end{pmatrix}
\cdot
\begin{pmatrix}
v_{h,1}^{(n)} \\
v_{h,2}^{(n)} \\
v_{h,3}^{(n)} \\
v_{h,4}^{(n)}
\end{pmatrix}
=
\begin{pmatrix}
f_1 - \tilde{K}_1^{far} v_h^{(n-1)} \\
f_2 - \tilde{K}_2^{far} v_h^{(n-1)} \\
f_3 - \tilde{K}_3^{far} v_h^{(n-1)} \\
f_4 - \tilde{K}_4^{far} v_h^{(n-1)}
\end{pmatrix}
, \quad n=1,2,...
\tag{8.13}
$$

Let us now work out the idea of panel clustering in more detail. To keep things as simple as possible let us assume that piecewise constant trial functions are used:

$$
b_i(x) = \begin{pmatrix} 1 & \text{for } x \in S_i \\ 0 & \text{otherwise} \end{pmatrix}, \quad i=1,...,N,
\tag{8.14}
$$

thus, supp $b_j = S_j$, $j=1,...,N$. The generalization to higher order trial functions is straightforward. For each fixed surface panel S_i one has to split the surface S into the so-called near-field part $S^{near}(S_i)$ consisting of panels located "near" S_i and the far-field part $S^{far}(S_i)$ consisting of clusters C_m being union of some panels:

$$S = (S_1 \cup S_2 \cup \ldots \cup S_{p_i}) \cup (C_1 \cup C_2 \cup \ldots \cup C_{c_i}) =: S^{near}(S_i) \cup S^{far}(S_i), \qquad (8.15)$$

where p_i denotes the number of panels in $S^{near}(S_i)$ and c_i the number of clusters in $S^{far}(S_i)$. Now we consider the i-th element of the matrix-vector product $\underline{K}^{far}\underline{v}$:

$$(\underline{K}^{far}\underline{v})_i := K_i^{far} v_h = \sum_{1 \le m \le c_i} K_{i,m}^{far} v_h, \qquad (8.16)$$

where $K_{i,m}^{far} v_h$ denotes the contribution of cluster C_m to the i-th element of $\underline{K}^{far}\underline{v}$. This contribtuion is

$$K_{i,m}^{far} v_h := \int_{S_i} \int_{C_m(S_i)} k(x, y-x) v_h(y) \, dS_y \, dS_x = \int_{S_i} \sum_{S_j \subset C_m(S_i)} \int_{S_j} k(x, y-x) v_h(y) \, dS_y \, dS_x. \quad (8.17)$$

Since the distance between $C_m(S_i)$ and S_i is large and $x \in S_i$, $y \in C_m(S_i)$, the kernel function $k(x,y-x)$ is small because of (8.3) and may be approximated by its Taylor series expansion up to order M around the centre z_m of the cluster $C_m(S_i)$:

$$k(x,z) \approx \sum_{|\delta| < M} k_\delta(x, z_m) F_\delta(y) =: k_{M,m}(x,z), \quad z := y-x. \qquad (8.18)$$

By carefully chosing M, the resulting error does not affect the asymptotic convergence order of the Galerkin scheme (cf. Section 10). Sauter (1992) has shown that for $M \in \mathbb{N}$ there always exists an $\eta \in (0,1)$ and a constant $\varepsilon > 0$ such that

$$|k(x,z) - k_{M,m}(x,z)| \le \epsilon(\eta, M)\big(\|z\|^{-s} + \|z_m\|^{-s}\big), \text{ for all } \|z - z_m\| \le \eta \|z_m\|. \qquad (8.19)$$

s denotes the order of singularity of the kernel function $k(x, \bullet)$. For sufficiently small η, ε can be estimated asymptotically by $\varepsilon(\eta, M) \le C\eta^M$. Thus, any desired accuracy can be achieved by a suitable choice of η and M. Estimate (8.19) implies that, when replacing $k(x,y-x)$ by its Taylor series expansion, the clusters have to be formed according to

$$\text{diam}\, C_m + \text{diam}\, S_i \le 2\eta \cdot |z_{C_m} - z_{S_i}|. \qquad (8.20)$$

z_m is the centre point of the cluster C_m (cf. Figure 8.2). Thus, the cluster C_m may be chosen larger if $\text{dist}(S_i, C_m)$ becomes larger due to the behaviour of the kernel function $k(x,y-x)$. The cluster is then called admissible with respect to the panel S_i. There are many such admissible coverings but only one which has a minimal number of clusters, and exactly this coverage is chosen. Now $K_{i,m}^{far} v_h$ is approximated by

$$\tilde{K}_{i,m}^{far} v_h = \sum_{|\delta| < M} J_\delta^m(v_h) \int_{S_i} k_\delta(x, z_m) \, dS_x, \quad J_\delta^m(v_h) := \sum_{S_j \subset C_m(S_i)} \int_{S_j} F_\delta(y) v_h(y) \, dS_y. \qquad (8.21)$$

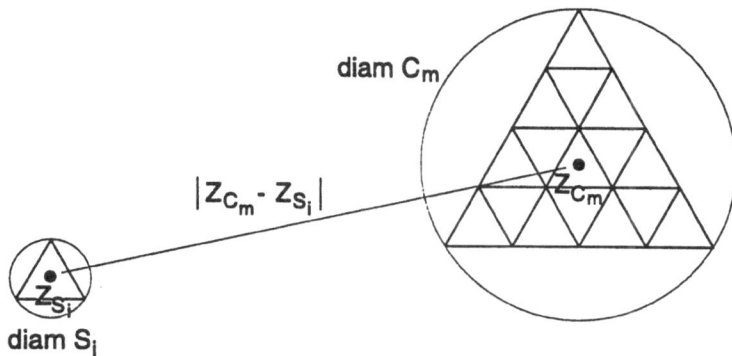

Fig. 8.2. Admissible cluster C_m w.r.t. panel S_i

Note that $J_\delta^m(v_h)$ is independent of x. Moreover, Sauter (1992) has shown that $c_i = O(\log N)$, i.e. there are at most $O(\log N)$ clusters in the far-field of each panel S_i. Observing (8.21), we finally obtain for the i-th element of the matrix-vector product $\underline{K}^{far}\underline{v}$ the approximation

$$(\tilde{\underline{K}}^{far}\underline{v})_i = \sum_{1 \leq m \leq c_i} \tilde{K}_{i,m}^{far} v_h = \sum_{1 \leq m \leq c_i} \sum_{|\delta| < M} J_\delta^m(v_h) \int_{S_i} k_\delta(x, z_m) dS_x .$$ (8.22)

The approximation error can be estimated by

$$|(\underline{K}^{far}\underline{v})_i - (\tilde{\underline{K}}^{far}\underline{v})_i| \leq C \epsilon(\eta, M) \|\underline{v}_h\|_\infty .$$ (8.23)

Taking the definition of the near-field contribution into account, we obtain the panel clustering algorithm:

Let $v_{h,j}^{(0)} = 0$, $1 \leq j \leq N$. For $n = 1, 2, \ldots$:

$$\sum_{1 \leq j \leq N} \left(\delta_{ij} \int_{S_i} \lambda(x) dS_x - \int_{S_i} \int_{S_j \cap S^{near}(S_i)} k(x, y-x) dS_y dS_x \right) \cdot v_{h,j}^{(n)} = \int_{S_i} f(x) dS_x$$ (8.24)

$$+ \sum_{1 \leq m \leq c_i} \sum_{|\delta| < M} \int_{S_i} k_\delta(x, z_m) dS_x \sum_{S_j \subset C_m(S_i)} v_{h,j}^{(n-1)} \int_{S_j} F_\delta(y) dS_y , \quad i = 1, \ldots, N .$$

The near field matrix entries are calculated as usual by using efficient numerical integration formulas. The matrix on the left-hand side is sparse with $O(N \cdot \log^r N)$ non-zero elements, and $r > 0$.

The numerical cost of the panel clustering depends on the parameters M and η. Sauter (1992) has proved that for the generation part, $O(NM^3)$ operations and for the evaluation of the matrix-vector product $O(NM^7)$ operations are needed. The storage requirements are about $O(NM^3)$ words. Note that $M=O(\log N)$, because the far-field contains about $O(\log N)$ clusters. For given M and η, the efficiency is determined by the decay of the kernel function $k(x,y-x)$ for large $|y-x|$, expressed by the order s of the singularity. Cauchy-singular (s=2) and hyper-singular (s=3) kernel functions reduce the number of panels in the near-field and allow to choose larger clusters in the far-field. Thus, the numerical costs are less than for weakly-singular kernel functions (s=1).

8.2 Multiscale Methods

An alternative to panel clustering which also provides a sparse equation matrix with $O(N\log^r N)$ non-zero elements for some positive r are multiwavelets. To keep things as simple as possible we want to restrict to zero-order boundary integral equations of the second kind. Moreover, we assume that the boundary integral equation is strongly elliptic (cf. Section 10). Generalizations to strongly elliptic integral operators of order ± 1 are possible provided that a suitable and stable multiwavelet basis is available.

Let us first recall the Galerkin method: given a dense sequence of finite dimensional subspaces $\{V_l: l=0,1,...\}$ of $L^2(S)$; solve $<Av_l,\varphi>=<f,\varphi>$ for all $\varphi \in V_l$. Till now we did not consider a whole sequence of subspaces but only one, namely the subspace V_h that provides the desired resolution. Let us assume that this corresponds to level n, i.e. $V_h \equiv V_n$. For our multiscale Galerkin method we have to consider the whole sequence of subspaces starting from the lowest level l=0 to the highest one l=n. Moreover, we need that the subspaces V_l are nested, i.e. $V_0 \subset V_1 \subset ... \subset V_n$. To construct such a sequence of nested subspaces we proceed as follows. Let us assume a subdivision of the boundary surface into a small number of N_S macro elements, e.g. curved triangles. N_S might be equal to 8, 12, or 20, depending on our subdivision algorithm. The mappings $\{\chi_i: i=1,...,N_S\}$ define them as smooth images of the standard triangle T. The N_S macro elements define the triangulation of level l=0. The triangulations of S at the higher levels are obtained by a certain subdivision procedure of the macro elements: first of all, we divide T into 4 congruent subtriangles $T_{1,j}$, j=1...4, by halving the sides. Each subtriangle is the image of the standard triangle under the bijective mapping $\psi_{1,j}$. We call the resulting triangulation the triangulation of level l=1. We then apply the same procedure to all triangles at level 1 and obtain the triangulation of level 2, consisting of 16 subtriangles $T_{2,j}$, j=1...16, per panel. They can also be defined as the image of the standard triangle under the mappings $\psi_{2,j}$. Again, with the mappings χ_i they are lifted to the boundary surface defining there a triangulation of the boundary surface into curved triangles. After n levels we have arrived at the final triangulation. Then, T has been subdivided into 4^n subtriangles, each macro element into the same number of (curved) triangles and the whole surface into $N_S \cdot 4^n$ curved triangles (cf. Fig. 8.3). They define the panels.

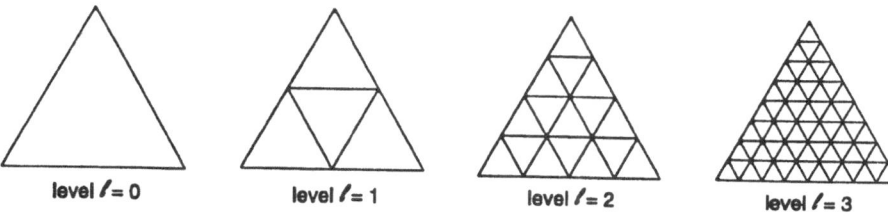

level $l = 0$ level $l = 1$ level $l = 2$ level $l = 3$

Fig. 8.3. Successive triangulation of a macro element

In the classical Galerkin approach, the space V_h is defined in the following way (cf. Section 5): On the standard triangle we define an algebraic polynomial of total degree d in local coordinates (u,v). With the bijective mappings $\psi_{n,j}$ they are transferred to the subtriangles $T_{n,j}$ defining there algebraic polynomials of total degree d in local coordinates (x,y). Finally, they are lifted to the surface S by the bijective affine mappings χ_i. The lifted polynomials define a basis of the finite-dimensional subspace V_h. Applied to the subspaces V_l, we obtain:

$$V_l = \{ v \in L^2(S) : (\gamma_i v) \circ \chi_{i_{|T_{l,j}}} \in \prod_d (T_{l,j}), \ i=1,...,N_S, \ j=1,...,4^l \} . \tag{8.25}$$

$\prod_d(T_{l,j})$ is the space of algebraic polynomials of total degree d on $T_{l,j}$. Obviously, $\dim V_l = \frac{1}{2}(d+1)(d+2)N_S 4^l$. Thus, the dimension N of V_h is

$$N = \dim V_h = \frac{1}{2}(d+1)(d+2) N_S 4^n . \tag{8.26}$$

The construction of an orthonormal basis of V_l is straightforward. Let

$$\{ \overline{b}_k : k=1,...,\tfrac{1}{2}(d+1)(d+2) \} \tag{8.27}$$

be such an L^2 orthonormal basis of $\prod_d(T)$. With $\psi_{l,j} : T \to T_{l,j}$ we define

$$\gamma_i \overline{\phi}_{l,\alpha} \circ \chi_i = \begin{cases} 2^{l-1} \overline{b}_k \circ \psi_{l,j}^{-1} & \text{on } T_{l,j} , \\ 0 & \text{else} \end{cases}$$

$$\overline{\phi}_{l,\alpha_{|S \backslash S_l}} = 0 , \tag{8.28}$$

$$\{ \overline{\phi}_{l,\alpha} | \ \alpha \in \overline{\Delta}_n = \{ (i,j,k) | \ i=1...N_S, j=1...4^l, k=1...\tfrac{1}{2}(d+1)(d+2) \} \} .$$

Then, the $\{ \overline{\psi}_{l,\alpha} \}$ set defines an orthonormal basis of V_l. γ_i is the restriction operator to the macro element S_i, and α a three-dimensional multi-index. In wavelet terminology, the basis functions at level n, $\overline{\phi}_{n,\alpha}$, define the so-called *fine scale basis* which spans V_h. Instead of representing v_h in the fine scale basis, we may represent it also in terms of a *multiscale expansion* using the wavelets. Since for zero-order integral operators the trial functions need not even be continuous, we can use a simple strategy to get appropriate multiresolution sequences defined on the surface S.

Because of the construction, we observe that the subspaces V_i, $i=0,1,...$ break up $L^2(S)$ in a sequence of nested subspaces; each coarser space V_{i-1} is a subspace of the finer space V_i:

$$V_0 \subset V_1 \subset ... \subset V_n \subset L^2 .$$ (8.29)

Therefore, we may introduce the orthogonal complement W_i of V_{i-1} in V_i:

$$W_i := \{v \in V_i : <v, \varphi> = 0, \text{ for all } \varphi \in V_{i-1}\} .$$ (8.30)

Since $V_i = V_{i-1} \oplus W_i$, we obtain the following multilevel splitting of V_h:

$$V_h = V_0 \oplus W_1 \oplus W_2 \oplus ... \oplus W_n .$$ (8.31)

Correspondingly, we obtain a multilevel decomposition of v_h:

$$v_h = v_0 \oplus w_1 \oplus w_2 \oplus ... \oplus w_n, \quad w_i \in W_i .$$ (8.32)

The term w_i may be viewed as the detail I have to add to a current approximation v_{i-1} to get the next finer approximation v_i: $v_i = v_{i-1} + w_i$. The representation (8.32) is called *multiscale expansion* of v_h. Note that the detail information w_i belong to the orthogonal complement W_i of V_{i-1} in V_i. In the wavelet terminology, the basis functions of W_i are called *wavelets* and the basis functions of the subspaces V_i are called *scaling functions*.

If we want to represent v_h in the form (8.32), we need to know a (stable) basis of W_i, $i=1...n$. This is the most difficult step, because we have to do that on a closed surface which excludes shift-invariance and corresponding Fourier techniques. Moreover, depending on the order α of the integral operator A, we need some minimal regularity of the wavelet basis functions. For operators of order $\alpha=0$, piecewise smooth discontinuous polynomials are sufficient but for $\alpha=1$, we need continuity.

To keep things simple we want to show how to construct a stable basis of $\{W_i : i=1,...,n\}$ for piecewise constant functions, i.e. for $d=0$. Schneider (1995) has shown how to do that for discontinuous piecewise linear functions ($d=1$). As mentioned before, piecewise discontinuous polynomials of degree d are consistent with operators of order $\alpha \leq 0$, but not with for instance $\alpha=1$ where $V_h \subset H^{1/2}(S)$ is needed. Since W_1 is the orthogonal complement of V_0 in V_1, its dimension is the difference of the dimension of V_1 and V_0, i.e. $4N_S - N_S = 3N_S$, where N_S is the number of macro elements of the initial triangulation. Therefore, we have to construct 3 wavelets per macro element. Following an idea of Mitrea (1994), we express the wavelets as some linear combination of the basis functions of V_1, which are the characteristic functions $h_{1,k}$ of the 4 subtriangles $T_{1,k}, k=1...4$:

$$b_k = c_k \cdot h_{1,k+1} - d_k \sum_{1 \le m \le k} h_{1,m}, \quad k=1...3 .$$ (8.33)

The coefficients c_k and d_k are determined by the two conditions

$$\int_T b_k dT = 0, \qquad \int_T |b_k| dT = |T| = \frac{1}{2} .$$ (8.34)

The first condition means that the wavelets have one vanishing moment, the second one is a simple normalization for stability reasons. The resulting basis functions are linear combinations of the characteristic functions and orthogonal to each other. We obtain for instance:

$$b_1 = 2(h_{1,2} - h_{1,1}), \quad b_2 = 2h_{1,3} - (h_{1,1} + h_{1,2}), \quad b_3 = \frac{9}{4}h_{1,4} - \frac{3}{4}(h_{1,1} + h_{1,2} + h_{1,3}).$$ (8.35)

A basis of W_1 is then defined by

$$\gamma_i \phi_{1,\alpha} \circ \chi_i = \begin{cases} b_k & \text{in } T \\ 0 & \text{else} \end{cases}, \quad \alpha \in \Delta_1 = \{(i,j,k) \mid i=1...N_S, j=1, k=1...3\} .$$ (8.36)

The detail space W_2 has dimension $12N_S$, i.e. there are 12 wavelet basis functions per macro element. They are derived from the previous 3 basis functions with help of the 4 bijective mappings $\psi_{1,j}$ and a scaling for stability reasons. The basis functions of the spaces $\{W_m: m=3,..,n\}$ are obtained in the same way, i.e. with help of the bijective mappings $\psi_{m-1,j}$ which map $T_{m-1,j}$ onto T. We obtain:

$$\gamma_i \phi_{m,\alpha} \circ \chi_i = \begin{cases} 2^{m-1} b_k \circ \psi_{m-1,j}^{-1} & \text{in } T_{m-1,j}, \\ 0 & \text{else} \end{cases}$$ (8.37)

$$\alpha \in \Delta_m = \{(i,j,k) \mid i=1...N_S, j=1...4^{m-1}, k=1...3\} .$$

The dimension of W_m is the difference between the dimension of V_m and V_{m-1}, i.e. $\frac{3}{2}(d+1)(d+2)N_S 4^{m-1}$. Finally, the mappings $\{\chi_i: i=1,...,N_S\}$ lift the wavelet basis functions on the boundary surface, yielding the multilevel wavelet basis of the approximating subspace V_h. Thus, the basis functions of V_h are

$$\bigcup_{r=0}^{n} \{\phi_{r,\alpha} \mid \alpha \in \Delta_r\}, \quad \phi_{0,\alpha} := \bar{\phi}_{0,\alpha}; \quad \Delta_0 := \bar{\Delta}_0 .$$ (8.38)

Note that the basis functions $\phi_{r,\alpha}$ are fully orthonormal w.r.t. the scalar product

$$<u,v> = \sum_{i=1}^{N_S} \int_T (\gamma_i u \circ \chi_i)(\gamma_i v \circ \chi_i) \, dT, \qquad (8.39)$$

which is equivalent to the usual $L^2(S)$ scalar product. Obviously, the constructed wavelets are not the translates and dilates of one mother function. However, it still holds that basis functions at a finer level can be used to write out basis functions at a coarser level, i.e. a multiresolution structure is maintained. Wavelets of this type are often called *second generation wavelets* (Schröder & Sweldens 1994).

Now we are prepared to introduce the *multiscale Galerkin algorithm*. The unknown function v_h is expressed in terms of the multiscale wavelet basis instead of the single scale basis as usually done:

$$v_h = \sum_{r=0}^{n} \sum_{\alpha \in \Delta_r} v_{r,\alpha} \phi_{r,\alpha} . \qquad (8.40)$$

The first index r runs over the different scales, the second index over the basis functions of the corresponding detail space W_r. The Galerkin principle then defines the multiscale coefficients $v_{r,\alpha}$ as solution of a linear system of algebraic equations

$$\sum_{r=0}^{n} \sum_{\alpha \in \Delta_r} <(\lambda I - K)\phi_{r,\alpha}, \phi_{s,\beta}> v_{r,\alpha} = <f, \phi_{s,\beta}>, \quad s=0...n; \quad \beta \in \Delta_s . \qquad (8.41)$$

The matrix

$$\underline{K} = \left(<K\phi_{r,\alpha}, \phi_{s,\beta}> \right) \qquad (8.42)$$

is still dense. But the decisive point is that the matrix elements show a characteristic decay in the chosen wavelet basis. This decay is based on two facts: firstly, the behaviour of the kernel function inverse proportional to 2, which implies the estimate

$$|D_x^p D_y^q k(x,y)| \leq \frac{C(p,q,S)}{|x-y|^{2+|p|+|q|}} \qquad (8.43)$$

for the higher order derivatives. Secondly, because our wavelet basis has vanishing moments up to order $\bar{d}$ (cf. 8.39):

$$\int_{T_{r,\alpha}} (\gamma_i \phi_{r,\alpha} \circ \chi_i) \cdot x^p \, dT = 0, \quad |p| \leq \bar{d} \qquad (8.44)$$

(cf. (8.35) for $\bar{d}=0$). For Haar-wavelets, $\bar{d}=0$, i.e. the wavelet basis has vanishing moments up to order 0. Both imply that each matrix entry can be estimated above by

$$|<(\lambda I - K)\phi_{r,\alpha},\phi_{s,\beta}>| \le C \cdot \text{dist}^{-(2\bar{d}+4)} \cdot 2^{-(r+s)\cdot(\bar{d}+2)}, \quad \text{dist}:= \text{dist}(S_{r,\alpha},S_{s,\beta}). \tag{8.45}$$

This estimate immediately follows after a Taylor series expansion of the integral kernel in local coordinates. Here, dist is the distance between the support of the basis functions involved in the definition of the corresponding matrix element. Obviously, each vanishing moment of the basis functions increases the rate of decay of the matrix element in terms of dist. That is the reason why we should make use of wavelets having many vanishing moments.

Based on the estimates (8.43) and (8.44) we can propose the following level-dependent *truncation strategy*: each matrix element for which the distance between the support of the corresponding basis functions is larger than a certain threshold $\delta_{r,s}$ is set equal to zero:

$$\tilde{K}_{ij} := \begin{cases} K_{ij} & \text{if } \text{dist}(S_{r,\alpha},S_{s,\beta}) \le \delta_{r,s} \\ 0 & \text{otherwise} \end{cases}. \tag{8.46}$$

The truncation parameters $\delta_{r,s}$ depends only on the resolution levels r,s of the two basis functions involved in the definition of the matrix entry. Dahmen, Prössdorf & Schneider (1995) and Schneider (1995) have shown that for the choice

$$\delta_{r,s} = \lambda \cdot n^{\frac{1}{2\bar{d}+2}} \cdot 2^{\frac{n(2d+2)-(r+s)(\bar{d}+d+2)}{2\bar{d}+2}} \tag{8.47}$$

we obtain a sparse matrix $\tilde{K}$ with a number of non-zero elements proportional to the number of unknowns up to a logarithmic term. The exponent of the logarithmic term depends on the order of vanishing moments $\bar{d}$ of the wavelet basis functions (cf. (8.48)). In addition, the truncation parameter depends on a certain constant $\lambda>1$, the refinement level n, the order of vanishing moments $\bar{d}$, the degree of the polynomial scaling functions d, and the levels r and s of the involved basis functions. The smaller $\delta_{r,s}$ the more matrix elements will be zero. The parameter λ plays an important role. However, its value can hardly be estimated but has to be find out numerically. When the truncation parameters $\delta_{r,s}$ fulfil (8.48) the compressed matrix $\tilde{K}$ has

$$O(N\log^{2+\frac{1}{\bar{d}+1}}N) \tag{8.48}$$

non-zero entries. The work estimates for the multiscale Galerkin scheme are slightly better than for panel clustering because the exponent of the logarithmic term is smaller. However, the obtained estimates are purely asympotic. Therefore, a realistic comparison of the panel clustering with the multiscale Galerkin scheme can only be done based on numerical results which are not yet available.

Let us finish this section with two remarks: (1) Schneider (1995) has proved that the numerical scheme which is based on the approximation of the original dense matrix $\underline{K}$ by a sparse matrix $\underline{\check{K}}$ is consistent, stable and does converge (cf. also Petersdorff & Schwab, 1995). (2) The given results hold if the entries of the compressed matrix $\underline{\check{K}}$ are computed *exactly*. In practise they have to be computed *numerically* by some suitable cubature formulas. Petersdorff & Schwab (1995) have presented and analyzed a cubature strategy which preserves the asymptotic convergence rates of the compressed scheme and essentially retains the efficiency (8.48) of the compressed scheme. We omit the details.

As an *example* let us consider an oblique BVP which has been transformed into a second kind integral equation with strongly singular kernel over a spherical surface. The corresponding integral operator has order $\alpha=0$. A triangulation of the spherical boundary surface has been derived from a recursive subdivision of an octahedron. We used Haar wavelets ($d=\bar{d}=0$) and applied the truncation scheme (8.46), (8.47). A resolution of 50 km at the equator corresponds to level 7 (131072 triangles) if Haar wavelets are chosen as basis functions.

Figure 8.4 shows as black dots the matrix entries which have to be computed for a refinement level $n=6$ (32768 panels) and smallest possible λ (one). Level $n=6$ corresponds to a gravity field resolution of about $1°$. It is easy to see that the lower scales interact with most other scales so we have to calculate all corresponding matrix entries. But with higher scales the interactions diminish more and more and most of the matrix entries can be set equal to zero. Thus, we get a sparse matrix containing roughly 32 times less coefficients than the original full matrix. Table 8.1 shows how the efficiency of the truncation scheme (8.46), (8.47) depends on the choice of the tuning parameter λ and the refinement level. With efficiency we mean the ratio of the number of elements of the original full matrix to that of the truncated sparse matrix, sometimes called sparsity factor or rate of compression. We observe to important things: firstly, for a given parameter λ, the efficiency increases exponentially with the refinement level. That means for high resolution gravity field approximations we gain much more than for low resolutions. For a 50 km resolution (level 7) and Haar wavelet basis functions ($d=\bar{d}=0$) we can expect a sparsity factor of up to 100. Secondly, higher values of λ reduce the efficiency considerably, i.e. less matrix entries can be set equal to zero. That is easy to understand, if we look at the definition of the truncation parameters (8.47). Higher values of λ increase the minimal distance between the support of the corresponding basis functions needed for truncation. Therefore more matrix entries have to be calculated. The correct value of λ must be find out by numerical experiments. First test calculations indicate that it is near to the optimal value 1. Table 8.2 shows how the rate of compression depends on the number of vanishing moments. An improvement of more than factor 2.5 can be expected at refinement level 7 when wavelets with ($d=\bar{d}=0$) are replaced by wavelets with ($d=\bar{d}=1$). Then, compression factors of more than 200 seem to be feasible.

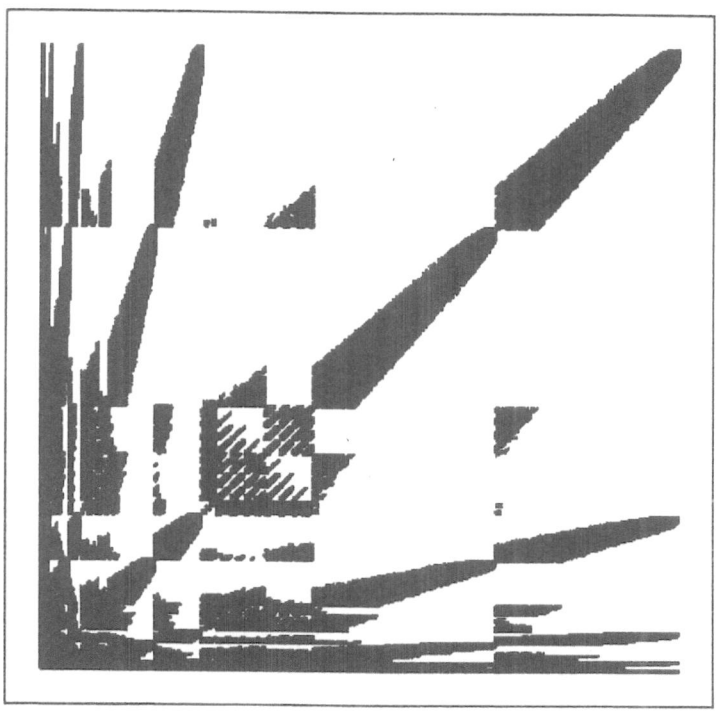

Fig. 8.4: Galerkin matrix in the wavelet basis (n=6, d=d̄=0, λ=1, N=32768, compression rate=34.3)

Table 8.1: Compression rate as function of the parameter λ and the refinement level (α=0, d=d̄=0); (·) extrapolated values

level n	resolution	# panels (unknowns N)	factor λ			
			1	2	4	8
3	~ 10°	512	5.9	3.0	1.6	1.0
4	~ 5°	2048	9.3	4.3	1.8	1.0
5	~ 2°	8192	16.0	7.3	3.3	1.6
6	~ 1°	32768	34.3	14.6	6.0	2.7
7	~ 0.5°	131072	(85.6)	(36.2)	(14.9)	(6.7)

Table 8.2: Compression rate as function of the number of vanishing moments and the refinement level ($\alpha=0$, $\lambda=1, d=\bar{d}$); ($\cdot$) extrapolated values

level n	resolution (d=0)	# panels	# unknowns (N)		compression rate	
			$\bar{d}=0$	$\bar{d}=1$	$\bar{d}=0$	$\bar{d}=1$
3	~ 10°	512	512	1536	5.9	9.4
4	~ 5°	2048	2048	6144	9.3	17.5
5	~ 2°	8192	8192	24576	16.0	36.2
6	~ 1°	32768	32768	98304	34.3	(83.9)
7	~ 0.5°	131072	131072	393216	(85.6)	(217.0)

8.3 Numerical Integration

It remains to consider the computation of the $O(N \log^\kappa N)$ non-zero elements of the Galerkin equation matrix. They are defined by

$$K_{ij} = <K b_j, b_i> = \int_{\text{supp} b_i} b_i(x) \int_{\text{supp} b_j} k(x, y-x) b_j(y) \, dS_y \, dS_x \,,$$

$$k(x, y-x) = \frac{\sum_{|\alpha| \geq t} c_\alpha(x)(y-x)^\alpha}{|y-x|^{s+t}}, \quad s \in \{1,2,3\}, \ s+t \text{ odd} .$$

(8.49)

Depending on the distance between the support of the basis functions b_i and b_j, special formulas have to be chosen because of the singular behaviour of the kernel function $k(x, y-x)$ in the neighbourhood of $y=x$. If $\text{supp}(b_i) \cap \text{supp}(b_j) \neq \emptyset$, the inner integral is singular and some regularization has to be introduced, depending on the order s of the singularity. This leads to the concept of Cauchy-singular and hyper-singular integrals (cf. Section 9). If $\text{supp}(b_i) \cap \text{supp}(b_j) = \emptyset$ but $\text{dist}(\text{supp}(b_i)$, $\text{supp}(b_j)) = O(h)$, the kernel function behaves similar as a singular function and special formulas have to be designed. Only when $\text{dist}(\text{supp}(b_i)$, $\text{supp}(b_j)) = O(1)$ standard cubature formulas can be used. But even then special care has to be taken otherwise the order of convergence will be affected. The latter is discussed in Section 10.3.

For multiscale methods, Petersdorff & Schwab (1995) have derived a cubature scheme for the entries of the compressed Galerkin scheme which preserves all asymptotic convergence properties to be established in Section 10. They showed that tensor product Gauß-Legendre quadrature formulas can be used to compute the matrix elements if the number of nodes are chosen appropriately and element subdivisions are applied. Some of their results are based on the numerical integration formulas of Sauter (1992) which have been developed to efficiently compute elements of the Galerkin matrix for boundary integral operators of any integer order and which cover all situations mentioned before. The technique of Sauter is the most powerful one available today and is also used in combination with panel clustering. It is based on the following statements which have been proved by Sauter (1992) for piecewise smooth S and smooth panels, e.g. for polynomial surface pieces:

(a) the integral (8.49) may be split into the sum over panels of the support of b_i and b_j:

$$K_{ij} = \sum_{S_x \subset \text{supp} \, b_i} \sum_{S_y \subset \text{supp} \, b_j} K_{ij}^{S_x, S_y}. \qquad (8.50)$$

(b) if $S_x \cap S_y = \varnothing$, $K_{ij}^{S_x, S_y}$ exist as weakly singular integrals

$$K_{ij}^{S_x, S_y} = \int_{S_x \times S_y} b_i(x) \, k(x, y-x) \, \phi_j(x,y) \, dS_y \, dS_x. \qquad (8.51)$$

(c) if $S_x \cap S_y \neq \varnothing$, the limit process can be interchanged with the outer integration:

$$K_{ij}^{S_x, S_y} = \lim_{\epsilon \to 0} \int_{\substack{S_x \times S_y \\ |x-y| > \epsilon}} b_i(x) \, k(x, y-x) \, \phi_j(x,y) \, dS_y \, dS_x \qquad (8.52)$$

Here, the function ϕ_j is defined by

$$\phi_j(x,y) := \begin{cases} b_j(y) - b_j(x), & \text{if k is hyper-singular (s=3)} \\ b_j(y), & \text{if k is Cauchy-singular (s=2)} \\ & \text{of weakly singular (s=1)} \end{cases} \qquad (8.53)$$

A detailed discussion of the problem of numerical integration w.r.t. the computation of the matrix entries K_{ij} is beyond the scope of the lecture notes. We refer to Sauter (1992), Hackbusch & Sauter (1993), and Lage (1995).

9. Solution in the domain

The solution of the linear system of equations gives the approximate solution of the boundary integral equation. If the integral equation was derived using Green-type representation formulae, this may already provide the desired information on the boundary. When following an indirect approach, e.g. single- and double-layer representation or one of Brovar's representation formulas, we have to evaluate another integral operator with singular kernel to get the solution on the boundary. When the solution of the BVP in the domain is required the representation formula gives the answer. Then, only regular surface integrals have to be evaluated. The integration has a smoothing effect and suppresses high-frequency errors. The same holds true if we want to calculate any partial derivative of arbitrary order of the potential in the domain.

Let us assume that we have solved the Galerkin equations and that we want to know the (approximate) solution of the BVP in the domain $\bar{D}^c$. Then, independently of the chosen representation formula, we have to compute integrals of type

$$\int_S k(x,y-x)\,v_h(y)\,dS_y, \quad x \in \bar{D}^{\,c} \tag{9.1}$$

with kernel function

$$k(x,y-x) = \frac{\displaystyle\sum_{|\alpha| \geq t} c_\alpha(x)(y-x)^\alpha}{|y-x|^{s+t}}, \quad s+t \text{ odd}. \tag{9.2}$$

Since S has been partitioned into panels S_i, and each of the S_i is the image of an analytic application $y = \chi_i(v)$ defined on the standard triangle T (see Section 5), we can restrict to consider integrals

$$I(x) := \int_{S_i} k(x,y-x)\,f(y)\,dS_y. \tag{9.3}$$

S_i is an analytic surface piece (panel) which is, without loss of generality, a curved triangle. f is usually a polynomial in local coordinates, thus a smooth function on S_i. Let us drop the index i from now on, i.e. we write simply $y=\chi(v)$, $S=\chi(T)$ and (9.3) becomes

$$I(x) = \int_S k(x,y-x)\,f(y)\,dS_y. \tag{9.4}$$

where S is a curved triangle. Mostly, $s \in \{1,2,3\}$, $t \in \{0,1\}$ with s+t odd. The numerical costs and the choice of the integration technique depend on the distance between calculation point and boundary surface. When $x \in D^c$, the integral is a usual Lebesgue integral. However, if $dist(x,S)=O(h)$ the numerical behaviour of the kernel function is complicated and quite similar to that of singular integrals. Therefore, standard cubature formulas such as Gauß-Legendre formulas loose their asymptotic accuracy because the derivatives of the kernel function appearing in the error functional produce negative powers of the discretization parameter h, cancelling the positive powers of h of higher order formulas. Therefore, special cubature formulas have been constructed and investigated, e.g. Klees (1992), Hackbusch & Sauter (1993), Hackbusch & Sauter (1994), Schwab (1994), Hayami & Matsumoto (1994), Rosen & Cormack (1994), Wu (1995). If dist(x,S)=O(1) the integral can be calculated efficiently using standard integration techniques. Both cases will not be treated here.

It reamains to consider the case that $x \in S$. Then the integral is singular. More precisely, if s<2 the integral is weakly singular but absolutely integrable. For $s \geq 2$, however, it has to be considered as a *part-fini* (or *finite part*) integral: Let

$$I(x,\epsilon):= \int_{S,\,|y-x|>\epsilon} k(x,y-x)f(y)\,dS_y, \quad x \in S .$$

(9.5)

$I(x,\epsilon)$ is regular for $\epsilon>0$. If x is located in the interior of the triangle S, and f is sufficiently smooth, and $\epsilon \to 0$, it has the representation

$$I(x,\epsilon) = c_0(x) + c_1(x)\log\epsilon + \sum_{2 \leq i \leq m} c_i(x)\epsilon^{1-i} + o(1), \quad \epsilon \to 0$$

(9.6)

with the estimates

$$|c_i(x)| \leq C, \quad 1 \leq i \leq s-1,$$

$$|c_0(x)| \leq \begin{cases} C \ln[\text{dist}(x,\partial S_y)] & \text{for } s=2 \\ C \text{ dist}^{2-s}(x,\partial S_y) & \text{for } s>2 \\ C & \text{for } s<2 \end{cases}$$

(9.7)

and constants C independent of x. Then, the part-fini of I(x) is defined as

$$\text{p.f. } I(x):= c_0(x) ,$$

(9.8)

and the Cauchy-principal value of I(x) is defined as

$$\text{p.v. } I(x):= \lim_{\epsilon \to 0} I(x,\epsilon)$$

(9.9)

if the limit exists. To calculate p.f. I(x) let us consider the *regular* integral $I(x,\epsilon)$. Using the parametrization $x=\chi(u)$, $y=\chi(v)$, we obtain

$$I(x,\epsilon) = \int_{S,\,|y-x|>\epsilon} k(x,y-x)f(y)\,dS_y = \int_{T,\,|v-u|>R_\epsilon} k(\chi(u),\chi(v)-\chi(u))f(\chi(v))J(v)\,dT_v .$$

(9.10)

R_ϵ denotes the image of $|y-x|=\epsilon$ in the parameter domain T, i.e. $\epsilon=\chi(R_\epsilon)$ and $J(v)$ the Jacobian of χ. Note that in general the image of the domain $|y-x|<\epsilon$ under the mapping χ^{-1} is not the domain $|v-u|<\epsilon$, but the domain $|v-u|<R_\epsilon$. R_ϵ is a function of $x=\chi(u)$ and defines a non-circular closed curve in the parameter domain. When (ρ,ϕ) denote the polar coordinates in the parameter

domain with respect to the point $u=\chi^{-1}(x)$, i.e.

$$v-u=\rho\begin{pmatrix}\cos\phi\\\sin\phi\end{pmatrix}, \tag{9.11}$$

we may write $R_\varepsilon=R_\varepsilon(u,\phi)$. Let us transform $I(x,\varepsilon)$ into polar coordinates. Using the Taylor series expansions

$$|y-x|^2=\rho^2\sum_{i\geq0}\rho^i\,F_i(u,\phi),\quad\sum_{|\alpha|\geq t}c_\alpha(x)(y-x)^\alpha=\rho^t\sum_{i\geq0}\rho_i\,f_i(u,\phi),$$

$$k(x,y-x)=\rho^{-s}\sum_{i\geq0}\rho^i\,k_i(u,\phi),\quad f(y)J(v)=\sum_{i\geq0}\rho^i\,\tilde{f}_i(u,\phi), \tag{9.12}$$

$$|y-x|=\epsilon\;\rightarrow\;|v-u|=R_\epsilon(u,\phi)=\epsilon F_0^{-1/2}-\tfrac{1}{2}\epsilon^2F_0^{-2}F_1+O(\epsilon^3),$$

we obtain

$$I(x,\epsilon)=\int_T\tilde{k}(u,v-u)\left(\tilde{f}(v)-\sum_{j=0}^{s-2}\rho^j\,\tilde{f}_j(u,\phi)\right)dT_v+\int_T\left(\tilde{k}(u,v-u)-\rho^{-s}\sum_{i=0}^{s-2}\rho^i\,k_i(u,\phi)\right)$$

$$\sum_{j=0}^{s-2}\rho^j\,\tilde{f}_j(u,\phi)dT_v\;+\int_{T,|v-u|>R_\epsilon}\rho^{-s}\sum_{i=0}^{s-2}\rho^i\,k_i(u,\phi)\sum_{j=0}^{s-2}\rho^j\,\tilde{f}_j(u,\phi)dT_v, \tag{9.13}$$

where

$$\tilde{k}(u,v-u):=k(\chi(u),\chi(v)-\chi(u)),\quad\tilde{f}(v):=f(\chi(v))J(v). \tag{9.14}$$

The two integrals over T are weakly singular. They will be treated later on. For the integral over $T\backslash\{v:\,|v-u|<R_\varepsilon\}$, we obtain

$$\int_{T,|v-u|>R_\epsilon}\rho^{-s}\sum_{i=0}^{s-2}\rho^i\,k_i(u,\phi)\sum_{j=0}^{s-2}\rho^j\,\tilde{f}_j(u,\phi)dT_v=$$

$$\sum_{i=0}^{s-2}\sum_{j=0}^{s-2}\int_0^{2\pi}k_i(u,\phi)\tilde{f}_j(u,\phi)\int_{R_\epsilon(u,\phi)}^{R(u,\phi)}\rho^{i+j+1-s}d\rho d\phi, \tag{9.15}$$

where $R(u,\phi)$ is the boundary of T in polar coordinates. It is

$$\int_{R_\epsilon(u,\phi)}^{R(u,\phi)} \rho^{i+j+1-s} d\rho = \begin{cases} \dfrac{\ln R(u,\phi)}{R_\epsilon(u,\phi)}, & \text{when } i+j=s-2 \\[2mm] \dfrac{1}{\kappa}\left(R^\kappa(u,\phi) - R_\epsilon^\kappa(u,\phi)\right), & \kappa = i+j+2-s \neq 0 \end{cases}, \tag{9.16}$$

and therefore,

$$\text{p.f.} \int_{0}^{R(u,\phi)} \rho^{i+j+1-s} d\rho = \begin{cases} \ln R(u,\phi) + \frac{1}{2}\ln F_0(u,\phi), & \text{for } i+j=s-2 \\[2mm] \frac{1}{\kappa}R^\kappa(u,\phi), & \kappa = i+j+2-s \neq 0 \end{cases}. \tag{9.17}$$

Finally, we obtain for p.f. I(x):

$$\text{p.f. } I(x) =$$

$$\int_T \tilde{k}(u,v-u)\left(\tilde{f}(v) - \sum_{j=0}^{s-2} \rho^j \tilde{f}_j(u,\phi)\right) dT_v + \int_T \left(\tilde{k}(u,v-u) - \rho^{-s}\sum_{i=0}^{s-2} \rho^i k_i(u,\phi)\right)$$

$$\sum_{j=0}^{s-2} \rho^j \tilde{f}_j(u,\phi) dT_v + \sum_{i=0}^{s-2}\sum_{j=0}^{s-2} \int_0^{2\pi} k_i(u,\phi)\tilde{f}_j(u,\phi) \begin{cases} \ln R(u,\phi) + \frac{1}{2}\ln F_0(u,\phi), & \text{for } i+j=s-2 \\[2mm] \frac{1}{\kappa}R^\kappa(u,\phi), & \kappa = i+j+2-s \neq 0 \end{cases}. \tag{9.18}$$

Thus, we have reduced the computation of the part-fini integral of a kernel function with singularity of order $s \geq 2$ over a curved panel $S \subset \mathbb{R}^3$ to the computation of two weakly singular integrals over the standard triangle $T \subset \mathbb{R}^2$ and the sum of $(s-1)^2$ regular, one-dimensional integrals.

It remains to show how to compute the weakly singular integrals efficiently. The integrals can be written as

$$\int_T g(u,v-u)p(v)\,dT_v, \quad g(u,v-u) = \rho^{-1}\sum_{i\geq 0} \rho^i g_i(u,\phi) = O(\rho^{-1}), \quad \rho \to 0, \tag{9.19}$$

where, without loss of generality we may assume that u is at $(0,0)$. Klees (1996) has studied extensively the computation of this type of integral. One of the methods that always works fine makes use of the polar coordinates with respect to u. Then, we obtain for (9.19)

$$\int_T g(u,v-u)p(v)\,dT_v = \sum_{i\geq 0} \int_0^{R(u,\phi)} \rho^i \int_0^{\phi_0} g_i(u,\phi)p(v(\rho,\phi))\,d\phi\,d\rho. \tag{9.20}$$

The integrand is analytic in ρ and ϕ and standard cubature formulas may be used to compute them efficiently, for instance Gauß-Legendre formulas. For alternative methods and a detailed comparison we refer to Klees (1996) and the references given there.

10. Error estimates

Usually the integral equations over general surfaces in $\mathbb{R}^3$ cannot be solved analytically. Boundary Element Methods like other numerical methods always lead to approximation errors. The main error sources are due to the discretization of the boundary integral equation, the approximation of the boundary, the numerical integration to set up the linear system of equations and to evaluate the representation formulae, and the round-off errors. Realistic a priori error estimates, however, are hard to get since the properties of the integral equation operators and the different steps of a boundary element application affect the errors in a very complicated way. Available are a-priori asymptotic error estimates for $h\to0$, i.e. when the maximum diameter of the panel tends to zero. In the following we want to give an overview over that what is known about the different error sources. More details can be found in Wendland (1983, 1985, 1987), Schwab & Wendland (1992), Nédélec (1976, 1977), and Sauter & Krapp (1995).

10.1 Discretization error

It is well-known that all boundary integral operators A to regular elliptic BVPs are continuous linear operators of integer order α, A: $H^{\alpha/2}(S)\to H^{-\alpha/2}(S)$, where for the examples given in Section 4, $\alpha\in\{-1,0,1\}$. But for solving the equation $Av=f$, continuity of A is not sufficient; we also need properties that provide the *existence* of A^{-1}. The corresponding key property is the *strong ellipticity* of A, which for continuous, linear A of order α is equivalent to the *coercitivity* of A in the form of a *Gårding inequality*:

Let A: $H^{\alpha/2}(S)\to H^{-\alpha/2}(S)$ be a continuous linear operator. Then, there exists a compact operator C: $H^{\alpha/2}(S)\to H^{-\alpha/2}(S)$ and a $\gamma>0$ such that for all $v\in H^{\alpha/2}(S)$

$$<(A+C)v,v> \geq \gamma \|v\|^2_{H^{\alpha/2}(S)}, \tag{10.1}$$

i.e. $A+C$ is $H^{\alpha/2}(S)$-elliptic on S. Gårding's inequality implies with the Lax-Milgram theorem the invertibility of $A+C$ in $H^{\alpha/2}(S)$. Hence, the classical Fredholm alternative holds: $Av=0$ has a finite number $n\geq0$ of linear independent solutions and there exists n linear independent solvability conditions. Moreover, uniqueness implies solvability.

For the Galerkin procedure, we have a sequence of finite dimensional subspaces $V_h\subset H^{\alpha/2}(S)$, $0<h<h_0$, and we look for $v_h\in V_h$ satisfying $<Av_h,g>=<f,g>$ for all

$g \in V_h$, where $f \in H^{-\alpha/2}(S)$. If for all $v \subset H^{\alpha/2}(S)$

$$\inf \{ \|v - \tilde{v}\|_{H^{\alpha/2}(S)} \mid \tilde{v} \in V_h \} \to 0, \ h \to 0 \tag{10.2}$$

then Gårding's inequality (10.1) implies stability and convergence of the Galerkin scheme: If A is bijective, i.e. $Av=f$ is uniquely solvable for any $f \in H^{-\alpha/2}(S)$, then

(a) there exists a $h_0 > 0$ such that for all $0 < h < h_0$ the Galerkin equation $<Av_h, g> = <f, g>$ is uniquely solvable for all $g \in V_h \subset H^{\alpha/2}(S)$;

(b) $v_h \to v$, for $h \to 0$;

(c) there exists constants C_1, C_2, independent of v and h with

$$\|v_h\|_{H^{\alpha/2}(S)} \le C_1 \cdot \|v\|_{H^{\alpha/2}(S)} \ \text{(stability)}$$

$$\|v - v_h\|_{H^{\alpha/2}(S)} \le C_2 \cdot \inf_{g \in V_h} \|v - g\|_{H^{\alpha/2}(S)} \ \text{(quasi-optimal convergence)} . \tag{10.3}$$

Equation (10.3) implies that the Galerkin discretization error $\|v - v_h\|$ depends on the approximation properties of the subspace V_h. For the boundary element trial spaces of class $S^{k+1,t}(S)$ (see Section 5) the following approximation properties in Sobolev spaces are known (cf. Babuska & Aziz 1972).

Let $s \le r \le k+1$ and $s \le t$. Then there exists a constant C, independent of v and h such that for any $v \in H^r(S)$ there exists an approximating $v_h \in V_h$ with

$$\|v - v_h\|_{H^s(S)} \le C \cdot h^{r-s} \|v\|_{H^r(S)} . \tag{10.4}$$

Here t measures the inter-panel differentiability and k is the degree of the polynomials. For the Lagrangian triangular elements discussed in Section 5, Table 10.1 gives an overview about their approximation properties in Sobolev spaces. This together with the quasi-optimal convergence (10.3) gives the estimate of the Galerkin discretization error in the energy norm $H^{\alpha/2}$:

$$\inf_{g \in V_h} \|v - g\|_{H^{\alpha/2}} \le \|v - v_h\|_{H^{\alpha/2}(S)} \le C \cdot h^{r-\alpha/2} \|v\|_{H^r(S)} \tag{10.5}$$

if $v \in H^r(S)$. Note that the *conformity condition* $V_h \in H^{\alpha/2}(S)$ implies $t \ge \alpha/2$. Therefore, for second kind integral equations of order $\alpha = 0$ we can use discontinuous trial functions. For $\alpha = 1$, the trial functions must be at least continuous. The approximating subspaces discussed in Section 5 provide also asymptotic error estimates in norms *different* from the energy norm $H^{\alpha/2}(S)$: Let $\alpha - (k+1) \le s \le t$, $\alpha - t \le r \le k+1$, $s \le r$. Then there exists a constant C, independent of v and h such that for any $v \in H^r(S)$ there exists an approximation $v_h \in V_h$ with

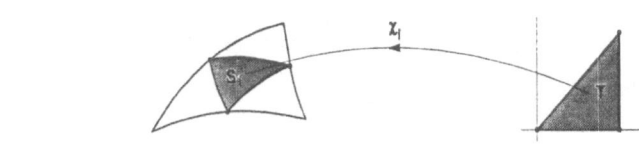

$V_h \in S^{k+1,t}$				
boundary element T	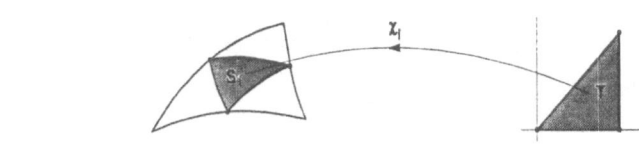			
$\Pi_h v\vert_T = p \in P_k$	$k=0$	$k=1$	$k=2$	$k=3$
$V_h \subset H^t(S)$	$t=0$	$t=1$	$t=1$	$t=1$
regularity of v on S $v \in H^\sigma(S)$	$\sigma=1$	$\sigma=2$	$\sigma=3$	$\sigma=4$
$\Vert v - \Pi_h v \Vert_{H^s(S)} \leq C h^{r-s} \Vert v \Vert_{H^r(S)}$	$s=0 : O(h)$	$s=0 : O(h^2)$ $s=1 : O(h)$	$s=0 : O(h^3)$ $s=1 : O(h^2)$	$s=0 : O(h^4)$ $s=1 : O(h^3)$

Table 10.1. Approximation properties of Lagrangian triangular elements in Sobolev spaces

$$\Vert v - v_h \Vert_{H^s(S)} \leq C \cdot h^{r-s} \Vert v \Vert_{H^r(S)} . \tag{10.6}$$

The highest order of convergence is obtained for $s=\alpha-(k+1)$ and $r=k+1$:

$$\Vert v - v_h \Vert_{H^{\alpha-(k+1)}(S)} \leq C \cdot h^{2k+2-\alpha} \Vert v \Vert_{H^{k+1}(S)} \tag{10.7}$$

if $v \in H^{k+1}(S)$. The results for Lagrangian triangular elements are shown in Table 10.2. Obviously, integral operators of negative order α give a higher order of convergence. However, it is known that the L^2-condition number of the linear system of equations blows up like $O(h^{-\vert\alpha\vert})$ (Wendland 1985). This reflects nothing else but the ill-posedness of such operators. The choice of the discretization parameter h and the machine precision determine whether the discretization error or the round-off error due to a large condition number of the equation matrix become dominant.

For the solution u of the BVP at points $x \in D_0$, $\bar{D}_0 \subset \mathbb{R}^3 \backslash S$, we obtain the pointwise error estimate

$$\vert \partial^p u - \partial^p u_h(x) \vert \leq C(p) \cdot \Vert v - v_h \Vert_{H^s(S)} \tag{10.8}$$

	α			$v \in H^r(S)$
	-1	0	1	
t = 0 k = 0	-	$O(h^2)$	-	r = 1
t = 1 k = 1	$O(h^5)$	$O(h^4)$	$O(h^3)$	r = 2
t = 1 k = 2	$O(h^7)$	$O(h^6)$	$O(h^5)$	r = 3

Table 10.2. Highest order of convergence for Lagrangian triangular elements

for *any* s. Therefore, away from the boundary S, u_h and its partial derivatives of any order p≥1 converge pointwise with order $O(h^{2k+2-\alpha})$:

$$|\partial^p u(x) - \partial^p u_h(x)| \leq C(p) \cdot h^{2k+2-\alpha} \|v\|_{H^{k+1}(S)}, \quad x \in D^c \qquad (10.9)$$

These results have been obtained using variational methods and the weak formulation of the boundary integral equation. For the GBVPs this holds for Galerkin but also for the least-squares method because it is identical to the Galerkin method when the operator A is replaced by A*A (A* is the adjoint operator) and when in Gårding's inequalitiy (10.1) α/2 is replaced by α. For collocation only Fredholm integral equations of the second kind with weakly singular kernels have been studied successfully. The highest order of convergence for the solution of the integral equation is $O((h^{k+1-\alpha})$, i.e. k+1 orders less than for Galerkin. The condition number of the linear system of equations is again $O(h^{-|\alpha|})$.

Example: We consider the linearized fixed gravimetric BVP which is a classical oblique BVP. Using the single-layer representation, we obtain a second kind integral equation with strongly singular kernel (cf. Table 3a). The corresponding integral operator maps continuously from $L^2(S)$ into $L^2(S)$. V_h is as subspace of $L^2(S)$; e.g. the space of continuous functions whose restriction to each panel is a polynomial of degree k. For k=0 discontinuous functions are allowed, as well. The Galerkin matrix has no special structure and the condition number is bounded if h→0. The error estimate for the single-layer density ε is

$$\|\varepsilon - \varepsilon_h\|_{L^2(S)} \leq c \cdot h^{k+1} \cdot \|\varepsilon\|_{H^{k+1}(S)} \tag{10.10}$$

and for any partial derivative of the potential

$$|\partial^p u(x) - \partial^p u_h(x))| \leq c \cdot h^{t+k+1} \cdot \|\varepsilon\|_{H^t(S)} \tag{10.11}$$

for $-(k+1) \leq t \leq k+1$ and $x \in D_0$, $\bar{D}_0 \subset \mathbb{R}^3 \backslash S$. The maximum order of convergence is $O(h^{2k+2})$. Using piecewise constant functions on each panel, wo obtain at most $O(h^2)$.

10.2 Integration error

Mostly, the elements of the Galerkin system of equations must be computed numerically using suitable integration techniques. The error estimates given in Section 10.1 only hold if the matrix elements are calculated *exactly*. In practise, however, this assumption is unrealistic since mostly the integrations have to be done numerically using suitable cubature formulas. Thus, the elements are not error-free. As a consequence, an additional consistency error is introduced and, besides the question of stability, we have to know what the minimal order of consistency is and what accuracy must be achieved. The most recent results for Galerkin Boundary Element Methods have been obtained by Sauter (1992) and Sauter & Krapp (1995). For the inner integration see Schwab & Wendland (1992), and for a weakly singular Fredholm integral equation of the second kind discretized by bilinear elements, see Johnson & Scott (1989).

Let us restrict to Galerkin Boundary Element Methods. Moreover, w.r.t. K_{ij} we only consider the far-field elements; for the near-field elements including those, which are defined by singular integrals, similar results are obtained if semi-analytic techniques as described in Section 9 are applied (see also Hackbusch & Sauter 1993). Let us define a cubature formula $Q_{S_x}^{\gamma_x}$ which integrates polynomials of total degree γ_x over the triangular panel S_x exactly:

$$Q_{S_x}^{\gamma_x}[g] := \sum_k w_k g(x_k) \text{ with } E_{S_x}^{\gamma_x}[g] := \int_{S_x} g(x) dS_x - Q_{S_x}^{\gamma_x}[g] = 0 \text{ for all } g \in P_{\gamma_x}(S_x) \tag{10.12}$$

where $P_{\gamma_x}(S_x)$ denotes the space of polynomials of total degree k on the panel S_x. Let us assume that the elements of the Galerkin system of equations (cf. equation (8.1) are approximated by

$$\hat{M}_{ij} := \sum_{S_x \in \, \text{supp} \, b_i} Q^{\gamma}_{S_x}[b_i, \lambda], \quad \hat{K}_{ij} := \sum_{S_x \in \, \text{supp} \, b_i} \sum_{S_y \in \, \text{supp} \, b_j} Q^{\gamma_x}_{S_x} Q^{\gamma_y}_{S_y}[b_i \, k b_j],$$

$$\hat{f}_i := \sum_{S_x \in \text{supp} \, b_i} Q^{\gamma}_{S_x}[b_i, f], \tag{10.13}$$

and let $\hat{v}_h$ denote the solution of the corresponding linear system of equations. Then, there exists a $h_0 > 0$ and a constant C independent of h such that for $h < h_0$:

$$\|v_h - \hat{v}_h\|_{H^s(S)} \le C h^{k+1-s} \|v\|_{H^{k+1}(S)}, \quad \alpha -(k+1) \le s \le t, \quad V_h \in S^{k+1,t}, \tag{10.14}$$

provided that

$$\gamma_y = \lceil k-s \rfloor, \; \gamma_x = \gamma_y + k - \lfloor \alpha -s \rfloor, \; \gamma = 2k-s-\lfloor \alpha -s \rfloor. \tag{10.15}$$

Here, $\lceil r \rceil$ denotes the smallest integer $\ge r$ and $\lfloor r \rfloor$ the largest integer $\le r$. Moreover, we assumed that $V_h \in S^{k+1,t}$ (S). Table 10.3 shows the required values (γ_x, γ_y) in order to get an optimal error estimate in the $H^s(S)$ norm with piecewise constant and piecewise linear trial functions, i.e. $V_h \in S^{1,0}(S)$ and $V_h \in S^{2,1}(S)$, respectively.

Table 10.3. (γ_x, γ_y) for k=0 (piecewise constant|linear trial functions) to get error estimate in $H^s(S)$

	s = 0	s = -1/2	s = -1	s = -3/2	s = -2
$\alpha = 1$	- | (1,1)	- | (2,2)	- | (1,2)	-	-
$\alpha = 0$	(0,0) | (2,1)	(1,1) | (3,2)	(0,1) | (2,2)	- | (3,3)	- | (2,3)
$\alpha = -1$	-	(1,1) | (2,2)	(1,1) | (3,2)	(2,2) | (4,3)	(1,2) | (3,3)

References

Adams, R.A. (1975): Sobolev Spaces. Academic Press, New York.
Babuška, I.; Aziz, A.K. (1972): Survey lectures on the mathematical foundations of the Finite Element Method. In: A.K. Aziz (ed.), The Mathematical Foundation of the Finite Element Method with Applications to Partial Differential Equations, Academic Press, New York, 3-359.

Baker, E.M. (1988): A finite element model of the earth's anomalous gravitational potential. Reports of the Department of Geodetic Science and Surveying, No. 391, The Ohio State University, Columbus, Ohio.

Ballani, L.; Barthelmes, F.; Klees, R. (1995): On the Use of Wavelets in Geodesy. To be published in Proceedings of the III. Hotine-Marussi Symposium, L´Aquila, Italy, May 29 - June 3, 1994.

Beylkin, G.; Coifman, R.; Rokhlin, V. (1991): Fast wavelet transforms and numerical algorithms I. Comm. Pure and Appl. Math. 44, 141-183.

Bosch, W. (1987): High degree spherical harmonics analysis. Paper presented at XIX. General Assembly of IUGG, Aug. 9-22, Vancouver, Canada.

Brandt, A.; Lubrecht, A.A. (1990): Multilevel matrix multiplication and fast solution of integral equations. Journal of Computational Physics 90, 348-370.

Brenner, S.C.; Scott, L.R. (1994): The mathematical theory of Finite Element Methods. Springer, New York.

Brovar, V.V. (1963): Solutions of the Molodensky boundary value problem. Geodesy and Aerophotography , 237-240.

Brovar, V.V. (1964): Fundamental harmonic functions with a singularity on a segment and solution of outer boundary problems. Geodesy and Aerophotography, 150-155.

Ciarlet, P.G. (1978): The Finite Element Method for Elliptic Problems. North-Holland, Amsterdam.

Ciarlet, P.G.; Lions, J.L. (1991): Handbook of Numerical Analysis, Volume II, Finite Element Methods (Part 1), North Holland, Amsterdam.

Costabel, M. (1987): Principles of boundary element methods. Computer Physics Reports 6, 243-274.

Courant, R.; Hilbert, D. (1962): Methods of Mathematical Physics II. Interscience Publ., New York.

Dahmen, W.; Prößdorf, S.; Schneider, R. (1995): Multiscale methods for pseudo-differential equations on manifolds. In: C.K. Chui (ed.), Wavelet Analysis and its Applications, 5, Academic Press.

Engels, J. (1991): Eine approximative Lösung der fixen gravimetrischen Randwertaufgabe im Innen- und Aussenraum der Erde. Deutsche Geodätische Kommission, Reihe C, Nr. 379, München.

Fabes, E. (1988): Layer potential methods for boundary value problems on Lipschitz domains. In: Král, J.; Lukeš, J.; Netuka, I.; Veselý, J. (eds.), Potential Theory - Surveys and Problems. Lecture Notes in Mathematics 1344, Springer, Berlin, 55-80.

Geers, N.; Klees, R. (1993): Out-of-core solver for large dense nonsymmetric linear systems. Manuscripta geodaetica 18, 331-342.

Geers, N.; Klees, R. (1995): Solution of large linear systems on pipelined SIMD machines. Paper submitted for publication in International Journal for Numerical Methods in Engineering.

Greengard, L.; Rokhlin, V. (1987): A fast algorithm for particle simulations. Journal of Computational Physics 73, 325-348.

Grafarend, E. (1989): The geoid and the gravimetric boundary value problem. The Royal Institute of Technology, TRITA GEOD series, Report no. 18, Stockholm.

Günter, N.M. (1957): Die Potentialtheorie und ihre Anwendung auf Grundaufgaben der mathematischen Physik. Teubner, Leipzig.

Guiggiani, M.; Gigante, A. (1990): A general algorithm for multidimensional Cauchy principal value integrals in the boundary element method. ASME J. Appl. Mech. 57, 907-915.

Hackbusch, W. (1989): Integralgleichungen. Teubner, Stuttgart.

Hackbusch, W.; Nowak, Z.P. (1989): On the fast matrix multiplication in the boundary element method by panel clustering. Numer. Math. 54, 463-491.

Hackbusch, W.; Sauter, St. (1992): On the efficient use of the Galerkin method to solve Fredholm integral equations. Institut für Praktische Mathematik, Bericht Nr. 92-18, Universität Kiel.

Hackbusch, W.; Sauter, St. (1993): On the efficient use of the Galerkin method to solve Fredholm integral equations. Applications of Mathematics 38, 301-322.

Hackbusch, W.; Sauter, St. (1993): On numerical cubatures of nearly singular surface integrals arising in BEM collocation. Institut für Praktische Mathematik, Bericht Nr. 93-4, Universität Kiel.

Hackbusch, W.; Sauter, St. (1994): On numerical cubatures of nearly singular surface integrals arising in BEM collocation. Computing 52, 139-159.

Harten, A.; Yad-Shalom, I. (1992): Fast multiresolution algorithm for matrix-vector multiplication. ICASE report no. 92-55.

Hayami, K.; Matsumoto, H. (1994): A numerical quadrature for nearly singular boundary element integrals. Engineering Analysis with Boundary Elements 13, 143-154.

Heck, B. (1991): On the linearized boundary value problems of Physical Geodesy. Department of Geodetic Science, Ohio State University, Report No. 407, Columbus, Ohio.

Hörmander, L. (1976): The boundary problems of physical geodesy. Archive for Rational Mechanics and Analysis 62, 1-52.

Johnson, C.G.L.; Scott, L.R. (1989): An analysis of quadrature errors in second-kind boundary integral methods. SIAM Journal of Numerical Analysis 26, 1356-1382.

Klees, R. (1992): Lösung des fixen geodätischen Randwertproblems mit Hilfe der Randelementmethode. Deutsche Geodätische Kommission, Reihe C, Nr. 382, München.

Klees, R. (1996): Efficient calculation of surface integrals with potential type singularities. Part I: Weakly singular integrals. Accepted for publication in Journal of Geodesy.

Koch, K.-R. (1967): Die Berechnung des Störpotentials und seiner Ableitungen aus den Integral- und Integrodifferentialgleichungen der Greenschen Fundamentalformel mit Hilfe schrittweiser Näherung. Deutsche Geodätische Kommission, Reihe C, Nr. 105, München.

Koch, K.-R. (1970): Reformulation of the geodetic boundary value problem in view of the results of geometric satellite geodesy. In: Kattner, W.T. (ed.), Advances in Dynamic Gravimetry, Pittsburg.

Koch, K.-R. (1971): Die geodätische Randwertaufgabe bei bekannter Erdoberfläche. Zeitschrift für Vermessungswesen 96, 218-224.

Koch, K.-R. (1972): Method of integral equations for geodetic boundary value problems. Mitteilungen aus dem Institut für Theoretische Geodäsie, Universität Bonn, Nr. 4, Bonn.

Král, J.; Wendland, W.L. (1988): On the applicability of the Fredholm-Radon method in potential theory and the panel method. In: Ballmann, J.; Eppler, R.; Hackbusch, W. (eds.), Panel Method in Fluid Mechanics with Emphasis on Aerodynamics. Notes on Numerical Fluid Mechanics 21, Vieweg, Braunschweig, 120-136.

Lage, C. (1995): Softwarentwicklung zur Randelementmethode: Analyse und Entwurf effizienter Techniken. PhD thesis, Christian-Albrechts-University Kiel.

Meissl, P. (1981): The use of finite elements in physical geodesy. Reports of the Department of Geodetic Science and Surveying, No. 313, The Ohio State University, Columbus, Ohio.

Miranda, C. (1970): Partial differential equations of elliptic type. Springer, Berlin.

Mitrea, M. (1994): Singular integrals, Hardy spaces and Clifford wavelets. Lecture Notes in Mathematics no. 1575, Springer New York.

Mizohata, S. (1973): The Theory of Partial Differential Equations. University Press, Cambridge.

Moritz, H. (1989): Advanced Physical Geodesy. 2nd edition, Wichmann, Karlsruhe.

Nédélec, J.C. (1976): Curved finite element methods for the solution of singular integral equations on surfaces in $\mathbb{R}^3$. Computer Methods in Applied Mechanics and Engineering 8, 61-80.

Nédélec, J.C. (1977): Approximation des equations intégrales en méchanique et en physique. Cours de l'école d'été d'analyse numérique. CEA-IRIA-EDF, Palaiseau.

Partridge, P.W.; Brebbia, C.A.; Wrobel, L.C. (1991): The dual reciprocity boundary element method. Computational Mechanics Publications, Southampton.

Petersdorff, T. von; Schwab, C.; Schneider, R. (1994): Multiwavelets for second kind integral equations. Preprint, University of Maryland.

Petersdorff, T. von; Schwab, C. (1995): Fully discrete multiscale Galerkin BEM. Preprint Seminar für Angewandte Mathematik, ETH Zürich.

Prasad, K.G.; Keyes, D.E.; Kane, J.H. (1993): Wavelet-based preconditioners for boundary element problems. Paper presented at IABEM 93, Braunschweig.

Rapp, R. (1993): Use of altimeter data in estimating global gravity models. In: Rummel, R.; Sansò, F. (eds.) 'Satellite Altimetry in Geodesy and Oceanography', Lecture Notes in Earth Sciences Vol. 50, Springer, Berlin, 374-420.

Rapp, R.; Pavlis, N.K. (1990): The development and analysis of geopotential coefficient models to spherical harmonic degree 360. J. Geophys. Res., 95, B13, 21885-21911.

Rokhlin, V. (1983): Rapid solution of integral equations of classical potential theory. Journal of Computational Physics 60, 187-207.

Rosen, D.; Cormack, D.E. (1994): The continuation approach for singular and nearly singular integration. Engineering Analysis with Boundary Elements 13, 99-113.

Sacerdote, F.; Sansò, F. (1986): The scalar boundary value problem of physical geodesy. Manuscripta Geodaetica 11, 15-28.

Sauter, St. (1991): Der Aufwand der Panel-Clustering-Methode für Integralgleichungen. Report Nr. 9115, Christian-Albrechts-Universität, Kiel.

Sauter, St. (1992): Über die effiziente Verwendung des Galerkinverfahrens zur Lösung Fredholmscher Integralgleichungen. Dissertation, Christian-Albrechts-Universität, Kiel.

Sauter, St.; Krapp, A. (1995): On the effect of numerical integration in the Galerkin boundary element method. Report no. 9504, Institute for informatics and practical mathematics, Christian-Albrechts-University, Kiel.

Schneider, R. (1995): Multiskalen- und Wavelet-Matrix Kompression: Analysis basierte Methoden zur effizienten Lösung großer vollbesetzter Gleichungssysteme. Habilitation Thesis, Technical University Darmstadt, Germany (in German).

Schröder, P.; Sweldens, W. (1994): Spherical wavelets: Efficiently representing functions on the sphere. Preprint, Department of Mathematics, University of South Carolina, Columbia SC 29208.

Schwab, C. (1994): Variable order composite quadrature of singular and nearly singular integrals. Computing 53, 173-194.

Schwab, C.; Wendland, W.L. (1992): On numerical cubatures of singular surface integrals in boundary element methods. Numer. Math. 57, 343-369.

Schwarz, H.R. (1980): Methode der finiten Elemente. Teubner, Stuttgart.

Wendland, W.L. (1983): Boundary Element Methods and their Asymptotic Convergence. In: Filippi, P. (ed.), Theoretical Acoustics and Numerical Techniques. CISM Courses 277, Springer, Wien, 135-216.

Wendland, W.L. (1985): On some mathematical aspects of boundary element methods for elliptic problems. In: Whiteman, J.R. (ed.), The Mathematics of Finite Elements and Applications V, Academic Press, London, 193-227.

Wendland, W.L. (1987): Strongly elliptic boundary integral equations. In: Iserles, A.; Powell, M. (eds.), The State of the Art in Numerical Analysis. Clarendon Press, Oxford, 511-561.

Wenzel, H.-G. (1985): Hochauflösende Kugel-funktionsmodelle für das Gravitationspotential der Erde. Wissenschaftliche Arbeiten der Fachrichtung Vermessungswesen der Universität Hannover, Nr. 137, Hannover.

Witsch, K.J. (1985): On a free boundary value problem of physical geodesy, I (Uniqueness). Mathematical Methods in the Applied Sciences 7, 269-289.

Wu, S. (1995): On the evaluation of nearly singular kernel integrals in boundary element analysis. Communications in Numerical Methods in Engineering 11, 331-337.

Zhdanov, M.S. (1988): Integral Transforms in Geophysics. Springer, Berlin.

Zienkiewicz, O. (1983): The Finite Element Method. McGraw-Hill, London, 3rd edition, reprinted.

Zienkiewicz, O.; Morgan, K. (1983): Finite Elements and Approximation. John Wiley & Sons.

Solving Geodetic Boundary Value Problems with Parallel Computers

R. Lehmann

1 Introduction

In the last years modern mathematical methods for the solution of boundary value problems (BVPs) attracted much interest in Geodesy, namely the finite difference method (FDM), the finite element method (FEM), and the boundary element method (BEM). They are able to solve general linear BVPs directly, i.e. up to any desired numerical precision. We could then abolish several doubtful classical geodetic approximation steps like spherical and constant radius approximation.

Investigations on the use of the FEM in Physical Geodesy are found in (MEISSL 1981, BAKER 1988). This method has been further applied to the solution of the geodetic BVP in spherical approximation by SHAOFENG & DINGBO (1991). Suggestions on the geodetic application of FDM came from KELLER (1995). So far the best investigated method of this kind in Geodesy is the BEM, which for a number of reasons is well suited for geodetic problems. Readers not familiar with the details of this method are referred to the contribution of KLEES in this volume and the citings therein. However, our contribution will not require a complete knowledge of its theoretical background.

The basic drawback of all those methods is that they are computationally much more demanding than classical geodetic techniques. Apart from some open questions, this has so far prevented their extensive application in Geodesy. Superscalar and vector computers have already failed to solve todays geodetic problems. On the one hand, the size of these problems will grow very rapidly in the future due to

- new observation techniques like airborne gravimetry or satellite gradiometry, and
- new accuracy requirements like the 1 cm geoid.

On the other hand, the power of classical sequential computers cannot be expected to grow much in the foreseeable future.

Several activities have been started to overcome this difficulty. Efficient numerical algorithms for the BEM have been discussed already in the contribution of KLEES in this volume. However, we also witness an immense progress in the branch of computer science and technology. Today, parallel computers achieve tremendous computer power (Gflops), and they will achieve even more in the near future (Tflops).

Example 1: Let us assume a moderate number of $n = 64000$ observations or unknowns. In several methods like the BEM, where we have to solve a linear system of equations with a dense $(n \times n)$ system matrix, we would then have to store a 16 Gbyte matrix (32 bit precision). This amount of storage will probably be available everywhere on the hard disk. But if one wants to apply iterative solvers of CG-type (conjugate gradients), the system must be stored in the core memory. The only computers available today, that can offer such a tremendous amount of core memory, are parallel computers. In the past, out-of-core solvers have been used in Geodesy (GEERS&KLEES 1993). Unfortunately, this is only possible for direct (non-iterative) solvers, typically requiring much more floating point operations.

The iterative solution with CG-solvers requires in each step mainly one matrix $\times$ vector multiplication. The number of floating point operations is $2n^2$, here $8 \cdot 10^9$. The question is: How much time can one afford to spend for one iteration step? Probably not more than one second, because perhaps several thousands of iteration steps are required for this size of problem, in order to achieve a reasonable accuracy. This requires a sustained performance of 8 Gflops in our example. Again, this requirement can only be met by parallel computers.

In other technical sciences, were BVPs occur, the use of parallel computers is standard nowadays. It is time to tap this giant resource of computer power also in Geodesy. However, supercomputers should *not* be used if problems can be solved in reasonable time with conventional computers. Up to now there is little reason to employ parallel computers in areas where fast classical methods (like FFT) can be applied.

2 Hardware issues

At present, parallel processing is the basic possibility to make supercomputers faster. If we talk about parallelism, we must distinguish between internal and external parallelism. *Internal parallelism* is already used extensively in conventional computer architectures:

- In superscalar processors there exist multiple floating point units that can perform a few floating point operations in parallel.
- In vector pipeline processors the technique of pipeline processing has a similar effect.

As a matter of fact, no major improvement of performance can be expected anymore from internal parallelism. Much better are the results and prospects using external parallelism, i.e. using multiple processing units. Hence, the scope of the subsequent discussion is *external parallelism*. It enables us to achieve *tremendous computing power at moderate prices*.

Subsequently, we will briefly discuss the most important parallel computer architectures:

1. SIMD[1] parallel computers. Such computers are usually equipped with an array processor ($\approx 2^{10} \ldots 2^{16}$ processing elements). All processing elements perform identical operations with different data in each step, except when some are masked and idle. This turns out to be a severe restriction for the user. Examples: Connection machines CM-1, CM-2, MasPar MP-1, MP-2. The MP-2 has a peak performance of 2.4 Gflops using 16384 RISC processors. This falls relatively short compared to the requirements in the example 1. In general, such computers are not well suited for engineering purposes.

2. MIMD[2] computers with shared memory. In order to overcome the strong limitation of SIMD machines we must allow each processing unit to run its own program, which is achieved with MIMD machines. The simplest approach is to connect each processing unit immediately with a single (shared) memory, i.e. to realize a global address space. Such computers are usually equipped with few ($\approx 2 \ldots 64$) processing units. We may distinguish between machines using

- inexpensive superscalar processors (known as SMP[3], examples: Sun SPARC-center 2000, Cray CS6400), or
- powerful vector processors (known as PVP[4], examples: Convex C3, Cray Y/MP, NEC SX-4)

Their basic problem is the scalability, i.e. the number of processors sharing a global memory cannot easily be enlarged without constricting the memory bottleneck.

3. MIMD computers with distributed memory. One might also give up the restriction of a shared memory, using a distributed memory instead (local

---

[1] Single Instruction stream / Multiple Data stream
[2] Multiple Instruction stream / Multiple Data stream
[3] Symmetric Multi Processor
[4] Parallel Vector Processor

address spaces), interconnected by a fast network. Again we may distinguish between machines using

- many ($\approx 32 \ldots 10000$) inexpensive superscalar processors (known as MPP[5], examples: IBM RS/6000 SP, Intel Paragon, Cray T3D, CM-5), or
- few powerful vector processors (known as scalable PVP, examples: Fujitsu/SNI VPP300)

Although not achieving highest performances, even interconnected workstations (workstation clusters) can be used as such a parallel computer. With distributed memory all (previously technical) problems are now shifted to the software side, i.e. to the user.

4. MIMD computers with virtually shared memory. Here the memory is physically local, but logically shared, i.e. the operating system *simulates* a global address space. This concept was meant to merge the best features of shared and distributed memory, but *failed so far*. Examples: KSR-1, KSR-2, Stanford Dash.

3 Programming issues

The basic approach to parallel programming is problem decomposition, i.e. to define subproblems. The following criteria should be met by the subproblems:

- uniform load distribution (in order to keep all processors busy)
- maximum independence of the subproblems (i.e. to avoid interactions)
- scalability (i.e. to avoid redundant storage and computation)

Usually, these goals are in *conflict* with each other. E.g. if subproblems are maximally independent then the load distribution might be poor, and vice versa. Even worse, the best *tradeoff* is extremely hardware dependent.

The best established parallel programming concepts are:

1. Data parallel programming, e.g. with High Performance Fortran (HPF) or with standard languages, supported by compiler options. This concept was introduced for SIMD parallel computers, but can also be used elsewhere.

2. Using message passing libraries like PVM, PARMACS, MPI, MPL, etc. They provide explicit data exchange commands (**send, receive, wait,** ...), which makes them best suited for distributed memory parallel computers.

---

[5]Massively Parallel Processor

Sometimes one chooses the number of subproblems (parallel tasks) larger than the number of processors available. The design of a mapping of the set of tasks onto the set of processors, minimizing the total runtime, is called load balancing. Static load balancing means that this mapping is a priori defined based on estimates of the computational complexity of the tasks. If such an estimate cannot be obtained very easily then dynamic load balancing will work better. Here the mapping is defined at runtime, depending on the actual progress of the different processors. Among the various algorithms available we only mention the task pool concept: While there are unprocessed tasks in a pool, each idling processor gets one and processes it.

More details can be found in textbooks like (KUMAR et al. 1994, FOSTER 1995).

4 Geodetic problems of parallel numerical integration

In Geodesy there appears the problem of computing integrals of type

$$F(x_1) \quad = \quad \int_{y \in D} f(x_1, y) dy$$

$$\quad \vdots \quad \vdots \quad \vdots \tag{1}$$

$$F(x_N) \quad = \quad \int_{y \in D} f(x_N, y) dy$$

This problem can be computationally very complex because typically

- D is (an approximation of) the surface of the Earth or a piece of this surface, which can be very complicated,
- f is a non-trivial (often singular) integrand, and
- $(x_1, \ldots, x_N)$ is a large set of evaluation points.

If we want to employ parallel computers in this area, we must decompose the problem (1) into subproblems. This is apparently easy: We may decompose it

- either with respect to the evaluation points

$$(x_1, \ldots, x_{N_1}), \ldots, (x_{N_{p-1}+1}, \ldots, x_N) \tag{2}$$

where

$$0 < N_1 < \ldots < N_{p-1} < N$$

Table 1: Parallel evaluation of the terrain attraction

decomposition of	evaluation points (2)	domain (3)
Is the load well balanced ?	in general NO (height dependent)	in general NO (e.g. roughness dependent)
Are the tasks independent ?	NO: special methods of orbit integration YES: otherwise	YES
Scalability with shared memory ?	YES	YES
Scalability with distributed memory ?	NO: terrain data must reside in each local memory	YES

– or with respect to the domain

$$D = \bigcup_{i=1}^{n} D_i, \quad \text{measure}(D_i \cap D_j) = 0 \text{ if } i \neq j \tag{3}$$

Example 2: Let us illustrate the effect for the problem of evaluating the terrain attraction (see table 1). The decomposition principle (2) seems to work slightly better.

At this point it is important to realize the following facts:

– Although easily expressed in a general form (1) numerical integration problems are highly individual, which makes it difficult to design parallel all-purpose algorithms.
– The load balancing problem is the most serious problem in geodetic applications of parallel numerical integration.
– The considerations shown in table 1 are the job of geodesists rather than of computer experts or (least of all) of compilers.

Example 3: Let us compute the single layer potential u and its gradient ∇u for a sphere D with constant layer density in points on D. The integral is weakly singular for u and strongly singular for ∇u. We do not take advantage of the existence of an analytical solution, but decompose D (as usual in the BEM) into 8142 nearly equilateral triangles, and sum over the partial integrals of the triangles. An accuracy of 6 decimal digits is ensured by comparison of the numerical results in all centers of the triangles with the analytical solution. This computation takes 356s on a single processor of IBM RS/6000 SP, i.e. 11ms per global integral. Using 16 times more globally well distributed evaluation points and 16 processors of the same computer takes 368s, i.e. 0.70ms per global integral. The subproblems are defined as in (3). The efficiency is $356/368 \approx 97\%$ because the load imbalance is minimal. However, if we restrict the evaluation points to the inside of a spherical cap $D_\theta \subset D$ of radius $\theta = 10°$, i.e. area$(D_\theta) \approx 0.01$ area(D), then the load imbalance is much worse. If by any chance for a processor i with an assigned surface piece D_i in (3) holds $D_\theta \subset D_i$ then this processor is employed with a multiple of the average workload (many singular integrals). Using the taskpool method mentioned in the previous section is a good remedy but still takes 438s on IBM RS/6000 SP, i.e. no more than $356/438 \approx 81\%$ efficiency is retained.

Example 4: Using the Galerkin discretization method for the single layer integral equation of the fixed gravimetric BVP there appears the problem of computing n^2 integrals of type

$$\int_{\pi_i} \mu_i(x) \int_{\pi_j} \mu_j(y) k(x,y) d\pi_i(y) d\pi_j(x)$$

$$k(x,y) := \frac{\partial}{\partial l(x)} \frac{1}{|y-x|}, \quad i,j := 1,\ldots,n$$

with μ_i, μ_j denoting some (constant or linear) shape functions on the boundary surface, $\pi_i := \text{supp}(\mu_i)$ the boundary elements or unions of boundary elements, and $l(x)$ the direction of normal gravity at x. It has been shown in (LEHMANN&KLEES 1996) that the maximum ratio of the computational complexities of these integrals may easily amount up to $100\ldots 1000$, which indicates a severe load imbalance in the parallelization. Moreover, it turned out that a more or less naive approach to load balancing gives poor results. In (LEHMANN&KLEES 1996) a special variant of a task pool method has been proposed and successfully implemented.

More details on general aspects of parallel numerical integration can be found in textbooks like (KROMMER&UEBERHUBER 1994).

5 Geodetic problems of parallel linear algebra

Fortunately, numerical linear algebra problems are less individual than numerical integration problems, such that in this field we may to a large extent rely on the experiences of computer experts.

With *shared memory*, the problems are thought to be easy, e.g. the inner loop is vectorized (if possible), and the outer loop is parallelized. Nevertheless, the chosen algorithm usually has a major impact on performance.

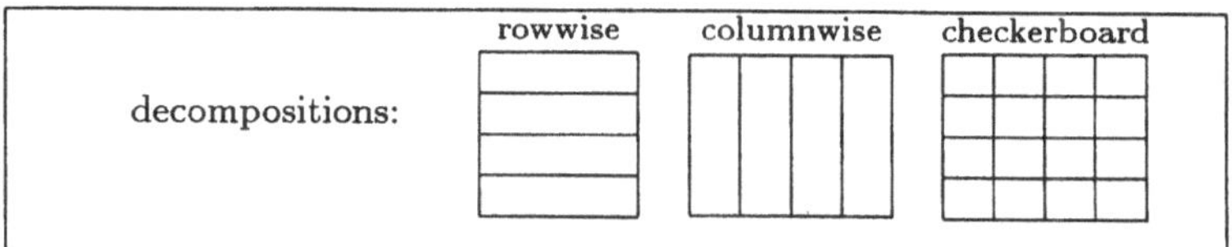

decompositions:	rowwise	columnwise	checkerboard

block mapping:	index	1	2	3	4	5	6	7	8
	processor	1	1	1	1	2	2	2	2
cyclic mapping:	index	1	2	3	4	5	6	7	8
	processor	1	2	1	2	1	2	1	2
block-cyclic mapping:	index	1	2	3	4	5	6	7	8
	processor	1	1	2	2	1	1	2	2

Fig. 1: Matrix decompositions and mappings

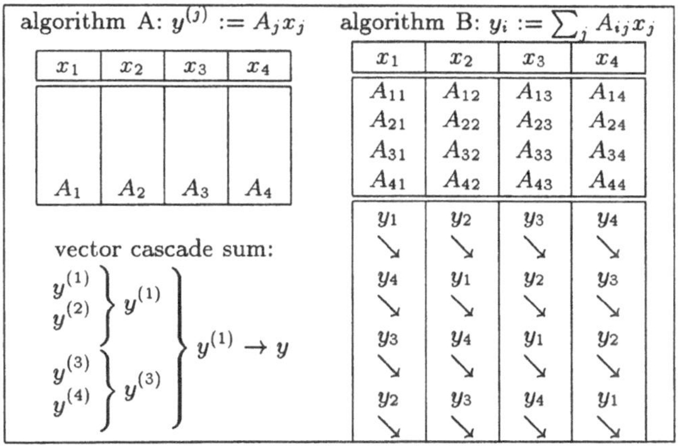

Fig. 2: Two dense matrix $\times$ vector multiplication ($y = Ax$) algorithms

With *distributed memory* matters are more difficult: matrices and vectors must be decomposed and *distributed* over the local memories (scalability!, see figure 1). This is always up to the user. Again, the chosen method may severely influence performance.

Example 5: Let us illustrate the problems with distributed memory for the basic linear algebra operation of the BEM. If we employ a CG-type solver for the solution of the linear system, this operation corresponds to a dense nonsymmetric matrix times vector multiplication $y := Ax, x, y \in \mathbb{R}^n$. Two different parallel realizations of this operation are shown in figure 2. In both cases a columnwise decomposition of the matrix and a block mapping is used. In algorithm A each processor computes its own contribution to y independently. Later the solution is obtained by a vector cascade sum. Algorithm B occupies slightly less storage because also y is decomposed. In each step the processors compute only their contribution to the local piece of y. Between the steps a global ring shift of these pieces takes place.

Which variant works better? It depends on the hardware we are going to use. Sometimes the ring shift operation is faster than the vector cascade sum (e.g. for a ring network). With a PVP algorithm A might work better because the vector length is maximum (n).

Complementary algorithms exist for rowwise decomposition.

Unlike in nontrivial applications of numerical integration, the runtime behaviour of linear algebra operations is to a large extent predictable, thus leaving little room for load imbalances.

6 Results

Today, parallel computers offer a tremendous computing power at moderate prices. However, parallel computing is not easy because straightforward implementations of "classical" sequential (or vectorized) codes are likely to fail. We must check our numerical algorithms and data structures if they are suitable for parallel computers.

In the near future, Geodesists cannot always solve their problems with classical computers anymore, least of all geodetic BVPs. On the one hand, the power of these computers will not increase as rapidly as the size of our numerical problems (number of data, desired resolution). On the other hand, the applicability of fast classical methods (like FFT) is based on approximations which may not meet all future accuracy requirements.

In the far future, parallel computing will probably become the normal way of scientific computing.

References

Baker, E.M. (1988): A finite element model of the Earth's anomalous gravitational potential. Dep. Geodetic Sci. and Surv., Rep. No. 391, The Ohio State University, Columbus.

Geers, N., R. Klees (1993): Out-of-core solver for large dense nonsymmetric linear systems. manuscr. geod. 18, 331–342.

Foster, I. (1995): Design and Building Parallel Programs: Concepts and Tools for Parallel Software Engineering. Addison-Wesley Publishing Company, Inc. Reading, Mass.

Keller, W. (1995): Finite differences schemes for elliptic boundary value problems. Section IV Bulletin IAG, No. 1.

Klees, R. (1996): Contribution in this volume.

Krommer, A.R., C.W. Ueberhuber (1994): Numerical Integration on Advanced Computer Systems. Lecture Notes in Computer Sciences, Vol. 848. Springer Verlag Berlin Heidelberg.

Kumar, V., A. Grama, A. Gupta, G. Karypis (1994): Introduction to Parallel Computing: Design and Analysis of Algorithms. The Benjamin/Cummings Publishing Company, Inc. Redwood City, Cal.

Lehmann, R., R. Klees (1996): Parallel Setup of Galerkin Equation System for a Geodetic Boundary Value Problem. To appear in: W. Hackbusch, G. Wittum (eds.): "Boundary Elements: Implementation and Analysis of Advanced Algorithms", Notes on Numerical Mechanics. Vieweg Verlag, Braunschweig.

Meissl, P. (1981): The use of finite elements in physical geodesy. Dep. Geodetic Sci. and Surv., Rep. No. 313, The Ohio State University, Columbus.

Shaofeng, B., Dingbo, C. (1991): The finite element method for the geodetic boundary value problem. manu geod. 16, 353-359

Application of Boundary Value Techniques to Satellite Gradiometry

W. Keller

1 Introduction

For a high resolution determination of the gravity field much hope is invested in a future satellite gradiometry mission. A number of possible mission scenarios have already been studied. They have the following features in common:

- The satellite flies at a sun-synchronous repeat orbit at a height of about 200 - 400 km with a repeat cycle of about 120 - 180 days.

- Due to the sun synchonous orbit, the orbit inclination differs slightly from a polar orbit. The consequence is that the satellite never flies over to spherical caps of radius 6 degrees around the poles.

- In the orbit the satellite measures the second order derivative V_{dd} of the gravitational potential in some direction d . The direction d and the accuracy of the gravity gradients depend on the mission under consideration. It can be the radial direction and an accuracy of 0.01 E.U as in the ARISTOTELES mission or the direction normal to the orbital plane and 0.0001 E.U. as in the STEP mission.

- The long repeat cycle and the high sampling rate produce an enormous amount of data which is (except of the polar gaps) distributed very regularly over the surface of the Earth.

- A good prior information about the unknown gravity field is available e.g in the form of gravity field models.

A gravity field model with an accuracy of about 10 cm in terms of geoid undulation and with a resolution up to degree and order 200 is to be determined from the combination of gradiometry and ground gravity data.

There are two completely different solution strategies : the so called *time-wise approach* and the *spacewise approach*.

In the timewise approach the unknown gravitational potential is assumed to have a series expansion in spherical harmonics. The unknown coefficients of this expansion are determined such that the evaluation of the second order derivatives of the series coincide as well as possible with the measured gravity gradients.

The advantage of the timewise approach is the great flexibility in the modelling of the measurement process. Its disadvantage is that the huge amount of data generates matrices of very high dimension, which are difficult to invert. Additionally, a a-priory forecast of the obtainable accuracy is not possible in the timewise approach.

The spacewise approach is complementary to the timewise approach. It takes advantage of the high data-rate and the dense distribution of data points on a surface Σ , which represents the orbit of the satellite. Because the data spacing is small compared to the desired resolution the data coverage can (except of the polar data gaps) be assumed to be continuous and the problem can be considered as the following overdetermined boundary value problem:

$$\Delta V(x) = 0, \qquad x \in \text{ext } S \tag{1}$$

$$\Gamma = \frac{\partial^2 V}{\partial d^2}\Big|_\Sigma + \epsilon \tag{2}$$

$$g = \frac{\partial V}{\partial r}\Big|_S + \eta \tag{3}$$

It will be shown later that this continuous approach has the advantage that it leads to an inversion free solution and that it automatically provides an a-priori error estimate.

The disadvantage of the spacewise approach is the assumption that the data is given in a surface Σ , representing the orbit of the satellite. In reality the data is measured directly in the orbit i.e. above or below the reference surface Σ. Hence, the measured data is to be interpolated from the location where the measuremnet took place to the closest point of the reference surface Σ. It has to be made sure that the interpolation error does not exceed the measurement error of the gradiometer.

In the following, it is assumed that the interpolation problem is solved and the discussion is completely restricted to the spacewise approach.

2 Mathematical Model

In order to extract the gravitational potential from the noisy data a mathematical model has to be built up, which properly handles the overdetermination of the problem and the influence of the noise, contained in the data. This model has to have several components:

- The **functional model**, describing the connection between the error-free data and the unknown solution.

- The **stochastic model**, describing the error behaviour of the data.

The functional model itself has again two components: The function spaces, describing the regularity of the given data and the unknown solution and the operators, acting between these spaces, and mapping the gravitational potential to the measured quantities.

2.1 Functional model

2.1.1 Sobolev spaces

Generally speaking, a function space is a set of functions, defined on the same set Ω, which have at least a certain degree of regularity. The different types of function spaces differ by the way regularity is measured.

For the considerations to be made here, the so called *Sobolev spaces* are an appropriate measure of regularity. A *Sobolev space* $W^{l,2}(\Omega)$ is defined as the set of all function which are

- defined on Ω,

- in the generalized sense l-times differentiable on Ω and

- which have generalized derivatives which are square integrable on Ω up to the order l

The higher l is the more regular the functions in $W^{l,2}(\Omega)$ are. The index l can also be negative, meaning that after l-times integration the functions are square integrable. Members of *Sobolev spaces* with negativ order are very irregular functions. Therefore, *Sobolev spaces* with negative order are well suited to model irregular signals as *white noise*.

Regularity or irregularity of signals can also be expressed in terms of their frequency content: regular signals have dominating long-wave components. In irregular signals dominate the high frequencies. Hence, it should be possible to describe *Sobolev space* also by the spectra of its members.

This characterization starts from the differentiation formula of the *Fourier transform*.

The differentation theorem of *Fourier transform* now states that the *Fourier transform* of a derivative equals the *Fourier transformation* of the original function times the corresponding powers of the frequency.

$$\mathcal{F}\{D^\alpha V(x)\}(k) = i^{|\alpha|} k_1^{\alpha_1} k_2^{\alpha_2} \ldots k_n^{\alpha_n} \mathcal{F}\{V(x)\}(k) \tag{4}$$

One remembers that a function V belongs to a *Sobolev space* $W^{l,2}(\Omega)$ if its derivatives up to the order l are square integrable. As a consequence of *Parsevals theorem* a function is square integrable if and only if its *Fourier transform* is square integrable. Hence, a function $D^\alpha V(x)$ is square integrable if and only if $\imath^{|\alpha|}k_1^{\alpha_1}k_2^{\alpha_2}\ldots k_n^{\alpha_n}\mathcal{F}\{V(x)\}(k)$ is square integrable. To be a member of $W^{l,2}(\Omega)$ the *Fourier transform* has to decay faster than any polynomial of degree l.

This leads to the equivalent definition of a *Sobolev space*

$$V \in W^{l,2}(\Omega) \longleftrightarrow \int_{\mathcal{R}^n} \left(1 + |k|^2\right)^l \left(\mathcal{F}\{V\}\right)^2 dk \tag{5}$$

This approach of describing the regularity of functions by the decay of their spectra has a natural extension from *Euklidian spaces* to the sphere. A *Sobolev space* on a sphere S_R of radius R is defined as the set of all functions having spherical harmonics expansions with a given asymptotic of their coefficients:

$$V \in W^{l,2}(S_R) \Longleftrightarrow \sum_{n=0}^{\infty} \sum_{m=-n}^{n} \left(\frac{2n+1}{2R}\right)^{2l} v_{nm} < +\infty \tag{6}$$

with

$$V = \sum_{n=0}^{\infty} \sum_{m=-n}^{n} v_{nm} \bar{Y}_{nm} \tag{7}$$

The similarity of the definitions of *Sobolev spaces* in *Euklidian spaces* and on the sphere is obvious. The spectrum $\mathcal{F}\{V\}(k)$ corresponds to the sequence of coefficients $\{v_{nm}\}$. In both cases a function belongs to $W^{l,2}$ if its spectrum decays faster than k^l or n^l respectively.

2.1.2 Pseudodifferential Operators

The next step in the construction of the mathematical model is the definition of operators which connect, in the noise-free case, the given data with the unknown solution. This connection can be given in the form of a differential operator, as in the case of gradiometry, or in the form of an integral operator as e.g. the *Stokes operator*. For both types of operators a unique concept has to be found. This will be the concept of *Pseudodifferential operators*. The point of departure is again the differentation formula of *Fourier transform*:

$$\mathcal{F}\{D^\alpha V(x)\}(k) = \imath^{|\alpha|}k_1^{\alpha_1}k_2^{\alpha_2}\ldots k_n^{\alpha_n}\mathcal{F}\{V(x)\}(k) \tag{8}$$

Applying the inverse *Fourier transform* to this relation the following representation of a differential operator is found

$$D^\alpha V(x) = \mathcal{F}^{-1}\{\imath^{|\alpha|}k_1^{\alpha_1}k_2^{\alpha_2}\ldots k_n^{\alpha_n}\mathcal{F}\{V(x)\}(k)\} \tag{9}$$

This representation of a differential operator can be interpreted as a filtering: The function is transformed in the frequency domain. Then the high frequencies are amplified and the modified spectrum is transformed back to the space domain.

The situation for integral operators is quite similar. For example, a convolution operator can be transformed in the frequency domain as follows

$$\mathcal{F}\{K \star V\}(k) = \mathcal{F}\{K\}(k) \cdot \mathcal{F}\{V\}(k) \tag{10}$$

If again the inverse *Fourier transform* is applied to this relation the following representation of a convolution operator is found

$$(K \star V)(x) = \mathcal{F}^{-1}\{\mathcal{F}\{K\} \cdot \mathcal{F}\{V\}\}(x) \tag{11}$$

Again, an interpretation of this representation as a filter is possible: The signal is transformed in the frequency domain. The high frequencies are damped and the modified spectrum is transformed back to the space domain. This leads to the definition of a certain class of filters as *Pseudodifferential operators*

Definition 2.1 *A mapping*

$$p : \begin{cases} W^{l,2}(\Omega) & \rightarrow & W^{l-s,2}(\Omega) \\ u & \mapsto & \mathcal{F}^{-1}\{a(k)\mathcal{F}\{u\}(k)\}(x) \end{cases} \tag{12}$$

with

$$|D^{\alpha}a(k)| < C \cdot (1 + |k|)^{s-\alpha}, \quad k \in \mathcal{R}^n \tag{13}$$

is called pseudodifferential operator of order s. *The filter characteristic* $a(k)$ *is called* symbol *of the PDO p.*

This definition of a pseudodifferential operator in a *Euclidian space* has an immediate extension to the definition of pseudodifferential operators on a sphere.

Definition 2.2 *(PDO on a sphere)*
A mapping

$$p : \begin{cases} W^{l,2} & \rightarrow & W^{l-s,2} \\ u & \mapsto & \sum_{n=0}^{\infty}\sum_{m=-n}^{n} a_n u_{n,m}\bar{Y}_{n,m}(\vartheta,\lambda), \quad a_n = O(n^s) \end{cases} \tag{14}$$

with

$$u(\vartheta,\lambda) = \sum_{n=0}^{\infty}\sum_{m=-n}^{n} u_{n,m}\bar{Y}_{n,m}(\vartheta,\lambda) \tag{15}$$

is called invariant pseudodifferential operator *on a sphere.*

The similarities of the definitions of pseudodifferential operators on *Euclidian spaces* and on the sphere are obvious:

- The spherical harmonics coefficients $u_{n,m}$ correspond to the spectrum $\mathcal{F}\{u\}$.

- The symbol $a(k)$ corresponds to the sequence $\{a_k\}$ of filter coefficients and

- the inverse *Fourier transform* corresponds to the superposition of multiples of spherical harmonics.

The most remarkable property of PDOs is the homomorphy of their algebra with the algebra of their symbols:

$$symb(p + q) = symb(p) + symb(q) \tag{16}$$
$$symb(p \cdot q) = symb(p) \cdot symb(q) \tag{17}$$
$$symb(\alpha p) = \alpha symb(p) \tag{18}$$

This means that it is possible to work with the symbol instead of the operators themselves. The replacement of operators by their symbols means a considerable simplification. It is much easier to handle symbols, because they are simple real functions or real sequences, than to handle the original operators.

2.2 Stochastic model

The stochstic model is meant to describe the behaviour of the measurement noise, contained in the data. Though the data is sampled at discrete points of time and consequently, due to the motion of the satellite, at discrete locations, a continuous model of measuring errors is needed in order to fit in the framework of boundary value techniques. Hence, the theory of random variables or random sequences is not longer applicable. One has to use random variables, having their values in some *Hilbert space* instead of the real numbers, the so called *Hilbert space* valued random variables.

Definition 2.3 *A (measurable) mapping*

$$\epsilon : [\Omega, \mathcal{A}, P] \to H \tag{19}$$

from a probability space to some Hilbert space H, is called a Hilbert space valued random variable.

The usual concepts of expectation, variance and covariance now have to be extended to *Hilbert spaces*. A direct generalization of the usual definition of expectation

$$E\{X(\omega)\} := \int_\Omega X(\omega)dP(\omega)$$

is not useful, because one would need a kind of *Lebesque integral* for *Hilbert space* valued functions.
Therefore, an indirect definition is preferred.

Let ϵ be a *Hilbert space* valued random variable. Then, for an arbitrary $x \in H$ the scalar product (x, ϵ) is a random variable in the usual sense and its expectation $E\{(x, \epsilon)\}$ can be computed.

Definition 2.4 *If there is a Element $m \in H$ such that $E\{(x, \epsilon)\} = (x, m)$ for any $x \in H$, m is called expectation of ϵ.*

$$m = E\{\epsilon\} \iff E\{(x, \epsilon)\} = (x, m), \quad \forall x \in H \tag{20}$$

Similarly, the covariance can be defined.

Definition 2.5 *Let ϵ be a Hilbert space valued random variable and m be its expectation.*
An operator $C_{\epsilon\epsilon} : H \rightarrow H$ is called covariance operator of ϵ if

$$E\{(x, \epsilon - m)(y, \epsilon - m)\} = (C_{\epsilon\epsilon}x, y), \quad \forall x, y \in H \tag{21}$$

Obviously, the covariance operator is positive, i.e.

$$(C_{\epsilon\epsilon}x, x) \geq 0, \quad (C_{\epsilon\epsilon}, x, x) = 0 \Leftrightarrow x = 0 \tag{22}$$

holds. Following a theorem of *Bochner - Schwartz* a positive function $S_{\epsilon\epsilon}(k) > 0$ exists such, that the covariance operator $C_{\epsilon\epsilon}$ has the representation

$$C_{\epsilon\epsilon}x = \mathcal{F}^{-1}\{S_{\epsilon\epsilon}(k)\mathcal{F}\{x\}\} \tag{23}$$

Consequently, the covariance operator $C_{\epsilon\epsilon}$ is a PDO of order zero. How can the positive function $S_{\epsilon\epsilon}$ be interpreted ?
In the case of the usual stationary stochastic processes the *Fourier transform* of the covariance function $C_{\epsilon\epsilon}$ is called *spectral density* of the process ϵ.

$$S_{\epsilon\epsilon}(k) := \mathcal{F}\{C_{\epsilon\epsilon}(\tau)\} \tag{24}$$

If for the *Hilbert space* valued process ϵ the *Fourier transform* of the covariance operator $C_{\epsilon\epsilon}$ is computed one obtains

$$\mathcal{F}\{C_{\epsilon\epsilon}\} = \mathcal{F}\{\mathcal{F}^{-1}\{S_{\epsilon\epsilon}\mathcal{F}\{\bullet\}\}\} = S_{\epsilon\epsilon} \tag{25}$$

Hence, the function $S_{\epsilon\epsilon}$ plays the role of the spectral density for stationary stochastic processes. Therefore, it is called *spectral density*, too.

This is the point of departure for the definition of *white noise*. White noise is a mathmathical model for a process which is **completely uncorrelated** at different locations. Because for a *Hilbert space* valued random variable ϵ with zero expectation and for highly concentrated test functions x, y the ordinary random variables $(x, \epsilon), (y, \epsilon)$ can be thought of as values of ϵ at the mass centres of x and y. For a white noise the covariance between different locations vanishes. Hence

$$(C_{\epsilon\epsilon}x, y) = 0$$

has to be fulfilled for functions x, y with disjunct support. For these functions also

$$(x, y) = 0$$

holds. Consequently, the only possible choice is $C_{\epsilon\epsilon} = \sigma^2 I$.

Definition 2.6 *A Hilbert space valued stochastic variable ϵ is called* white noise, *if*

$$E\{\epsilon\} \;=\; 0 \tag{26}$$
$$C_{\epsilon\epsilon} \;=\; \sigma^2 I \tag{27}$$

What can be said about the regularity of realizations of a white noise random variable ?
To find an answer to this question the severe assumption of *ergodicity* of the random variable has to be made. Ergodicity means that all stochastic properties of a random variable can be reconstructed already from one single realization. For a ergodic stochastic variable the spectral density can be computed in the following way

$$S_{\epsilon\epsilon}(k) = |\mathcal{F}\{\epsilon(\omega_0)\}|^2 \tag{28}$$

From this relation

$$P(\epsilon \in W^{l,2}) = 1 \Longleftrightarrow S_{\epsilon\epsilon} \in W^{2l,2} \tag{29}$$

can be deduced.
For white noise the spectral density is identical to σ^2

$$S_{\epsilon\epsilon}(k) = \sigma^2$$

A constant function is a member of $W^{l,2}$, $l \leq -1$. Hence,

$$P(\epsilon \in W^{l,2}) = 1 \Longleftrightarrow l \leq -\frac{1}{2} \tag{30}$$

holds.

2.3 Overdetermined problems

It was already mentioned that the determination of the gravitational potential V from two different data - gravity gradients Γ at satellite altitude and
modulus of gravitation g at the Earth's surface -is an overdetermined problem.
Geodesists are very familiar with overdetermined problems if one only things
of network adjustment. Opposite to the network adjustment in the gradiometry problem the given data and the unknown solution can only be described
by an **infinite** number of coefficients. Hence, familiar adjustment theory,
which is well known for finite dimensional spaces $\mathcal{R}^n$ has to be extended to
infinite dimensional *Hilbert spaces H*.

Let H_1, H_2 be *Hilbert spaces*. The space H_2 is the space which contains
the given data and the space H_1 is the space where we look for a solution.
The definition of the spaces H_2, H_1 reflects the knowledge about the regularity of the given data and the regularity requirements for the unknown
solution.
The mathematical model representing the connection between the given data
and the unknown solution is therefore given by $A \in \mathcal{L}(H_1, H_2)$ a linear continuous operator A which maps the solution space H_1 in the data space H_2.
Let $y \in H_2$ be the measured data. The data y inavoidably contains measurement noise ϵ, which is assumed to be an ergodic stochastic process with
$S_{\epsilon\epsilon} \in H_2$, i.e. its spectral density contained in the data space H_2.
Hence the mathematical model for an abstract overdetermined problem in
infinite dimensional *Hilbert spaces* is:

$$y = A\beta + \epsilon, \qquad C_{\epsilon\epsilon} = \mathcal{F}^{-1}\{S_{\epsilon\epsilon}\mathcal{F}\{\bullet\}\} \tag{31}$$

The first idea for the estimation of the unknown solution β would probably
be the extension of the usual least squares principle to infinite dimensions.
Unfortunately, this extension fails.

Beside any mathematical formalism this result can be explained as follows. In the finite dimensional case the expectation of the weighed sum of
the squares of the residuals equals the number of observations. Hence, in
an infinite dimensional space the expectation of this sum has to be infinite.
Because it is impossible to minimize a quantity, which itself is infinite an
extention of the least squares principle to infinite dimensional spaces is impossible.
The solution of this dilemma again comes from the finite dimensional case.
There the least squares principle is equivalent to some other estimation principles as e.g. the best linear unbiased estimation (BLUE). Hence, an estimation principle in infinite dimensional spaces is obtained from a generalization
of the BLUE principle instead of the least squares principle.
What is meant with BLUE ? For a given $c \in H_1$, which has to fulfill some

technical requirements - c has to be a so called *estimable function* - one looks for a $d \in H_2$ such that

- $\quad (c, \beta) = E\{(d, y)\},$ $\qquad\qquad$ unbiasedness, linearity $\qquad$ (32)

- $\quad E\{[(d, y) - E\{(d, y)\}]^2\} \to min,$ $\quad$ minimal variance $\qquad$ (33)

In the following the question is addressed, how the data filter d can be obtained from the given signal filters c and A.

This data filter has to meet two requirements *unbiasedness* and *minimal variance*. First, the unbiasdness condition shall be considered. Because of

$$(c, \beta) = E\{(d, y)\} = E\{(d, A\beta + \epsilon)\} = (d, A\beta) = (A^*d, \beta)$$

the estimation can only be unbiased, if d fulfills the linear constraint

$$A^*d = c \tag{34}$$

The condition of minimal variance simplifies to

$$E\{[(d, y) - E\{(d, y)\}]^2\} = E\{(d, \epsilon)^2\} = (C_{\epsilon\epsilon}d, d) \to min \tag{35}$$

Hence, the BLUE estimation is equivalent to the solution of the following linear constrained minimization problem

$$
\begin{aligned}
(C_{\epsilon\epsilon}d, d) &\to min & (36) \\
A^*d &= c & (37)
\end{aligned}
$$

It can be solved by the usual *Lagrangian multiplier* technique.

The solution of the system of equations gives

$$d = C_{\epsilon\epsilon}^{-1} A (A^* C_{\epsilon\epsilon}^{-1} A)^{-1} c \tag{38}$$

This is the general solution. It describes how for an estimable signal filter c the inverse data filter d has to be designed. Beside the condition that c has to be *estimable* it can be chosen arbitrarily. It is interesting to consider special choices for c as e.g. $c = e^{ik^T x}$:

For this choice the scalar product (c, β) equals the *Fourier transform* of the signal β

$$
\begin{aligned}
\mathcal{F}\{\beta\} &= (c, \beta) \\
&= (d, y) \\
&= (C_{\epsilon\epsilon}^{-1} A (A^* C_{\epsilon\epsilon}^{-1} A)^{-1} c, y) \\
&= (c, (A^* C_{\epsilon\epsilon}^{-1} A)^{-1} A^* C_{\epsilon\epsilon}^{-1} y) \\
&= \mathcal{F}\{(A^* C_{\epsilon\epsilon}^{-1} A)^{-1} A^* C_{\epsilon\epsilon}^{-1} y\}
\end{aligned}
$$

This means that the *Fourier transform* of the signal β equals the *Fourier transfom* of the solution of the normal equations. Hence, using $c = e^{ik^{\mathsf{T}}x}$ as an estimable function, this leads to an estimation in the frequency domain. This is important due to the fact that the operators $A, C_{\epsilon\epsilon}$ are PDOs and consequently their combination

$$p := (A^* C_{\epsilon\epsilon}^{-1} A)^{-1} A^* C_{\epsilon\epsilon}^{-1} \tag{39}$$

is a PDO too. As a PDO it has the representation

$$py = \mathcal{F}^{-1}\{symb\ p\ \mathcal{F}\{y\}\} \tag{40}$$

and the symbol of p can easily be computed from the symbols of A and $C_{\epsilon\epsilon}$

$$symb\ p = \left[(symb\ A)^{\mathsf{T}} S_{\epsilon\epsilon}^{-1}(symb\ A)\right]^{-1} (symb\ A)^{\mathsf{T}} S_{\epsilon\epsilon}^{-1} \tag{41}$$

This leads to the following solution of the overdetermined problem in the frequency domain

$$\mathcal{F}\{\beta\} = \left[(symb\ A)^{\mathsf{T}} S_{\epsilon\epsilon}^{-1}(symb\ A)\right]^{-1} (symb\ A)^{\mathsf{T}} S_{\epsilon\epsilon}^{-1} \mathcal{F}\{y\} \tag{42}$$

This solution has the form of a filter. The data spectrum $\mathcal{F}\{y\}$ is multiplied by a real valued function $symb\ p$, representing the filter characteristic, and the spectrum $\mathcal{F}\{\beta\}$ of the unknown signal is obtained.
This change from the space domain to the frequency domain was made possible by the circumstance that all operators involved are PDO. The advantage of this change is that the operations in the frequency domain are much simpler than in the space domain. Instead of evaluation singular integral operators in the space domain only multiplications with their symbols, which are real valued functions, have to be performed in the frequency domain.
Finally, if the result is really needed in the space domain it easily can be transformed back using FFT.
In many cases, it is advantageous to stay in the frequency domain. For error consideration it is necessary to know which parts of the spectrum are significantly disturbed from the data noise and which parts are not. Hence, a a-priori error estimate in the frequency domain is desirable. Under the severe assumption that the error process ϵ is ergodic this error estimate can be obtained by simple error propagation:

$$S_{\beta\beta}(k) = \sigma^2 \left[(symb\ A)^{\mathsf{T}} S_{\epsilon\epsilon}^{-1}(symb\ A)\right]^{-1} \tag{43}$$

The equation (43) indicates the estimation error variance as a function of the wave number k. It only needs information about the error behaviour of the data and the symbols of the operator, which connects the data and the unknown solution. The data y itself is not needed.

Therefore, the relation (43) can be used for an a-priori error estimate, in order to investigate the influence of the various mission parameters on the obtainable accuracy of the solution. Because this error estimate is easy to evaluate several mission scenarios can be simulated and a mission optimization can be achieved.

3 Satellite gradiometry

At this point all mathematical tools for the description and the solution of the gradiometry problem are compiled. They now shall be applied to the ARISTOTELES mission, one of two different satellite gradiometry missions intensively discussed during the last years.

3.1 ARISTOTELES mission

The earliest and probably best prepared proposal for a satellite gradiometry mission was the so called ARISTOTELES project of ESA. It was a combined gravity field and magnetic field mission. Here, only the gravitational part of the mission will be discussed.

It was proposed to launch a satellite in an almost polar orbit at an altitude of about $h = 200$ km. The parameter semi major axis a and orbital plane i were tuned such that the satellite had a sun-synchronous repeat orbit with a repeat period of about 180 days.

The satellite was supposed to carry a tri-axial gradiometer with axis oriented in radial, along track and cross track direction, measuring the second order derivatives in these directions with an accuracy of about 0.01 E.U. .

Because the radial component of the *Eötvös tensor* carries the main information about the gravitational signal, only the second order radial derivative $\Gamma = V_{rr}$ will be analysed in the sequel.

The gradiometer signal is given at satellite altitude but the solution, the gravitational potential V, is needed at the surface of the Earth. Hence, the solution of the gradiometry problems somehow includes the harmonic downward continuation. This process is known to be unstable. Therefore, additional ground gravity data $g = -V_r$ has to be included in the data in order to stabilize the harmonic downward continuation. Because these ground gravity data is needed for stabilization only and not for the recovery of Earth's gravity field they can be rather inaccurate. If nothing better is available the ground gravity data can be computed from a global gravity field model such as OSU91A, providing a global accuracy of about 50 mGal.

In a spherical approximation, i.e. under the assumption that both the surface Σ, representing the orbit of the satellite and S representing the Earth's surface are spheres of radius $R + h$ and R respectively, the gradiometry problem can be formulated as the following overdetermined problem

$$\Delta V(x) = 0, \qquad x \in \text{ext } S \tag{44}$$

$$\Gamma = \frac{\partial^2 V}{\partial r^2}\Big|_\Sigma + \epsilon \tag{45}$$

$$g = -\frac{\partial V}{\partial r}\Big|_S + \eta \tag{46}$$

The error processes ϵ, η are assumed to be white noise i.e. having the spectral densities

$$S_{\epsilon\epsilon}(k) = \sigma_\Gamma^2, \qquad S_{\eta\eta}(k) = \sigma_g^2 \tag{47}$$

with $\sigma_\Gamma = 10^{-11}s^{-2}$ and $\sigma_g = 5 \cdot 10^{-4} ms^{-2}$.

Following the general formalism presented above, the following steps have to be carried out:

- Fixing the regularity of data and solution by specification of the used *Hilbert spaces* H_1, H_2,

- formulation of the connection between data and solution in the form of PDOs and

- application of the BLUE estimator in the frequency domain.

First the specification of the *Hilbert spaces* has to be considered.
For the restriction of the unknown solution on the surface of the Earth S only square integrability has to be required. Consequently, H_1 is chosen as

$$H_1 := W^{0,2}(S)$$

Both data taypes contain white noise. Hence,

$$\Gamma \in W^{-\frac{1}{2},2}(\Sigma), \quad g \in W^{-\frac{1}{2},2}(S)$$

holds. The data space H_2 is defined as the *Cartesian product* of these two spaces

$$H_2 := W^{-\frac{1}{2},2}(\Sigma) \times W^{-\frac{1}{2},2}(S)$$

The next step is the definition of suitable PDOs. This cannot be done directly but with the help of some auxiliary operators.
The first of these auxiliary operators is the operator u of harmonic upward continuation. From the well known *Poisson formula* we have

$$u : \begin{cases} H_1 & \to \quad W^{-1/2,2}(\Sigma) \\ \\ V & \mapsto \quad \frac{R^2 - r^2}{4\pi} \int_S \frac{V}{l^3} dF \end{cases} \tag{48}$$

Lemma 3.1 *The operator u defined by (48) is an invariant PDO on the sphere S with the spherical symbol $\{(R/(R+h))^{n+1}\}$.*

With a similar technique the following two lemmata can be proved

Lemma 3.2 *Let S_r be a sphere of radius r. The mapping*

$$q: \begin{cases} W^{l,2}(S_r) & \longrightarrow & W^{l+1,2}(S_r) \\ \\ V & \longmapsto & \frac{1}{2\pi} \int_{S_r} \frac{V}{l} dF \end{cases} \tag{49}$$

is an invariant PDO on S_r with the spherical symbol

$$\left\{ \frac{2R}{(2n+1)} \left(\frac{R}{r} \right)^{n+1} \right\} \tag{50}$$

Lemma 3.3 *Let S_r be a sphere of radius r. The mapping*

$$t: \begin{cases} W^{l,2}(S_r) & \longrightarrow & W^{l+1,2}(S_r) \\ \\ V & \longmapsto & \frac{1}{2\pi} \int_{S_r} \frac{\partial l^{-1}}{\partial n} V dF \end{cases} \tag{51}$$

is an invariant PDO on S_r with the spherical symbol

$$\left\{ \frac{-1}{(2n+1)} \left(\frac{R}{r} \right)^{nn+1} \right\} \tag{52}$$

The three auxiliary operators u, q, t now can be used to build the PDOs, which connect the unknown solution V with the data g, Γ. First, the construction for the ground gravity data g is carried out.
This construction starts from the well known *Greens representation theorem*

$$2\pi V + \int_S \left(l^{-1} \frac{\partial V}{\partial n} - \frac{\partial l^{-1}}{\partial n} V \right) dF = 0$$

With the help of the operators u, q, t this theorem can be rewritten

$$IV + q \left(\frac{\partial V}{\partial n} \right) - tV = 0 \tag{53}$$

Solving this relation for $\frac{\partial V}{\partial n}$ one obtains

$$\frac{\partial V}{\partial r} |s = \frac{\partial V}{\partial n} = -q^{-1}(I - t)V \tag{54}$$

If one defines $A_2 := q^{-1}(I - t)V$, the operator is an invariant PDO with the spherical symbol $\{(n + 1)/R\}$ and it connects the ground gravity data g with the unknown solution V in the following way

$$g = A_2V + \eta \tag{55}$$

The next step is the representation of the gravity gradients Γ with the help of u, q, t. Because the gravity gradient is the second order radial derivative, one could try to apply the operator A_2 twice. Unfortunately, this is not possible. The function A_2V represents the first order radial derivative of V **only**, if V is harmonic. The first order radial derivative of a harmonic function in general is **not** harmonic. Hence, the repeated application of A_2 does not represent the second order radial derivative.

Nevertheless, there is something like an integrating factor. Though, for a harmonic function V the radial derivative $\partial V/\partial r$ is not harmonic the product $r \cdot (\partial V/\partial r)$ of the radial and the radial distance is harmonic . Hence, it is possible to represent the second order derivative at satellite altitude by the evaluation of

$$A_2(r \cdot A_2(uV))$$

The result is

$$\Gamma = A_1V + \epsilon \tag{56}$$

with A_1 beeing a PDO with the spherical symbol

$$\left\{ \frac{(n + 1)(n + 2)}{R^2} \left(\frac{R}{R + h} \right)^{n+3} \right\}$$

The two equations (56) and (55) are now combined in a matrix notation

$$y := \begin{bmatrix} \Gamma \\ g \end{bmatrix}, \ A := \begin{bmatrix} A_1 \\ A_2 \end{bmatrix}, \ \varepsilon := \begin{bmatrix} \epsilon \\ \eta \end{bmatrix}, \ S_{\epsilon\epsilon} := \begin{bmatrix} \sigma_\Gamma^2 & 0 \\ 0 & \sigma_g^2 \end{bmatrix}, \ \beta := V \tag{57}$$

leading to the compact notation of the overdetermined gradiometry problem

$$y = A\beta + \varepsilon \tag{58}$$

with the spherical symbol

$$symb\ A = \begin{bmatrix} \left\{ \frac{(n+1)(n+2)}{R^2} \left(\frac{R}{R+h} \right)^{n+3} \right\} \\ \{(n + 1)/R\} \end{bmatrix} \tag{59}$$

The only thing, which is left is to insert this spherical symbol into the estimation formula (42). For the denominator of (42)

$$(symb\ A)^\mathsf{T} S_{\epsilon\epsilon}^{-1}(symb\ A) = \frac{(n + 1)^2(n + 2)^2}{\sigma_\Gamma^2 R^4} \left(\frac{R}{R + h} \right)^{2n+6} + \frac{(n + 1)^2}{\sigma_g^2 R^2} \tag{60}$$

is obtained and the nominater has the form

$$(symb\ A)^{\top} S_{\varepsilon\varepsilon}^{-1} \mathcal{F}\{y\} = \frac{(n+1)(n+2)}{\sigma_\Gamma^2 R^2} \left(\frac{R}{R+h}\right)^{n+3} \Gamma_{nm} + \frac{(n+1)}{\sigma_g^2 R} g_{nm} \quad (61)$$

, with Γ_{nm}, g_{nm} being the spherical harmonics coefficients of the data Γ and g, respectively.

Inserting (60) and (61) in (42) the final estimation formula in the frequency domain is obtained.

$$v_{nm} = \frac{\frac{(n+1)(n+2)}{\sigma_\Gamma^2 R^2} \left(\frac{R}{R+h}\right)^{n+3} \Gamma_{nm} + \frac{(n+1)}{\sigma_g^2 R} g_{nm}}{\frac{(n+1)^2(n+2)^2}{\sigma_\Gamma^2 R^4} \left(\frac{R}{R+h}\right)^{2n+6} + \frac{(n+1)^2}{\sigma_g^2 R^2}} \quad (62)$$

This formula has a simple interpretation: The spherical harmonics coefficients of the solution are the weighed means of the corresponding spherical harmonics coefficients of the data. The weights are indirectly proportional to the error variances of the data. Hence, the more accurate one data type is the more it influences the final solution.

The summary of the solution theory is a three step procedure

1. Determination of the spherical harmonics coefficients Γ_{nm}, g_{nm} of the data using some numerical efficient analysis tool like FFT or *Gaussian quadrature*.

2. Determination of the spherical harmonics coefficients v_{nm} of the solution as a weighed mean of the spherical harmonics coefficients of the data.

3. Synthethis of the solution $V(\vartheta, \lambda)$ from its spherical harmonics coefficients

Beside the solution itself its accuracy is also of interest. The variance σ_{nm}^2 of the estimation of the spherical harmonics coefficient v_{nm} is given by

$$\sigma_{nm}^2 = \left[\frac{(n+1)^2(n+2)^2}{\sigma_\Gamma^2 R^4} \left(\frac{R}{R+h}\right)^{2n+6} + \frac{(n+1)^2}{\sigma_g^2 R^2}\right]^{-1} \quad (63)$$

Unlike the estimation of the solution the estimation of the acccuracy *does not* need the data itself. It is sufficient to have knowledge about the error spectrum of the data. Consequently, it is possible to discuss several mission scenarios and to investigate the influence of different mission parameters on the accuracy of the solution.

For such a simulation study first the information content of the two data types is investigated. In the following two figures the spectra of the gravity and the gradiometry signal are plotted in red color. These spectra are compared to the spectra of the noise contained in the corresponding data. The noise spectra are plotted in green. The figures clearly show that the gravity information can only contribute to the solution up to degree and order 40. For higher frequency the noise level exeeds the signal power.

The high frequency information can only be extracted from the gradiometry data, where the signal to noise ratio is better than 1 up to degree and order 180.

These investigations only show which parts of the unknown solution are visible in the different data types. It does not show whether this information really can be recovered from the data.

In order to answer this question two single solutions were computed first. Single solution means that the solution is computed from one data type only. The accuracy of such a single solution can be estimated by setting the variance of the complementary data type equal to infinity. It turns out that the error of the gravity solution exceeds the signal for degrees larger than about 40. The error spectrum for the gradiometry solution for the long and medium wavelength is much smaller than the signal itself. For higher frequencies the error of the gradiometry solution increases very fast. This reflects the instability of the harmonic downward computation.

So far the gravity data does not contribute to the solution. The gradiometry solution has much higher resolution than the gravity solution.

The errors of the combined solution increase much slower for increasing frequencies than the errors of the single gradiometry solution. This means, despite the fact that the gravimetry data do not directly contribute to the solution they stabilize the harmonic downward continuation. This moderated increase of the estimation error also increases the resolution. Instead of a resolution up to degree 180 the combined solution makes a resolution up to degree 220 possible.

References

[Sve] *Svensson, L*: Pseudodifferential Operators - a New Approach to the Boundary Problems of Physical Geodesy. Manus. geod. 8(1983), pp 1 -40

[Rum] *Rummel, R. et al.*: Spherical Harmonics Analysis of Satellite Gradiometry. Nethl. Geod. Comm. New Series Nr. 39

[RuSw] *Rummel, R and Schwintzer, P. (eds)*: A Major STEP for Geodesy. Report 1994 of the STEP Geodesy Working Group.

The Polar Gap

N. Sneeuw and M. van Gelderen

1. Introduction

For geodetic purposes the ideal satellite orbit would be polar, simply because in that case the pattern of groundtracks covers the entire earth. For several reasons, especially engineering ones, a sun-synchronous orbit is favourable, though. The inclination for low earth orbiters would be around 97°. This will leave caps around the poles without data coverage, denoted as *polar gaps*.

In the literature, the polar gap problem has hardly been studied, presumably because only polar orbits were assumed for geodetic missions. A further reason might be, that most simulation models employ spherical harmonics as base functions. Since these are global functions, they are awkward for taking into account local data gaps. Moreover, simulation results used to be presented in terms of error degree variances, which do not reveal the polar gap effects on the full spherical harmonic (SH hereinafter) error spectrum. The first attempts of such a description in the 2-D spectral domain were made in the course of investigations of the STEP gradiometry mission, cf. (Belikov et al. 1994) and (Sneeuw 1994).

Although the polar gaps are relatively small, about 7° radius for sun-synchronous orbits, they do influence gravity field recovery. Considered as a spherical harmonic analysis problem, it is the task to derive the SH spectrum from a function, multiplied by a rectangular window. In Fourier terms this causes leakage. In analogy the polar gap causes leakage of the SH spectrum.

Several strategies are possible to find the potential coefficients from the (incomplete) data. The problem can be formulated in the time domain (see section 4.) or in the space domain. Although they are essentially identical, they each have their peculiarities (Rummel et al. 1993). Then we can either use a quadrature approach to solve for the potential coefficients, which implies that the inverse relation is available, or minimize some error (e.g., least-squares). The latter is often preferred as it has, obviously, minimal propagated variances. The next problem is that the system of equations can be unstable. If it needs to be regularized, bias will be introduced, cf. (Xu 1992).

Finally, we only try to determine the SH coefficients up to a certain degree. Thus aliasing will play a role as well. This aliasing is not only due to undersampling, as it is defined classically. In general it is signal that will be estimated on the wrong frequencies during the inversion.

2. Distortion of the Spherical Harmonic Spectrum

The polar gap causes distortion of the SH spectrum. In order to see which part of the spectrum is affected, a number of error simulations have been performed with increasing polar gap radius $\theta_0 = |\pi/2 - I|$. The second radial gradient T_{zz} was simulated with white noise assumption, which should give a homogeneous SH error spectrum, that only depends on the degree l. The simulation results are displayed in fig. 2.1 with growing polar gap. Note that only the spectral domain centered around the zonal coefficients is displayed.

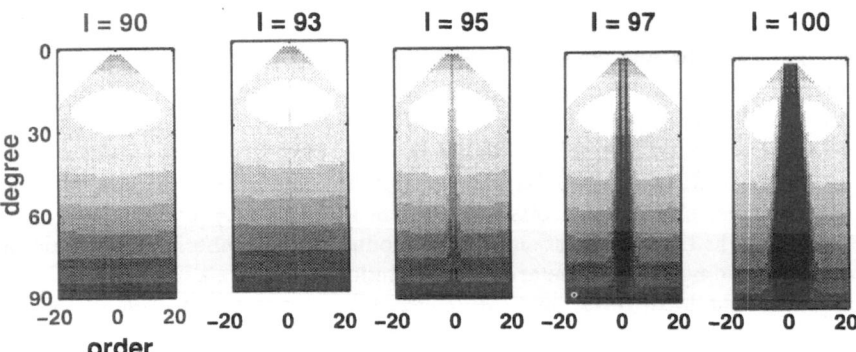

Figure 2.1. Polar gap wedge from least-squares error analysis of T_{zz} tensor component under increasing gap radius.

Without going into details about computation methods, see (Rummel et al. 1993), about meaning of the gray values and about actual results, the figures make clear that only a polar orbit assures a homogeneous error spectrum (for the T_{zz}-observable, to be precise). At 93° inclination, or $\theta_0 = 3°$, the zonals show slightly reduced performance already. At $I = 95°$ the domain of affected coefficients shows up as a narrow wedge. This *polar gap wedge* grows wider for increasing gap radius.

So the low order part of the SH spectrum is seen to be affected by the polar gap. This fact would not be clear if only the degree error variances would be given. They are horizontal averages, loosely speaking, in the figures 2.1. The effect of a polar gap would be invisible, though present. Thus special care must be taken, when representing the results from a non-polar satellite mission by degree error measures. Such a 1-D error representation is not representative for the full σ_{lm}-spectrum anymore.

The wedge shape is explained partly by analyzing the definition, e.g. (Heiskanen and Moritz 1967), of the associated Legendre function:

$$P_{lm}(x) = (1 - x^2)^{m/2} \frac{\mathrm{d}^m P_l(x)}{\mathrm{d}x^m}$$

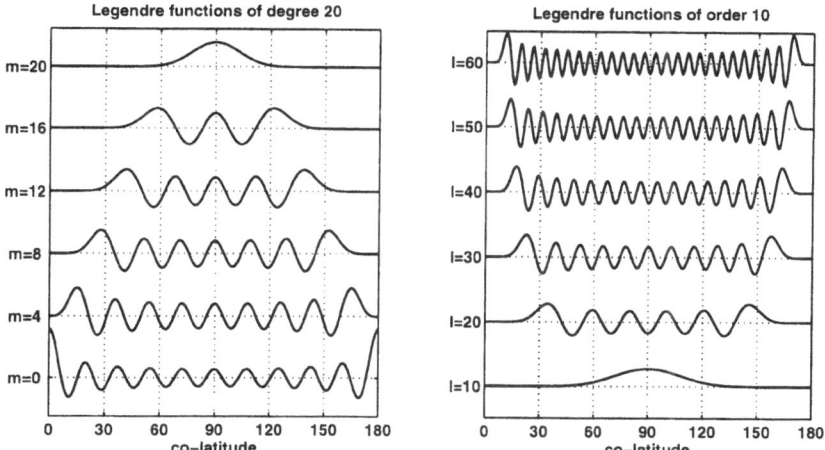

Figure 2.2. Squeezing effects for changing degree or order.

The Legendre polynomial P_l is of degree l, and has l roots on $x \in (-1, 1)$. The derivative $\mathrm{d}^m P_l(x)/\mathrm{d}x^m$ is a polynomial of degree $l - m$, having $l - m$ roots. The latter polynomial gets multiplied by the function $(1 - x^2)^{m/2}$. Substituting $x = \cos\theta$, the factor becomes $\sin^m\theta$, which will give more and more weight to the area around the equator for increasing order m. The left hand figure in fig. 2.2 show this effect for $P_{20,m}(\cos\theta)$. The only Legendre functions - and thus SH-coefficient – that contribute to the function at and near the poles are the zonal and low-order ones. Vice versa it must then be concluded that the only coefficients, likely to be affected by missing data around the poles, are the near-zonals. Besides the order m, the degree l also plays a role, though less pronounced. For fixed order and increasing degree, the support of the Legendre functions grows, cf. the right hand side of fig. 2.2, displaying $P_{l,10}(\cos\theta)$.

3. A Simple Approach

From a simplified point of view, the polar gap problem is that of estimating the SH spectrum of a function $f(\theta, \lambda)$ on the sphere, multiplied by the rectangular window $h(\theta, \lambda)$, which equals zero in the polar gaps and 1 for $\theta \in [\theta_0, \pi - \theta_0]$. If the potential coefficients are estimated by means of a quadrature approach from potential observations on the the earth's surface, the estimator reads:

$$C_{lm} = \frac{1}{4\pi} \int_\sigma f(\sigma) Y_{lm}^*(\sigma)\mathrm{d}\sigma$$

in which $Y_{lm}(\sigma)$ are the fully normalized surface spherical harmonics:

$$Y_{lm}(\sigma) = Y_{lm}(\theta, \lambda) = e^{im\lambda} N_{l,|m|} P_{l,|m|}(\cos\theta).$$

Here, N_{lm} denotes the normalization factor. Due to the window function, the above quadrature is now replaced by, cf. (v. Gelderen & Koop 1996):

$$\hat{C}_{lm} = \frac{1}{4\pi} \int_{\sigma} f(\sigma)h(\sigma)Y_{lm}^*(\sigma)\mathrm{d}\sigma = \frac{1}{4\pi} \int_{\sigma - \sigma_0} f(\sigma)Y_{lm}^*(\sigma)\mathrm{d}\sigma. \qquad (3.1)$$

Here σ_0 stands for both polar gaps. The relative error in the coefficients reads $\epsilon_{lm} = (\hat{C}_{lm} - C_{lm})/C_{lm}$, whose error measure:

$$M\{\epsilon_{lm}^2\} \approx \frac{1}{4\pi} \int_{\sigma_0} Y_{lm}^*(\sigma)\, Y_{lm}(\sigma)\mathrm{d}\sigma \qquad (3.2)$$

can be considered the *loss of power* of the spherical harmonic base functions in the polar gaps σ_0. The error ϵ_{lm} represents a signal error alone. No stochastics have been included yet.

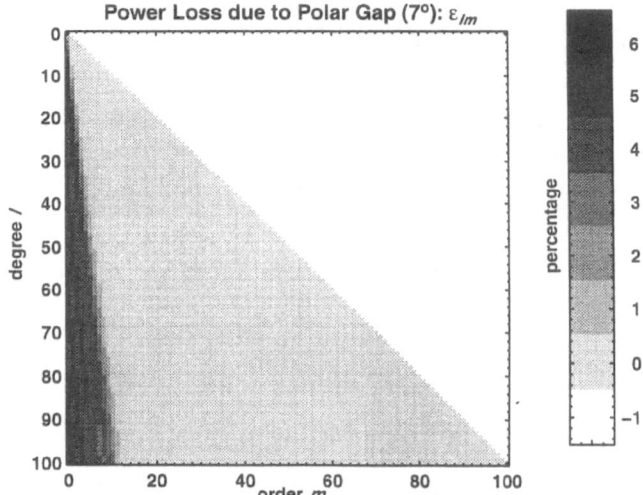

Figure 3.1. Relative error ϵ_{lm} in percent.

The power loss is displayed for a 7° polar gap in fig. 3.1, revealing a wedge shape indeed, delimited rather sharply. In (v. Gelderen & Koop 1996) the following rule of thumb is derived for the maximum order $m_{\max}$ (at a specific degree) that delimits the polar gap wedge:

$$m_{\max} = |\pi/2 - I| \cdot l = \theta_0 \cdot l \qquad , \theta_0 \text{ in radians.} \qquad (3.3)$$

Remark 3.1. Inspection of fig. 2.1 shows that the rule of thumb does not strictly apply to those wedges. They are in fact narrower. An explanation in terms of power loss, based on the quadrature approach, is apparently too simplistic for more realistic calculations. As a rule of thumb it performs well enough, though.

4. A Time-wise Approach: Lumped Coefficients

The polar gap problem is inherent to any type of satellite-based observations of functionals of the geopotential, e.g. specific gradiometer components or GPS derived orbit perturbations. Each functional has its own specific transfer and therefore its own gravity field recovery characteristics. Consequently the way that the polar gap problem interferes with the coefficient recovery must depend on observable type. E.g. the cross-track gradient tensor component yields a gravity field that is weaker in the lower orders than one from radial gradiometry. So a polar gap cannot inflict much harm there (Sneeuw 1994).

In order to separate the behaviour of the basic linear system from other particularities (observable type, orbit height), we will investigate now a simplified situation. Suppose the earth were a sphere and a satellite would revolve around it at zero height. The observable would be the geopotential directly, instead of its functionals. The along-orbit time-series of this observable can be expressed as the following Fourier series (Wagner & Klosko 1977):

$$V(t) = \sum_{m=-L}^{L} \sum_{k=-L}^{L} A_{mk} e^{i\psi_{mk}(t)} \qquad (4.1)$$

whose coefficients A_{mk} are the so-called *lumped coefficients*. The angular argument $\psi_{mk} = k(\omega + M) + m(\Omega - \theta)$ is composed of the argument of latitude $\omega + M$ (the in-orbit angle, counted from the ascending node) and the longitude of the ascending node $\Omega - \theta$. The spectral lines of the signal $V(t)$ are thus defined by $\dot{\psi}_{mk}$. The lumped coefficients are a linear combination of the potential coefficients:

$$A_{mk} = \frac{GM}{R_{\mathrm{E}}} \sum_{l} F_{lmk}(I) C_{lm}. \qquad (4.2)$$

Here the functions $F_{lmk}(I)$ are complex *inclination functions*. Every functional can similarly be expressed as a lumped coeffficient series, based on eq. (4.2). Examples of equations for gradiometer components and of orbit perturbations can be found in (Sneeuw 1994).

Now each order m gives a separate linear system, yielding a block system. Apart from the constant GM/R_{E} the linear systems consist only of the inclination functions $F_{lmk}(I)$. They build matrices F, again for each order m separately. The lumped coefficients of this order are stored in a vector a, and the potential coefficients in c. Now the linear system (4.2) simply becomes $a = Fc$. The extent to which this linear system can be solved, solely depends on the conditioning of the matrix F.

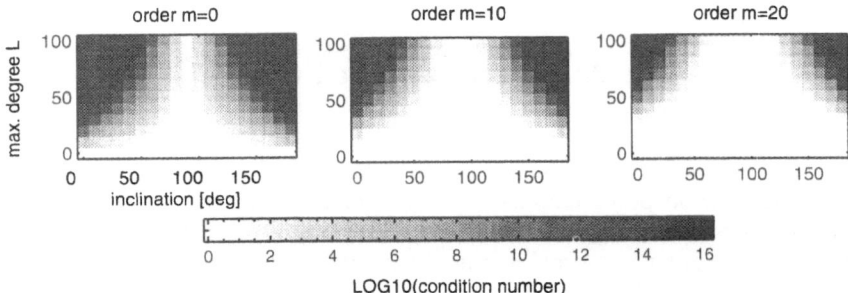

Figure 4.1. Conditioning of inclination function matrices, depending on inclination I, maximum degree L and order m.

Figure 4.1 displays the condition numbers of several such F-matrices, i.e. the ratio of the maximum and minimum singular values, depending on inclination, order and maximum degree. To be precise, the logarithm of the condition numbers has been rendered. This is a measure for the number of valid digits that can be lost during inversion of $a = Fc$, due to round-off and/or measurement errors (Strang 1986). Thus the gray value white (logarithm equals zero) represents the optimal situation; black denotes that F cannot be inverted.

With the inclination changing from $0°$ towards $180°$, the conditioning moves from black to white (at $I = 90°$) to black again. A polar orbit is optimal for inverting the linear system. Thus a good ground-track coverage, i.e. sampling of the earth, is translated in numerical terms as stability of the inclination function matrices F. The speed by which the conditioning changes with inclination depends on the order m. The smaller the order, the narrower the white band of good invertibility around $I = 90°$.

The maximum degree L determines the size of the matrix F, which has size $(2L+1) \times (L$-$m+1)$, since $-L \leq k \leq L$ and $m \leq l \leq L$. Figure 4.1 shows that for smaller L, conditioning gets better, which is a general behaviour: the smaller the matrix, the better its conditioning. A 1×1 matrix has condition number 1. This fact will be of importance to processing GPS-type observables, which typically are included up till degree $L = 50$ or $L = 75$ at most. As can be seen from the plot. even for the case $m = 0$ the F-matrices for $L = 50$ are still stable when the orbit is not strictly polar. Thus, loosely speaking, GPS observables can always be inverted (for reasonable inclination).

5. A Least-Squares Formulation

A more thorough analysis of the polar gap problem must also include a bias consideration. Figure 4.1 explains that for low orders the matrices F become unstable. Their inversion will require regularization by means of adding a priori knowledge of the unknown potential coefficients, usually in the form of a signal power spectrum, such as Kaula's rule. Xu (1992) argues that this type of regularization gives rise to biased estimation of the unknowns.

Another aspect to be treated is aliasing. Due to limitation of the SH domain, only a limited number of coefficients C_{lm} will be estimated. The following general model is able to treat these aspects:

$$y = [A_1 \ A_2] \begin{bmatrix} x_1 \\ x_2 \end{bmatrix} \tag{5.1}$$

in which y denotes the vector of observations, x_1 the vector of coefficients to be solved, i.e. up to degree L, and x_2 the unresolved coefficients, in principle up to infinity. The vector x_2 (and thus the matrix A_2) have been added for analysis of the aliasing phenomenon. The design matrices A_1, A_2 may come from a space-wise or from a time-wise formulation, which is irrelevant at this point. In the latter case they consist basically of the inclination functions, eq. (4.2), and of the specific transfer, due to the observable type. Least-squares inversion of the linear system (5.1) yields the estimate $\hat{x}_1 = By$, where the matrix B stands for:

$$B = (A_1^T P A_1)^{-1} A_1^T P \qquad \text{in the unbiased case, or}$$
$$B = (A_1^T P A_1 + R)^{-1} A_1^T P \qquad \text{with regularization matrix } R.$$

P denotes the weight matrix of the vector of observations y, which is the inverse of the covariance matrix Q_{yy}. Now three types of errors must be investigated in order to assess a realistic error for the polar gap:

1. A straightforward *noise propagation* gives the covariance matrix of the estimated unknowns:

$$Q_{x_1 x_1} = B Q_{yy} B^T = B P^{-1} B^T.$$

In the unbiased case, this reduces to the inverse of the normal matrix: $Q_{x_1 x_1} = (A_1^T P A_1)^{-1}$. In the biased case this will only approximately be true: $Q_{x_1 x_1} \approx (A_1^T P A_1 + R)^{-1}$. Also note that $(A_1^T P A_1 + R)^{-1}$ will be smaller than $(A_1^T P A_1)^{-1}$ (Xu 1992).

2. For the *bias* the following expression holds: $x_1 - \hat{x}_1 = (I - BA_1)x_1$. Without regularization and with full invertibility of the normal matrix (polar orbit or global coverage), one has $BA_1 = I$. This expresses unbiasedness. Since the unknowns are unknown indeed, the above expression cannot be evaluated. In a mean square sense one would have:

$$M\{(x_1 - \hat{x}_1)^2\} = (I - BA_1)C_{xx}(I - BA_1)^T.$$

The matrix C_{xx} represents a variance model, e.g. from Kaula's rule or from an actual gravity field model (OSU91a). The sum of propagated measurement noise $Q_{x_1 x_1}$ and bias variance is called MSE (mean square error), cf. (Xu 1992).

3. Considering the full linear model (5.1), where A_2 is not a zero matrix, part of the signal in vector x_2 will map onto the estimate $\hat{x}_1$. This *aliasing* is expressed by the second term in:

$$x_1 - \hat{x}_1 = (I - BA_1)x_1 + BA_2 x_2.$$

The latter part has the following error measure: $BA_2 C_{xx} A_2 B$. Again, for the evaluation of it, one can insert a signal model of SH coefficients for the matrix C_{xx}.

In fig. 5.1 the three effects are displayed separately.

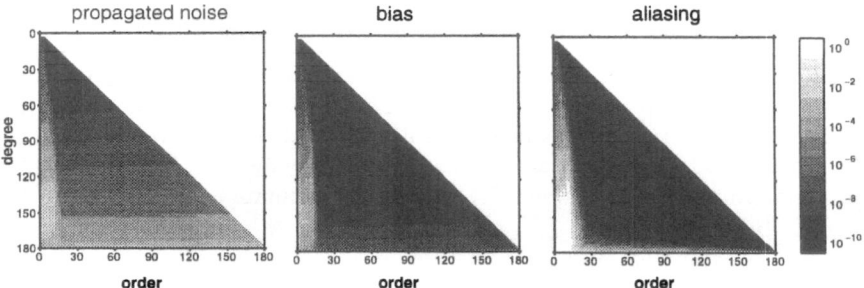

Figure 5.1. The three errors (an example with V_{zz} observed).

6. Error Propagation onto the Sphere

The polar gaps distort the low-order SH coefficients in the polar gap wedge. Although spherical harmonics are functions with *global* support, the polar gaps themselves are well-localized functions. This paradoxical situation will now further be investigated by a formal covariance propagation. The *full* covariance block matrix will be propagated onto geoid errors on the sphere. In this case the propagation result can be shown to be longitude-independent on the sphere. Moreover the propagated variance will be symmetric in the equator. It suffices therefore to represent the propagated geoid errors as a function of co-latitude θ only, on the interval $\theta \in [0; \pi/2]$.

The propagation results are separated in non-regularized (left hand side) and regularized (right hand side) simulations, cf. figure 6.1. Simulation #1 corresponds to a polar orbit (left picture in fig. 2.1) and #2 to a non-regularized simulation with $I = 97°$. The corresponding simulations with

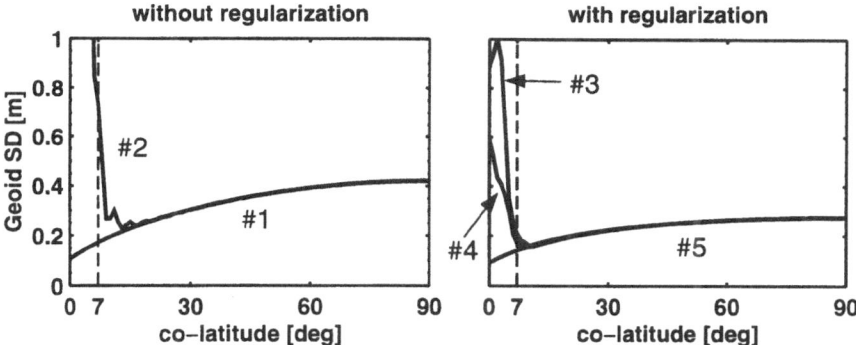

Figure 6.1. Propagated geoid standard deviations.

regularization (by Kaula's rule) are #5 and #3 respectively. For simulation #4 GPS-type observables are included. First compare the geoid standard deviations of the two polar simulations. It is seen that $\sigma_N^{\#5}(\theta)$ remains below $\sigma_N^{\#1}(\theta)$. This is an artefact, due to the regularization. Since the curves come from error propagation of the inverted normal matrices $Q_{x_1 x_1}$, it is basically a comparison of $(A_1^T P A_1 + R)^{-1}$ vs. $(A_1^T P A_1)^{-1}$, see item 1. above.

The full polar gap effect gets visible in the unregularized #2. The $\sigma_N^{\#2}(\theta)$ curve stretches far beyond the plot margin of 1 m (to around 10 m). At the edge of the polar gap however, at $\theta = 7°$, the curve falls down towards the reference curve of simulation #1. From $\theta \approx 20°$ onwards, both geoid error curves are equal. The lack of data within the polar gap does hardly propagate to the area outside of it. Only a rather small area around shows increased errors. Outside the gap, and the surrounding areas, the errors remain well-behaved, at the level of a polar orbit simulation. So even though the SH error spectrum, representing errors in the coefficients of global functions, the errors as a spatial function are confined to the gaps.

The error curve $\sigma_N^{\#3}(\theta)$ shows the influence of regularization. The maximum standard deviation within the gap is reduced to 1 m. At the same time, the radius of the transition zone at the polar gap edge is reduced. The curve conforms to the $\sigma_N^{\#3}(\theta)$-curve directly beyond $\theta = 7°$, whereas $\sigma_N^{\#3}(\theta)$ fluctuates till about $\theta = 20°$. The GPS-like simulation has the same effect. The geoid standard deviation within the gap is even further reduced.

7. Conclusions

The effects of a slightly non-polar orbit on gravity field recovery, both in the spectral and in the spatial domain, are summarized by the following conclusions:

- In the SH *spectral* domain the low order coefficients are affected, which can be concluded from both a space-wise and a time-wise view. The spectral band of these coefficients has a wedge shape, whose upper limit in m-direction is represented by the rule of thumb $m_{max} = |\theta_0|l$, (v. Gelderen & Koop 1996), which follows from a quadrature. For least-squares inversions the gap is even smaller than that, approximately by a factor 2.
- This wedge can be understood in numerical terms by ill-conditioned normal matrices, which are based on inclination function matrices F. The fact that the smaller the matrix (through smaller L), the better its condition, explains why GPS-type observations hardly suffer from the polar gap.
- Analytically, the wedge shape is explained by the shape of Legendre functions, which quickly tend towards zero at the higher latitudes when the order m increases. Thus the polar gap can only interfere with the low order Legendre functions. To a lesser extent, the degree l plays a similar role.
- The polar gap has relatively little influence on observation types which have reduced performance at the lower orders anyway, e.g. V_{yy} or V_{yz}.
- Irrespective of the global nature of spherical harmonics, the polar gap error in the *spatial* domain is delimited to the polar gap itself and a small surrounding transition area. Outside this area, the error characteristic is the same as for polar simulations. Thus regional gravity field recovery methods do not suffer from a polar gap.
- Regularization has a dual effect. It shrinks the transition area outside the gaps, and it reduces the error within the gaps, cf. fig. 6.1.

References

Belikov, M., Van Gelderen, M., Koop, R. (1994): Determination of the gravity field from satellite gradiometry. in: A Major STEP for Geodesy, Report 1994 of the STEP Geodesy Working, pp 55–64, Group, eds. R. Rummel and P. Schwintzer, GFZ Potsdam, Germany

Gelderen, M. van, Koop, R. (1996): The use of degree variances in satellite gradiometry. submitted to J. of Geodesy

Rummel, R., Van Gelderen, M., Koop, R., Schrama, E., Sansò, F., Brovelli, M., Migliaccio, F., Sacerdote, F. (1993): Spherical harmonic analysis of satellite gradiometry. Netherlands Geodetic Commission, New Series, **39**, Delft, The Netherlands

Sneeuw, N. (1994): Global Gravity Field Error Simulations for STEP–Geodesy. in: A Major STEP for Geodesy, Report 1994 of the STEP Geodesy Working, pp 45–54, Group, eds. R. Rummel and P. Schwintzer, GFZ Potsdam, Germany

Strang, G. (1986): Introduction to applied mathematics. Wellesley-Cambridge Press, Wellesley, Massachusetts

Wagner, C.A., Klosko, S.M. (1977) Gravitational Harmonics from Shallow Resonant Orbits. Celest. Mech. **16**:143-163

Xu, P.L. (1992): The Value of Minimum Norm Estimation of Geopotential Fields. Geophys. J. Int. **111**, 170–178.

European Capabilities and Prospects for a Spaceborne Gravimetric Mission

M. Schuyer

1 Introduction

The purpose of this seminar is to provide information on studies which are ongoing at the European Space Agency (ESA) on the theme of a spaceborne gravimetry mission. The Agency regroups 14 Member States and has a well established Earth Observation programme which until recently has been illustrated by the well-known series of Meteosat geostationary satellites for meteorology, and the 2 European Remote Sensing Satellites (ERS-1 and -2) placed in nearly polar low altitude orbits. These satellites reaped a wealth of new data with optical and Synthetic Aperture Radar (SAR) observation of the Earth, and many other original observations. The continuation of this successful programme is marked by the larger Meteosat Second Generation (MSG), the very large Envisat polar satellite and a series of operational meteorological satellites, Metop (Meteorological Operational Satellite Programme) – also in near polar orbit– which together with MSG will be operated by the Eumetsat organisation. This programme will be completed by dedicated research satellites and precursor operational satellites, respectively designated as Earth Explorer and Earth Watch missions. Among the nine candidate research themes proposed to be implemented as the first project in the Earth Explorer series, a spaceborne gravimetry mission reflects the long-lasting expectation of the geodesy science community, but also the needs of other scientists such as oceanographers and solid Earth geophysicists. If retained in the first step of the Earth Explorer selection process towards the end of 1996, the mission would – together with another 2-3 missions – be subjected to industrial feasibility studies in the course of 1997, preparing for the final selection in 1998 of the first Earth Explorer mission which should be launched in the 2003-2004 time frame.

2 Science Background and Mission Objectives

In some areas of geophysical research such as ocean circulation and transport, physics of the interior of the Earth, insufficient knowledge and modelling of the Earth's gravity field and/or its zero-potential surface, the geoid, is a significant hurdle to further advances. For instance it has been recognised that a very accurate geoid up to a high spatial resolution is needed to fully exploit sea surface topographic maps as obtained from spaceborne radar altimetry missions, whether past or future; in "solid Earth" research, seismic tomography data need to be constrained by independent measurements such as gravity field mapping, in order to remove ambiguities in determining the structure and dynamics of the Earth's interior. Moving now to more "operational" applications, the use of a very accurate high resolution geoid as a reference surface over emerged areas would greatly improve the new efficient technique of terrain levelling by Global Positioning System (GPS).

In regard of these scientific and applications needs, what is the present situation of our knowledge of the gravity field? So far, the only global information beyond the traditional gravimetric surveys (including airborne) has been the processing of satellite orbit perturbations data, and of sea surface heights from satellite radar altimetry (using techniques such as time-averaging and orbit cross-overs).

Using these data sets for gravimetry, even to the best of possibilities, has strong limitations in performance, one reason being the too high altitude by gravimetric standards, of nearly all satellites, another one being the contribution of non gravitational forces to orbit perturbations, which need to be separated in the processing of measurements.

Satellite missions were proposed in the recent past to try and make some progress towards a better knowledge of the gravity field and geoid. The ESA-NASA Aristoteles (also addressing magnetometry) had been studied for many years but was eventually terminated for budgetary reasons. Concept such as BRIDGE (Mini-Satellite Concept to Bridge the Past and Future in Gravity Field Research), a mission combining precise orbit determination of a low altitude satellite supplemented by accelerometric measurements (to account for nongravitational forces) have been studied in France, while the CHAMP (Catastrophes and Hazard Monitoring and Prediction) gravity and magnetic field project is now being funded in Germany. However, the resulting gravity field and geoid performance would lie nearly an order of magnitude beyond the requirements from the research needs which have been discussed above, and which are summarised in Table 2.1.

Table 2.1. Required resolution and accuracy of the gravity field and geoid

	Accuracy		Spatial Resolution
	Geoid	Gravity	(half wavelength)
Ocean Circulation			
- Small scale	2 cm		60-250 km
- Bassin scale	<1 cm		1000 km
Geodynamics			
- Continental lithosphere (thermal structure, post-glacial rebound)		1-2 mgals	50-400 km
- Mantle composition, rheology		1-2 mgals	100-5000 km
- Ocean lithosphere and interaction with asthenosphere (subduction processes)		5-10 mgals	100-200 km
Geodesy			
- Ice and land vertical movements	2 cm		100-200 km
- Rock basement under polar ice sheets		1-5 mgals	50-100 km
- World-wide height system	<5 cm		50-100 km

The satellite mission responding to these requirements has been studied in 1995-1996 as a candidate Earth Explorer mission (cf. supra, Introduction) and will now be presented in more details.

3 The Gravity Field and Steady-State Ocean Circulation Explorer (GOCE) Space Mission Concept

The requirements on the gravimetric performance of the GOCE mission, as illustrated in Table 2.1, lead to the following typical mission parameters:

- orbit altitude 270 km , circular
- orbit inclination 96.5° (heliosynchronous)
- equatorial mode crossing time 6hrs - 18 hrs
- mission duration 8 months

The chosen altitude is a compromise between the need to achieve a good sensitivity of the in situ gravimetric measurements, and satellite safety considerations. Indeed, examination of the gravity field expansion, e.g. for the radial diagonal component of the gravity gradient:

$$g_r = (GM/_{r^3}) \sum_i \sum_l \sum_m (l+1)(l+2)(R/_r)^l C^*_{ilm} Y^*_{ilm} \qquad (i=1.2)$$

shows that the sensitivity of measurements dramatically decreases when altitude increases, an effect which worsens for the higher harmonics. Conversely, low altitudes imply high atmospheric densities, hence high aerodynamic forces and torques. It has been shown, cf. Figure 3.1, that a satellite can only be safely left unattended (from the ground station) if it is above a certain altitude.

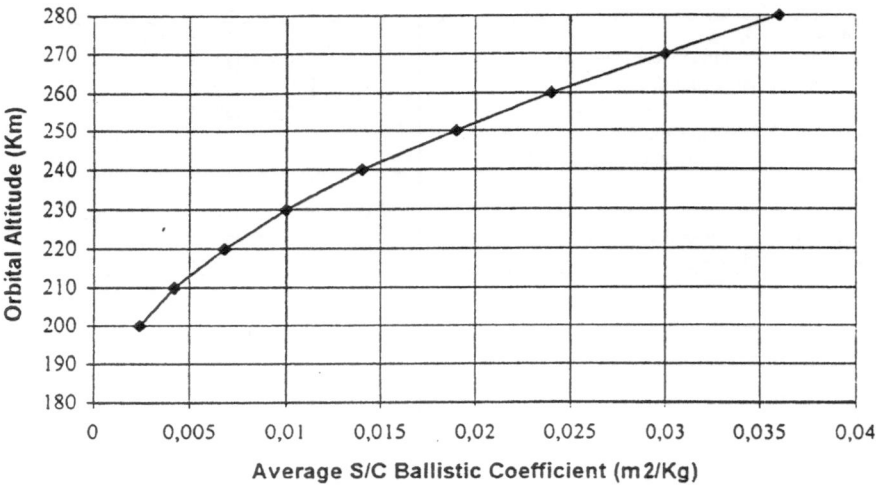

Fig. 3.1.   Minimum Orbital Altitude Compatible with the 7-Day Survival Requirement

The inclination of the orbit should ideally be 90 degrees to ensure that all areas on the Earth are covered by measurements. However, engineering considerations, which heavily impact on the satellite costs, lead to strongly prefer a heliosynchronous orbit.

The selected equatorial node crossing times allow the satellite to have a fixed solar generator, facing the Sun at right angles for maximum efficiency of energy conversion.

The planned mission duration of 8 months has been determined so that there are at least two complete coverages of the Earth's surface by the satellite's tracks.

Having fixed the mission duration and orbit parameters, it still remains to choose an attitude stabilisation law. Here, the choice would be between a fully inertially stabilised satellite ("3-axes control") and an Earth pointing satellite. The latter has been finally retained, despite the incurred precession at the orbital rate of 0.0012 rad/sec (which impacts the accuracy of gravity gradient measurements), because it allows to have:

- a low satellite cross-section with respect to the orbital velocity thus minimising aerodynamic drag and torques
- a simple configuration of thrusters for the compensation of aerodynamic forces

4 Scientific Instruments

4.1 Principles of Gravimetric Measurements from Space

In practical terms there are two main techniques for the determination of the gravity field by a space mission:
- satellite-to-satellite tracking (SST) and
- spaceborne gravity gradiometry (SGG)

Furthermore, depending on the altitudes of the mutually tracking satellites, one usually distinguishes between high-low SST and low-low SST (in the latter, satellites would remain quite close to each other).

The various techniques are illustrated in Figure 4.1.

While NASA proposals often favoured low-low SST, which calls for the launch of 2 satellites, ESA with the past Aristoteles project, and currently with the GOCE proposal has firmed up its preference for a single satellite combining high-low SST and gradiometry. In the ESA approach, the high-satellites will be the existing constellations of GPS and/or Global Navigation Satellite System (GLONASS), with the sole implication that the lower satellite (GOCE) carries a high performance GPS/GLONASS receiver. For low-low SST the major difficulty lies with the mutual tracking of the two satellites, which has to be extremely accurate, whereas the alternative technique which will be used on GOCE, calls for a complex instrument, the gradiometer.

Given the Signal to Noise Ratio (SNR) achievable by the best GNSS receivers, the high-low SST alone cannot properly determine the gravity field beyond harmonics 70-80. Gravity gradiometers will take care of measurements up to, say, harmonic 200. Comparative inspection of the harmonic expansions of the gravity field and of the radial diagonal components of its gradient:

$$\Delta g = (GM/_{r^2}) \sum_{i,l,m} (l-1) C_{ilm}^* Y_{ilm} \qquad (i=1.2)$$

$$g_r = (GM/_{r^3}) \sum_{i,l,m} (l+1)(l+2) C_{ilm}^* Y_{ilm} \qquad (i=1.2)$$

suggests a better sensitivity of the higher harmonics in the gravity gradient expansion. Conversely, the gradiometer alone cannot properly determine the lower harmonics. This is due to the limited stability of its response over time. If, for instance, it cannot be guaranteed that the change in properties of the instrument (scale factor, bias) stays within tolerances for times longer than 200 seconds, then it will not only poorly determine harmonics below 27 (assuming an orbital period of about 5400 sec), but recent computer analyses at Technical University of München also showed that degradation will already start below harmonic 120.

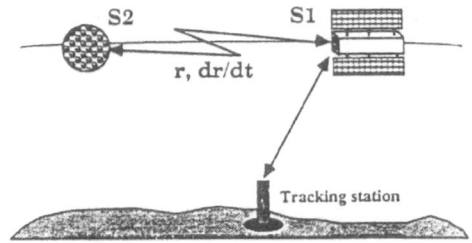

Schematic of low-low SST. One measures the distance r between the two co-orbiting spacecraft, or the range-rate dr/dt. One spacecraft can be also tracked from the ground

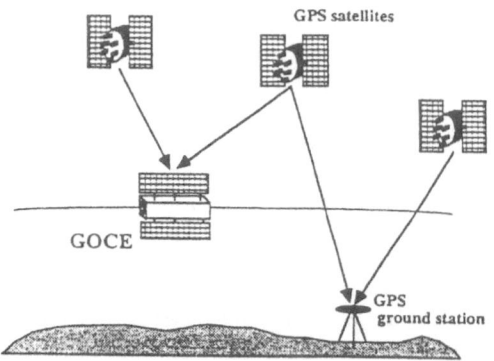

High-low SST using GPS satellites which orbits are monitored by a network of ground stations

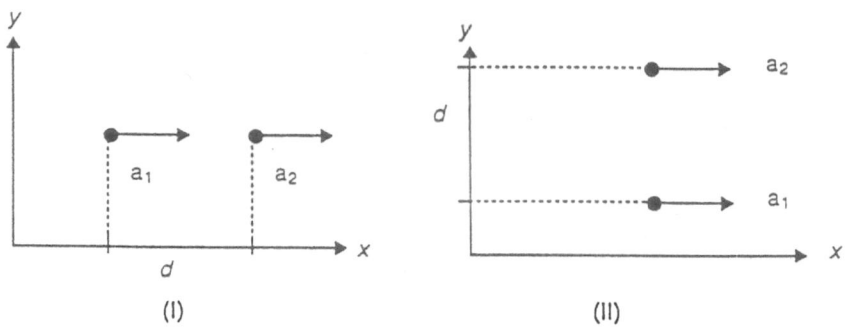

Principle of gradiometry based on pairs of accelerometers. In (I), one measures : $(a_2 - a_1) / d = V_{xx}$ and in (II) : $(a_2 - a_1) / d = V_{yy}$

Fig. 4.1.   Techniques for the determination of gravity field from space

Hence the single satellite in low orbit of the GOCE mission has to use in combination a GNSS receiver and a gradiometer, both instruments having high performance.

4.2 The GNSS Receiver

Development of very high performance GNSS receiver has been initiated at ESA. As one of its main applications will be the production of altitude profiles of atmospheric density and other atmospheric properties by the technique of limb sounding from a low orbiting satellite, it is labelled GNSS Receiver for Atmospheric Sounding (GRAS).

The GRAS receiver is intended to allow the simultaneous ranging of the GOCE satellite from 12 satellites of the GPS/GLONASS constellations. The necessary ranging accuracy imposes 2 frequency carrier phase tracking as the ranging method and differential observations between the GOCE satellite and a ground network of GPS/GLONASS geodetic receivers. The already operational International GPS Geodynamics Service (IGS) is well adapted to this role.

The carrier phase noise on GRAS should be less than 1 mm (1 σ) with a 10 Hz sampling rate. To obviate for encryption of the ranging code on one of the frequencies "codeless" operation is possible, with a phase noise of 2mm with 1 Hz sampling. Errors on pseudoranges, either with GPS or GLONASS, are all specified to be 1m or less.

Figure 4.2 shows a very simplified block diagram of the GRAS breadboard already built and tested.

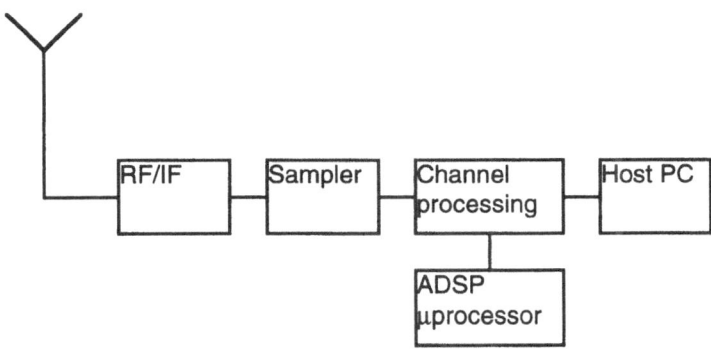

Fig. 4.2. Block diagram of the GNSS dual-frequency L1/L2 receiver (GRAS)

Test results with the preliminary 8-channel receiver breadboard showed that the parallel tracking with several GPS/GLONASS satellites was working, with acquisition times not exceeding 60 seconds. Table 4.1 demonstrates that adequate performance was achieved with the first breadboard model.

Table 4.1. Results of "zero baseline" GRAS test results

System	Double difference observable standard deviation				
	L1 C/A code (m)	L1 P code (m)	L2 P code (m)	L1 carrier (mm)	L2 carrier (mm)
GPS (codeless)	0.288	1.478	1.109	0.39	2.96[†]
GLN (coded)	0.986	0.323	0.359	0.719	0.475[†]

[†]The L1 carrier loop aids the L2 carrier loop
so there is a common component of noise
which is removed to produce this
measurement

The estimated mass and power consumption of the GRAS receiver (with 12 channels) are 3.2 kg and 24.

4.3   The Gravity Gradiometer

4.3.1   General

The gravity gradiometer is designed to measure the diagonal components of the local gravity at the satellite's altitude. Off-diagonal components will also be measured, but with a much degraded accuracy. The principle used in this type of instrument is differential accelerometry, for instance 2 accelerometers separated by 0.5 m, each with a noise figure of $10^{-12} m.s^{-2}/\sqrt{Hz}$ in the bandwidth would determine the gradient to $2.8 \times 10^{-12} s^{-2}/\sqrt{Hz} = 2.8 \times 10^{-3} E/\sqrt{Hz}$ (it is reminded that the Eötvös unit, E, equals $10^{-9} s^{-2}$).

For the sake of comparison, the radial gravity gradient signal at 270km above a minimum frequency of 5 mHz is of the order of 0.1E.

The common mode of each pair of accelerometers serves to measure the non-gravitational forces.

In the current studies two possible European designs are being considered for the gradiometer. Both are based on the response of a proof mass-spring system:

- one from ONERA (France) uses capacitive measurement of the small shift of the proof mass, which is kept in place by a force feedback servoloop using the same electrodes as for detection. The operating temperature is ambient.
- the other from Oxford Instruments (Great Britain) uses inductive detection of the proof mass uncontrolled motion. The supersensitive position detection is performed by a supraconducting loop coupled to a Supraconducting Quantum Interference Detector (SQUID) measuring changes in loop currents caused by the shift. Supraconductivity requires cooling of the instrument by liquid helium in superfluid state at 1.5 K.

4.3.2 Electrostatic Ambient Temperature Gradiometer

In this option, the building block of the gradiometer is the ONERA ultra-high sensitivity accelerometer designated as GRADIO. The instrument is made up of 3 pairs of accelerometers along each of the measurement axes, mounted on a structure of high dimensional stability.

To allow sufficient gradiometric sensitivity, the separation of two accelerometers in a pair is 50 cm.

The proof mass is a square slab of a platinum-rhodium alloy, with a mass of the order of 320 grammes. The less sensitive axis, with a thinner gap between the proof mass and the electrodes, is perpendicular to the faces of the slab.

The gradiometer assembly and an exploded view of the GRADIO accelerometer are shown respectively in Figures 4.3 and 4.4.

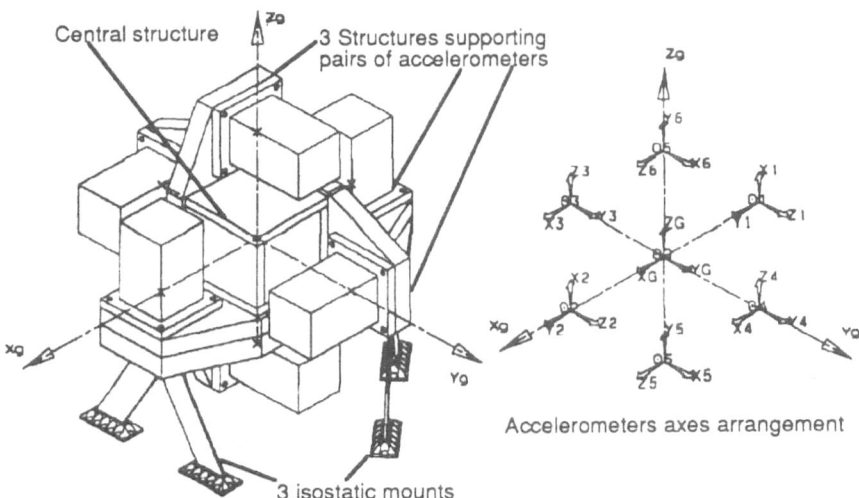

Fig. 4.3.   Schematic of the gradiometer assembly (6 capacitive 3-axis accelerometers)

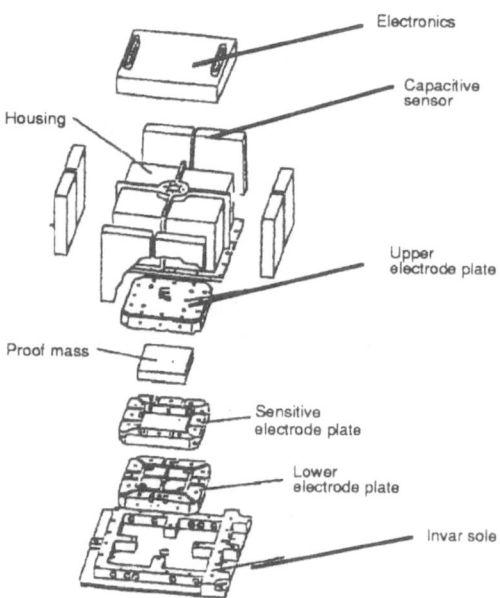

Fig. 4.4.   Exploded view of the capacitive accelerometer

The GRADIO accelerometer has been in the last ten years extensively tested, in preparing for the Aristoteles mission. Since the termination of this programme, a simplified, much less performant model, labelled ASTRE, has been flown in June 1996 on a Shuttle Spacelab mission and performed according to specified noise level.

4.3.3   Inductive Supraconductive Gradiometer

Conversely to the electrostatic gradiometer, the building block here is a pair of accelerometers along each sensitive axis, with a separation of 12 cm between accelerometers.

Proof masses are made of niobium, have the shape of a belted cylindrical shell and weigh 150 grams each.

The principle of the measurement circuittry for differential measurements is shown in Figure 4.5 and a view of the 3-axis gradiometer assembly is shown in Figure 4.6

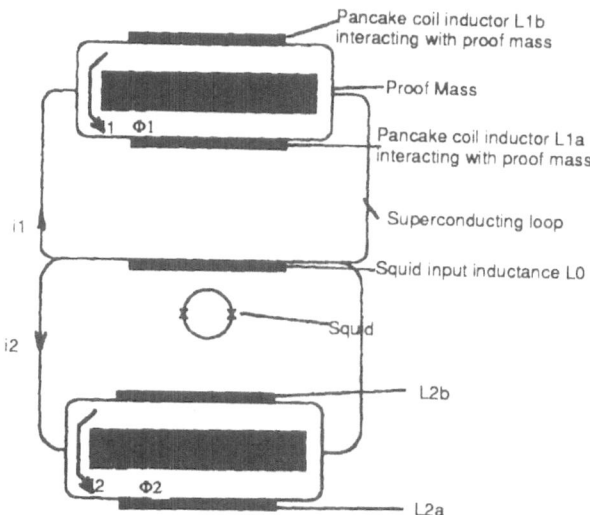

Fig. 4.5 Differents measurement circuity of the cryogenic inductive gradiometer (1-channel)

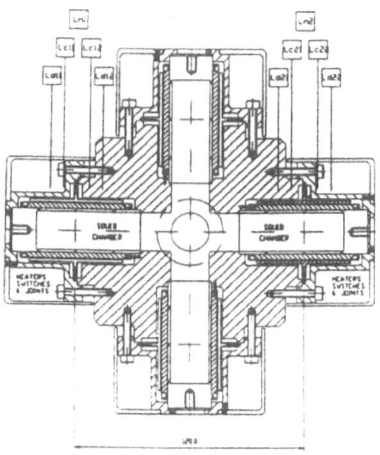

Fig. 4.6 Schematic of the 3-axis cryogenic gradiometer

A key feature of this accelerometer is the wiring arrangement in a pair of accelerometers, which allows for a lower stiffness of the differential mode (for gradiometry) than for the common mode (for non-gravitational forces determination).

In addition, built in electrodynamic shakers provide a calibration facility. Like the electrostatic suspension accelerometer, the inductive gradiometer benefits from a long background of laboratory testing, which demonstrated a differential accelerometric noise level of 10^{-11} ms^{-2} to 10^{-12} ms^{-2} using a "weak" mechanical suspension instead of the proposed suspension by magnetic levitation.

4.3.4 Comparison of the two options

A theoretical performance comparison of the two kinds of gradiometers has been done by Prof. S. Vitale, University of Trento, concerning all various sources of noise: thermal, back-action of the displacement detector, additive noise of SQUIDs (inductive device) or amplifiers (capacitive servoloop device).

In brief, Table 4.2 summarises the predicted theoretical performance of key parameters for both gradiometers, under realistic assumptions.

Table 4.2 Comparative performance of capacitive and inductive gradiometers

	Capacitive		Inductive (cryogenic)	
noise level	$7 \times 10^{-5} - 10^{-4}$ ms^{-2} . Hz$^{-\frac{1}{2}}$		$7 \times 10^{-4} - 10^{-3}$ ms^{-2} . Hz$^{-\frac{1}{2}}$	
Common Mode Rejection Ratio (CMRR)	$10^{6} - 10^{7}$	(after differential mode balancing)	$10^{5} - 10^{6}$	(after in orbit calibration)

The performance of the capacitive gradiometer has been found, as will be seen later, to be acceptable for the GOCE mission while that of the cryogenic is much better. The capacitive GRADIO accelerometer was extensively tested until 1991. On the other hand, a 1-axis breadboard model of the inductive magnetically suspended gradiometer will be completed and tested toward the end of 1996. In conclusion, the selection of the gradiometer instrument is still open and should take place in 1997.

5 The GOCE Satellite

5.1 Specific Requirements

To satisfy the gravimetric performance, it is required that some characteristics of the GOCE satellite match the very high sensitivity of the science payload instruments, the GRAS receiver and more critically the gravity gradiometer.

The most important perturbation is due to the non-gravitational forces and moments, partly due to solar and terrestrial radiation pressure but chiefly due to the relatively high atmospheric density at the GOCE altitude of 270 km.

The shape of the satellite shall be as slender as possible in order to minimise the propellant mass needed for orbit maintenance during the entire mission.

For the use of the GRAS data it is sufficient that the acceleration be determined by the common mode of the accelerometer pairs in the gradiometer, to accuracies better than 10^{-8} ms^{-2}.

Turning now to the operation of the gradiometer, it will be optimally placed near the centre of mass of the satellite, where it has been verified that gravity induced by the remainder of the satellite ("self-gravity") is negligible. The non-gravitational perturbation shall be minimised by a continuous operation of the propulsion system which maintains the orbit altitude. Ideally a 3-axes Drag Free Control (DFC) propulsion system should be implemented, whereby the propulsion system's thrust continuously matches the fluctuating non-gravitational forces to a level of 0.1% - 1%. One can see · that with a typical drag deceleration along orbital velocity vector of 10^{-5} ms^{-2}, the DFC will reduce it to 10^{-8} ms^{-2} - 10^{-7} ms^{-2}, values which are compatible with an accelerometric noise of 10^{-3} ms^{-2} and a gradiometer CMRR of 10^6.

The specification on pointing control accuracy is fairly loose, the usual satellite pointing error of 0.1 degree being tolerable provided that it is measured, for processing on ground, with an accuracy of a few arcseconds.

Conversely stringent requirements shall be put on angular rates and angular accelerations of the satellite. This can be seen by observing the gravity gradient matrix as affected by rotational terms, denoting by ω_o the constant pitch rate at the orbital angular velocity of 1.2×10^{-3} rad.sec^{-1} and by $\omega_{x,y,z}$ respectively the fluctuations of angular rates:

$$U' \cong \begin{vmatrix} U_{xx} - 2\omega_o\omega_y & U_{xy} + \omega_o\omega_x - \dot{\omega}_z & U_{xz} + \omega_z\omega_x + \dot{\omega}_y \\ U_{yz} + \omega_o\omega_x + \dot{\omega}_z & U_{yy} - \omega_z^2 - \omega_x^2 & U_{yz} + \omega_o\omega_z - \dot{\omega}_x \\ U_{zx} + \omega_z\omega_x - \dot{\omega}_y & U_{zy} + \omega_z\omega_o + \dot{\omega}_x & U_{zz} - 2\omega_o\omega_y \end{vmatrix}$$

Given the orders of magnitude of gravity gradient components in the typical measurement bandwidth of 5×10^{-3} Hz to 0.1 Hz it results that angular velocities are assumed to be typically 10^{-7} - 10^{-6} rad.sec^{-1} Hz$^{-\frac{1}{2}}$ and angular accelerations 10^{-8} - 10^{-7} rad.sec-2.Hz$^{-\frac{1}{2}}$, with actual values measured to a few percent.

With tight requirements on its orbit and attitude control functions, the GOCE satellite needs specific designs which are hereafter briefly reviewed.

5.2 Orbit Maintenance and DFC

The propulsion system has to cope with two major constraints:
- to minimise the propellant mass needed for orbit maintenance during 8 months
- to precisely and continuously match the fluctuating non-gravitational force
  Fluctuations of the non-gravitational deceleration for a satellite weighing 700 kg with
a cross-section of 1 m² against the orbital velocity, are illustrated in Figures 5.1 and 5.2.

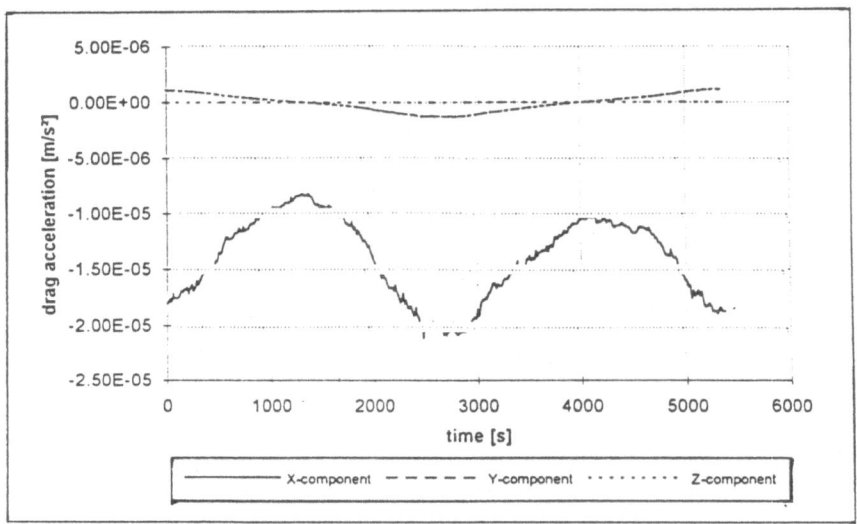

Fig. 5.1. Fluctuation of GOCE non-gravitational accelerations

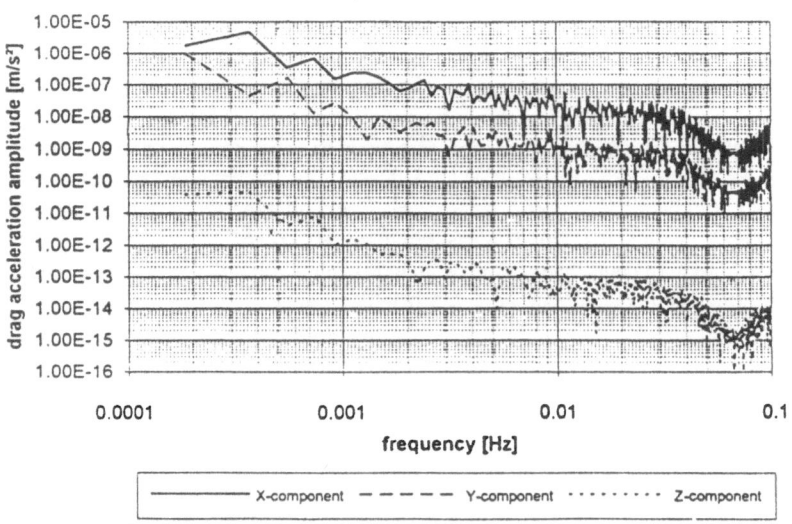

Fig. 5.2. Amplitude spectrum of GOCE non-gravitational accelerations

The propellant mass needed for orbit maintenance depends on the type of propulsion chosen as shown hereafter:

	Propellant mass (kg)
Hydrazine	85
Ion thruster (xenon)	6
Helium	150

Catalytic decomposition of hydrazine is the classical means of propulsion for satellites. However, ion propulsion provides a considerable gain in propellant mass, but at the expense of a high power consumption, of the order of 200 Watts.

Only ion and helium propulsion allow the thrust to be modulated, by throttling the thrusters, so that it can continuously match the non-gravitational force. Hence the hydrazine propulsion is not considered for GOCE. Propulsion using the gaseous helium exhaust from the cryostat is of interest only if the cryogenic gravity gradiometer is selected.

Pending the selection of either capacitive or inductive (cryogenic) gradiometer, electric propulsion is adopted. The operation of an ion thruster is schematised in Figure 5.3.

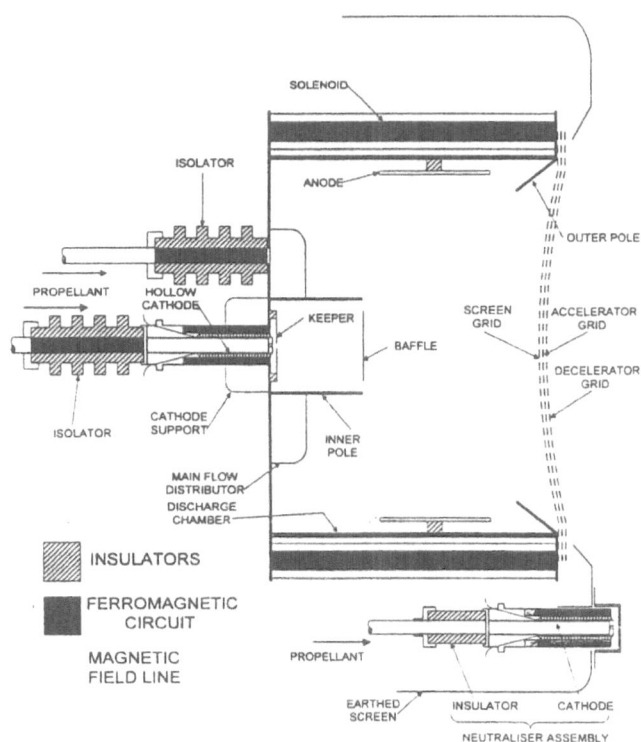

Fig. 5.3.   Schematic diagram of the T5 Ion Thruster

Xenon gas flows through the cathode produces positive ions in the presence of a magnetic field. The ions are then accelerated to a very high exhaust velocity by a set of high voltage grids. At the exhaust the outcoming flux is neutralised by an auxiliary flow of propellant through another cathode, in order to avoid charge build up in the satellite.

The simplest configuration of the GOCE mission is to have one (or two, for redundancy) throttleable ion thrusters along the longitudinal axis of the satellite. More ion thrusters would allow in addition to balance non-gravitational forces in all directions. This is presently considered as an optional possibility, as transverse forces on the satellite are 1 or even 2 order(s) of magnitude smaller than along the orbital velocity. Throttling of the ion thrusters has been experimentally tested and proven feasible with an acceptable response in time (less than 100 seconds).

5.3   Attitude Control

As it has already been explained, satellite angular velocities and accelerations strongly impact the gravity gradient measurements. Hence a very smooth motion of the satellite is indispensable.

Design solutions to this problem can be:
-   electric ion thrusters, which have a very low thrust, but a complex multi-thruster configuration is needed, and a very deep modulation of the thrust
-   proportional helium thrusters, if the cryogenic gradiometer is used. A design of such thrusters exists in the US
-   reaction wheels, although any noise in their operation may affect the gradiometer measurement bandwidth e.g. by aliasing. Special testing of reaction wheels noise is being performed in industry
-   low thrust nitrogen thrusters. They operate in on-off mode at a fixed thrust level, typically 20mN. Drawback is that gradiometric measurements are unacceptably degraded during operation, bu the time lost in this manner is only about 1% of the total mission duration

Currently, the preferred design would rely on nitrogen impulsive or helium (cryogenic exhaust) proportional thrusters, although results of reaction wheel tests – if positive – would lead to prefer this last approach.

5.4   The GOCE Satellite Subsystems

The Drag Free Control and attitude control subsystems are very specific to GOCE and have already been briefly described. Other subsystems in the satellite are more classical, to various degrees.
-   Structure and thermal control: To allow accommodation of the gradiometer instrument near the satellite's centre of mass, the structure design uses the external panels as load-carrying members, thus departing from the usual approach calling for a central cylindrical tube. A possible design appears on Figure 5.4.

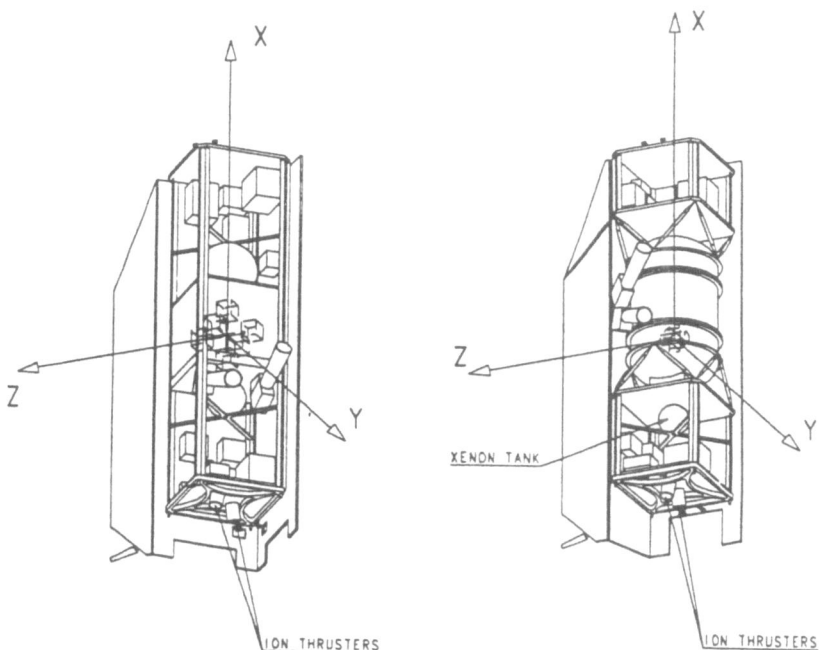

Fig. 5.4. Structure of the GOCE satellite

The overall configuration of the satellite is displayed in Figure 5.5.

This classical design will feature a larger radiator on the face opposite to the Sun direction, and the thermal control of the gradiometer will be autonomous.

- Power generation: The overall power demand is driven by electric propulsion. With the simplest configuration featuring only 2 ion thrusters, the maximum power demand will reach 850 Watts which will be satisfied with an output of 1100 Watts from the solar array. To reduce the total area of solar cells, the more efficient gallium-arsenide cells have been proposed to replace the silicon cells which are commonly produced in Europe, resulting in a total area of 8.5 m² which can be accommodated on body-fixed panels. The energy storage of nearly 700 Ah could use classical nickel-cadmium batteries.

- Data handling and communications: The satellite and its science instruments will continuously generate a data flow of 8 Kbps. With one ground station at a high latitude it will then be sufficient to have on-board a mass memory of 0.3 Gbps which is very much within reach of modern Dynamic Random Access Memory (DRAM) solid state memory units. During each ground station visibility pass, the satellite's data will be telemetered at 1 Mbps. while commands will be up linked at a rate of 4 Kbps.

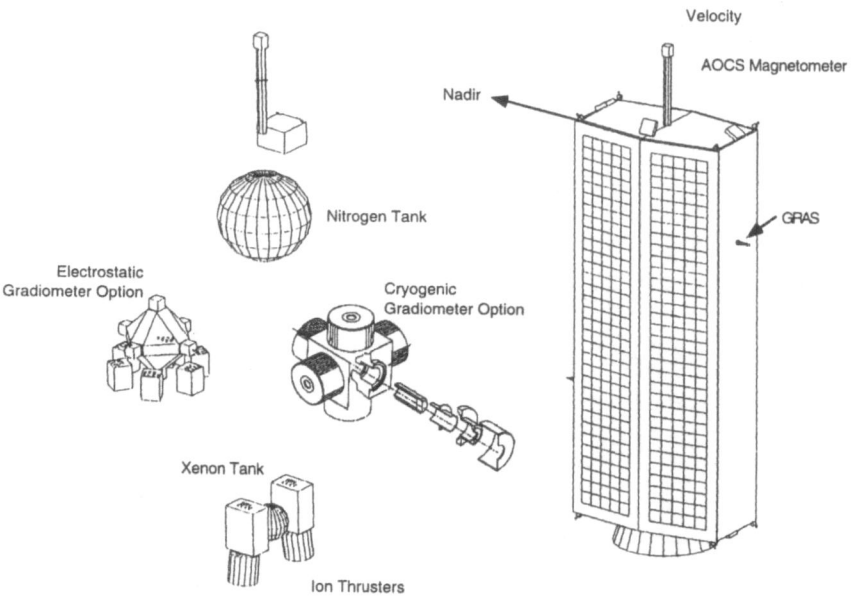

Fig. 5.5.   Overall configuration of the GOCE satellite

5.5   Launch and Orbit Operations

- Launch: Being a satellite of relatively modest mass and dimensions, GOCE could be best accommodated in one of the "small satellite" launch vehicles now becoming available, e.g. Eurorockot (Russia) or Taurus (US). The matter will be further addressed in a full feasibility study.
- Operations: The satellite operations will be supported by a conventional ground segment the main elements of which are one ground station for telemetry and telecommand, located at a high latitude (e.g. Kiruna, Sweden) so as to have many satellite passes in a day. Operations will be remotely managed from a satellite control centre, which will dispatch the satellite raw data, after decommutation and telemetry calibration, to other centres in charge of archiving and further processing.

6 Evaluation of the ExpectedMission Performance

There are two major steps in evaluating if the GOCE mission will perform according to its demanding requirements as recalled in Table 21:

- determination of the in situ measurement accuracy via an error budget, accounting for the instruments' intrinsic accuracies, and the impact of the instruments' environment assuming that the satellite performs as required
- evaluation by scientific institutes of the final performance, using the measurement accuracies as inputs to covariance analyses. In a few cases (such as the study at GFZ, Postdam, of GNSS tracking without gradiometry), computer simulations were called for and validated the covariance analyses.

The performance of the GRAS receiver has already been mentioned in Chapter 4.2, and is in accordance with the requirements.

For what concerns gradiometry, major contributions to the error budget originate from the satellite angular rates and accelerations, and – only for the electrostatics gradiometer – from its dimensional stability (even with very low CTE materials).

The outcome is a measurement performance slightly above 10^{-2} E.Hz$^{-1/2}$ for the capacitive gradiometer and about 10^{-3} E.Hz$^{-1/2}$ for the (cryogenic) inductive device.

Scientific covariance analyses show that when supplemented by measurements from the GRAS receiver, both options appear compatible with the requirements of Table 2.1. An example of such analyses, from the Technical University of München, is given in Figure 6.1.

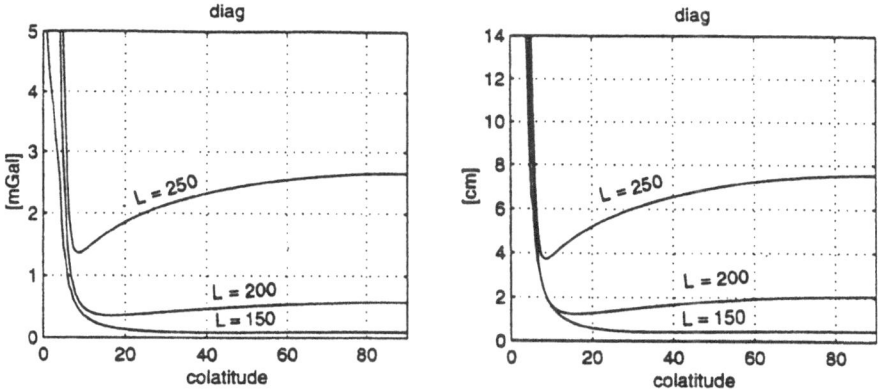

Fig. 6.1. Typical result of GOCE scientific performance analyses

It assumes a gradiometric performance of 5×10^{-3} E.Hz$^{-1/2}$ in the measurement bandwidth from a satellite flying at 260 km altitude in an orbit inclined by 96.5 degrees, with GRAS receiver performance from 1 cm (radial) to 3 cm (along-track). The mission duration is 8 months.

7 Conclusion

Studies of a dedicated single-satellite mission dedicated to gravimetry (GOCE), in which would be flown a high-accuracy GNSS receiver and a high accuracy gradiometer, have resulted in realistic concepts. The achievable gradiometric performance appears to be compatible with the needs of many scientific users in the areas of oceanography, geodesy and solid Earth physics, as well as with those of a few peculiar applications.

A fully detailed system study will allow among other things to select the gradiometer concept and more generally to firm up the satellite design and establish its technical and financial feasibility, in view of a later decision on the project funding.

List of Acronyms

ADSP	(Analog Devices) Digital Signal Processor
BRIDGE	Mini-Satellite Concept to Bridge the Past and Future in Gravity Field Research
CHAMP	Catastrophes and Hazard Monitoring and Prediction
CMRR	Common Mode Rejection Ratio
CTE	Coefficient of Thermal Expansion
DFC	Drag Free Control
DRAM	Dynamic Random Access Memory
ERS	(ESA) European Remote Sensing Satellite
ESA	European Space Agency
GFZ	Geo Forschungs Zentrum
GLONASS	(R) Global Navigation Satellite System
GNSS	Global Navigation Satellite System (based on GPS and GLONASS)
GOCE	Gravity Field and Steady-State Ocean Circulation Explorer
GPS	(US) Global Positioning System
GRAS	GNSS Receiver for Atmospheric Sounding
IF	Intermediate Frequency
IGS	International GPS Geodynamics Service
METOP	Meteorological Operational Satellite Programme
MSG	Meteosat Second Generation
PC	Personal Computer
RF	Radio Frequency
SAR	Synthetic Aperture Radar
SGG	Spaceborne Gravity Gradiometry
SNR	Signal to Noise Ratio
SQUID	Supraconducting Quantum Interference Detector
SST	Satellite to Satellite Tracking

Index

Springer
and the
environment

At Springer we firmly believe that an international science publisher has a special obligation to the environment, and our corporate policies consistently reflect this conviction.
We also expect our business partners – paper mills, printers, packaging manufacturers, etc. – to commit themselves to using materials and production processes that do not harm the environment. The paper in this book is made from low- or no-chlorine pulp and is acid free, in conformance with international standards for paper permanency.

 Springer

Lecture Notes in Earth Sciences